SpringerBriefs in Applied Sciences and Technology

T0236522

More information about this series at http://www.springer.com/series/8884

Luben Cabezas-Gómez · Hélio Aparecido Navarro
José Maria Saíz-Jabardo

Thermal Performance Modeling of Cross-Flow Heat Exchangers

 Springer

Luben Cabezas-Gómez
Departamento de Ciências Térmicas e dos
    Fluidos
Universidade Federal de São João del-Rei
São João del-Rei, Minas Gerais
Brazil

José Maria Saíz-Jabardo
Escuela Politécnica Superior
Universidad de la Coruña
La Coruña
Spain

Hélio Aparecido Navarro
Departamento de Engenharia Mecânica
Escola de Engenharia de São Carlos
Universidade de São Paulo
São Carlos, São Paulo
Brazil

Additional material to this book can be downloaded from http://extras.springer.com

ISSN 2191-530X   ISSN 2191-5318   (electronic)
ISBN 978-3-319-09670-4   ISBN 978-3-319-09671-1   (eBook)
DOI 10.1007/978-3-319-09671-1

Library of Congress Control Number: 2014946187

Springer Cham Heidelberg New York Dordrecht London

Printed on acid-free paper

Springer is part of Springer Science+Business Media (www.springer.com)

Acknowledgments

The first author acknowledges the financial support from the CNPq (Conselho Nacional de Desenvolvimento Científico e Tecnológico), process 307141/2011-0, and the FAPEMIG (Fundação de Amparo à Pesquisa do Estado de Minas Gerais), process PPM-00639/11.

Contents

Notations

A	Exchanger outer total heat transfer area, m^2
C	Heat capacity rate, W/K
C^*	Heat capacity rate ratio, C_{min}/C_{max}, dimensionless
F	LMTD correction factor, dimensionless
G	Designation of flow arrangement configuration
i	Number of elements in each tube
j	Number of tubes per row
k	Number of tube rows
N	Number of data points
N_c	Number of tube fluid circuits
N_e	Number of elements per tube
N_r	Number of rows in the heat exchanger
N_t	Number of tubes per row
NTU	Number of transfer units, UA/C_{min}, dimensionless
P	Temperature effectiveness, $(T_{c,o} - T_{c,i})/(T_{h,i} - T_{c,i})$, dimensionless
q	Heat transfer rate, W
R	Temperature ratio, $(T_{h,i} - T_{h,o})/(T_{c,o} - T_{c,i})$, dimensionless
T	Temperature, K
U	Overall heat transfer coefficient, $W/(m^2K)$

Greek Symbols

δ	Relative differential
δp	Established relative tolerance
ε	Conventional heat exchanger effectiveness, q/q_{max}, dimensionless
λ	Relative error
Γ	Dimensionless parameter defined by Eq. (2.30) (local effectiveness)
Λ	Constant, $C_c^e \Gamma^e / C_h^e$, dimensionless

Subscripts

∞	Infinite
A	Mixed fluid side
air	Air side (external fluid)
av	Average value
B	Unmixed fluid side
c	Cold fluid side of heat exchanger (external fluid, air side)
d	Total differential or derivative
e	Element
fr	Frontal face
h	Hot fluid side of heat exchanger (in-tube fluid)
i	Inlet conditions, or cold fluid entrance, $i = 0$, and exit, $i = 1$ (EES program)
j	Number of elements in each tube (EES program)
l	Number of in-tube fluid passes
m	Mean value or number of tube rows per pass of the in-tube fluid
max	Maximum value
min	Minimum value
n	Number of in-tube fluid rows
new	New value
o	Outlet conditions
r	Row
t	In-tube fluid side of heat exchanger
th	Theoretical

Superscripts

cc	Overall counter cross-flow configuration
cf	Cross-flow configuration
e	Element
k	Designates either cf or cc or pc superscripts
pc	Overall parallel cross-flow configuration

Abbreviations

DAES	Differential algebraic equation systems
LMTD	Logarithm mean temperature difference
HETE	Heat exchanger thermal effectiveness

Abstract

This monograph introduces a numerical computational methodology for thermal performance modeling and evaluation of cross-flow heat exchangers, which may find application in contemporary chemical, refrigeration, and automobile industries. According to this methodology, the heat exchanger is discretized into small elements following the tube side fluid circuits. Each element constitutes itself in a one pass mixed-unmixed cross-flow heat exchanger. The algebraic governing equations obtained for each element are solved iteratively for the whole heat exchanger using local values of the element's temperature, assuming constant physical properties, and heat transfer coefficients. This computational methodology has been used by the authors for simulating several flow arrangement configurations of cross-flow heat exchangers. The methodology allows obtaining effectiveness-number of transfer units (ε-NTU) data for several standard and complex flow arrangements. Simulated results have been validated through comparisons with results from analytical solutions for one pass cross-flow heat exchangers with one and more rows, and for 1–4 passes, parallel and counter cross-flow arrangements. In addition, comparisons have been performed with complex flow arrangements approximate solutions. Very accurate results have been obtained over wide ranges of NTU and C* values in all cases. New effectiveness data for some of the aforementioned configurations have been found, along with data for a complex flow configuration proposed elsewhere. The proposed procedure constitutes a useful research tool both for theoretical and experimental studies of cross-flow heat exchangers thermal performance. A computational program based on the proposed algorithm was implemented and named HETE, which is an acronym for *Heat Exchanger Thermal Effectiveness*. A version of this code will be presented in the appendix A of the monograph and will be available for downloading. Solved examples are also presented showing in details how the HETE code works for classical flow arrangements. Tables and plots for (ε, NTU) data are provided in appendix C for several flow arrangements configurations. In addition to the aforementioned capabilities, the proposed procedure can be used extensively for computing the classical log-mean temperature difference correction factor (F), the so-called "heat exchanger reversibility norm" (*HERN*), and the newly introduced concepts of heat exchanger efficiency, and dimensionless entransy dissipation number. Other heat exchanger effectiveness definitions could also be computed.

Keywords Cross-flow heat exchangers · Heat exchanger effectiveness · HETE code · Iterative computational algorithm · Number of transfer units · Numerical simulation

Chapter 1
Introduction

Cross-flow heat exchangers find a widespread application in contemporary chemical, petrochemical, refrigeration, air conditioning, food storage, automobile, and other industries. The main reasons are their wide range of design options, simple manufacturing process, less maintenance requirements, low cost and good thermal-hydraulic characteristics for gas, in general air, cooling and heating. The widespread use of this type of heat exchangers has prompted the need for computer tools that accurately could predict their thermal performance, Bes (1996). The purpose of the present monograph is the introduction of a numerical computational methodology for thermal performance modeling and evaluation of cross-flow heat exchangers. The modeling procedure relies on the use of the effectiveness—number of transfer units (ε-NTU) method for computing the thermal performance of several standard and complex flow arrangements of cross-flow heat exchangers. The computational methodology is fully described in the text which is expected to be a useful research and teaching tool in the field of cross flow heat exchangers. The complete code, written in C, designated by the acronym **HETE** (Heat Exchanger Thermal Efficiency), is listed in Chap. 5. This code is also available for downloading at the following electronic address http://extras.springer.com.

1.1 A Concise Literature Survey

According to Pignotti and Shah (1992) and Sekulič et al. (1999), the design and analysis of two-fluid heat exchangers, including cross-flow exchangers, can be accomplished by the following procedures: (1) the effectiveness-number of transfer units method (ε-NTU); (2) the logarithmic mean temperature difference method (LMTD); (3) the temperature effectiveness-number of transfer units method (P-NTU); and (4) a modified version of one of these methods such as the Mueller and Roetzel charts. For a complete listing and discussion of different

© The Author(s) 2015
L. Cabezas-Gómez et al., *Thermal Performance Modeling of Cross-Flow Heat Exchangers*, SpringerBriefs in Applied Sciences and Technology, DOI 10.1007/978-3-319-09671-1_1

methods see publications by Taborek (Taborek 1983), Shah and Mueller (1985) and ESDU (**E**ngineering **S**ciences **D**ata **U**nit) (ESDU 1991) reports.

Differences among procedures and statements regarding the advantages of one over the other have been reported in the open literature. Kays and London (1998) stressed arguments in favor of the ε-*NTU* method with respect to the *LMTD* method. The main advantage of the former is related to the rating problem since it allows a straightforward solution for the heat exchanger thermal performance evaluation without the need of successive approximations as required by the log-mean approach. In addition of being very useful for heat exchangers rating and design, as a general rule, ε-NTU relationships are helpful in treating experimental data for the determination of compact heat exchangers external heat transfer coefficients, Kays and London (1998), Wang et al. (2000).

Another procedure introduced before the ε-NTU one is the P-NTU, where P is an "efficiency" defined for either the hot or cold fluid (fluid 1 or fluid 2) in contrast with ε which is defined in terms of the fluid with minimum heat capacity, C_{min}. Pignotti and Shah (1992) emphasized the fact that the *P-NTU* method avoids the need for two expressions for a stream asymmetrical two-fluid heat exchanger required by the ε-*NTU* procedure, each expression depending on which of the fluids is the one with the minimum heat capacity. Hewitt et al. (1994) argued that *P-NTU* plots present rather compressed curves which make difficult an adequate interpolation. Instead, these authors opted for the iterations free Taborek (1983) procedure which is a combination of the *LMTD* method, with the correction factor F, with the θ-NTU one, the parameter θ being defined as $\theta = P/NTU$. It must be noted that all the procedures are interrelated and must be applied knowing the temperature field distribution in the heat exchanger, leading to the same results for the same set of input data. The present monograph will focus on the ε-*NTU* method due to its widespread application in heat exchangers along with its extensively use in computer programs, as stated by Shah and Sekulić (2003).

Sekulić et al. (1999) presented a comprehensive review of solution methods for the determination of ε-*NTU* relationships for two-fluid heat exchangers with simple and complex flow arrangements. The available methods were separated into the following categories: exact analytical methods, approximate (analytical, curve-fit and analog) methods, numerical methods, matrix formalism methods, and methods based on heat exchanger configuration properties. Despite their comprehensive investigation, Sekulić et al. (1999) claim that new ε-*NTU* correlations not reported in the literature are required to comply with continuing efforts to design more efficient systems, more compact exchangers, and specific operating conditions.

Using both matrix formalism and methods based on heat exchanger configuration, Pignotti and Shah (1992) developed ε-*NTU* expressions for 18 new complex flow arrangements, 16 of which were cross-flow heat exchangers. They used methods such as the matrix algebra treatment introduced by Domingos (1969), also known as Domingos' rules, the chain rule and the rules for exchangers with one fluid mixed, previously published by Pignotti (1988). In this paper, Pignotti introduced a matrix formalism used for the evaluation of the thermal effectiveness of complex heat exchanger configurations that can be broken into simple constitutive parts, linked to each other by unmixed streams.

Shah and Pignotti (1993) dealt with complex heat exchanger flow arrangements relating them to simple forms for which either an exact analytical solution is available or an approximate solution could be obtained. They provided ε-NTU correlations for tubular cross-flow exchangers having seven different flow arrangements that could be manufactured from the same six-rows tube bundle.

It is worth mentioning the studies by Taborek (1983), Bacliç (1990), and Stevens et al. (1957) dealing with the evaluation of the heat exchanger effectiveness of multi-pass parallel and counter cross-flow configurations. Bacliç (1990) presented a compilation of closed-form effectiveness expressions for a large number of new flow arrangements used in compact heat exchangers. Stevens et al. (1957) determined numerically the temperature distribution for one, two and three passes parallel and counter cross-flow heat exchangers with refrigerant (in-tube fluid) side mixed and air side unmixed. This procedure was used by Chen et al. (1998) to develop a closed form expression for the effectiveness of a four rows counter cross-flow heat exchanger. According to Chen et al., the study performed by Stevens et al. (1957) is a significant contribution to the theory of the multi-pass parallel and counter cross-flow heat exchangers. In fact, Sekuliç et al. (1999) claim that the first extensive numerical results for various cross-flow arrangements were obtained by Karst (1952) and Stevens et al. (1957).

Concerning cross-flow heat exchangers, the investigations carried out by Nusselt (1911, 1930) and Mason (1955) must be mentioned. They provided analytical solutions for unmixed-unmixed single-pass cross-flow heat exchangers. According to Sekuliç et al. (1999), in his paper of (1911), Nusselt used a Riemann's integration method to solve a partial differential equation of the second order to obtain the temperature distribution in a heat exchanger. Thereafter, in his second attempt in (1930), Nusselt solved the same problem by transforming the analytical model into a Volterra integral equation. This equation was solved by assuming a trial solution in the form of a power series, from which he obtained a complicated explicit expression for the effectiveness. Mason (1955) used Laplace transforms for the same problem to obtain a solution that is often included in the heat transfer textbooks. The analytical solution of the problem that Nusselt dealt with motivated the search for other procedures along with adjustments of the complex expression of the initial analytical solution in order to simplify it. The following are must reading publications regarding the unmixed-unmixed flow arrangement solutions: Binnie and Poole (1937), Bacliç (1978), Bacliç and Heggs (1985), Hansen (1983), and Li (1987). The Bacliç and Heggs (1985) paper presents a very detailed explanation of several employed procedures and their equivalence.

Several correlations for the heat exchanger effectiveness can be found in Chap. 2 of the book by Martin (1990), who developed these correlations from the solution of two differential equations corresponding to the dimensionless temperature distribution for each fluid. Martin (1990) applied the Nusselt (1930) solution with some a posteriori simplifications to an unmixed-unmixed single-pass cross-flow heat exchanger. In combined flow configurations, such as the parallel and counter-cross-flow arrangements, the cell method was applied. This method consists in dividing the heat exchanger in cells corresponding to a section covering the entire length of one pass of the mixed in-tube fluid and the corresponding portion of the unmixed external fluid. Each cell provides a pair of algebraic equations for the dimensionless temperature distribution of each fluid. Thus, the whole heat exchanger is simulated by a system of algebraic equations whose solution allows the evaluation of the overall thermal effectiveness.

The computer code HETE presented herein deals with the procedure described in the preceding paragraph having the potential of application to different heat exchangers configurations. However, it must be emphasized that in the HETE code the in-tube fluid is divided into several elements in each pass. An excellent summary of effectiveness relations for different flow arrangements is also presented in details in the German VDI Atlas.

1.2 The Development of the HETE Code

The HETE code was developed as part of a comprehensive theoretical/experimental investigation of compact heat exchangers. The code development was prompted by the need to apply the ε-NTU method in experimental data processing aiming at the determination of external heat transfer coefficients and the corresponding fitting of correlations. Initially, a numerical code was developed to compute the heat exchanger effectiveness as a function of the flow arrangement and the dimensionless parameters, Number of Transfer Units, NTU, and heat capacities ratio, C^*. Other capabilities were incorporated to the code afterwards, including such procedures to compute (1) NTU as a function of the effectiveness, C^*, and flow arrangement; (2) the P temperature effectiveness in terms of NTU, C^*, and the flow arrangement; and (3) finally the $LMTD$ correction factor F, in terms of the arguments P and C^*. The last version of the HETE code presents the capability of providing data for several cross-flow heat exchangers configurations using the ε-NTU, P-NTU and $LMTD$ methods, including the comparison with several closed form and approximate solutions. In addition, the code can be used to obtain simulation data for plots purposes along with polynomial and curve fitting capabilities. The version of the HETE code presented in this monograph only includes the ε-NTU method.

The first external report based on an investigation performed with the help of the HETE code was the paper by Navarro and Cabezas-Gómez (2005). The HETE code was validated through a comparison of simulation results with numerical values obtained from closed form effectiveness analytical relationships for one-pass cross-flow heat exchangers with one to four rows (see forthcoming Table 2.1). Comparisons were also made with approximate infinite series solution for an unmixed-unmixed cross-flow heat exchanger. Very small maximum errors were obtained in all the cases considered in that paper. Results were also very satisfactory when compared with analytical solutions in terms P-NTU from the paper by Shah and Pignotti (1993) for two flow arrangements.

Results involving the thermal performance of multipass parallel and counter-cross-flow heat exchangers with external fluid unmixed were reported in a second paper by Cabezas-Gómez et al. (2007). Numerically obtained effectiveness results were compared with both analytical solutions for one to four rows and pure counter and parallel cross flows (see forthcoming Table 2.2). Small relative errors were obtained in all the comparisons. Finally, ε-NTU data for a complex flow arrangement proposed by Guo et al. (2002) were also computed and plotted. In both publications, Navarro and Cabezas-Gómez (2005) and Cabezas-Gómez et al. (2007), the computational and numerical procedures implemented into a HETE software were explained.

Recently Cabezas-Gómez et al. (2009, 2012) provided thermal effectiveness data for special new flow arrangements. These flow arrangements consist of two fluid circuits displayed in two up to six tube rows (passes), and two tube lines. The new characteristic of the proposed arrangements consisted in the inversion of flowing direction of the in-tube fluids, with respect to each other, in each pass. This distinct feature allowed the increment of the thermal effectiveness in comparison with its counterpart, consisting in counter-cross-flow arrangements with the same number of tube rows and lines. The HETE code facilitated the simulation of these new configurations. In these two papers, the following evaluation norms were also computed numerically with the HETE code: the heat exchanger efficiency recently defined by Fahkeri (2007), the heat exchanger reversibility norm (HERN), introduced by Sekulić (1990), and the entransy dissipation number, as defined by Guo et al. (2009).

In a last contribution, Cabezas-Gómez et al. (2014) published new effectiveness data and closed form relations for recently proposed flow arrangements used in industries and studied in research laboratories (see Pongsoi et al. 2011, 2012). The effectiveness data determined by applying the proposed correlations were equal to the effectiveness data computed with the HETE code, considering the convergence error of about 10^{-8} provided by the HETE code between the last two consecutive iterations of the external mean outlet temperature. Applying both, either the proposed closed form expressions or the HETE code, it is possible to use the ε-NTU method to compute experimentally the external fluid heat transfer coefficient for new heat transfer surfaces and flow arrangements.

References

Baclíç BS (1978) A simplified formula for cross-flow heat exchanger effectiveness. ASME J Heat Transf 100:746–747

Baclíç BS (1990) ε-NTU analysis of complicated flow arrangements. In: Shah RK, Kraus AD, Metzger D (eds) Compact heat exchangers—A Festschrift for AL London. Hemisphere Publishing, New York, pp 31–90

Baclíç BS, Heggs PJ (1985) On the search for new solutions of the single-pass crossflow exchanger problem. Int J Heat Mass Transf 28(10):1965–1975

Bes TH (1996) Thermal performance of codirected cross-flow heat exchangers. Heat Mass Transf 31:215–222

Binnie AM, Poole EGC (1937) The theory of single-pass cross-flow heat exchanger. Proc Cambr Phyl Soc 33:404–411

Cabezas-Gómez L, Navarro HA, Saiz-Jabardo JM (2007) Thermal performance of multi-pass parallel and counter cross-flow heat exchangers. ASME J Heat Transf 129:282–290

Cabezas-Gómez L, Navarro HA, Godoy SM, Campo A, Saiz-Jabardo JM (2009) Thermal characterization of a cross-flow heat exchanger with a new flow arrangement. Int J Therm Sci 48:2165–2170

Cabezas-Gómez L, Navarro HA, Saiz-Jabardo JM, Hanriot SM, Maia CB (2012) Analysis of new cross flow heat exchanger flow arrangement—Extension to several rows. Int J Therm Sci 55:122–132

Cabezas-Gómez L, Saiz-Jabardo Navarro HA, Barbieri PEL (2014) New thermal effectiveness data and formulae for some cross-flow arrangements of practical interest. Int J Heat Mass Transf 69:237–246

Chen TD, Conklin JC, Baxter VD (1998) Comparison of analytical and experimental effectiveness of four-row plate-fin-tube heat exchangers with water, R-22 and R-410A, ASME Winter Annual Meeting, Anaheim, California

Domingos JD (1969) Analysis of complex assemblies of heat exchangers. Int J Heat Mass Transf 12:537–548

ESDU 86018 (1991) Effectiveness-NTU relations for the design and performance evaluation of two-stream heat exchangers, Engineering Science Data Unit 86018 with amendment, London ESDU International plc, pp 92–107

Fakheri A (2007) Heat exchanger efficiency. ASME J Heat Transf 129:1268–1276

Guo Z-Y, Zhou S-Q, Li Z-X, Chen L-G (2002) Theoretical analysis and experimental confirmation of the uniformity principle of temperature difference field in heat exchanger. Int J Heat Mass Transf 45:2119–2127

Guo JF, Cheng L, Xu MT (2009) Entransy dissipation number and its application to heat exchanger performance evaluation. Chin Sci Bull 29:2708–2713

Hausen H (1983) Heat transfer in counterflow, parallel flow, and cross flow. McGraw-Hill, New York

Hewitt G, Shires G, Bott T (1994) Process heat transfer. CRC Press, Florida

Karst HH (1952) Mean temperature difference in multi-pass crossflow heat exchangers. In: Proceedings of 1st US National congress of applied mechanics, ASME, pp 949–955

Kays WM, London AL (1998) Compact heat exchangers, 3rd edn. McGraw Hill, New York

Li CHA (1987) New simplified formula for crossflow heat exchanger effectiveness. ASME J Heat Transf 109:521–522

Martin H (1990) Heat exchangers. Hemisphere Publishing Corp, Washington, DC

Mason JL (1955) Heat transfer in cross flow. In: Proceedings of 2nd US National congress of applied mechanics, New York, pp 801–803

Navarro HA, Cabezas-Gómez L (2005) A new approach for thermal performance calculation of cross-flow heat exchangers. Int J Heat Mass Transf 48:3880–3888

Nusselt W (1911) Der Wärmeübertragung in Kreuztrom. VDI-Zeitschr 55:2021–2024

Nusselt W (1930) Eine neue Formel für den Wärme-durchgang in Kreuztrom. Tech Mech Thermodyn, Berl 1:417–422

Pignotti A (1988) Linear matrix operator formalism for basic heat exchanger thermal design. ASME J Heat Transf 110:297–303

Pignotti A, Shah RK (1992) Effectiveness-number of transfer units relationships for heat exchanger complex flow arrangements. Int J Heat Mass Transf 35:1275–1291

Pongsoi P, Pikulkajorn S, Wang CC, Wongwises S (2011) Effect of fin pitches on the air side performance of crimped spiral fin-and-tube heat exchangers with a multipass parallel and counter cross-flow configuration. Int J Heat Mass Transf 54:2234–2240

Pongsoi P, Pikulkajorn S, Wang CC, Wongwises S (2012) Effect of number of tube rows on the air side performance of crimped spiral fin-and-tube heat exchangers with a multipass parallel and counter cross-flow configuration. Int J Heat Mass Transf 55:1403–1411

Sekulić DP (1990) The second law quality of energy transformation in a heat exchanger. ASME J Heat Transf 112:295–300

Sekulić DP, Shah RK, Pignotti A (1999) A review of solution methods for determining effectiveness-NTU relationships for heat exchangers with complex flow arrangements. Appl Mech Rev 52(3):97–117

Shah RK, Mueller AC (1985) Heat exchanger basic thermal design methods. In: Roshenow WM, Hartnett JP, Ganić EN (eds) Handbook of Heat Transfer, 2 edn. McGraw Hill, NewYork (Chapter 4, part 1)

Shah RK, Pignotti A (1993) Thermal analysis of complex cross-flow exchangers in terms of standard configurations. ASME J Heat Transf 115:353–359

Shah RK, Sekulić DP (2003) Fundamentals of heat exchanger design. Wiley, New York

Stevens RA, Fernandez J, Woolf JR (1957) Mean-temperature difference in one, two, and three-pass crossflow heat exchangers. Trans ASME 79:287–297

Taborek J (1983) Charts for mean temperature difference in industrial heat exchanger configuration. In: Schlünder EW (ed) Heat exchanger design Handbook. Hemisphere Publishing, New York (Chapter 1.5)

Wang C-C, Webb RL, Chi K-Y (2000) Data reduction for air-side performance of fin-and-tube heat exchanger. Exp Thermal Fluid Sci 21:218–226

Chapter 2
Theoretical Development

The ε-*NTU* method was formally introduced in a 1942 unpublished paper by London and Seban (A generalization of the methods of heat exchangers analysis. Stanford University, Stanford, California, 1980). Later on, Kays and London in 1952 used extensively the procedure in their well-known book "Compact Heat Exchangers", Kays and London (Compact heat exchangers. McGraw Hill, New York, 1998), where data for different geometries and flow arrangements can be found. From that time on, applications of the ε-*NTU* method have been growing to the point that in the present this could be considered as the most accepted heat exchangers design and analysis procedure.

2.1 Fundamentals of the ε-*NTU* Method

The ε-*NTU* method was formally introduced in a 1942 unpublished paper by London and Seban (1980). Later on, Kays and London in 1952 used extensively the procedure in their well-known book "Compact Heat Exchangers", Kays and London (1998), where data for different geometries and flow arrangements can be found. From that time on, applications of the ε-*NTU* method have been growing to the point that in the present this could be considered as the most accepted heat exchangers design and analysis procedure.

In the ε-*NTU* method the heat exchanger effectiveness, ε, plays a central role, though its concept is relatively simple. The idea of effectiveness of the heat exchanger has to do with the energy conservation, since it is defined as the ratio between the actual rate of heat transfer and the maximum rate of heat transfer for given inlet temperatures of the hot and cold streams. It is defined as:

$$\varepsilon = \frac{q}{q_{\max}} = \frac{C_h\left(T_{h,i} - T_{h,o}\right)}{C_{\min}\left(T_{h,i} - T_{c,i}\right)} = \frac{C_c\left(T_{c,o} - T_{c,i}\right)}{C_{\min}\left(T_{h,i} - T_{c,i}\right)} \tag{2.1}$$

© The Author(s) 2015
L. Cabezas-Gómez et al., *Thermal Performance Modeling of Cross-Flow Heat Exchangers*, SpringerBriefs in Applied Sciences and Technology, DOI 10.1007/978-3-319-09671-1_2

The rate of heat transfer between the hot and cold streams is given by the numerators of Eq. (2.1):

$$q = C_h\left(T_{h,i} - T_{h,o}\right) = C_c\left(T_{c,o} - T_{c,i}\right) \tag{2.2}$$

Indices "h" and "c" stand for hot and cold whereas "i" and "o" refer to the inlet and outlet sections. Note that the maximum rate of heat transfer is the one that would be obtained in a counter-current heat exchanger with the same inlet temperatures of the hot and cold fluids when the fluid with the lower heat capacity, C_{min}, would attain the same temperature as the inlet one of the hot (or cold) fluid.

The actual heat transfer rate can be written as:

$$q = \varepsilon q_{max} = \varepsilon C_{min}\left(T_{h,i} - T_{c,i}\right) \tag{2.3}$$

It is interesting to note that when one of the streams is changing phase at constant pressure, the heat capacity numerically tends to infinity. Thus the other stream is the C_{min} one. In general, it is possible to express the thermal effectiveness as a function of the number of transfer units NTU, the heat capacity rate ratio, C^*, and the flow arrangement of the heat exchanger,

$$\varepsilon = \Phi\left(NTU,\ C^*,\ flow\ arrangement\right) \tag{2.4}$$

The dimensionless parameters NTU and C^* are defined as:

$$NTU = \frac{UA}{C_{min}} \tag{2.5}$$

and

$$C^* = \frac{C_{min}}{C_{max}} \tag{2.6}$$

U and A stand for the overall heat transfer coefficient and the heat transfer surface area. According to its definition, Eq. (2.6), the heat capacity ratio is a number less or equal to one. According to Shah and Sekuliç (2003), the exchanger heat transfer rate can be written in terms of the mean temperature difference between the hot and cold streams as:

$$q = UA\Delta T_m \tag{2.7}$$

Thus, combining Eqs. (2.3) and (2.7), an alternate expression for the thermal effectiveness of the heat exchanger can be obtained.

$$\varepsilon = \left(\frac{UA}{C_{min}}\right)\left(\frac{\Delta T_m}{\Delta T_{max}}\right) = NTU\left(\frac{\Delta T_m}{\Delta T_{max}}\right) \tag{2.8}$$

Note that ΔT_m is the effective mean temperature difference also known as the "mean temperature driving potential", Shah and Sekuliç (2003). Note also that ΔT_m is related to the log men temperature difference, $LMTD$, by the following expression:

$$\Delta T_m = F(LMTD) \tag{2.9}$$

F is the well-known correction factor of the $LMTD$ procedure.

According to Shah and Sekulič (2003), *NTU* could be considered as a dimensionless size or, in other words, the thermal size of the heat exchanger. Thus it is a design parameter. Another point of view regarding the physical meaning of this dimensionless parameter could be devised in terms of a temperature difference ratio. In fact, the following expression results by introducing Eqs. (2.2) and (2.7) into Eq. (2.5):

$$NTU = \frac{q/\Delta T_m}{q/\Delta T_{max}} = \frac{\Delta T_{max}}{\Delta T_m} \tag{2.10}$$

Note that

$$NTU = \begin{cases} \frac{\Delta T_c}{\Delta T_m} \to C_{min} = C_c \\[2mm] \frac{\Delta T_h}{\Delta T_m} \to C_{min} = C_h \end{cases}$$

Thus, *NTU* could be considered as the ratio between the maximum temperature variation and the mean temperature difference between the streams in the heat exchanger. High values of *NTU* would correspond, for example, to low temperature difference between the hot and cold streams. Figure 2.1 illustrates operational conditions corresponding to high and low values of *NTU*. Note that in the first case the value of the mean temperature difference is relatively high, corresponding to low *NTU*, whereas in the second is low, high *NTU*.

The functional relation expressed by Eq. (2.4) could have been obtained by dimensional analysis, the Buckinham's π theorem. This approach just proves that the effectiveness could be expressed as a function of *NTU*, C^*, and the particular geometry of the heat exchanger. However, the correlation that relates these dimensionless parameters is yet to be determined. This will be done for the counter flow heat exchanger of Fig. 2.1 assuming that the heat capacity of the hot fluid is lower than the cold one, corresponding to the right plot. To start, let us consider the expression for the log mean temperature difference:

$$LMTD = \frac{\left(T_{h,i} - T_{c,o}\right) - \left(T_{h,o} - T_{c,i}\right)}{\ln\left(\frac{T_{h,i} - T_{c,o}}{T_{h,o} - T_{c,i}}\right)}$$

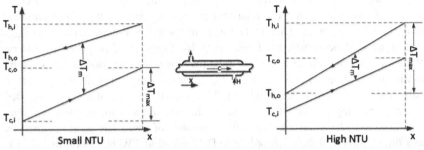

Fig. 2.1 Plots of temperatures of the hot and cold fluids along a counter flow heat exchanger illustrating conditions of high and low *NTU*

The left and right hand sides of Eq. (2.7) can be transformed into the following expressions:

$$\frac{q}{UA} = \frac{\left(T_{h,i} - T_{h,o}\right) - \left(T_{c,o} - T_{c,i}\right)}{\ln\left(\frac{T_{h,i}-T_{c,o}}{T_{h,o}-T_{c,i}}\right)} = \frac{\frac{q}{C_h} - \frac{q}{C_c}}{\ln\left(\frac{T_{h,i}-T_{c,o}}{T_{h,o}-T_{c,i}}\right)}$$

Combining the left and the right hand side and canceling out q, results:

$$\frac{\frac{UA}{C_h}\left(1 - \frac{C_h}{C_c}\right)}{\ln\left(\frac{T_{h,i}-T_{c,o}}{T_{h,o}-T_{c,i}}\right)} = 1 \Rightarrow \frac{T_{h,o} - T_{c,i}}{T_{h,i} - T_{c,o}} = \exp\left[-\frac{UA}{C_h}\left(1 - \frac{C_h}{C_c}\right)\right]$$

By summing and subtracting $T_{h,i}$ in the numerator of the right hand side and $T_{c,i}$ in the denominator of the left hand side of the above equation and introducing the definition equation of the thermal effectiveness, Eq. (2.1), remembering that $C_{min} = C_h$, results:

$$\frac{-\varepsilon + 1}{1 - \varepsilon C^*} = \exp\left[-NTU\left(1 - C^*\right)\right]$$

Finally,

$$\varepsilon = \frac{1 - \exp\left[-NTU(1 - C^*)\right]}{1 - C^* \exp\left[-NTU(1 - C^*)\right]} \tag{2.11}$$

Equation (2.11) was obtained for a geometry corresponding to the counter flow heat exchanger of Fig. 2.1. Expressions for other geometries can be found in heat transfer text books and in the general literature. Data for advanced geometries will be presented herein as case studies applications of the code HETE.

The functional relation, Eq. (2.4), is applied for rating the heat exchanger by determining the thermal effectiveness for a given geometry. In designing, ε-NTU correlations are used in sizing the heat exchanger based on the inlet and outlet temperatures of the streams. In this case explicit NTU correlations are needed in terms of ε and C^* for a given geometry.

Table 2.1 presents a summary of correlations for one-pass cross-flow configurations (ESDU 86018 1991; Stevens et al. 1957; Bacliç and Heggs 1985). Details regarding the derivation of these relationships can be found in the aforementioned publications and others cited herein. Correlations for unmixed-unmixed flow arrangements deserve some comments. Equation (2.16) (Table 2.1), proposed by Mason (1955), valid for an infinite number of tube rows, was the one used as reference by Navarro and Cabezas-Gómez (2005) for comparison purposes with the numerical results from the HETE code. In addition, this correlation is one of several suggested by Bacliç and Hegss (1985) for computing the effectiveness for this flow arrangement. Equation (2.17), also for an infinite number of rows, has been obtained by curve fitting numerical data. According to DiGiovanni and Webb (1989) its origin is uncertain, though it appears in a footnote on page 483 of the book by Eckert (1959). Navarro and Cabezas-Gómez (2005) claim that the application of this correlation can

Table 2.1 ε-NTU relationships for one pass cross-flow configurations with one or more rows

N_r	C_{min} side	Relation	Equations
1	A	$\varepsilon_A = 1 - e^{-(1-e^{NTU_A \cdot C_A^*})/C_A^*}$	(2.12a)
	B	$\varepsilon_B = \frac{1}{C_B^*}\left[1 - e^{-C_B^*(1-e^{NTU_B})}\right]$	(2.12b)
2	A	$\varepsilon_A = 1 - e^{-2K/C_A^*}\left(1 + \frac{K^2}{C_A^*}\right),\ K = 1 - e^{-NTU_A \cdot C_A^*/2}$	(2.13a)
	B	$\varepsilon_B = \frac{1}{C_B^*}\left[1 - e^{-2KC_B^*}\left(1 + K^2 C_B^*\right)\right],\ K = 1 - e^{-NTU_B/2}$	(2.13b)
3	A	$\varepsilon_A = 1 - e^{-3K/C_A^*}\left(1 + \frac{K^2(3-K)}{C_A^*} + \frac{3K^4}{2(C_A^*)^2}\right),\ K = 1 - e^{-NTU_A \cdot C_A^*/3}$	(2.14a)
	B	$\varepsilon_B = \frac{1}{C_B^*}\left[1 - e^{-3KC_B^*}\left(1 + K^2(3-K)\cdot C_B^* + \frac{3K^4(C_B^*)^2}{2}\right)\right],\ K = 1 - e^{-NTU_B/3}$	(2.14b)
4	A	$\varepsilon_A = 1 - e^{-4K/C_A^*}\left(1 + \frac{K^2(6-4K+K^2)}{C_A^*} + \frac{4K^4(2-K)}{(C_A^*)^2} + \frac{8K^6}{3(C_A^*)^3}\right),\ K = 1 - e^{-NTU_A \cdot C_A^*/4}$	(2.15a)
	B	$\varepsilon_B = \frac{1}{C_B^*}\left[1 - e^{-4KC_B^*}\left(1 + K^2(6-4K+K^2)C_B^* + 4K^4(2-K)(C_B^*)^2 + \frac{8K^6(C_B^*)^3}{3}\right)\right],\ K = 1 - e^{-NTU_B/4}$	(2.15b)
∞	Both fluids unmixed	$\varepsilon = \frac{1}{C^*NTU}\sum_{m=0}^{\infty}\left\{\left[1 - e^{-NTU}\sum_{m=0}^{n}\frac{(NTU)^m}{m!}\right]\left[1 - e^{-C^*NTU}\sum_{m=0}^{n}\frac{(C^*NTU)^m}{m!}\right]\right\}$	(2.16)
∞	Both fluids unmixed	$\varepsilon = 1 - e^{\left[NTU^{0.22}\left(e^{-C^*\, NTU^{0.78}}-1\right)/C^*\right]}$	(2.17)
∞	Both fluids unmixed	$\varepsilon = 1 - e^{-NTU} - e^{[-(1+C^*)NTU]}\sum_{n=1}^{\infty}C^{*n}P_n(NTU),\ P_n(y) = \frac{1}{(n+1)!}\sum_{j=1}^{n}\frac{(n+1-j)}{j!}y^{n+j}$	(2.18)

Equations (2.12)–(2.15) from Kays and London (1998), ESDU 86018 (1991), Stevens et al. (1957); Eq. (2.15) from Stevens et al. (1957) and Baclic and Heggs (1985); Eq. (2.16) from ESDU 86018 (1991); Eq. (2.18) from Shah and Sekulic (2003)
Fluid A mixed, Fluid B unmixed, $C_A^* = 1/C_B^*$ $\varepsilon_B = \varepsilon_A C_A^*,\ NTU_B = NTU_A C_A^*,\ C_A^* = C_A/C_B$

result in deviations of the order of 4 % for certain values of NTU and C^* when compared with more accurate solutions such as the HETE code and Eq. (2.16). Shah and Sekuliç (2003) suggest the use of Eq. (2.18) for the same flow arrangement.

Table 2.2 presents correlations for different multi-pass parallel and counter-cross-flow arrangements with widespread applications, Kays and London (1998), Taborek (1983), ESDU 86018 (1991), and Cabezas-Gómez et al. (2007). Equations (2.22) and (2.26) can be found in heat transfer and specialized heat exchanger textbooks such as the ones by Incropera et al. (2008) and Nellis and Klein (2009).

The development of the HETE code and its numerical solution will be the subject of the text that follows.

2.2 Differential Governing Equations

The numerical procedure of the HETE code is based on a mathematical model whose governing equations will be presented in this section. Based on the procedure by Kays and London (1998) (Appendix C), the governing equations consist of the Energy Conservation applied to a one tube row cross-flow heat exchanger with one fluid mixed and other unmixed. The procedure presented herein can be found in previous publications as in Navarro and Cabezas-Gómez (2005, 2007) and Cabezas-Gómez et al. (2007). The governing equations of the HETE code are based on the following basic assumptions, which can also be found in the books by Kays and London (1998) and Shah and Sekuliç (2003): (1) the heat exchanger operates under steady state conditions; (2) heat losses to the surroundings are neglected, that is, the heat exchanger is assumed to be adiabatic; (3) there are no thermal energy sources or sinks in the heat exchanger walls or fluids; (4) the tube side fluid is perfectly mixed in the tube cross section, its temperature varying linearly along the tube axis; (5) the external fluid is unmixed, i.e., there are fins in the external side; (6) heat transfer coefficients and transport properties of the fluids and heat exchanger walls are constant; (7) axial heat transfer in the solid walls and fluids are neglected; (8) there is no phase change in both streams; (9) for analysis purposes, the in-tube fluid is assumed the hot fluid though governing equations could be applied otherwise, that is, the in-tube being the cold one, by just interchanging the subscripts "c" and "h".

Figure 2.2 illustrates the temperature distribution of both fluids along the transversal and longitudinal direction with respect to the in-tube fluid for a one pass cross-flow mixed-unmixed heat exchanger having one tube row with one circuit. Along the differential strip of length "dx" shown in Fig. 2.2, the mass flow rate of the external fluid (cold one) is small and since the heat transfer rate is also small, one might expect that the in-tube fluid temperature remains essentially constant, as suggested in the figure. An energy balance in the differential strip length for the hot and cold fluids can be written as:

$$\delta q = -C_h dT_h \tag{2.27}$$

$$\delta q = (dC_c)(\Delta T_c) \tag{2.28}$$

Table 2.2 ε-NTU relationships for multi-pass parallel and counter cross-flow configurations

N_r	C_{min} side	Relation	Equation
Multi-pass parallel cross-flow			
2	A	$\varepsilon_A = (1 - \frac{K}{2})(1 - e^{-2K/C_A^*})$, $K = 1 - e^{-NTU_A \cdot C_A^*/2}$	(2.19a)
	B	$\varepsilon_B = \frac{1}{C_B^*}(1 - \frac{K}{2})(1 - e^{-2KC_B^*})$, $K = 1 - e^{-NTU_B/2}$	(2.19b)
3	A	$\varepsilon_A = 1 - (1 - \frac{K}{2})^2 e^{-3K/C_A^*} - K(1 - \frac{K}{4} + \frac{K}{C_A^*}(1 - \frac{K}{2}))e^{-K/C_A^*}$, $K = 1 - e^{-NTU_A \cdot C_A^*/3}$	(2.20a)
	B	$\varepsilon_B = \frac{1}{C_B^*}\left[1 - (1 - \frac{K}{2})^2 e^{-3KC_B^*} - K(1 - \frac{K}{4} + KC_B^*(1 - \frac{K}{2}))e^{-KC_B^*}\right]$, $K = 1 - e^{-NTU_B/3}$	(2.20b)
4	A	$\varepsilon_A = 1 - \frac{K}{2}(1 - \frac{K}{2} + \frac{K^2}{4}) - K(1 - \frac{K}{2})(1 + 2\frac{K}{C_A^*}(1 - \frac{K}{2}))e^{-2K/C_A^*} - (1 - \frac{K}{2})^3 e^{-4K/C_A^*}$, $K = 1 - e^{-NTU_A \cdot C_A^*/4}$	(2.21a)
	B	$\varepsilon_B = \frac{1}{C_B^*}\left[1 - \frac{K}{2}(1 - \frac{K}{2} + \frac{K^2}{4}) - K(1 - \frac{K}{2})(1 + 2KC_B^*(1 - \frac{K}{2}))e^{-2KC_B^*} - (1 - \frac{K}{2})^3 e^{-4KC_B^*}\right]$, $K = 1 - e^{-NTU_B/4}$	(2.21b)
∞	Parallel flow A or B	$\varepsilon = \frac{1 - e^{-NTU(1+C^*)}}{1+C^*}$	(2.22)
Multi-pass counter cross-flow			
2	A	$\varepsilon_A = 1 - \left(\frac{K}{2} + (1 - \frac{K}{2})e^{2K/C_A^*}\right)^{-1}$, $K = 1 - e^{-NTU_A \cdot C_A^*/2}$	(2.23a)
	B	$\varepsilon_B = \frac{1}{C_B^*}\left[1 - \left(\frac{K}{2} + (1 - \frac{K}{2})e^{2KC_B^*}\right)^{-1}\right]$, $K = 1 - e^{-NTU_B/2}$	(2.23b)
3	A	$\varepsilon_A = 1 - \left((1 - \frac{K}{2})^2 e^{3K/C_A^*} + (K(1 - \frac{K}{4}) - (1 - \frac{K}{2}\frac{K^2}{C_A^*})e^{K/C_A^*}\right)^{-1}$, $K = 1 - e^{-NTU_A \cdot C_A^*/3}$	(2.24a)
	B	$\varepsilon_B = \frac{1}{C_B^*}\left[1 - \left((1 - \frac{K}{2})^2 e^{3KC_B^*} + (K(1 - \frac{K}{4}) - (1 - \frac{K}{2})K^2 C_B^*)e^{KC_B^*}\right)^{-1}\right]$, $K = 1 - e^{-NTU_B/3}$	(2.24b)
			(continued)

Table 2.2 (continued)

N_r	C_{min} side	Relation	Equation
4	A	$\varepsilon_A = 1 - \left(\dfrac{K}{2}\left(1 - \dfrac{K}{2} + \dfrac{K^2}{4}\right) + K\left(1 - \dfrac{K}{2}\right)\left(1 - 2\dfrac{K}{C_A^*}\left(1 - \dfrac{K}{2}\right)\right)e^{2K/C_A^*} + \left(1 - \dfrac{K}{2}\right)^3 e^{4K/C_A^*}\right)^{-1}$, $K = 1 - e^{-NTU_A \cdot C_A^*/4}$	(2.25a)
	B	$\varepsilon_B = \dfrac{1}{C_B^*}\left[1 - \left(\dfrac{K}{2}\left(1 - \dfrac{K}{2} + \dfrac{K^2}{4}\right) + K\left(1 - \dfrac{K}{2}\right)\left(1 - 2KC_B^*\left(1 - \dfrac{K}{2}\right)\right)e^{2KC_B^*} + \left(1 - \dfrac{K}{2}\right)^3 e^{4KC_B^*}\right)^{-1}\right];$ $K = 1 - e^{-NTU_B/4}$	(2.25b)
∞	Counterflow A or B	$\varepsilon = \dfrac{1 - e^{-NTU(1-C^*)}}{1 - C^* e^{-NTU(1-C^*)}}$	(2.26)

Equations (2.20) and (2.24) from Kays and London (1998); Eqs. (2.19) from ESDU 86018 (1991); Eqs. (2.21)–(2.23) from Taborek (1983) and ESDU 86018 (1991)

Fluid A mixed, Fluid B unmixed, $C_A^* = 1/C_B^*$, $\varepsilon_B = \varepsilon_A\, C_A^*$, $NTU_B = NTU_A\, C_A^*$, $C_A^* = C_A/C_B$

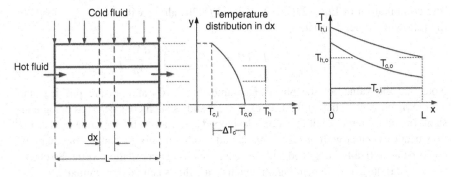

Fig. 2.2 Temperature variation along the transversal and longitudinal directions in a one pass cross-flow mixed-unmixed heat exchanger

Note that ΔT_c is the temperature variation of the cold fluid in the differential strip, that is, $\Delta T_c = T_{c,o} - T_{c,i}$, where the subscripts stand for the outlet and inlet temperatures of the external fluid (cold one) in the strip.

Given the fact that the mass flow rate of the cold fluid in the differential strip is small, it can be concluded that the heat capacity ratio for the strip heat exchanger is given by the following expression:

$$dC^* = \frac{dC_c}{C_h} \rightarrow 0 \tag{2.29}$$

This is a differential heat capacity ratio, which tends to zero since the mass flow rate of the cold fluid also tends to zero. Physically this result is equivalent as considering the differential strip heat exchanger a condenser, since the temperature of the hot fluid remains essentially constant, as shown in Fig. 2.2. The "local thermal effectiveness" of the strip exchanger, Γ, can thus be determined from Eq. (2.11) which assumes the following expression for the heat capacity ratio tending to zero:

$$\Gamma = \frac{\Delta T_c}{T_h - T_{c,i}} = 1 - \exp\left[-\frac{UdA}{dC_c}\right] \tag{2.30}$$

Assuming that both the frontal (approaching) area of the cold fluid side, A_{fr}, and the surface heat transfer area, A, are uniform over the heat exchanger width, the following expressions can be written:

$$\frac{dC_c}{dA_{fr}} = \frac{C_c}{A_{fr}} = \text{constant} \tag{2.31}$$

$$\frac{dC_c}{dA} = \frac{C_c}{A} = \text{constant} \tag{2.32}$$

Thus, introducing Eq. (2.32) into Eq. (2.30), the following local thermal effectiveness expression results, valid over the width L of the heat exchanger:

$$\Gamma = 1 - e^{-\frac{UA}{C_c}} = \text{constant} \tag{2.33}$$

The combination of Eqs. (2.27), (2.28), and (2.30) along with Eq. (2.31), leads to the following general equation:

$$\frac{dT_h}{T_h - T_{c,i}} = -\Gamma dC^* = -\Gamma \left(\frac{C_c}{C_h}\right)\left(\frac{dA_{fr}}{A_{fr}}\right) \tag{2.34}$$

Note that the values of the physical parameters C_c, C_h and A_{fr} are physical and geometric characteristics of the heat exchanger and as such considered as being constant. In addition, the set of equations developed so far are valid for a one pass cross-flow heat exchanger with one fluid mixed and another unmixed. This set of equations could be analytically integrated and the resulting ε-NTU correlation would be one of Eqs. (2.12a) or (2.12b) depending on which of the fluids is mixed or unmixed.

Cross-flow heat exchangers for engineering applications are commonly characterized by complex flow arrangements with several circuits and rows (see Chap. 6 for several examples). The previous set of equations Eqs. (2.27)–(2.34) does not have a trivial solution for those complex configurations since the conditions under which Eq. (2.34) has been derived are no longer valid due to: (1) application of Eqs. (2.32) and (2.33) is questionable in those cases; (2) the cold fluid inlet temperature at each tube row is not uniform as in Fig. 2.2. An adequate and accurate solution for heat exchanger complex geometries can be sought through a numerical computational procedure as the HETE code. This procedure is based on a set of algebraic equations, amenable to a numerical solution, whose development and analysis will be dealt with in the next section.

2.3 Finite Elements Governing Equations of the HETE Code

A differential point of view of the heat exchanger was considered in the last section. Though solutions of the differential set of equations for simple flow arrangements have been found, more complex flow arrangements are not amenable to simple solutions. In those cases, numerical solutions of a discretized set of equations could be the answer. This is the case of the HETE code whose finite elements governing equations will be developed in the present section.

The element in the HETE code is a finite volume heat exchanger involving a finite length of the in-tube fluid with its corresponding external fluid counterpart, as illustrated in Fig. 2.3. The governing equations of Sect. 2.2 are based on the assumption that the cold fluid mass flow rate is small in comparison with the hot fluid one, as suggested by Eq. (2.29). This is not the case for the element of Fig. 2.3. Thus the hot fluid temperature variation along the element cannot be neglected. Instead, it is assumed that it varies linearly so that its average temperature can be written as:

$$T_h^e = 0.5\left(T_{h,i}^e + T_{h,o}^e\right) \tag{2.35}$$

The superscript "e" has been introduced to designate an element whose identification, characterized by its space location, will be designated by the subscripts (i, j, k).

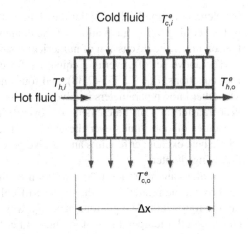

Fig. 2.3 Illustration of an element with associated inlet and outlet temperatures

By integrating the energy conservation for the hot fluid along the element, Eq. (2.27), the following expression results:

$$q^e = -C_h^e(T_{h,o}^e - T_{h,i}^e) \tag{2.36}$$

The integration of the energy conservation applied to the cold fluid, Eq. (2.28), followed by the introduction of Eq. (2.31), results in the following expression:

$$q^e = \Delta T_c^e \int_e dC_c = \Delta T_c^e \int_e \frac{C_c}{A_{fr}} dA_{fr} = \Delta T_c^e C_c^e \tag{2.37}$$

As previously observed, C_c and A_{fr} are the heat capacity of the cold fluid and the frontal area of the heat exchanger. At his point it is important to stress that, since the inlet as well as the outlet temperatures of the cold stream might vary along width of the element, the cold fluid variation along the element, $\Delta T_c^e = T_{c,o}^e - T_{c,i}^e$, is determined in terms of the average inlet and outlet temperatures of the cold fluid, that is, $T_{c,i}^e$ and $T_{c,o}^e$.

The closure of the set of algebraic equations requires an additional equation which is the one of the thermal effectiveness, given by integration of Eq. (2.30) over the total area of the element. For that purpose, Eq. (2.32) must be taken into account in such a way to introduce the relation $dA^e/dC_c^e = A^e/C_c^e$ into Eq. (2.30) with the following result:

$$\Gamma^e = \frac{\Delta T_c^e}{(T_h^e - T_{c,i}^e)} = 1 - e^{-\frac{UA^e}{C_c^e}} \tag{2.38}$$

Note that the temperature of the hot fluid is the average over the element instead of the inlet one. The right hand side of the equation above is formally identical to that of Eq. (2.33) though in this case the area and the heat capacity refer to the element. The result expressed by Eq. (2.38) is physically equivalent to admit that the size of the element is small. The smaller the element the more accurate will be the solution of the set of governing equations of the element. This aspect will be addressed further on under the discussion of numerical results from the HETE code.

The above set of governing equations for each element is a closed system with four equations, Eqs. (2.35)–(2.38), and four unknowns namely $T_{h,o}^e$, $T_{c,o}^e$, q^e, and Γ^e. The other known parameters involve the element inlet temperatures, $T_{h,i}^e$ and $T_{c,i}^e$, the heat transfer area of the element, the overall heat transfer coefficient, and the heat capacities of both fluids over the element. The numerical solution extended to the whole heat exchanger requires an iterative procedure which will be introduced in the following chapter.

An alternate approach to the solution of the set of governing equations would be through the reduction of the number of unknowns (and equations) by rearranging the four available equations in such a way to obtain explicit expressions for the element outlet temperatures of the hot and cold streams which assume the following expressions:

$$T_{c,o}^e = \frac{\Lambda^e + 2(1 - \Gamma^e)}{2 + \Lambda^e} T_{c,i}^e + \frac{2\Gamma^e}{2 + \Lambda^e} T_{h,i}^e \tag{2.39}$$

$$T_{h,o}^e = \frac{2 - \Lambda^e}{2 + \Lambda^e} T_{h,i}^e + \frac{2\Lambda^e}{2 + \Lambda^e} T_{c,i}^e \tag{2.40}$$

The parameter Λ^e is given by the following expression

$$\Lambda^e = C_c^e \Gamma^e / C_h^e \tag{2.41}$$

Note that the outlet temperatures can be directly determined from Eqs. (2.39) and (2.40) since the element thermal effectiveness can be determined from Eq. (2.38). The explicit expressions for the element outlet temperatures have been implemented into the HETE code, as will be shown in the next chapter.

References

Bacliç BS, Heggs PJ (1985) On the search for new solutions of the single-pass crossflow heat exchanger problem. Int J Heat Mass Transf 28(10):1965–1975

Cabezas-Gómez L, Navarro HA, Saiz-Jabardo JM (2007) Thermal performance of multi-pass parallel and counter cross-flow heat exchangers. ASME J Heat Transf 129:282–290

DiGiovanni MA, Webb RL (1989) Uncertainty in effectiveness-NTU calculations for cross flow heat exchangers. Heat Transf Eng 10(3):61–70

Eckert ERG (1959) Heat transfer, 2nd edn. McGraw-Hill, New York

ESDU 86018 (1991) Effectiveness-NTU relations for the design and performance evaluation of two-stream heat exchangers, Engineering Science Data Unit 86018 with amendment, London ESDU International plc, 92–107

Incropera FP, DeWitt DP, Bergman TL, Lavine AS (2008) Fundamentals of heat and mass transfer, 6th edn. Wiley, Hoboken, New York

Kays WM, London AL (1998) Compact heat exchangers, 3rd edn. McGraw Hill, New York

London AL, Seban RA (1980) A generalization of the methods of heat exchangers analysis, TR No. NTU-1, Mech. Engineering Department, Stanford University, Stanford, California, (1942): reprinted in International Journal of Heat and Mass Transfer 23:5–16

Mason JL (1955) Heat transfer in cross flow. In: Proceedings of 2nd US National congress of applied mechanics, New York, pp 801–803

Navarro HA, Cabezas-Gómez L (2005) A new approach for thermal performance calculation of cross-flow heat exchangers. Int J Heat Mass Transf 48:3880–3888

Navarro HA, Cabezas-Gómez L (2007) Effectiveness-NTU computation with a mathematical model for cross-flow heat exchangers. Braz J Chem Eng 24(4):509–521

Nellis G, Klein S (2009) Heat transfer. Cambridge University Press, New York

Shah RK, Sekuliç DP (2003) Fundamentals of heat exchanger design. Wiley, New York

Stevens RA, Fernandez J, Woolf JR (1957) Mean-temperature difference in one, two, and three-pass crossflow heat exchangers. Trans. ASME 79:287–297

Taborek J (1983) Charts for mean temperature difference in industrial heat exchanger configuration. In: Schlünder EW (ed) Heat exchanger design handbook, Hemisphere Publishing, New York (Sect. 1.5)

Chapter 3
Computational Procedure

The computational procedure of the HETE code will be developed in this chapter. This procedure has been introduced in several papers among them the ones by Navarro and Cabezas-Gómez (2005), Cabezas-Gómez et al. (2007) and Navarro et al. (2010). It constitutes a useful numerical methodology for computing the thermal performance parameters for, among other, cross-flow heat exchangers with diverse flow arrangements.

3.1 Numerical Procedure of the HETE Code

The numerical methodology employed by the HETE code to solve the set of element governing equations, Eqs. (2.35)–(2.38) or equivalently Eqs. (2.39) and (2.40), and their extension to the whole heat exchanger, aiming at determining the effectiveness, is based on the following steps along with some explanatory comments regarding other aspects of the code, Navarro and Cabezas-Gómez (2005).

(i) Initially the heat exchanger is divided into a set of finite, three dimensional control volumes, herein designated by elements, each one being a small size mixed-unmixed cross-flow heat exchanger, as shown in Fig. 3.1. Each element is identified by the triplet (i, j, k) which characterizes the spatial position of the element in the heat exchanger. Thus, the components of the triplet vary in the following ranges: $1 \leq i \leq N_e$; $1 \leq j \leq N_t$; and $1 \leq k \leq N_r$, where N_e is the number of elements along each tube, N_t the number of tubes per row, and N_r the number of tube rows. In the case of Fig. 3.1, there are 8 tubes per row and 4 tube rows. The number of elements along each tube is chosen by the user. The total number of elements in the heat exchanger is equal to the product $N_e \times N_t \times N_r$, which are part of the input data of the HETE code.

(ii) The next step involves the solution of the set of Eqs. (2.35)–(2.38), or equivalently Eqs. (2.39) and (2.40), for each element. The set of equations for the whole heat exchanger is solved in an iterative procedure whenever required, the convergence being achieved when a "steady" temperature distribution

© The Author(s) 2015
L. Cabezas-Gómez et al., *Thermal Performance Modeling of Cross-Flow Heat Exchangers*, SpringerBriefs in Applied Sciences and Technology,
DOI 10.1007/978-3-319-09671-1_3

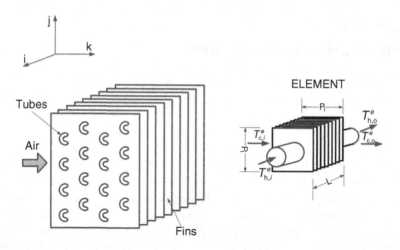

Fig. 3.1 Illustration of a cross flow mixed-unmixed heat exchanger and a tube element

over the heat exchanger is obtained. The procedure is performed step-by-step, starting from the first element of the first circuit, following the in-tube fluid along each circuit, through the range of the triplet (i, j, k).

(iii) Finally, with all the needed parameters being available, as, for example, the determined exit temperatures, the heat exchanger effectiveness is evaluated by applying Eq. (2.1).

(iv) The number of elements determines de accuracy of the results. In the present version of the code 1,000 elements for the whole heat exchanger are used though this number can be changed by the operator. Due to the small size of each element, the external fluid is the one with the minimum heat capacity, C_{min}. Physically, each element works as if it were a heat exchanger connected to others by the in-tube fluid along the circuits. It must be emphasized that the in-tube fluid is perfectly mixed in each element cross-sectional area whereas the external fluid is perfectly unmixed.

(v) From the heat exchanger point of view, the proposed algorithm could be used for an unmixed/unmixed configuration, see the configuration for 100 circuits in the internet site http://extras.springer.com. Cross-flow heat exchangers with other flow arrangements as, for example, multi-pass parallel and counter-cross-flow configurations with different in-tube fluid mixing arrangements can also be simulated. However, the present version of the code does not contemplate the mixture of in-tube fluid between passes. Mixture of the external fluid between passes, corresponding to the identical order coupling distribution Shah and Sekuliç (2003), can be simulated through a modification of the heat exchanger as suggested by Cabezas-Gómez et al. (2009) and Cabezas-Gómez et al. (2012). However, the present version of the code does not contemplate the complete mixture of the external fluid between passes.

Details of the procedure described above will be developed in the next section. The following is a summary of the steps involved in the procedure.

(1) The first step consists in reading the configuration input data file, where flow arrangement details are provided to the code (see Table 3.1).
(2) Then thermal input data are read.
(3) The iterative solution of the governing equations follows next. This is the step where the code algorithm is applied ending with the computation of the desired thermal characteristics.
(4) The log of obtained data in the results file ends the code process.

3.1.1 Reading the Configuration Input Data

The list of input data needed for performing the simulation of a particular heat exchanger flow arrangement is displayed in Table 3.1. In this table input data are shown for a parallel-cross-flow arrangement with four in-tube fluid passes regarding the external air flow. The input data for this configuration can be found in Chap. 6 under the name ("1tube4row4pass1cir(par).txt"). They can also be found on the internet site http://extras.springer.com. The flow arrangement shown schematically in Fig. 3.2 consists of four rows and one tube line (one tube per row).

Table 3.1 Input data for the flow arrangement of Fig. 3.2

Total number of tube rows (N_r)		4			
Total number of tube lines (horizontal lines) (N_l)		1			
Total number of tubes per row (N_t)		1			
Total number of in-tube fluid circuits (N_c)		1			
Data for each in-tube fluid circuit (*Provided separately for each circuit in the same file*)					
Inlet tube line (i circuit)	Inlet row (i circuit)	1	1		
End tube line (i circuit)	End row (i circuit)	1	4		
Number of data per circuit		4			
i Circuit trajectory data		A	B	C	D
		1	1	1	2
		2	0	1	3
		2	1	1	4
		3	0	0	0

A: Type of tube; 1 (Inlet tube), 2 (Intermediary tube), 3 (Ending tube)

B: Flow direction; 1 (Into the page), 0 (Out of the page)

C: Next tube line; 0 (indicates end of data for i circuit)

D: Next tube row; 0 (indicates end of data for i circuit)

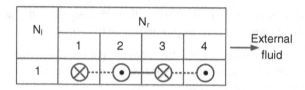

Fig. 3.2 Schematic representation of a parallel cross-flow heat exchanger with one circuit (one tube line) performing four passes and four rows (one tube per row). ("1tube4row4pass1cir(par).txt")

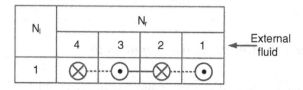

Fig. 3.3 Schematic representation of a counter cross-flow heat exchanger with one circuit (one tube line) performing four passes and four rows (one tube per row). ("1tube4row4pass1cir(cc).txt")

The following comments might be useful in the comprehension of Table 3.1:

- **Inlet tube line and inlet row**. Indicates where the ith in-tube fluid circuit starts (line and row). They are labeled by the particular tube line and row numbers, ordered sequentially.
- **Outlet tube line and outlet row**. Indicates where the ith in-tube fluid circuit ends (line and row). They are labeled by the particular tube line and row numbers.
- **Number of data per ith circuit**. Indicates how many data lines will be set for each in-tube fluid circuit. Each line of four data (A, B, C, and D) in the file corresponds to only one tube. This is the reason for using four data lines in Table 3.1.
- **In-tube fluid ith circuit path data**. Indicates the in-tube fluid circuit flow arrangement, starting from the inlet line and row down to the end line and row. The first column, column A in Table 3.1, indicates what kind of tube is being used. There are three types of tubes: **1** (inlet tube); **2** (intermediate tube); and **3** (end tube). The second column, column B in Table 3.1, indicates the flow direction of the fluid inside the tube. Two directions are considered: **1** (into the page); and **0** (out of the page). The third and fourth columns, columns C and D in Table 3.1, stand for the next tube line and row for a particular circuit. In both cases a zero value indicates the end of data for the ith circuit.

It is important to note that the air flow always starts from the first row of tubes ($N_r = 1$). In order to simulate a counter-cross-flow heat exchanger the in-tube fluid circuit must start from the last row, which, in the case of Fig. 3.3, is the fourth. This can be done in this case by changing the values of N_r as shown in Fig. 3.3. The input data for this flow arrangements are displayed in the file ("1tube4row4pass1cir(cc).txt") shown in Chap. 6.

Table 3.2 Thermal input data used for computing the thermal effectiveness for a particular configuration

Input data	User	Code
$C*$	X	
NTU	X	
C_{min} side	X	
$T_{h,i}$		X
$T_{c,i}$		X
UA		X

3.1.2 Thermal Input Data

The necessary input thermal characteristics for computing the heat exchanger effectiveness, $\varepsilon(C*, NTU)$ are shown in Table 3.2. The user provides values of $C*$, NTU, and the fluid with the minimum heat capacity rate, C_{min}. The other thermal variables necessary for computing the heat exchanger effectiveness are set inside the code. These variables are the inlet mean temperatures of both fluids, as well as the conductance, UA, of the heat exchanger. These data are constants in all the simulations.

The name of the configuration input data file should be given to the code before entering thermal input data. This file should be built according to Table 3.1. The heat capacity rate ratio, $C*$ can be set as either a single or a range of values. In this case, the different values of $C*$ in the established range are set in a loop. As for the heat capacity, the NTU number can be set as either a single or a range of values, in which case the different values are introduced through a loop. In order to obtain (ε, NTU) plots with good resolution it is recommended to use small intervals for the NTU values. Finally, the user should set the fluid with minimum heat capacity rate, C_{min}. The code version shown in Chap. 5 is programmed for the computation of a single effectiveness value, that is, for single values of $C*$ and NTU.

3.1.3 Algorithm for Computing the Heat Exchanger Effectiveness ε

The algorithm for solving the governing equations corresponding to the set of elements in which the heat exchanger is divided can be accomplished by an iterative procedure. The algorithm is shown in the block diagrams of Figs. 3.4, 3.5, 3.6, and 3.7.

In the block diagram of Fig. 3.4, the configuration data is initially read from a text file, see Sect. 3.1.1. Then the user enters the values of the thermal input parameters in the main program: NTU, $C*$ and $C_{min} = (C_c \, or \, C_h)$. The following variables are set by default in the program: $T_{c,i}$, $T_{h,i}$, and UA. The actual values of these variables are read in the main program, Fig. 3.4.

After reading and setting the input data, the subroutine "ComputeE", Fig. 3.5, is called to compute the heat exchanger effectiveness. The last operation in the main program is the print out of the computed effectiveness.

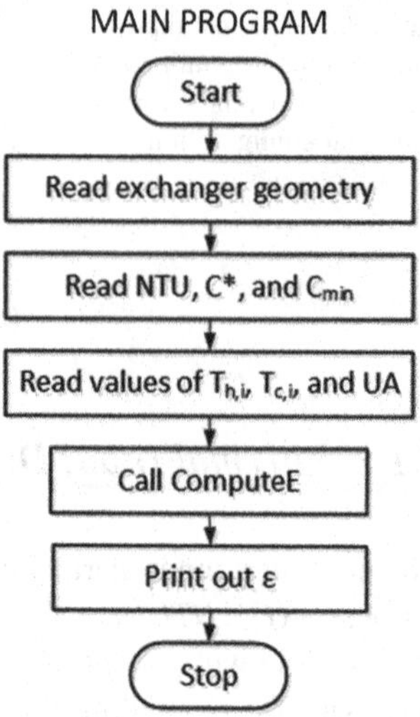

Fig. 3.4 Block diagram
of the main program of the
HETE code

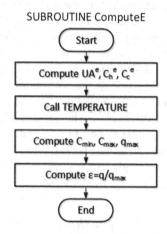

Fig. 3.5 Block diagram of
the "ComputeE" numerical
subroutine

The details of the effectiveness calculation performed by the "ComputeE" subroutine are shown in Fig. 3.5. After reading all the necessary input data, the parameters UA^e, C_c^e, and C_h^e are computed for a single element. The element conductance UA^e is calculated using the total number of elements, that is,

$$UA^e = \frac{UA}{N_e N_t N_r} \tag{3.1}$$

The hot and cold fluids heat capacities, C_h^e and C_c^e, are computed considering the mass flow rate corresponding to each element,

(i)
$$C_{\min} = C_c$$

Fig. 3.6 Block diagram
of the "TEMPERATURE"
numerical subroutine

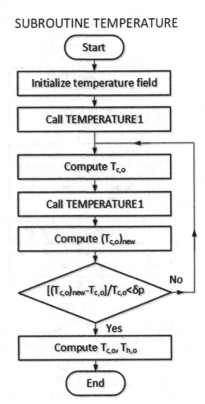

$$C_c^e = \frac{UA}{NTU \cdot N_e N_t} \tag{3.2}$$

$$C_h^e = \frac{UA}{NTU \cdot C^* N_r} \tag{3.3}$$

(ii) $$C_{\min} = C_h$$

$$C_c^e = \frac{UA}{NTU \cdot C^* N_e N_t} \tag{3.4}$$

$$C_h^e = \frac{UA}{NTU \cdot N_r} \tag{3.5}$$

Note that the computation of the element parameters UA^e, C_c^e, and C_h^e is performed in the program subroutines "ComputeUACminCa" and "ComputeUACminCr" called out by the "ComputeE" routine depending on the $C_{\min}$ fluid side, see code

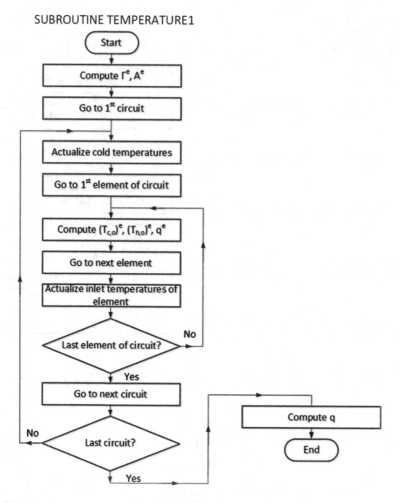

Fig. 3.7 Block diagram of the "TEMPERATURE1" numerical subroutine

in Chap. 5. Since the developed procedure assumes the element heat capacity of the unmixed fluid as being the minimum, that is, $C^e_{unmixed}/C^e_{mixed} \leq 1$ (see Sect. 2.2); the number of elements, N_e, should be chosen to satisfy this condition, Navarro and Cabezas-Gómez (2005); and Cabezas-Gómez et al. (2007).

The heat exchanger temperature distribution is determined iteratively in the next step following the in-tube fluid circuit by applying Eqs. (2.39) and (2.40) in the "TEMPERATURE" subroutine shown in details in Fig. 3.6. The overall numerical iterative process is explained in the following paragraphs. It must be noted that the temperature distribution is determined assuming constant both the thermodynamic properties and the overall heat transfer coefficient. As a result, the parameter Γ^e must remain constant over the whole heat exchanger. However, the code could handle variable thermodynamic properties by just evaluating the parameters UA^e, C_h^e, C_c^e, and Γ^e by using local properties and overall heat transfer coefficient. In any case, the element outlet hot and cold temperatures are determined from Eqs. (2.39) and (2.40), Cabezas-Gómez et al. (2007).

The subroutine "ComputeE" ends by the calculation of the heat exchanger effectiveness. Once the converged temperature field distribution has been obtained in the "TEMPERATURE" subroutine, thermal performance characteristics of the heat exchanger can be computed through their corresponding definition equations. These are the last steps displayed in the block diagram of Fig. 3.5. Note that in the HETE code, thermal performance parameters are computed in routine "ComputeE", external to the "TEMPERATURE" subroutine (see Chap. 5). The heat exchanger effectiveness is computed according to the following set of equations:

$$T_{c,o} = \frac{1}{N_t N_e} \sum_{e,r,N_r} T_{c,o}^e \tag{3.6}$$

$$T_{h,o} = \frac{1}{N_c} \sum_{\substack{last\ circuit \\ elements}} T_{h,o}^e \tag{3.7}$$

$$q = -C_h\left(T_{h,o} - T_{h,i}\right) = C_c\left(T_{c,o} - T_{c,i}\right) \tag{3.8}$$

$$\varepsilon = \frac{q}{q_{max}} \tag{3.9}$$

The convergence criterion is based on the cold fluid average outlet temperature though the global heat transfer rate, q, could also be used for that purpose.

The overall iterative computation is performed through the algorithm programmed in the main subroutine called "TEMPERATURE", shown in the block diagram of Fig. 3.6. Note that before calling for the first time the subroutine "TEMPERATURE1", displayed in Fig. 3.7, the whole heat exchanger temperature field, i.e. the temperatures of all elements, must be initialized once using the known inlet values of cold and hot fluids, $T_{c,i}$ and $T_{h,i}$ respectively. At the beginning of the iterative process, after obtaining the temperature distribution over the heat exchanger by applying subroutine "TEMPERATURE1", the cold fluid (external) average exit temperature, $T_{c,o}$, can be determined from the following expression and stored as$(T_{c,o})_{new}$:

$$\left(T_{c,o}\right)_{new} = \frac{1}{N_t N_e} \sum_{e,r,N_r} T_{c,o}^e \tag{3.10}$$

The iterative process starts in the following step by calling the subroutine "TEMPERATURE1" repeatedly until the relative error between the new, $(T_{c,o})_{new}$, and the previous value, $T_{c,o}$, of the average outlet temperature of the cold fluid (external and unmixed) is less or equal to a pre-established tolerance, that in this case is set equal to $\delta p = 10^{-8}$.

$$\frac{\left| (T_{c,o})_{new} - T_{c,o} \right|}{T_{c,o}} \leq \delta p \qquad (3.11)$$

In order to perform the iterative procedure, at the beginning of each iterative loop, the temperature T is given the value T_{new}. Then, the subroutine "TEMPERATURE1" is run again computing the new current value of T_{new} and the relative error is computed. The process goes on until the set tolerance is achieved.

As previously mentioned, the temperatures in each circuit are computed following the in-tube flowing direction. The temperature computations in each element are performed according to the algorithm programmed in the subroutine "TEMPERATURE1" shown in the block diagram of Fig. 3.7. The first operation is the computation of the element effectiveness Γ^e, and Λ^e, according to Eqs. (2.38) and (2.41), respectively. Since these parameters are constant, they are computed only once. Then, following the in-tube flow direction, the outlet temperatures for each element are computed from Eqs. (2.39) and (2.40), while the inlet temperatures for both fluids are set equal to the outlet temperatures of the upstream element in contact with the element. This procedure is repeated for all elements of the first circuit. Then the same procedure is performed for the following circuits to complete the whole heat exchanger temperature distribution. Since the order of elements in a circuit not always follows an order where the actual values of the inlet temperatures for a particular element are known in advance, the computation of the overall heat exchanger temperature distribution must be performed iteratively. Note that the subroutines "AirUpdate" and "FluidUpdate" are called in "TEMPERATURE1" to actualize the element inlet temperatures, as shown in Chap. 5.

The HETE code allows the simulation of several cross-flow arrangements. In the case of pure and parallel-cross-flow arrangements, the temperature distribution initially assumed corresponds to the actual inlet temperatures of both fluids and the iterative procedure is avoided. However, for other flow arrangements as, for example, the counter-cross-flow one, the initial temperature distribution does not correspond to the actual one, and the iterative procedure is mandatory. In counter-cross-flow arrangements the initial temperatures of the external fluid are not known at the inlet of the in-tube fluid, as the cold fluid enters the heat exchanger at the other side. To make it easier the comprehension of the HETE code computational procedure, two cases are illustrated corresponding to: (i) a pure cross-flow arrangement with one row and one circuit, as in Fig. 3.8 ("1tube1row1pass1cir.txt" configuration from Chap. 6); and (ii) a counter-cross-flow arrangement with four tube rows and one circuit performing four passes regarding the external fluid, as in Fig. 3.9 ("1tube4row4pass1cir(cc).txt" configuration from Chap. 6). The cold fluid is considered as unmixed in both cases.

Note that for any element in the tube of Fig. 3.8, the value of the inlet temperature of the cold (external) fluid, $T_{c,i}$, is uniform over the tube extension and known for all elements whereas the inlet temperature of the in-tube fluid in each element is the one corresponding to an outlet temperature of the upstream element. For the first element, the hot fluid inlet temperature is equal to $T_{h,i}$, which is one of the

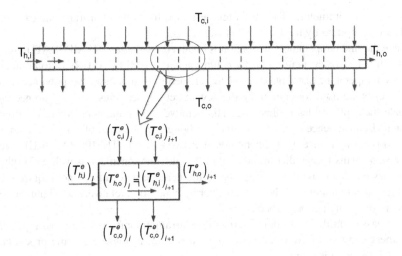

Fig. 3.8 A pure cross-flow arrangement with one row and one circuit, illustration of division in elements

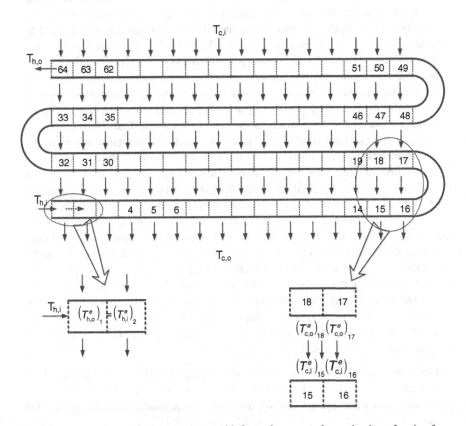

Fig. 3.9 A counter-cross-flow arrangement with four tube rows and one circuit performing four passes, illustration of division in elements

thermal input parameters. The outlet temperatures for both the in-tube and external fluids are computed by the Eqs. (2.39) and (2.40).

Once the first temperature distribution is obtained, by calling "TEMPERATURA1", the iterative procedure is restarted by calling again "TEMPERATURE1" to compute the new temperature distribution in all the elements. In this case, the previous computed values are used for computing the new outlet temperatures. There is no need to evaluate the input values in this case. The iterative procedure stops when the aforementioned convergence criterion is satisfied, though, in the case of Fig. 3.8, a single iteration is enough since the temperatures after two calls of "TEMPERATURE1" will be the same. This means that in the case of this flow arrangement as well as in other multi-pass parallel-cross-flow arrangements the iterative procedure is not required. In fact, these configurations can be treated either by hand, for a relatively small number of elements, or through a spreadsheet.

One-pass or multi-pass counter-cross-flow arrangements, as the one of Fig. 3.9, and others show a different temperature distribution. In these cases the procedure of the HETE code is required.

As noted in the first circuit elements of Fig. 3.9, the in-tube fluid distribution follows the same trend shown in Fig. 3.8. Except for the first element, in all the downstream elements, the inlet temperatures of the elements are equal to the exit temperatures of the upstream ones. In the first element, this temperature is equal to the inlet one in the heat exchanger of the in-tube fluid, $T_{h,i}$. The complexity in this flow arrangement is related to the temperature distribution of the cold fluid. Contrary to the pure cross-flow and the parallel-cross flow arrangements, where the inlets of the in-tube and the external fluids geometrically coincide, in counter-cross-flow arrangements, the cold fluid enters the heat exchanger at the opposite side of the in-tube fluid entrance or, in other words, at the exit of the in-tube fluid. Thus, in the first elements of the hot fluid (first in-tube pass), the inlet element temperatures of the cold fluid are unknown and change with each iteration up to attain the convergence.

In the first iteration, the cold fluid inlet temperature in each element, $T_{c,i}^e$ is assumed equal to the cold fluid heat exchanger inlet temperature, $T_{c,i}$. Upon calling "TEMPERATURE1", the values of this temperature change according to the simulated flow arrangement. Each element of the arrangement of Fig. 3.9 receives the cold fluid from the outlet of the element located in the tube upstream in the cold fluid flow direction. This is shown in Fig. 3.9, where the 15 and 16 elements, located in the fourth tube row, receive cold fluid respectively from the elements 18 and 17. Upon concluding iteration, the convergence criterion is applied involving the average temperature of the cold fluid. The error between iterations diminishes with the number of iterations. The computation process is stopped when the required error, δ_p, is attained.

The following are some explanatory remarks relative to the HETE code and the block diagrams of Figs. 3.4, 3.5, 3.6 and 3.7.

- It must be stressed that the code aims at developing (ε, NTU) or other relationships for heat exchangers of known geometry and flow configuration of both fluids.
- The assumed input parameters are the NTU of the heat exchanger along with one of the two following pairs: C^* and C_{min} or C_c and C_h.

- Since (ε, NTU) relationships are independent on the inlet temperatures of both fluids, they are arbitrarily assumed.
- The size of each element and its associated parameters such as UA^e, C_h^e and C_c^e, and NTU^e can be determined from the assumed number of elements, N_e, and the number of tubes and tube rows of the heat exchanger, N_t, and N_r. The number of elements must be chosen so that $C_{min}^e = C_c^e \leq C_h^e$, where the cold fluid is assumed to be the unmixed external fluid.
- The element effectiveness, Γ^e, can then be determined from Eq. (2.38) and is assumed to be constant for the heat exchanger effectiveness computation performed in this text.
- The first element is the one at the in-fluid tube entrance. For a parallel flow configuration, the inlet temperatures of both fluids are known and the exit temperatures can be obtained from Eqs. (2.39) and (2.40) without an iterative procedure. By applying a similar procedure to the succeeding elements along the path of the tube fluid, the corresponding exit temperatures can be determined, allowing for the mapping of the heat exchanger temperature. The average exit temperatures of both fluids can then be determined.
- The inlet temperature of the cold fluid in the first element is not known in the case of a counter-cross-flow configuration and others similar to it. This temperature must be assumed so that an iterative procedure must be followed in the determination of the temperature distribution, as suggested in the block diagram of Fig. 3.6. The proposed convergence criterion is given in terms of the average outlet temperature of the cold (external) fluid.

3.2 Examples Solved with the EES Software

EES® (Engineering Equation Solver, Klein and Alvarado 1995; Klein 2004) codes are introduced in this section. They were developed to compute the heat exchanger effectiveness for five different configurations. These codes are shown in Chap. 8 and are based on the discrete mathematical model of Sect. 2.3. The heat exchanger configurations to be analyzed are the following: (i) three one-pass pure cross-flow with one, two and three tube rows, respectively; (ii) two-passes parallel-cross-flow; and (iii) two-passes counter-cross-flow. In the configurations (ii) and (iii), there is just one in-tube fluid circuit. These configurations are shown schematically in Chap. 6. ("1tube1row1pass1cir.txt", "1tube2row1pass2cir.txt", "1tube3row1pass3cir.txt", "1tube2row2pass1cir(par).txt", "1tube2row2pass1cir(cc).txt").

The EES platform was chosen due to its extensive use in universities and research laboratories across the world. The EES is a powerful program for solving non-linear implicit algebraic equations or non-linear differential algebraic equation systems (DAES). The software has also many other functionalities which allow its use for solving many different thermodynamics and heat transfer problems, (Nellis and Klein 2009; and Klein and Nellis 2012). EES is used in the present monograph due to its capability in the solution of the set of discrete equations of Sect. 2.3. The most difficult task is the programming of the circuit distribution that

Fig. 3.10 Schematic diagram of one-pass pure cross-flow configuration

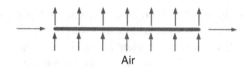

Air

is performed in somewhat different way with respect to that of the HETE code. Two characteristics of these configurations prompted their choice for analysis: (i) easy to program in EES; and (ii) the availability of effectiveness correlations for these flow arrangements (see Chap. 8).

The flow configuration illustrated in Fig. 3.10 corresponds to the one-pass pure cross-flow one which can be simulated with the HETE code using the input file named "1tube1row1pass1cir.txt". The EES program for this configuration starts with the standard code lines using the symbol $. These two commands allow choosing the unit system to be employed and the space tabulation of the text written in the command window of the EES code. The user comments are written in red and blue. Please, refer to the EES manual, the published books by the EES authors, Nellis and Klein (2009) and Klein and Nellis (2012), among others, and other material published in "http://www.fchart.com" site for assistance in the EES® use and manipulation.

```
"!One Pass Pure Cross-flow heat exchanger"
$UnitSystem SI MASS RAD PA  K J
$TABSTOPS   0.2 0.4 0.6 0.8 3.5 in

"This program allows to compute the HE effectiveness with the EES® code.
 The code is programmed following the book material and it is not optimized from the
 computational point of view"
```

Input of flow arrangement and thermal data comes next, as noted below. The flow configuration will be set in subsequent code lines. Note that the same variable symbols used in the monograph are used in the EES code.

```
"Heat exchanger flow arrangement data"
N_e = 100[-] "Number of elements in the heat exchanger"
N_t = 1[-]      "Number of tubes per row"
N_r = 1[-]      "Number of rows"
N_c = 1[-]      "Number of in-tube fluid circuits"

"Thermal input data - user"
NTU = 1 [-]   "Heat exchanger number of transfer of units"
Cr = 0.5[-]   "Heat capacity rate ratio Cr=C_min/C_max"
K = 1            "Chose of the fluid with the C_min heat capacity rate. K =1 -> C_min = C_c
(air), K = 2 -> C_min = C_h (fluid)"

"Thermal input data - code"
UA = 12 [W/K]      "Heat exchanger conductance"
T_h_i = 70 [°C]     "Mean in-tube fluid inlet temperature"
T_c_i = 35 [°C]     "Mean external fluid inlet temperature"
```

The computation of the following element parameters comes next: UA^e, C_c^e, C_h^e, NTU, and C^*. This part of the code can be improved by programming an EES function in order to compute these quantities as a function of some variable K that will indicate which is C_{min} fluid. This code stands for $C_c = C_{min}$. The last line of this part corresponds to the element effectiveness calculation.

```
"Compute the quantities C_c, C_h, UAle, C_cle and C_hle, C_min = C_c"
UAle = UA/(N_e*N_t*N_r)     "Element heat exchanger conductance"
C_c = UA/NTU                "Heat exchanger cold fluid heat capacity rate"
C_h = C_c/Cr               "Heat exchanger hot fluid heat capacity rate"
C_cle = UA/(NTU*N_e*N_t)    "Element cold fluid heat capacity rate"
C_hle = UA/(NTU*Cr*N_c)     "Element hot fluid heat capacity rate"

"Element effectiveness Gammale"
GAMMAle = 1-exp(-UAle/C_cle)
```

The inlet temperatures of the cold and hot fluids are introduced in the next EES code lines followed by the evaluation of the elements exit temperatures. All the temperatures are computed considering the flow configuration programming.

The hot fluid temperature only changes along the in-tube fluid circuit, which in the case of Fig. 3.10 is only one. Thus, a complicated matrix for computing this temperature is not required. A one-dimensional matrix can be programmed in EES as, for example, $T_h[j]$. The index j in the present case varies from 1 to N_e. The hot fluid (in-tube) inlet temperature for the first element is the one with $j = 0$ and it is given by the heat exchanger hot fluid inlet temperature, that is, $T_h[0] = T_{h,i}$ Two notes regarding EES variable edition: (i) the code variables and equations can be visualized in equation format in the formatted equation entry; and (ii) sub-indices are written after the symbol "_" as, for example, T_h which is written as T_h. Another example: the inlet and outlet temperatures of the element j would be written as $T\_h[j - 1]$ and $T\_h[j]$.

The external cold fluid temperature distribution is two-dimensional because this fluid is unmixed in the finned external side of the heat transfer surface. Thus, a two dimensional matrix in T_c is required, being designated by $T\_c[j, i]$ in EES, with j varying from 1 to N_e. The subscript i designates the cold fluid entrance, $i = 0$, and exit, $i = 1$. In the case of Fig. 3.10, the cold fluid entrance temperature is uniform over the length of the heat exchanger so that: $T\_c[j, 0] = T_{c,i}$ with $1 \leq j \leq N_e$. The cold fluid exit temperatures would be written as $T\_c[j, 1]$ with $1 \leq j \leq N_e$. The outlet temperatures in each element are determined according to Eqs. (2.39) and (2.40) as shown in the partial EES code shown below. The following are two additional characteristics of EES: (i) EES does not require programing an iterative procedure in case it would be needed since the program solves the set of algebraic equations, see, for example, the EES code for the two-pass counter-cross-flow arrangement in Chap. 8; and (ii) the DUPLICATE command is another important EES tool since it allows working with two sets of the same equations just by changing the indices of each matrix.

```
"Inlet values of temperatures for both fluids"
T_h[0] = T_h_i          "hot fluid"
duplicate j=1,N_e
   T_c[j,0]=T_c_i        "cold fluid"
end
```

```
"Heat exchanger temperature field distribution. Use of Eqs. (2.39) and (2.40)"
A = (C_cle*GAMMAle)/C_hle
duplicate j=1, N_e
   T_h[j] = (2-A)/(2+A)*T_h[j-1] + (2*A)/(2+A)*T_c[j,0]                          "hot fluid"
   T_c[j,1] = (A+2*(1-GAMMAle))/(2+A)*T_c[j,0] + (2*GAMMAle)/(2+A)*T_h[j-1] "cold
fluid"
end
```

Performance data can be computed once the heat exchanger temperature field distribution is computed. The following EES code fragment shows the set of commands needed for the effectiveness computation by the direct application of Eq. (2.1). The first step is the evaluation of the average outlet temperature of the cold fluid by applying Eq. (3.6) which requires the use of the internal EES function SUM. The same function must be used in the calculation of the heat transfer rates in both sides of the heat exchanger (internal and external) by applying Eqs. (2.36) and (2.37). Note that function SUM allows only the sum of elements of one dimensional array. Having the total rate of heat transfer, the effectiveness might be determined by dividing it by the maximum heat transfer rate, which is calculated in the last command line of the fragment below. Note that the effectiveness has been previously determined with the available heat exchangers inlet and outlet temperatures. Both values must coincide with each other. The result can be compared with the one that would be obtained from Eq. (2.12), Table 2.1.

```
T_bar_c_o=sum(T_c[k,1],k=1,N_e)/N_e       "mean outlet cold fluid temperature"

epsilon=(T_bar_c_o-T_c_i)/(T_h_i-T_c_i)   "element effectiveness"
epsilon_teste = Q_h_total/Q_max           "heat exchanger effectiveness"
"theoretical effectiveness. One-pass cross-flow with one row"
epsilon_th = (1/Cr)*(1-exp(-(Cr)*(1-exp(-NTU))))

duplicate j=1, N_e
   Q_dot_hle[j]=-C_hle*(T_h[j]-T_h[j-1])   "hot fluid"
   Q_dot_cle[j]=C_cle*(T_c[j,1]-T_c[j,0])  "cold fluid"
end

Q_h_total=sum(Q_dot_hle[k],k=1,N_e)        "total heat transfer rate, hot fluid"
Q_c_total=sum(Q_dot_cle[k],k=1,N_e)        "total heat transfer rate, cold fluid"
Q_max = min(C_c,C_h)*(T_h_i-T_c_i)         "maximum heat transfer rate"
```

The EES program fragment above is based on the application of the set of equations Eqs. (2.39) and (2.40) of Sect. 2.3. The same results could be obtained if instead Eqs. (2.35)–(2.38) were used, as in the program fragment below. However, the convergence of the EES program could be affected depending on the set of equations and the heat exchanger arrangement. In the present case study, any of the approaches would pose no problem of convergence. However, if the set Eqs. (2.35)–(2.38), corresponding to the next program fragment, were applied to a one-pass cross-flow configuration with three rows the EES program would present convergence problems. In such cases, the use of the first set, program fragment above, would be preferable since a faster convergence would be attained. The whole programs for these two cases can be found in the Chap. 8.

```
"Heat exchanger temperature distribution. Use of Eqs. (2.35 - 2.38)"
duplicate j=1, N_e
    Q_dotle[j]=-C_hle*(T_h[j]-T_h[j-1])                    "hot fluid"
    Q_dotle[j]=C_cle*(T_c[j,1]-T_c[j,0])                   "cold fluid"
    (T_c[j,1]-T_c[j,0])/(T_h_m[j]-T_c[j,0]) = GAMMAle      "element effectiveness"
    T_h_m[j] = (T_h[j] + T_h[j-1])/2                       "element mean hot fluid temperature"
end

T_bar_c_o=sum(T_c[k,1],k=1,N_e)/N_e       "mean outlet cold fluid temperature"
Q_total = sum(Q_dotle[k],k=1,N_e)         "total heat transfer rate"
Q_max = min(C_c,C_h)*(T_h_i-T_c_i)        "maximum heat transfer rate"

epsilon=(T_bar_c_o-T_c[1,0])/(T_h[0]-T_c[1,0])  "heat exchanger effectiveness"
epsilon_teste = Q_total/Q_max                   "heat exchanger effectiveness"
"theoretical effectiveness. One-pass cross-flow with one row"
epsilon_th = (1/Cr)*(1-exp(-(Cr)*(1-exp(-NTU))))
```

As an example of application of the EES program to the solution of the present case study, a one-pass pure cross-flow configuration of Fig. 3.10, the effect of the number of elements on the effectiveness is presented in the plot of Fig. 3.11 for the following thermal parameters: $NTU = 1$; $C^* = 0.5$; and $C_{min} = C_c$. The precision of the solution can be assessed by comparison with Eq. effectiveness. It can be noted that differences diminish with the number of elements tending to disappear when the number of elements is high enough. Note that for 20 elements the numerical error is minimal though the use of higher number of elements, say 100, would be advisable in this case. Deviations might vary depending on the values of thermal

Fig. 3.11 Plot illustrating the effect of the number of elements, N_e, on the EES obtained effectiveness of the flow arrangement of Fig. 3.10. The broken line corresponds to Eq. (2.12a) effectiveness

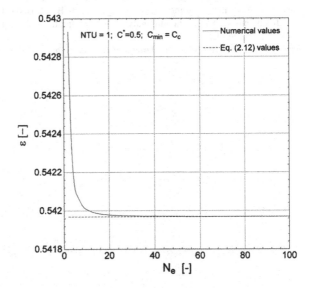

parameters such as NTU and C^*. The number of elements can be set easily in the HETE code and can be changed in an equally easy manner when working with EES though there is a limit of variables that the program can handle. The Professional version of EES can handle 12,000 variables and 6,000 the Academic one.

Similar analysis to the one of the case study presented herein with the EES program could be performed for other flow arrangements such as the ones included in Chap. 8. These codes could be changed in such a way to make them applicable to other flow arrangements. The reader is encouraged to consult Chaps. 6 and 8, where HETE and EES codes for five flow arrangements are presented, and run the programs as an exercise of comparison between both approaches.

References

Cabezas-Gómez L, Navarro HA, Godoy SM, Campo A, Saiz-Jabardo JM (2009) Thermal characterization of a cross-flow heat exchanger with a new flow arrangement. Int J Therm Sci 48:2165–2170

Cabezas-Gómez L, Navarro HA, Saiz-Jabardo JM (2007) Thermal performance of multi-pass parallel and counter cross-flow heat exchangers. ASME J Heat Transf 129:282–290

Cabezas-Gómez L, Navarro HA, Saiz-Jabardo JM, Hanriot SM, Maia CB (2012) Analysis of new cross flow heat exchanger flow arrangement—extension to several rows. Int J Therm Sci 55:122–132

Klein SA, Alvarado FL (1995) EES—engineering equation solver. F-Chart Software, Middleton, Wisconsin

Klein SA (2004) Engineering equation solver (EES®), v. 7.187-3D, F-Chart Software, Madison, USA

Klein S, Nellis G (2012) Thermodynamics. Cambridge University Press, New York

Nellis G, Klein S (2009) Heat transfer. Cambridge University Press, New York

Navarro HA, Cabezas-Gómez L (2005) A new approach for thermal performance calculation of cross-flow heat exchangers. Int J Heat Mass Transf 48:3880–3888

Navarro HA, Cabezas-Gómez L Zoghbi Filho JRB, Ribatski G, Saíz-Jabardo JM (2010) Effectiveness - NTU data and analysis for air conditioning and refrigeration air coils. J Braz Soc Mech Sci Eng 32:218–226

Shah RK, Sekuliç DP (2003) Fundamentals of heat exchanger design. Wiley, New York

Chapter 4
Numerical Results and Discussions

In the previous sections fundamentals of the HETE code were introduced. The objective of this section is to demonstrate the capabilities of the code through several nonstandard applications which will allow the reader to assess its potential. These applications will be introduced as an overview of several journal papers published by the authors namely Navarro and Cabezas-Gómez (2005); Cabezas-Gómez et al. (2007); Cabezas-Gómez et al. (2009); and Cabezas-Gómez et al. (2012). Thus, in Sect. 4.1 the model will be tested and validated against well-known theoretical relations from the open literature such as one-pass cross-flow exchangers, with one up to n rows for which theoretical correlations are provided in Table 2.1. Then some flow arrangements from Table 2.2 will be considered namely parallel and counter cross-flow heat exchanges with one, two, three, four, and more rows. More complex configurations, such as some flow configurations whose effectiveness is not known in a close form correlation, are introduced in Sect. 4.2.

In comparing the numerical code results with those provided by a theoretical correlation, deviations or errors will be introduced whose definition is as follows:

$$\lambda = 100 \left| \frac{\varepsilon - \varepsilon_{th}}{\varepsilon_{th}} \right| \tag{4.1}$$

and

$$\lambda_{av} = 100 \left[\frac{1}{N} \sum_{1}^{N} \left| \frac{\varepsilon - \varepsilon_{th}}{\varepsilon_{th}} \right| \right] \tag{4.2}$$

In the above equations, ε and ε_{th} are the HETE code effectiveness and the theoretically evaluated one by an available analytical correlation. In the evaluation of the average deviation, λ_{av}, N is the number of data points considered in the evaluation of a particular flow arrangement.

The notation from Shah and Pignotti (1993) will be used to identify heat exchanger geometries. According to this notation, the geometry is designated

© The Author(s) 2015
L. Cabezas-Gómez et al., *Thermal Performance Modeling of Cross-Flow Heat Exchangers*, SpringerBriefs in Applied Sciences and Technology, DOI 10.1007/978-3-319-09671-1_4

by $G_{l,m}^k$ where superscript "k" designates the set of heat exchangers known as "pure cross-flow", in which case $k \equiv c$, the "parallel cross-flow", $k \equiv pc$, and the "counter cross-flow", $k \equiv cc$. The first subscript, "l", stands for the number of passes of the in-tube fluid, and "m" for the number of rows per pass of the in-tube fluid.

4.1 Model Tests and Validations

The results of the HETE code are tested and validated against closed form correlations in the present section. The aforementioned correlations are the ones listed in Table 2.1, for a one pass and n rows cross-flow heat exchangers, and Table 2.2, for multipass parallel and counter-cross-flow heat exchangers with one up to n passes of the in-tube fluid.

Navarro and Cabezas-Gómez (2005) computed the maximum relative error, λ, between the effectiveness values computed with the HETE code, ε, and those from the closed form correlations from Table 2.1, Eqs. (2.12)–(2.15),for the corresponding four configurations $G_{1,1}^c$, $G_{1,2}^c$, $G_{1,3}^c$, and $G_{1,4}^c$ ("1tube1row1pass1cir", "1tube2row1pass2cir", "1tube3row1pass3cir", "1tube4row1pass4cir", see Chap. 6 for HETE input files). The maximum relative error was of the order of 10^{-6} % among 1,111 effectiveness values computed in the ranges $0 \le C_i^* \le 1$ and $0 \le NTU \le 10$, with 0.1 increments. The observed deviations from the correlations do confirm the accuracy of the results provided by the HETE code.

Results for the unmixed-unmixed pure cross-flow heat exchanger, $G_{1,\infty}^c$, were also compared with the Mason series solution and the curve fitting solution from Table 2.1, Eqs. (2.16) and (2.17), respectively. Table 4.1 presents the convergence history of the simulated results from 5 rows to 100 rows per in-tube pass with the deviations (relative errors) referring to the ones with respect to the infinite series

Table 4.1 Comparison between model prediction and infinite series solution, Eq. (2.16), Navarro and Cabezas-Gómez (2005)

N_r	Configuration	λ_{av} (%) $C_{min} = C_{air}$		λ_{max} (%) $C_{min} = C_h$	
5	$G_{1,5}^c$	0.63	2.88	0.45	2.89
6	$G_{1,6}^c$	0.44	2.10	0.32	2.10
7	$G_{1,7}^c$	0.33	1.56	0.24	1.56
8	$G_{1,8}^c$	0.25	1.22	0.18	1.22
9	$G_{1,9}^c$	0.20	0.97	0.14	0.97
10	$G_{1,10}^c$	0.16	0.79	0.12	0.79
20	$G_{1,20}^c$	0.04	0.20	0.03	0.20
50	$G_{1,50}^c$	0.006	0.033	0.005	0.033
100	$G_{1,100}^c$	0.0016	0.0082	0.0012	0.0082

solution, Eq. (2.16). According to Navarro and Cabezas-Gómez (2005) the following conclusions can be drawn from Table 4.1 results:

1. The maximum relative error decreases with the number of tube rows, attaining a value of 0.0082 % for $N_r = 100$, a result that confirms once again the accuracy of the HETE code.
2. The rather low deviation observed for a 100 rows indicates that a high number is needed of tube rows in order to achieve the unmixed fluid concept for the in-tube fluid.
3. The closed form correlation, Eq. (2.16), is rigorously valid only for an infinite number of tube rows. Analytical expressions for more than five tube rows are either scarce or not available in the literature. The developed methodology could be very useful in those cases providing accurate thermal effectiveness values.

It is interesting to note that the lack of accuracy in the determination of the thermal effectiveness can cause a significant error in cases where the effectiveness is used in the determination of the NTU value as in experimental analysis of compact heat exchangers. In fact, Navarro and Cabezas-Gómez (2005) analyzed the one tube row heat exchanger, $G_{1,1}^c$, for $C^* = 0.5$ and $NTU = 5$. They concluded that, even using an adequate (ε, NTU) correlation, as Eq. (2.12a) or (2.12b) in the present case, an effectiveness error of the order of 0.1 % caused a NTU error of the order of 4.0 %.

Equation (2.17) from Table 2.1 is extensively used to approximate (ε, NTU) relations for cases where the number of tube rows is higher than 5. However, according to DiGiovanni and Webb (1989), this correlation yields unphysical results in the range $NTU < 1$. In addition, they suggested that a maximum error of the order of 3.7 % would result if this correlation were used for the unmixed-unmixed configuration. Those are examples of applications where the HETE code could be useful.

The HETE code was further validated in the paper by Cabezas-Gómez et al. (2007) where the results from the code were compared with the correlations of Table 2.2 for the flow arrangements of Fig. 4.1. Input data for the simulation of these flow arrangements are provided in Chap. 6 (see the files "1tube1row1pass1cir","1tube2row2pass1cir(par)","1tube3row3pass1cir(par)", "1tube4row4pass1cir(par)", "1tube2row2pass1cir(cc)", "1tube3row3pass1cir(cc)", and "1tube4row4pass1cir(cc)", respectively). The heat exchanger configurations of Fig. 4.1 include a "pure cross-flow" heat exchanger, designated by $G_{1,1}^c$ and those of two, three and four in-tube passes, with one row per pass, parallel and counter cross-flow heat exchangers, designated, respectively, by $G_{l,1}^{pc}$ and $G_{l,1}^{cc}$, with "l" varying from 1 to 4. Effectiveness data comparisons were performed for C^* and NTU values varying in the following ranges: $0 \leq C_i^* \leq 1$ and $0 \leq NTU \leq 10$. Equation (2.12) from Table 2.1, and Eqs. (2.19)–(2.21) and (2.23)–(2.25), from Table 2.2, were applied depending on the heat exchanger configuration.

Table 4.2 presents the maximum relative errors obtained for the different configurations of Fig. 4.1 over the aforementioned ranges of C^* and NTU. It can be noted that the obtained errors are rather small for all cases, with upper limits being of the order of 10^{-4} and 10^{-6} %, respectively, for the counter cross-flow

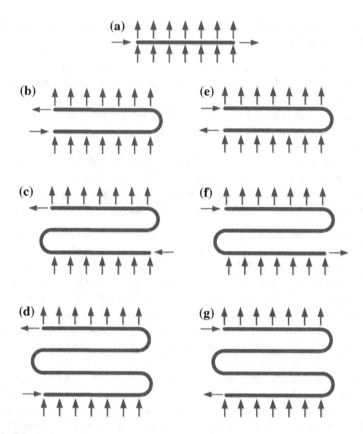

Fig. 4.1 Schematic representations of pure cross-flow (**a**) and several parallel-cross-flow with two, three and four passes (**b**), (**c**), and (**d**); and counter-cross flow with two, three and four passes (**e**), (**f**), (**g**) arrangements

Table 4.2 Maximum relative error λ_{max}, as a function of the number of tube rows, Cabezas-Gómez et al. (2007)

N_r	Geometry	$C_{min} = C_{air}$	$C_{min} = C_{hot}$
1	$G^c_{1,1}$ versus Eq. (2.12)	1.1×10^{-6}	1.4×10^{-6}
2	$G^{pc}_{2,1}$ versus Eq. (2.19)	8.7×10^{-7}	9.0×10^{-7}
3	$G^{pc}_{3,1}$ versus Eq. (2.20)	6.2×10^{-7}	8.9×10^{-7}
4	$G^{pc}_{4,1}$ versus Eq. (2.21)	6.3×10^{-7}	8.0×10^{-7}
2	$G^{cc}_{2,1}$ versus Eq. (2.23)	2.6×10^{-5}	1.4×10^{-5}
3	$G^{cc}_{3,1}$ versus Eq. (2.24)	5.7×10^{-5}	2.4×10^{-5}
4	$G^{cc}_{4,1}$ versus Eq. (2.25)	1.2×10^{-4}	3.8×10^{-5}

and parallel cross-flow arrangements. These are rather small errors to be considered significant for any practical purpose. As a general rule, the parallel cross flow arrangement errors are lower than the counter cross-flow errors. This trend probably is related to the iterative procedure required in counter cross-flow as compared with its counterpart arrangement. However, this does not mean that the effectiveness computation is not accurate in all cases. Considering the four passes, counter cross-flow arrangement, for conditions (C^* and NTU) corresponding to the maximum relative error, 1.2×10^{-4} %, according to Table 4.2, effectiveness values from simulation and Eq. (2.16) are equal to 0.996189 and 0.996188, respectively. This minimal difference implies that, even for an iteratively computed effectiveness, as for the counter cross-flow arrangements, the results are very accurate, Cabezas-Gómez et al. (2007).

The correlations of Table 2.2 for cross-flow arrangements are limited to 4 tube rows. Note that, since the number of tubes per tube row in the present case is equal to one, the use of rows or in-tube passes is indifferent. Up to the authors' knowledge, no correlations for more than six tube passes are available. The general practice has been to assume the correlations for pure parallel and counter flows, as these are the limit arrangements when the number of tube passes tends to infinity. However, this practice might lead to significant errors in the thermal effectiveness determination. Table 4.3 has been raised in order to determine the order of magnitude of these errors. The actual effectiveness in this table, ε, evaluated by applying the HETE code, is compared with the effectiveness from either the parallel or the counter flow arrangements, ε_{th}, for cross-flows with a higher number of tube passes ($N_r \geq 4$). The range and increment intervals of C^* and NTU considered in the development of Table 4.3 are the same as those of Table 4.2. Table 4.3 results are presented in terms of the maximum and the average relative errors.

	N_r	Geometry	$\lambda_{av}, \lambda_{max}$ (%) $C_{min} = C_{air}$	$\lambda_{av}, \lambda_{max}$ (%) $C_{min} = C_{hot}$
Table 4.3 Comparison of simulated effectiveness with pure parallel and counter flow ones from Eqs. (2.22) and (2.26) of Table 2.2, respectively, Cabezas-Gómez et al. (2007)	4	$G_{4,1}^{pc}$	0.2, 1.4	0.6, 3.1
	5	$G_{5,1}^{pc}$	$7.4 \times 10^{-2}, 0.34$	0.2, 1.2
	6	$G_{6,1}^{pc}$	$3.8 \times 10^{-2}, 0.16$	0.1, 0.6
	7	$G_{7,1}^{pc}$	$2.8 \times 10^{-2}, 0.12$	0.06, 0.3
	8	$G_{8,1}^{pc}$	$1.9 \times 10^{-2}, 9.2 \times 10^{-2}$	0.04, 0.2
	9	$G_{9,1}^{pc}$	$1.6 \times 10^{-2}, 7.4 \times 10^{-2}$	0.03, 0.1
	10	$G_{10,1}^{pc}$	$1.2 \times 10^{-3}, 5.9 \times 10^{-2}$	0.02, 0.08
	20	$G_{20,1}^{pc}$	$3.2 \times 10^{-3}, 1.5 \times 10^{-2}$	0.003, 0.01
	50	$G_{50,1}^{pc}$	$5.1 \times 10^{-4}, 2.4 \times 10^{-3}$	0.0005, 0.002
	100	$G_{100,1}^{pc}$	$1.3 \times 10^{-4}, 5.9 \times 10^{-4}$	0.0001, 0.0006
	4	$G_{4,1}^{cc}$	1.5, 6.7	1.4, 6.7
	5	$G_{5,1}^{cc}$	1.0, 4.7	0.9, 4.7
	6	$G_{6,1}^{cc}$	0.7, 3.5	0.7, 3.5
	7	$G_{7,1}^{cc}$	0.5, 2.7	0.5, 2.7
	8	$G_{8,1}^{cc}$	0.4, 2.1	0.4, 2.1
	9	$G_{9,1}^{cc}$	0.3, 1.7	0.3, 1.7
	10	$G_{10,1}^{cc}$	0.26, 1.4	0.2, 1.4
	20	$G_{20,1}^{cc}$	0.06, 0.37	0.06, 0.4
	50	$G_{50,1}^{cc}$	0.01, 0.06	0.01, 0.06
	100	$G_{100,1}^{cc}$	0.002, 0.01	0.002, 0.01

An overview of Table 4.3 allows one to draw the following general conclusions:

- The first arrangement for both cross-flow configurations is the one with 4 tube passes. As expected, the errors are rather high attaining the order of 3.1 %, with an average of 0.6 %, for the parallel cross-flow arrangement, and 6.7 % and average of 1.5 %, for the counter cross-flow arrangement.
- The maximum and average errors diminish with the number of tube passes. Negligibly small deviations are attained for a number of tube passes of the order of 50, though, for practical purposes, this number could be much smaller, as the results of Table 4.1 clearly indicate.
- Errors found in counter cross-flow arrangements are significantly higher than their parallel cross-flow counterparts. In fact, whereas the latter arrangement presents a maximum error of the order of 0.2 % $C_{min} = C_{hot}$ for 8 tube passes, the maximum deviation in the case of the counter cross-flow arrangement is of the order of 2.1 %.

It has been a common practice to assume that the heat exchanger effectiveness of a cross-flow heat exchanger with more than six tube passes can be approximated by either the pure parallel or the pure counter flow arrangements for industrial applications, Stevens et al. (1957) and Taborek (1983). The analysis of Table 4.3 data demonstrates that this is not the case, at least for the counter cross-flow arrangements in heat exchangers with high NTU and C^* values (not shown in Table 4.3). Bes (1996) has also called attention to this fact. He found out that the absolute error between his proposed analytical relation for a counter cross-flow arrangement of twenty tube passes and Eq. (2.17), for NTU > 5, tends to 2.5 %. For a similar flow arrangement, according to the present approach, the maximum relative error is of the order of 0.4 %, for $C^* \geq 0,7$ and $NTU > 5$. This is a significantly lower value than that found by Bes (1996), but still too high to be neglected. It could be argued that these errors are rather low for practical purposes. However, if one considers that in design analysis, the effectiveness is known and the NTU value must be determined from it in order to evaluate the product UA, this conclusion might be questionable. As previously suggested, Navarro and Cabezas-Gómez (2005) have shown that a very small error in the effectiveness can cause NTU errors of more than one order of magnitude higher in the high values of NTU range.

Effectiveness data for configurations involving five to ten tube passes has been determined by applying the HETE code. Results can be found in Tables 4.4 and 4.5 for parallel cross-flow and counter cross-flow configurations. It must be noted that effectiveness data for flow arrangements of more than six tube passes have not been published so far in the open literature. The effectiveness of the limit configurations of pure parallel and pure counter flow has also been included in the last column of each table as a reference. The effectiveness considered in these tables is the so called "P temperature effectiveness". The P temperature effectiveness for each fluid is defined as the ratio between the temperature variation in this fluid (hot or cold) and the inlet temperature difference:

$$P_h = \frac{T_{h,i} - T_{h,o}}{T_{h,i} - T_{c,i}} \quad P_c = \frac{T_{c,o} - T_{c,i}}{T_{h,i} - T_{c,i}} \tag{4.3}$$

Table 4.4 Temperature effectiveness P for a parallel cross-flow arrangement for five to ten rows, with one tube per row, respectively

R	NTU	Temperature effectiveness P						
		Multipass parallel-cross-flow						Parallel-flow
		Five passes	Six passes	Seven passes	Eight passes	Nine passes	Ten passes	
0.2	0.2	0.1778	0.1778	0.1778	0.1778	0.1778	0.1778	0.1778
0.2	0.4	0.3177	0.3177	0.3177	0.3177	0.3177	0.3177	0.3177
0.2	0.6	0.4278	0.4278	0.4278	0.4278	0.4278	0.4277	0.4277
0.2	0.8	0.5145	0.5144	0.5144	0.5144	0.5143	0.5143	0.5143
0.2	1.0	0.5827	0.5826	0.5825	0.5825	0.5825	0.5824	0.5823
0.2	2.0	0.7587	0.7584	0.7582	0.7581	0.7580	0.7580	0.7577
0.2	3.0	0.8116	0.8113	0.8111	0.8109	0.8109	0.8108	0.8106
0.2	4.0	0.8272	0.8270	0.8268	0.8267	0.8267	0.8267	0.8265
0.2	5.0	0.8317	0.8315	0.8315	0.8314	0.8314	0.8314	0.8313
0.2	6.0	0.8329	0.8329	0.8328	0.8328	0.8328	0.8328	0.8327
0.2	7.0	0.8332	0.8332	0.8332	0.8332	0.8332	0.8332	0.8331
0.2	8.0	0.8333	0.8333	0.8333	0.8333	0.8333	0.8333	0.8333
0.2	9.0	0.8333	0.8333	0.8333	0.8333	0.8333	0.8333	0.8333
0.2	10.0	0.8333	0.8333	0.8333	0.8333	0.8333	0.8333	0.8333
0.4	0.2	0.1745	0.1745	0.1744	0.1744	0.1744	0.1744	0.1744
0.4	0.4	0.3064	0.3063	0.3063	0.3063	0.3063	0.3063	0.3063
0.4	0.6	0.4062	0.4061	0.4060	0.4060	0.4060	0.4060	0.4059
0.4	0.8	0.4817	0.4815	0.4815	0.4814	0.4814	0.4813	0.4812
0.4	1.0	0.5388	0.5386	0.5385	0.5384	0.5383	0.5383	0.5381
0.4	2.0	0.6722	0.6718	0.6715	0.6714	0.6713	0.6712	0.6708
0.4	3.0	0.7047	0.7043	0.7041	0.7040	0.7039	0.7038	0.7036
0.4	4.0	0.7123	0.7121	0.7120	0.7119	0.7118	0.7118	0.7116
0.4	5.0	0.7139	0.7138	0.7138	0.7137	0.7137	0.7137	0.7136
0.4	6.0	0.7143	0.7142	0.7142	0.7142	0.7142	0.7142	0.7141
0.4	7.0	0.7143	0.7143	0.7143	0.7143	0.7143	0.7143	0.7142
0.4	8.0	0.7143	0.7143	0.7143	0.7143	0.7143	0.7143	0.7143
0.4	9.0	0.7143	0.7143	0.7143	0.7143	0.7143	0.7143	0.7143
0.4	10.0	0.7143	0.7143	0.7143	0.7143	0.7143	0.7143	0.7143
0.6	0.2	0.1712	0.1712	0.1712	0.1712	0.1712	0.1712	0.1712
0.6	0.4	0.2956	0.2955	0.2955	0.2955	0.2955	0.2955	0.2954
0.6	0.6	0.3860	0.3859	0.3859	0.3858	0.3858	0.3858	0.3857
0.6	0.8	0.4518	0.4516	0.4515	0.4514	0.4514	0.4514	0.4512
0.6	1.0	0.4996	0.4994	0.4992	0.4991	0.4991	0.4990	0.4988
0.6	2.0	0.6009	0.6004	0.6002	0.6000	0.5999	0.5998	0.5995
0.6	3.0	0.6208	0.6204	0.6203	0.6202	0.6201	0.6201	0.6199
0.6	4.0	0.6244	0.6242	0.6242	0.6241	0.6241	0.6241	0.6240

(continued)

Table 4.4 (continued)

R	NTU	Temperature effectiveness P						Parallel-flow
		Multipass parallel-cross-flow						
		Five passes	Six passes	Seven passes	Eight passes	Nine passes	Ten passes	
0.6	5.0	0.6250	0.6248	0.6249	0.6248	0.6248	0.6248	0.6248
0.6	6.0	0.6251	0.6249	0.6250	0.6250	0.6250	0.6250	0.6250
0.6	7.0	0.6251	0.6250	0.6250	0.6250	0.6250	0.6250	0.6250
0.6	8.0	0.6251	0.6250	0.6250	0.6250	0.6250	0.6250	0.6250
0.6	9.0	0.6252	0.6250	0.6250	0.6250	0.6250	0.6250	0.6250
0.6	10.0	0.6252	0.6250	0.6250	0.6250	0.6250	0.6250	0.6250
0.8	0.2	0.1680	0.1680	0.1680	0.1680	0.1680	0.1680	0.1680
0.8	0.4	0.2853	0.2853	0.2852	0.2852	0.2852	0.2852	0.2851
0.8	0.6	0.3673	0.3672	0.3671	0.3670	0.3670	0.3670	0.3669
0.8	0.8	0.4246	0.4244	0.4243	0.4242	0.4241	0.4241	0.4239
0.8	1.0	0.4646	0.4643	0.4642	0.4641	0.4640	0.4639	0.4637
0.8	2.0	0.5416	0.5412	0.5410	0.5408	0.5407	0.5407	0.5404
0.8	3.0	0.5538	0.5534	0.5534	0.5533	0.5533	0.5532	0.5530
0.8	4.0	0.5555	0.5552	0.5553	0.5552	0.5552	0.5552	0.5551
0.8	5.0	0.5557	0.5555	0.5556	0.5555	0.5555	0.5555	0.5555
0.8	6.0	0.5558	0.5555	0.5556	0.5555	0.5556	0.5555	0.5555
0.8	7.0	0.5559	0.5555	0.5556	0.5555	0.5556	0.5556	0.5556
0.8	8.0	0.5560	0.5555	0.5556	0.5555	0.5556	0.5556	0.5556
0.8	9.0	0.5561	0.5554	0.5556	0.5555	0.5556	0.5556	0.5556
0.8	10.0	0.5563	0.5554	0.5556	0.5555	0.5556	0.5556	0.5556
1.0	0.2	0.1649	0.1649	0.1649	0.1649	0.1649	0.1648	0.1648
1.0	0.4	0.2755	0.2755	0.2754	0.2754	0.2754	0.2754	0.2753
1.0	0.6	0.3498	0.3497	0.3496	0.3496	0.3495	0.3495	0.3494
1.0	0.8	0.3998	0.3995	0.3994	0.3993	0.3993	0.3992	0.3991
1.0	1.0	0.4333	0.4330	0.4328	0.4327	0.4326	0.4326	0.4323
1.0	2.0	0.4919	0.4915	0.4914	0.4912	0.4911	0.4911	0.4908
1.0	3.0	0.4993	0.4990	0.4990	0.4989	0.4989	0.4989	0.4988
1.0	4.0	0.5002	0.4998	0.4999	0.4999	0.4999	0.4999	0.4998
1.0	5.0	0.5004	0.4999	0.5000	0.5000	0.5000	0.5000	0.5000
1.0	6.0	0.5005	0.4998	0.5001	0.5000	0.5000	0.5000	0.5000
1.0	7.0	0.5007	0.4998	0.5001	0.5000	0.5000	0.5000	0.5000
1.0	8.0	0.5010	0.4998	0.5001	0.5000	0.5000	0.5000	0.5000
1.0	9.0	0.5013	0.4997	0.5001	0.5000	0.5000	0.5000	0.5000
1.0	10.0	0.5017	0.4996	0.5001	0.5000	0.5000	0.5000	0.5000
2.0	0.2	0.1505	0.1504	0.1504	0.1504	0.1504	0.1504	0.1504
2.0	0.4	0.2332	0.2331	0.2331	0.2330	0.2330	0.2330	0.2329
2.0	0.6	0.2787	0.2786	0.2785	0.2784	0.2784	0.2784	0.2782

(continued)

Table 4.4   (continued)

R	NTU	Temperature effectiveness P						Parallel-flow
		Multipass parallel-cross-flow						
		Five passes	Six passes	Seven passes	Eight passes	Nine passes	Ten passes	
2.0	0.8	0.3037	0.3035	0.3034	0.3033	0.3033	0.3032	0.3031
2.0	1.0	0.3174	0.3172	0.3171	0.3170	0.3169	0.3169	0.3167
2.0	2.0	0.3331	0.3325	0.3327	0.3326	0.3326	0.3326	0.3325
2.0	3.0	0.3341	0.3329	0.3335	0.3332	0.3334	0.3333	0.3333
2.0	4.0	0.3349	0.3327	0.3336	0.3332	0.3334	0.3333	0.3333
2.0	5.0	0.3360	0.3323	0.3337	0.3331	0.3334	0.3333	0.3333
2.0	6.0	0.3375	0.3318	0.3339	0.3331	0.3335	0.3333	0.3333
2.0	7.0	0.3392	0.3311	0.3341	0.3330	0.3335	0.3333	0.3333
2.0	8.0	0.3410	0.3303	0.3344	0.3329	0.3335	0.3332	0.3333
2.0	9.0	0.3428	0.3294	0.3348	0.3327	0.3336	0.3332	0.3333
2.0	10.0	0.3445	0.3285	0.3353	0.3326	0.3336	0.3332	0.3333
3.0	0.2	0.1377	0.1377	0.1377	0.1377	0.1377	0.1377	0.1377
3.0	0.4	0.1998	0.1997	0.1997	0.1996	0.1996	0.1996	0.1995
3.0	0.6	0.2277	0.2276	0.2275	0.2275	0.2274	0.2274	0.2273
3.0	0.8	0.2403	0.2401	0.2400	0.2400	0.2399	0.2399	0.2398
3.0	1.0	0.2459	0.2456	0.2456	0.2455	0.2455	0.2455	0.2454
3.0	2.0	0.2507	0.2496	0.2501	0.2498	0.2500	0.2499	0.2499
3.0	3.0	0.2521	0.2490	0.2504	0.2497	0.2501	0.2499	0.2500
3.0	4.0	0.2541	0.2481	0.2508	0.2496	0.2502	0.2499	0.2500
3.0	5.0	0.2566	0.2469	0.2513	0.2493	0.2503	0.2498	0.2500
3.0	6.0	0.2594	0.2454	0.2520	0.2490	0.2505	0.2497	0.2500
3.0	7.0	0.2622	0.2437	0.2529	0.2486	0.2507	0.2497	0.2500
3.0	8.0	0.2648	0.2419	0.2539	0.2481	0.2509	0.2495	0.2500
3.0	9.0	0.2672	0.2401	0.2549	0.2475	0.2512	0.2494	0.2500
3.0	10.0	0.2692	0.2384	0.2561	0.2468	0.2516	0.2492	0.2500
5.0	0.2	0.1165	0.1165	0.1165	0.1165	0.1165	0.1165	0.1165
5.0	0.4	0.1518	0.1517	0.1516	0.1516	0.1516	0.1516	0.1515
5.0	0.6	0.1624	0.1622	0.1622	0.1622	0.1622	0.1622	0.1621
5.0	0.8	0.1656	0.1653	0.1654	0.1653	0.1654	0.1653	0.1653
5.0	1.0	0.1666	0.1662	0.1664	0.1662	0.1663	0.1663	0.1663
5.0	2.0	0.1686	0.1656	0.1672	0.1663	0.1668	0.1665	0.1667
5.0	3.0	0.1715	0.1639	0.1680	0.1658	0.1671	0.1664	0.1667
5.0	4.0	0.1749	0.1616	0.1692	0.1651	0.1675	0.1661	0.1667
5.0	5.0	0.1781	0.1591	0.1707	0.1642	0.1680	0.1658	0.1667
5.0	6.0	0.1808	0.1565	0.1724	0.1630	0.1687	0.1654	0.1667
5.0	7.0	0.1830	0.1541	0.1741	0.1618	0.1695	0.1649	0.1667
5.0	8.0	0.1848	0.1518	0.1757	0.1604	0.1704	0.1642	0.1667

(continued)

Table 4.4   (continued)

R	NTU	Temperature effectiveness P						
		Multipass parallel-cross-flow						Parallel-flow
		Five passes	Six passes	Seven passes	Eight passes	Nine passes	Ten passes	
5.0	9.0	0.1862	0.1499	0.1772	0.1591	0.1714	0.1635	0.1667
5.0	10.0	0.1873	0.1481	0.1785	0.1578	0.1724	0.1628	0.1667
7.0	0.2	0.0998	0.0998	0.0998	0.0998	0.0998	0.0998	0.0998
7.0	0.4	0.1201	0.1200	0.1200	0.1199	0.1199	0.1199	0.1199
7.0	0.6	0.1242	0.1240	0.1240	0.1240	0.1240	0.1240	0.1240
7.0	0.8	0.1251	0.1247	0.1249	0.1248	0.1248	0.1248	0.1248
7.0	1.0	0.1255	0.1247	0.1251	0.1249	0.1250	0.1249	0.1250
7.0	2.0	0.1280	0.1231	0.1259	0.1244	0.1253	0.1247	0.1250
7.0	3.0	0.1312	0.1206	0.1273	0.1234	0.1259	0.1244	0.1250
7.0	4.0	0.1340	0.1178	0.1290	0.1221	0.1267	0.1238	0.1250
7.0	5.0	0.1361	0.1152	0.1307	0.1207	0.1276	0.1231	0.1250
7.0	6.0	0.1376	0.1128	0.1323	0.1192	0.1287	0.1223	0.1250
7.0	7.0	0.1386	0.1108	0.1337	0.1177	0.1298	0.1214	0.1250
7.0	8.0	0.1393	0.1090	0.1349	0.1163	0.1308	0.1205	0.1250
7.0	9.0	0.1399	0.1074	0.1359	0.1150	0.1318	0.1195	0.1250
7.0	10.0	0.1402	0.1061	0.1366	0.1138	0.1326	0.1186	0.1250

The method (P, NTU) is based on the P effectiveness for each fluid, hot or cold, in such a way that:

$$q = P_h C_h (\Delta T)_h = P_c C_c (\Delta T)_c \tag{4.4}$$

The P effectiveness of Tables 4.4 and 4.5 is given in terms of NTU and R, the latter being defined as:

$$R = \frac{C_c}{C_h} \tag{4.5}$$

The following conclusions can be drawn from the effectiveness data of Tables 4.4 and 4.5, (Cabezas-Gómez et al. 2007):

- The effectiveness of parallel cross-flow and counter cross-flow configurations tend to that of the pure parallel and pure counter flow, respectively, as the number of tube passes increases.
- It is interesting to note that the effectiveness varies differently with the number of tube passes in the parallel and counter cross-flows. The effectiveness increases with the number of tube passes in the counter cross-flow configuration with the limit value (rounded off to four decimals) being attained at different number of passes depending on the values of R and NTU.

Table 4.5 Temperature effectiveness P values for counter cross-flow arrangement for five to ten rows, with one tube per row, respectively

R	NTU	Temperature effectiveness P						
		Multipass counter-cross-flow						Counterflow
		Five passes	Six passes	Seven passes	Eight passes	Nine passes	Ten passes	
0.2	0.2	0.1782	0.1782	0.1782	0.1782	0.1782	0.1782	0.1782
0.2	0.4	0.3203	0.3203	0.3204	0.3204	0.3204	0.3204	0.3204
0.2	0.6	0.4349	0.4350	0.4350	0.4350	0.4350	0.4350	0.4351
0.2	0.8	0.5281	0.5282	0.5283	0.5283	0.5283	0.5284	0.5284
0.2	1.0	0.6046	0.6047	0.6048	0.6049	0.6049	0.6049	0.6050
0.2	2.0	0.8302	0.8306	0.8309	0.8311	0.8312	0.8313	0.8317
0.2	3.0	0.9239	0.9246	0.9250	0.9252	0.9254	0.9255	0.9261
0.2	4.0	0.9648	0.9655	0.9659	0.9662	0.9664	0.9666	0.9671
0.2	5.0	0.9832	0.9839	0.9843	0.9845	0.9847	0.9848	0.9853
0.2	6.0	0.9917	0.9923	0.9926	0.9928	0.9929	0.9930	0.9934
0.2	7.0	0.9957	0.9962	0.9964	0.9966	0.9967	0.9968	0.9970
0.2	8.0	0.9977	0.9981	0.9982	0.9984	0.9984	0.9985	0.9987
0.2	9.0	0.9987	0.9990	0.9991	0.9992	0.9992	0.9993	0.9994
0.2	10.0	0.9992	0.9994	0.9995	0.9996	0.9996	0.9997	0.9997
0.4	0.2	0.1752	0.1752	0.1752	0.1752	0.1752	0.1753	0.1753
0.4	0.4	0.3112	0.3113	0.3113	0.3113	0.3113	0.3113	0.3113
0.4	0.6	0.4191	0.4192	0.4192	0.4192	0.4193	0.4193	0.4194
0.4	0.8	0.5061	0.5062	0.5063	0.5064	0.5064	0.5065	0.5066
0.4	1.0	0.5772	0.5775	0.5777	0.5778	0.5778	0.5779	0.5781
0.4	2.0	0.7916	0.7925	0.7930	0.7934	0.7936	0.7938	0.7945
0.4	3.0	0.8889	0.8904	0.8913	0.8919	0.8923	0.8926	0.8938
0.4	4.0	0.9374	0.9393	0.9404	0.9412	0.9417	0.9420	0.9435
0.4	5.0	0.9630	0.9651	0.9663	0.9670	0.9676	0.9679	0.9695
0.4	6.0	0.9771	0.9791	0.9803	0.9811	0.9816	0.9819	0.9834
0.4	7.0	0.9851	0.9871	0.9882	0.9889	0.9893	0.9896	0.9909
0.4	8.0	0.9898	0.9917	0.9927	0.9933	0.9937	0.9940	0.9950
0.4	9.0	0.9927	0.9944	0.9953	0.9958	0.9962	0.9964	0.9973
0.4	10.0	0.9945	0.9961	0.9969	0.9973	0.9976	0.9978	0.9985
0.6	0.2	0.1723	0.1723	0.1723	0.1723	0.1723	0.1723	0.1723
0.6	0.4	0.3024	0.3024	0.3025	0.3025	0.3025	0.3025	0.3025
0.6	0.6	0.4037	0.4038	0.4039	0.4039	0.4040	0.4040	0.4041
0.6	0.8	0.4845	0.4848	0.4849	0.4850	0.4851	0.4851	0.4853
0.6	1.0	0.5503	0.5507	0.5509	0.5510	0.5511	0.5512	0.5515
0.6	2.0	0.7497	0.7510	0.7517	0.7522	0.7526	0.7529	0.7539
0.6	3.0	0.8453	0.8476	0.8490	0.8499	0.8506	0.8510	0.8529
0.6	4.0	0.8975	0.9007	0.9027	0.9039	0.9048	0.9054	0.9081
0.6	5.0	0.9283	0.9322	0.9345	0.9361	0.9371	0.9379	0.9411

(continued)

Table 4.5 (continued)

R	NTU	Temperature effectiveness P						Counterflow
		Multipass counter-cross-flow						
		Five passes	Six passes	Seven passes	Eight passes	Nine passes	Ten passes	
0.6	6.0	0.9473	0.9517	0.9544	0.9561	0.9573	0.9581	0.9616
0.6	7.0	0.9595	0.9643	0.9672	0.9690	0.9702	0.9711	0.9748
0.6	8.0	0.9677	0.9727	0.9756	0.9775	0.9788	0.9796	0.9833
0.6	9.0	0.9733	0.9784	0.9814	0.9833	0.9845	0.9854	0.9889
0.6	10.0	0.9772	0.9824	0.9854	0.9873	0.9885	0.9893	0.9926
0.8	0.2	0.1694	0.1695	0.1695	0.1695	0.1695	0.1695	0.1695
0.8	0.4	0.2938	0.2939	0.2939	0.2939	0.2939	0.2940	0.2940
0.8	0.6	0.3888	0.3890	0.3891	0.3891	0.3892	0.3892	0.3893
0.8	0.8	0.4636	0.4639	0.4641	0.4642	0.4643	0.4643	0.4645
0.8	1.0	0.5239	0.5244	0.5247	0.5248	0.5249	0.5250	0.5254
0.8	2.0	0.7057	0.7073	0.7082	0.7088	0.7093	0.7096	0.7109
0.8	3.0	0.7946	0.7975	0.7993	0.8004	0.8013	0.8018	0.8043
0.8	4.0	0.8454	0.8496	0.8522	0.8540	0.8551	0.8560	0.8597
0.8	5.0	0.8771	0.8826	0.8860	0.8882	0.8898	0.8909	0.8957
0.8	6.0	0.8980	0.9046	0.9087	0.9114	0.9133	0.9147	0.9206
0.8	7.0	0.9124	0.9200	0.9247	0.9279	0.9301	0.9317	0.9386
0.8	8.0	0.9226	0.9310	0.9363	0.9399	0.9423	0.9441	0.9518
0.8	9.0	0.9300	0.9392	0.9450	0.9488	0.9515	0.9535	0.9619
0.8	10.0	0.9354	0.9453	0.9515	0.9556	0.9585	0.9606	0.9696
1.0	0.2	0.1666	0.1666	0.1666	0.1667	0.1667	0.1667	0.1667
1.0	0.4	0.2855	0.2856	0.2856	0.2856	0.2856	0.2857	0.2857
1.0	0.6	0.3744	0.3746	0.3747	0.3748	0.3748	0.3749	0.3750
1.0	0.8	0.4434	0.4437	0.4439	0.4440	0.4441	0.4442	0.4444
1.0	1.0	0.4984	0.4989	0.4992	0.4994	0.4995	0.4996	0.5000
1.0	2.0	0.6609	0.6626	0.6637	0.6644	0.6649	0.6652	0.6667
1.0	3.0	0.7393	0.7424	0.7444	0.7457	0.7466	0.7393	0.7500
1.0	4.0	0.7841	0.7887	0.7916	0.7935	0.7949	0.7841	0.8000
1.0	5.0	0.8123	0.8183	0.8221	0.8246	0.8264	0.8123	0.8333
1.0	6.0	0.8312	0.8385	0.8431	0.8462	0.8484	0.8312	0.8571
1.0	7.0	0.8444	0.8528	0.8582	0.8619	0.8645	0.8444	0.8750
1.0	8.0	0.8538	0.8633	0.8695	0.8737	0.8767	0.8538	0.8889
1.0	9.0	0.8608	0.8712	0.8780	0.8827	0.8861	0.8608	0.9000
1.0	10.0	0.8661	0.8772	0.8847	0.8898	0.8935	0.8661	0.9091
2.0	0.2	0.1534	0.1534	0.1534	0.1534	0.1534	0.1534	0.1535
2.0	0.4	0.2476	0.2477	0.2478	0.2478	0.2478	0.2476	0.2479
2.0	0.6	0.3102	0.3104	0.3105	0.3106	0.3107	0.3102	0.3109
2.0	0.8	0.3539	0.3542	0.3545	0.3546	0.3547	0.3539	0.3551
2.0	1.0	0.3855	0.3860	0.3864	0.3866	0.3867	0.3855	0.3873

(continued)

Table 4.5 (continued)

R	NTU	Temperature effectiveness P						Counterflow
		Multipass counter-cross-flow						
		Five passes	Six passes	Seven passes	Eight passes	Nine passes	Ten passes	
2.0	2.0	0.4598	0.4609	0.4617	0.4621	0.4625	0.4598	0.4637
2.0	3.0	0.4828	0.4841	0.4849	0.4854	0.4858	0.4828	0.4872
2.0	4.0	0.4914	0.4926	0.4934	0.4938	0.4942	0.4914	0.4954
2.0	5.0	0.4952	0.4962	0.4968	0.4971	0.4974	0.4952	0.4983
2.0	6.0	0.4970	0.4978	0.4983	0.4986	0.4988	0.4970	0.4994
2.0	7.0	0.4979	0.4986	0.4990	0.4992	0.4994	0.4979	0.4998
2.0	8.0	0.4985	0.4991	0.4994	0.4995	0.4996	0.4985	0.4999
2.0	9.0	0.4988	0.4993	0.4996	0.4997	0.4998	0.4988	0.5000
2.0	10.0	0.4990	0.4995	0.4997	0.4998	0.4999	0.4990	0.5000
3.0	0.2	0.1414	0.1415	0.1415	0.1415	0.1415	0.1414	0.1415
3.0	0.4	0.2155	0.2156	0.2157	0.2158	0.2158	0.2155	0.2159
3.0	0.6	0.2582	0.2584	0.2586	0.2587	0.2587	0.2582	0.2589
3.0	0.8	0.2842	0.2845	0.2847	0.2848	0.2849	0.2842	0.2852
3.0	1.0	0.3006	0.3009	0.3012	0.3013	0.3014	0.3006	0.3018
3.0	2.0	0.3280	0.3284	0.3286	0.3287	0.3288	0.3280	0.3292
3.0	3.0	0.3321	0.3323	0.3325	0.3325	0.3326	0.3321	0.3328
3.0	4.0	0.3330	0.3331	0.3331	0.3332	0.3332	0.3330	0.3333
3.0	5.0	0.3332	0.3333	0.3333	0.3333	0.3333	0.3332	0.3333
3.0	6.0	0.3333	0.3333	0.3333	0.3333	0.3333	0.3333	0.3333
3.0	7.0	0.3333	0.3333	0.3333	0.3333	0.3333	0.3333	0.3333
3.0	8.0	0.3333	0.3333	0.3333	0.3333	0.3333	0.3333	0.3333
3.0	9.0	0.3333	0.3333	0.3333	0.3333	0.3333	0.3333	0.3333
3.0	10.0	0.3333	0.3333	0.3333	0.3333	0.3333	0.3333	0.3333
5.0	0.2	0.1209	0.1209	0.1210	0.1210	0.1210	0.1209	0.1210
5.0	0.4	0.1660	0.1661	0.1662	0.1662	0.1662	0.1660	0.1663
5.0	0.6	0.1848	0.1849	0.1850	0.1851	0.1851	0.1848	0.1852
5.0	0.8	0.1930	0.1931	0.1932	0.1933	0.1933	0.1930	0.1934
5.0	1.0	0.1967	0.1968	0.1969	0.1969	0.1969	0.1967	0.1971
5.0	2.0	0.1999	0.1999	0.1999	0.1999	0.1999	0.1999	0.1999
5.0	3.0	0.2000	0.2000	0.2000	0.2000	0.2000	0.2000	0.2000
5.0	4.0	0.2000	0.2000	0.2000	0.2000	0.2000	0.2000	0.2000
5.0	5.0	0.2000	0.2000	0.2000	0.2000	0.2000	0.2000	0.2000
5.0	6.0	0.2000	0.2000	0.2000	0.2000	0.2000	0.2000	0.2000
5.0	7.0	0.2000	0.2000	0.2000	0.2000	0.2000	0.2000	0.2000
5.0	8.0	0.2000	0.2000	0.2000	0.2000	0.2000	0.2000	0.2000
5.0	9.0	0.2000	0.2000	0.2000	0.2000	0.2000	0.2000	0.2000
5.0	10.0	0.2000	0.2000	0.2000	0.2000	0.2000	0.2000	0.2000
7.0	0.2	0.1042	0.1043	0.1043	0.1043	0.1043	0.1042	0.1043

(continued)

Table 4.5 (continued)

R	NTU	Temperature effectiveness P						
		Multipass counter-cross-flow						Counterflow
		Five passes	Six passes	Seven passes	Eight passes	Nine passes	Ten passes	
7.0	0.4	0.1314	0.1315	0.1315	0.1315	0.1315	0.1314	0.1316
7.0	0.6	0.1393	0.1394	0.1394	0.1394	0.1394	0.1393	0.1395
7.0	0.8	0.1417	0.1418	0.1418	0.1418	0.1418	0.1417	0.1418
7.0	1.0	0.1425	0.1425	0.1425	0.1425	0.1425	0.1425	0.1426
7.0	2.0	0.1429	0.1429	0.1429	0.1429	0.1429	0.1429	0.1429
7.0	3.0	0.1429	0.1429	0.1429	0.1429	0.1429	0.1429	0.1429
7.0	4.0	0.1429	0.1429	0.1429	0.1429	0.1429	0.1429	0.1429
7.0	5.0	0.1429	0.1429	0.1429	0.1429	0.1429	0.1429	0.1429
7.0	6.0	0.1429	0.1429	0.1429	0.1429	0.1429	0.1429	0.1429
7.0	7.0	0.1429	0.1429	0.1429	0.1429	0.1429	0.1429	0.1429
7.0	8.0	0.1429	0.1429	0.1429	0.1429	0.1429	0.1429	0.1429
7.0	9.0	0.1429	0.1429	0.1429	0.1429	0.1429	0.1429	0.1429
7.0	10.0	0.1429	0.1429	0.1429	0.1429	0.1429	0.1429	0.1429

- For the parallel cross-flow arrangement, the effectiveness alternates between larger and lower than the limit values depending on the number of tube pass. It can be noted that for an odd number of passes, the effectiveness is higher than the limit whereas for an even number the effectiveness is lower. In any case, the difference from the limit effectiveness diminishes with the number of tube passes.

4.1.1   Cross-Flow Heat Exchangers with More Complex Configurations

The good predictive capacity of the HETE code was demonstrated in the paper by Navarro and Cabezas-Gómez (2005) regarding the simulation of two of the configurations studied by Shah and Pignotti (1993), cases 4 and 7 of their paper. Those authors examined complicated heat exchanger flow arrangements considering seven possible flow arrangements for a cross-flow heat exchanger of 60 tubes, arranged in six rows of ten tubes each. The authors related the analyzed flow arrangements to standard cross-flow configurations using the Domingos' rules, Domingos (1969). These standard configurations were described by exact or approximate solutions using, when needed, the recursive algorithms presented in Pignotti and Cordero (1983).

The equivalent two dimensional view of geometries (4) and (7) from Shah and Pignotti (1993) are shown in Fig. 4.2. These geometries were also simulated with

Fig. 4.2 Equivalent two dimensional geometries according to Shah and Pignotti (1993) and simulated with HETE. **a** case 4 and **b** case (7). Configurations "1tube6row3pass2cir(cc)" and "1tube6row6pass1cir(cc)", respectively

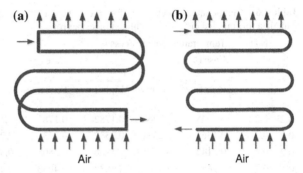

the HETE code. The term "equivalent geometry" designates an approximation of the real heat exchanger configuration considering the influence of the temperature distribution on the effectiveness. This approximation consists in obtaining schematically the flow arrangement distribution of the two fluids streams in the heat exchanger. HETE code simulated P temperature effectiveness data can be found in Table 4.6 for the configurations (4) and (7) from Shah and Pignotti (1993) for different values of the R and NTU parameters. The simulated effectiveness values match exactly those of Table 1 from Shah and Pignotti (1993) for all the range of R values.

Data are also presented in Table 4.6 for similar geometries as those of cases (4) and (7) from Shah and Pignotti (1993) (Fig. 4.2) though for four and eight rows with ten tubes per row.

It is very important to note that Shah and Pignotti (1993) computed the temperature effectiveness data through a very strong and useful procedure consisting in the use of the Domingos' rules and the recursive algorithms of Pignotti and Cordero (1983). Through this recursive algorithm these authors were capable of obtaining the P temperature effectiveness of a counter-cross-flow heat exchanger with six passes of the mixed in-tube fluid, represented schematically in Fig. 4.2b, which corresponds to case (7). The HETE code allows the direct simulation of these configurations without applying procedures from the literature and their associated potential difficulties. More details about the P temperature effectiveness computation and simulation with the aforementioned procedure can be found in Shah and Pignotti (1993) and Navarro-Cabezas-Gómez (2005).

The HETE code has also been applied by Cabezas-Gómez et al. (2007) to the evaluation of the heat exchanger effectiveness of a complex cross flow arrangement proposed by Guo et al. (2002) whose schematic representation is shown in Fig. 4.3.

Guo et al. (2002) compared this flow arrangement with three other configurations, namely, the pure cross-flow, the one pass cross-flow with four tube rows, and the two passes counter cross-flow, Fig. 4.1e. They found that their proposed configuration was the best performing. In their paper from 2007, Cabezas-Gómez et al. (2007) provided new (ε, NTU) plotted data for the Guo et al. (2002)

Table 4.6 P temperature effectiveness for cases (4) and (7) from Shah and Pignotti (1993) along with data for four and eight rows, with ten tubes per row

R	NTU	Temperature effectiveness P						
		Case (4)			Case (7)			Counter flow
		Four rows	Six rows	Eight rows	Four rows	Six rows	Eight rows	
0.2	0.2	0.0938	0.1782	0.1782	0.1782	0.1782	0.1782	0.1782
0.2	0.3	0.1365	0.2531	0.2532	0.2532	0.2532	0.2532	0.2532
0.2	0.5	0.2142	0.3805	0.3806	0.3806	0.3807	0.3807	0.3807
0.2	0.7	0.2832	0.4836	0.4838	0.4838	0.4839	0.4840	0.4841
0.2	1.0	0.3729	0.6039	0.6044	0.6043	0.6047	0.6049	0.6050
0.2	1.5	0.4927	0.7414	0.7423	0.7421	0.7429	0.7432	0.7436
0.2	2.0	0.5847	0.8285	0.8299	0.8293	0.8306	0.8311	0.8317
0.2	3.0	0.7126	0.9217	0.9236	0.9226	0.9246	0.9252	0.9261
0.2	5.0	0.8453	0.9814	0.9831	0.9819	0.9839	0.9845	0.9853
0.2	7.0	0.9041	0.9947	0.9958	0.9948	0.9962	0.9966	0.9970
0.2	10.0	0.9410	0.9988	0.9993	0.9987	0.9994	0.9996	0.9997
0.3	0.2	0.0933	0.1767	0.1767	0.1767	0.1767	0.1767	0.1767
0.3	0.3	0.1355	0.2502	0.2502	0.2502	0.2503	0.2503	0.2503
0.3	0.5	0.2118	0.3741	0.3743	0.3743	0.3744	0.3744	0.3745
0.3	0.7	0.2791	0.4738	0.4742	0.4741	0.4744	0.4745	0.4746
0.3	1.0	0.3659	0.5900	0.5906	0.5905	0.5911	0.5913	0.5915
0.3	1.5	0.4812	0.7231	0.7245	0.7241	0.7253	0.7257	0.7263
0.3	2.0	0.5692	0.8088	0.8108	0.8101	0.8120	0.8127	0.8136
0.3	3.0	0.6914	0.9040	0.9070	0.9055	0.9086	0.9096	0.9110
0.3	5.0	0.8193	0.9714	0.9745	0.9722	0.9759	0.9772	0.9787
0.3	7.0	0.8774	0.9894	0.9919	0.9896	0.9928	0.9937	0.9948
0.3	10.0	0.9151	0.9966	0.9981	0.9962	0.9984	0.9989	0.9994
0.5	0.2	0.0924	0.1737	0.1738	0.1738	0.1738	0.1738	0.1738
0.5	0.3	0.1335	0.2444	0.2444	0.2444	0.2445	0.2445	0.2445
0.5	0.5	0.2071	0.3617	0.3620	0.3619	0.3621	0.3622	0.3623
0.5	0.7	0.2711	0.4548	0.4553	0.4552	0.4556	0.4558	0.4560
0.5	1.0	0.3525	0.5623	0.5633	0.5631	0.5640	0.5643	0.5647
0.5	1.5	0.4589	0.6858	0.6879	0.6872	0.6892	0.6899	0.6908
0.5	2.0	0.5391	0.7669	0.7702	0.7689	0.7721	0.7732	0.7746
0.5	3.0	0.6495	0.8621	0.8672	0.8645	0.8700	0.8719	0.8744
0.5	5.0	0.7657	0.9404	0.9474	0.9421	0.9506	0.9535	0.9572
0.5	7.0	0.8197	0.9681	0.9752	0.9684	0.9779	0.9810	0.9847
0.5	10.0	0.8561	0.9835	0.9896	0.9817	0.9913	0.9940	0.9966
0.7	0.2	0.0915	0.1708	0.1709	0.1709	0.1709	0.1709	0.1709
0.7	0.3	0.1316	0.2387	0.2388	0.2388	0.2389	0.2389	0.2389
0.7	0.5	0.2026	0.3497	0.3500	0.3500	0.3502	0.3503	0.3504

(continued)

Table 4.6 (continued)

R	NTU	Temperature effectiveness P						
		Case (4)			Case (7)			Counter flow
		Four rows	Six rows	Eight rows	Four rows	Six rows	Eight rows	
0.7	0.7	0.2634	0.4363	0.4370	0.4369	0.4374	0.4376	0.4379
0.7	1.0	0.3397	0.5352	0.5366	0.5363	0.5374	0.5378	0.5384
0.7	1.5	0.4376	0.6481	0.6508	0.6500	0.6525	0.6534	0.6545
0.7	2.0	0.5103	0.7227	0.7268	0.7252	0.7293	0.7308	0.7326
0.7	3.0	0.6090	0.8127	0.8196	0.8159	0.8233	0.8260	0.8295
0.7	5.0	0.7118	0.8936	0.9043	0.8961	0.9094	0.9143	0.9207
0.7	7.0	0.7597	0.9274	0.9399	0.9276	0.9452	0.9516	0.9598
0.7	10.0	0.7923	0.9499	0.9633	0.9462	0.9676	0.9752	0.9845
1.0	0.2	0.0901	0.1666	0.1666	0.1666	0.1666	0.1667	0.1667
1.0	0.3	0.1287	0.2305	0.2306	0.2306	0.2307	0.2307	0.2308
1.0	0.5	0.1959	0.3324	0.3328	0.3328	0.3331	0.3332	0.3333
1.0	0.7	0.2523	0.4099	0.4107	0.4105	0.4112	0.4115	0.4118
1.0	1.0	0.3214	0.4962	0.4978	0.4974	0.4989	0.4994	0.5000
1.0	1.5	0.4077	0.5923	0.5955	0.5945	0.5975	0.5986	0.6000
1.0	2.0	0.4699	0.6549	0.6597	0.6578	0.6626	0.6644	0.6667
1.0	3.0	0.5521	0.7301	0.7380	0.7336	0.7424	0.7457	0.7500
1.0	5.0	0.6348	0.7995	0.8121	0.8020	0.8183	0.8246	0.8333
1.0	7.0	0.6724	0.8306	0.8461	0.8304	0.8528	0.8619	0.8750
1.0	10.0	0.6977	0.8532	0.8713	0.8483	0.8772	0.8898	0.9091
1.5	0.2	0.0879	0.1598	0.1598	0.1598	0.1599	0.1599	0.1599
1.5	0.3	0.1242	0.2175	0.2177	0.2177	0.2178	0.2178	0.2179
1.5	0.5	0.1855	0.3055	0.3060	0.3060	0.3064	0.3065	0.3067
1.5	0.7	0.2351	0.3690	0.3700	0.3698	0.3707	0.3709	0.3713
1.5	1.0	0.2936	0.4361	0.4379	0.4375	0.4391	0.4396	0.4404
1.5	1.5	0.3630	0.5057	0.5089	0.5078	0.5109	0.5120	0.5134
1.5	2.0	0.4104	0.5474	0.5518	0.5500	0.5545	0.5561	0.5584
1.5	3.0	0.4693	0.5929	0.5989	0.5954	0.6023	0.6049	0.6084
1.5	5.0	0.5237	0.6287	0.6357	0.6297	0.6390	0.6426	0.6474
1.5	7.0	0.5467	0.6420	0.6488	0.6414	0.6514	0.6551	0.6598
1.5	10.0	0.5614	0.6502	0.6565	0.6479	0.6581	0.6615	0.6652
2.0	0.2	0.0857	0.1533	0.1534	0.1534	0.1534	0.1534	0.1535
2.0	0.3	0.1199	0.2054	0.2056	0.2056	0.2057	0.2058	0.2058
2.0	0.5	0.1759	0.2811	0.2816	0.2816	0.2820	0.2822	0.2824
2.0	0.7	0.2194	0.3325	0.3335	0.3333	0.3342	0.3345	0.3348
2.0	1.0	0.2689	0.3832	0.3849	0.3845	0.3860	0.3866	0.3873
2.0	1.5	0.3244	0.4308	0.4334	0.4325	0.4351	0.4360	0.4372
2.0	2.0	0.3601	0.4560	0.4591	0.4577	0.4609	0.4621	0.4637

(continued)

Table 4.6 (continued)

R	NTU	Temperature effectiveness P						Counter flow
		Case (4)			Case (7)			
		Four rows	Six rows	Eight rows	Four rows	Six rows	Eight rows	
2.0	3.0	0.4016	0.4793	0.4824	0.4805	0.4841	0.4854	0.4872
2.0	5.0	0.4365	0.4930	0.4952	0.4932	0.4962	0.4971	0.4983
2.0	7.0	0.4499	0.4966	0.4982	0.4964	0.4986	0.4992	0.4998
2.0	10.0	0.4581	0.4983	0.4993	0.4978	0.4995	0.4998	0.5000
3.0	0.2	0.0816	0.1413	0.1414	0.1414	0.1415	0.1415	0.1415
3.0	0.3	0.1119	0.1836	0.1838	0.1838	0.1839	0.1840	0.1841
3.0	0.5	0.1585	0.2389	0.2394	0.2393	0.2398	0.2400	0.2402
3.0	0.7	0.1922	0.2716	0.2724	0.2723	0.2730	0.2733	0.2736
3.0	1.0	0.2275	0.2990	0.3002	0.2999	0.3009	0.3013	0.3018
3.0	1.5	0.2629	0.3191	0.3203	0.3199	0.3211	0.3215	0.3221
3.0	2.0	0.2831	0.3268	0.3278	0.3273	0.3284	0.3287	0.3292
3.0	3.0	0.3035	0.3316	0.3321	0.3318	0.3323	0.3325	0.3328
3.0	5.0	0.3178	0.3331	0.3332	0.3331	0.3333	0.3333	0.3333
3.0	7.0	0.3224	0.3333	0.3333	0.3332	0.3333	0.3333	0.3333
3.0	10.0	0.3250	0.3333	0.3333	0.3333	0.3333	0.3333	0.3333
5.0	0.2	0.0743	0.1208	0.1209	0.1209	0.1209	0.1210	0.1210
5.0	0.3	0.0980	0.1482	0.1484	0.1484	0.1486	0.1486	0.1487
5.0	0.5	0.1306	0.1769	0.1772	0.1772	0.1775	0.1776	0.1777
5.0	0.7	0.1509	0.1892	0.1896	0.1895	0.1898	0.1900	0.1902
5.0	1.0	0.1690	0.1963	0.1966	0.1965	0.1968	0.1969	0.1971
5.0	1.5	0.1838	0.1993	0.1994	0.1993	0.1995	0.1995	0.1996
5.0	2.0	0.1906	0.1998	0.1999	0.1998	0.1999	0.1999	0.1999
5.0	3.0	0.1959	0.2000	0.2000	0.2000	0.2000	0.2000	0.2000
5.0	5.0	0.1986	0.2000	0.2000	0.2000	0.2000	0.2000	0.2000
5.0	7.0	0.1992	0.2000	0.2000	0.2000	0.2000	0.2000	0.2000
5.0	10.0	0.1995	0.2000	0.2000	0.2000	0.2000	0.2000	0.2000
7.0	0.2	0.0678	0.1041	0.1042	0.1042	0.1043	0.1043	0.1043
7.0	0.3	0.0866	0.1217	0.1219	0.1219	0.1220	0.1221	0.1221
7.0	0.5	0.1095	0.1362	0.1364	0.1364	0.1366	0.1366	0.1367
7.0	0.7	0.1219	0.1406	0.1408	0.1408	0.1409	0.1410	0.1410
7.0	1.0	0.1316	0.1424	0.1425	0.1424	0.1425	0.1425	0.1426
7.0	1.5	0.1382	0.1428	0.1428	0.1428	0.1428	0.1428	0.1428
7.0	2.0	0.1406	0.1429	0.1429	0.1429	0.1429	0.1429	0.1429
7.0	3.0	0.1421	0.1429	0.1429	0.1429	0.1429	0.1429	0.1429
7.0	5.0	0.1427	0.1429	0.1429	0.1429	0.1429	0.1429	0.1429
7.0	7.0	0.1428	0.1429	0.1429	0.1429	0.1429	0.1429	0.1429
7.0	10.0	0.1428	0.1429	0.1429	0.1429	0.1429	0.1429	0.1429

Fig. 4.3 Schematic representation of the Guo et al. (2002) flow arrangement. Configuration 34 "2tube4row2pass4cir.txt" from Chap. 6

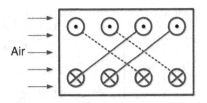

configuration and compared the effectiveness of this configuration with that of the four passes counter cross-flow one of Fig. 4.1g. The numerical results for limited ranges of C^* and NTU are presented in Table 4.7. These results are complementary to those from Guo et al. (2002) since they include extended performance data to wider ranges of C^* and NTU.

The following trends can be devised from the observed differences between the different configurations of Table 4.7, (Cabezas-Gómez et al. 2007).

(i)   The effectiveness of the four passes counter cross-flow configuration, Fig. 4.1g, is higher than that of Fig. 4.3 for certain ranges of C^* and NTU.

(ii) As expected, the thermal effectiveness increases with NTU for both configurations. However, the effectiveness of the Fig. 4.1g arrangement is higher for low NTU values, the difference diminishing with NTU.

(iii) The effectiveness of the Fig. 4.1g arrangement is higher over the whole range of NTU for the C^* values of 0.2 and 0.4. However, for higher values of C^* a cross over is noted at a given NTU. The cross over NTU diminishes with C^*.

(iv) The thermal effectiveness of both configurations tend to that of the pure counter flow one for lower values of C^* and high values of NTU, as should be expected. It is interesting to note that the best performance of the Fig. 4.3 configuration with respect to that of Fig. 4.1g occurs at high C^* and NTU values, for which the deviations with respect to the pure counter flow are higher (see shadow areas in Table 4.7). A more uniform temperature difference along the heat exchanger could explain the above trend for this configuration at high values of C^* and NTU, as suggested by Guo et al. (2002).

(v)   It could be stated that the conclusions from Guo et al. (2002) with respect to the thermal performance of the Fig. 4.3 configuration are limited to the flow arrangements considered for comparison in their paper. These conclusions cannot be generalized as the data of Table 4.7 clearly demonstrate.

It is important to emphasize at this point that the choice of any flow configuration should not be limited to the theoretical characteristics of the heat exchanger. Parameters such as the allowed pressure drop and heat transfer coefficients should also be considered in design and rating of a heat exchanger along with manufacturing and durability aspects, Shah and Pignotti (1993).

Similar configurations to that proposed by Guo et al. (2002) and numerically simulated by Cabezas-Gómez et al. (2007), shown in Fig. 4.3, were proposed and simulated with the HETE code as reported in the papers by Cabezas-Gómez et al. (2009) and Cabezas-Gómez et al. (2012). These were two and three tube rows

Table 4.7 Heat exchanger effectiveness for three configurations: (i) Fig. 4.3 (Guo et al. 2002); (ii) Fig. 4.1g; and counter-flow (Cabezas-Gómez et al. 2007)

C^{*}	NTU	Heat Exchanger Effectiveness				
		$C_{min} = C_c$		$C_{min} = C_h$		
		Figure 4.3	Figure 4.1 (g)	Figure 4.3	Figure 4.1 (g)	Counter-flow
0.2	2	0.8252	0.8293	0.8258	0.8294	0.8317
0.2	5	0.9785	0.9819	0.9799	0.9825	0.9853
0.2	8	0.9958	0.9969	0.9968	0.9976	0.9987
0.2	9	0.9974	0.9981	0.9982	0.9987	0.9994
0.2	10	0.9982	0.9987	0.9989	0.9992	0.9997
0.2	11	0.9988	0.9991	0.9994	0.9996	0.9999
0.2	12	0.9991	0.9993	0.9996	0.9997	0.9999
0.4	2	0.7830	0.7899	0.7838	0.7900	0.7945
0.4	5	0.9527	0.9591	0.9551	0.9606	0.9695
0.4	8	0.9835	0.9860	0.9863	0.9886	0.9950
0.4	9	0.9874	0.9890	0.9902	0.9919	0.9973
0.4	10	0.9900	0.9910	0.9928	0.9940	0.9985
0.4	11	0.9918	0.9923	0.9945	0.9954	0.9992
0.4	12	0.9930	0.9932	0.9958	0.9964	0.9996
0.6	2	0.7389	0.7473	0.7395	0.7474	0.7539
0.6	5	0.9130	0.9211	0.9155	0.9230	0.9411
0.6	8	0.9554	0.9583	0.9594	0.9628	0.9833
0.6	9	0.9619	0.9635	0.9664	0.9688	0.9889
0.6	10	0.9666	0.9671	0.9714	0.9731	0.9926
0.6	11	0.9700	0.9696	0.9753	0.9763	0.9951
0.6	12	0.9725	0.9715	0.9782	0.9788	0.9967
0.8	2	0.6939	0.7028	0.6942	0.7029	0.7109
0.8	5	0.8593	0.8674	0.8609	0.8688	0.8957
0.8	8	0.9060	0.9082	0.9089	0.9118	0.9518
0.8	9	0.9138	0.9144	0.9173	0.9188	0.9619
0.8	10	0.9197	0.9190	0.9236	0.9240	0.9696
0.8	11	0.9241	0.9223	0.9285	0.9279	0.9757
0.8	12	0.9274	0.9247	0.9323	0.9310	0.9804
1.0	2	0.6490	0.6578	0.6490	0.6578	0.6667
1.0	5	0.7952	0.8020	0.7952	0.8020	0.8333
1.0	8	0.8373	0.8384	0.8373	0.8384	0.8889
1.0	9	0.8445	0.8442	0.8445	0.8442	0.9000
1.0	10	0.8499	0.8483	0.8499	0.8483	0.9091
1.0	11	0.8539	0.8514	0.8539	0.8514	0.9167
1.0	12	0.8570	0.8537	0.8570	0.8537	0.9231

configurations, as shown in Fig. 4.4, though numerical results were also obtained for configurations of up to six rows. The numerical results were compared against the standard counter-cross-flow arrangements of two to six passes which find widespread applications in refrigeration, automotive and air conditioning systems due to its high thermal effectiveness. These flow arrangements are shown in Fig. 4.1e–g for two, three and four rows, respectively. Arrangements with more tube rows are an extension of these ones.

In order to compare the performance of those configurations, Cabezas-Gómez et al. (2012) used several criteria, including the heat exchanger effectiveness, ε, the heat exchanger reversibility norm, Sekuliç (1990), the heat exchanger efficiency, Fakheri (2007), and the entransy dissipation number, Guo et al. (2009). The

Fig. 4.4 Proposed flow arrangements considering two and three tube rows

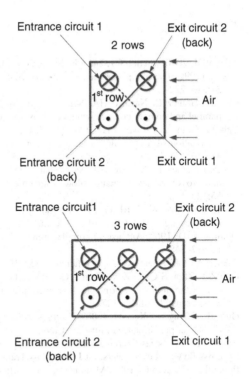

proposed flow arrangements, Fig. 4.4, were the ones with the best performance according to all the criteria considered in the comparison. In addition to excellent thermal performance, these configurations present good applicability for single phase applications. This behavior is due to the better local mean temperature difference distribution obtained due to the redistribution of the in-tube fluid streams. In fact, the configurations of Fig. 4.4 behave as a heat exchanger with an identical order coupling distribution of the external fluid. This kind of heat exchanger provides an increment in the thermal performance with respect to heat exchangers that have an inverse order coupling distribution as those with the standard multi-pass counter-cross-flow configurations, Figs. 4.1e–g. According to Shah and Sekuliç (2003) the identical order coupling leads to a lower local temperature difference distribution in comparison with the inverse order coupling. Correlations for the effectiveness of heat exchangers with identical order coupling distributions can be found in Stevens et al. (1957), Shah and Sekuliç (2003) and VDI Atlas (2010), among others. This kind of flow arrangements are also called by the VDI Atlas (2010) as "co-directed countercurrent cross-flow" heat exchangers.

From the constructive point of view, the configurations shown in Fig. 4.4 represent other practical method of obtaining the so called identical order coupling distribution. Two in-fluid circuits are used in this case, with an overall counter-flow arrangement between them. Other way is to redirect the air flow between each tube pass, as suggested by Shah and Sekuliç (2003).

References

Atlas VDI (2010) Section C1, 2nd edn. Springer, Berlin, pp 41–43

Bes Th (1996) Thermal performance of codirected cross-flow heat exchangers. Heat Mass Transf 31:215–222

Cabezas-Gómez L, Navarro HA, Saiz-Jabardo JM (2007) Thermal performance of multi-pass parallel and counter cross-flow heat exchangers. ASME J Heat Transf 129:282–290

Cabezas-Gómez L, Navarro HA, Godoy SM, Campo A, Saiz-Jabardo JM (2009) Thermal characterization of a cross-flow heat exchanger with a new flow arrangement. Int J Therm Sci 48:2165–2170

Cabezas-Gómez L, Navarro HA, Saiz-Jabardo JM, Hanriot SM, Maia CB (2012) Analysis of new cross flow heat exchanger flow arrangement—extension to several rows. Int J Therm Sci 55:122–132

DiGiovanni MA, Webb RL (1989) Uncertainty in effectiveness-NTU calculations for crossflow heat exchangers. Heat Transf Eng 10(3):61–70

Domingos JD (1969) Analysis of complex assemblies of heat exchangers. Int J Heat Mass Transf 12:537–548

Fakheri A (2007) Heat exchanger efficiency. ASME J Heat Transf 129:1268–1276

Guo Z-Y, Zhou S-Q, Li Z-X, Chen L-G (2002) Theoretical analysis and experimental confirmation of the uniformity principle of temperature difference field in heat exchanger. Int J Heat Mass Transf 45:2119–2127

Guo JF, Cheng L, Xu MT (2009) Entransy dissipation number and its application to heat exchanger performance evaluation. Chin Sci Bull 29:2708–2713

Navarro HA, Cabezas-Gómez L (2005) A new approach for thermal performance calculation of cross-flow heat exchangers. Int J Heat Mass Trans 48:3880–3888

Pignotti A, Cordero G (1983) Mean temperature difference in multipass crossflow. ASME J Heat Transf 105:584–591

Sekuliç DP (1990) The second law quality of energy transformation in a heat exchanger. ASME J Heat Transf 112:295–300

Shah RK, Pignotti A (1993) Thermal analysis of complex cross-flow exchangers in terms of standard configurations. ASME J Heat Transf 115:353–359

Shah RK, Sekuliç DP (2003) Fundamentals of heat exchanger design. Wiley, New York

Stevens RA, Fernandez J, Woolf JR (1957) Mean-temperature difference in one, two, and three-pass crossflow heat exchangers. Trans ASME 79:287–297

Taborek J (1983) Charts for mean temperature difference in industrial heat exchanger configuration, In: Schlünder EW (ed), Heat exchanger design handbook, Hemisphere Publishing, New York

Chapter 5
Computational Code HETE (Standard C Programming Language)

The computational code HETE, written in standard C programming language, is displayed in this chapter. Though its executable version could be downloaded from the Springer site http://extras.springer.com, the printed version presented in this chapter has been included as a helpful tool in the reader comprehension which can be complemented with the detailed development of Chap. 3. For further assistance related with the HETE code, the users are encouraged to contact the authors by e-mail. The current e-mails of the authors are printed right after the first line of the HETE code.

```
/*  HETE (Heat Exchanger Thermal Efficiency) software
        Helio Aparecido Navarro. email: han@sc.usp.br
        Luben Cabezas-Gomez. email: cab35ezas@yahoo.com.br
        Jose Maria Saiz Jabardo. email: jose.saiz.jabardo@udc.es
*/

#include <math.h>
#include <stdio.h>
#include <stdlib.h>
#include <string.h>


#define TAM_ARQ 81

#define NMAX_Ne 2001
#define NMAX_Nt 101
#define NMAX_Nr 101
#define NMAX_Nl 202
#define NMAX_Nc 101

typedef struct {double Tae,Tas,Tre,Trs,Qa,Qr;} elementtype;
typedef struct {
        int ET; /*Exit tube 1:yes, 0:no*/
        int tt; /*Tube type 0: without tube, 1: inlet, 2: intermediary, 3: exit*/
        int s;/*Flow direction 1: into the page, 0: out of the page }*/
        int proxl; /* next tube line */
        int proxk; /* next tube row */
 } elementtypegeometria;
typedef struct  {int Nd; int inicl, inick, endl, endk; } circuittype;
typedef struct  {
        int Nr; /* number of rows */
        int Nl; /* number of lines */
        int Nt; /* number of tubes per row*/
        int Nc; /* circuit number */
        elementtypegeometria g[NMAX_Nl][NMAX_Nr]; /* geometry elements */
 } geometrytype;
```

© The Author(s) 2015
L. Cabezas-Gómez et al., *Thermal Performance Modeling of Cross-Flow Heat Exchangers*, SpringerBriefs in Applied Sciences and Technology, DOI 10.1007/978-3-319-09671-1_5

```
elementtype e[NMAX_Ne][NMAX_Nt][NMAX_Nr];
geometrytype ge;
circuittype ci[NMAX_Nc];

/*Input data*/
int Ne; // number of elements per tube

double Tae; // [C] heat exchanger inlet temperature of air (external unmixed fluid)
double Tre; // [C] heat exchanger inlet temperature of in-tube fluid

/*Output data*/
double Tas; // [C] heat exchanger air outlet temperature
double Trs; // [C] heat exchanger outlet refrigerant temperature
double Ca; // [W/K] element thermal capacity of air
double Cr; // [W/K] element thermal capacity of refrigerant
double UA; // [W/K] element UA: overall heat transfer coefficient x area
double UAt; // [W/K] heat exchanger UA: global heat exchanger coefficient x area
double Cat; // [W/K] heat exchanger thermal capacity of air
double Crt; // [W/K] heat exchanger thermal capacity of refrigerant
double Cmin; // [W/k] minimum thermal capacity
double Cmax; // [W/k] maximum thermal capacity
double CR; // heat capacity rate ratio Cmin/Cmax
double Qr,Qa; // [W] heat transfer rate for (refrigerant and air) in the HE
double Qta,Qtr; // [W] heat transfer rate (refrigerant and air) in the elements
double Qmax; // [W] heat exchanger maximum heat transfer rate
double NUT; // number of transfer units
int CminCa; // Cmim=1 (air side)   Cmim=0 (tube side)
double E; // heat exchanger thermal effectiveness

FILE *file_geom;
char name_geometry_file[TAM_ARQ];

/* Procedures */

double dmin(double x, double y)
/* minimum value*/
{
      return ((x<=y) ? x : y);
}

double dmax(double x, double y)
/*maximum value*/
{
      return ((x>=y) ? x : y);
}

void ReadGeometry(void)

/* Procedure for reading heat exchanger geometry from the input file*/
{
int l,k,m,n,lc,kc;
fscanf(file_geom,"%d",&ge.Nr);
fscanf(file_geom,"%d",&ge.Nl);
fscanf(file_geom,"%d",&ge.Nt);
fscanf(file_geom,"%d",&ge.Nc);
for(l=1;l<=ge.Nl;l++) for(k=1;k<=ge.Nr;k++) {
      ge.g[l][k].ET=0;ge.g[l][k].tt=0;ge.g[l][k].s=0;
   ge.g[l][k].proxl=0;ge.g[l][k].proxk=0;
}
/* For each of Nc circuits */
for(m=1;m<=ge.Nc;m++) {
   fscanf(file_geom, "%d %d",&ci[m].inicl,&ci[m].inick);
   fscanf(file_geom,"%d %d",&ci[m].endl,&ci[m].endk);
   fscanf(file_geom,"%d",&ci[m].Nd);
   lc=ci[m].inicl;kc=ci[m].inick;
   for(n=1;n<=ci[m].Nd;n++) {
     ge.g[lc][kc].ET=1;
     fscanf(file_geom,"%d %d %d %d",&ge.g[lc][kc].tt,&ge.g[lc][kc].s,
        &ge.g[lc][kc].proxl,&ge.g[lc][kc].proxk);
     l=ge.g[lc][kc].proxl;k=ge.g[lc][kc].proxk;
     lc=l;kc=k;
   }
}
}
```

```
int NextElement(int m, int *l, int *i, int *j, int *k)
/*search next element in the geometry following path*/
 {
int end, stop,lc,kc;
end=0;  stop=0;

lc=*l;kc=*k;
if ((! stop)&&(*i>=1)&&(*i<Ne)&&(ge.g[lc][kc].s==1)) {*i=*i+1;  stop=1; }
if ((! stop)&&(*i>1)&&(*i<=Ne)&&(ge.g[lc][kc].s==0)) {*i=*i-1;  stop=1; }
if ((! stop)&&(((*i==Ne)&&(ge.g[lc][kc].s==1))||((*i==1)&&(ge.g[lc][kc].s==0)))) {
  if (ci[m].Nd==1) {
  end=1;
  }
  else {
   *l=ge.g[lc][kc].proxl; *k=ge.g[lc][kc].proxk;
   *j=*l;
   if (ge.g[lc][kc].tt==3) end=1;
  }

}
if ((end)&&(m<ge.Nc)) {
     *l=ci[m+1].inicl;*k=ci[m+1].inick;
  if (ge.g[*l][*k].s==1) *i=1; else *i=Ne;
  *j=*l;
}
return(end);
 }

void AirUpdate(int i, int j, int k)
/* Update air-side temperature in the circuits */
{
   if (k>1) {
   e[i][j][k].Tae=e[i][j][k-1].Tas;
   }
 }

void FluidUpdate(int i, int j, int k, int iant, int jant, int kant)
/* Update fluid-side temperature in the circuits */
 {
      e[i][j][k].Tre=e[iant][jant][kant].Trs;
 }

void ComputeUACminCa(double NUT,double CR)
 { // case: Cat<=Crt; Cmin=Cat
Cat=UAt/NUT;
Crt=Cat/CR;
Ca=UAt/(NUT*Ne*ge.Nt);
Cr=UAt/(NUT*CR*ge.Nc);
UA=UAt/(Ne*ge.Nt*ge.Nr);
 }

void ComputeUACminCr(double NUT,double CR)
 { // case: Cat>=Crt; Cmin=Crt
Crt=UAt/NUT;
Cat=Crt/CR;
Ca=UAt/(NUT*CR*Ne*ge.Nt);
Cr=UAt/(NUT*ge.Nc);
UA=UAt/(Ne*ge.Nt*ge.Nr);
 }

void ComputeTemperature1(void)
/* Procedure for computing Heat Exchanger temperature only once*/
 { // compute temperature in the element, Kays & London
int i,j,k,l,m,iant,jant,kant;
int end; // last element of the circuit
double G;
double A;
double NUTe,CRe;

G=1.-exp(-UA/dmin(Ca,Cr));
A=Ca*G/Cr;

Qta=0.;Qtr=0.;
l=ci[1].inicl;k=ci[1].inick;
if (ge.g[l][k].s==1) i=1; else i=Ne;
j=ci[1].inicl;
for(m=1;m<=ge.Nc;m++) {
     end=0;
  if (!end) AirUpdate(i,j,k);
```

```
                /*computing the outlet air temperature*/

            while (!end)  {
        e[i][j][k].Trs=e[i][j][k].Tae*2.*A/(2.+A)+e[i][j][k].Tre*(2.-A)/(2.+A);
        e[i][j][k].Tas=e[i][j][k].Tae*(1.-2.*G/(2.+A))+e[i][j][k].Tre*(2.*G)/(2.+A);

            /* computing heat transfer rates*/
        e[i][j][k].Qr=-Cr*(e[i][j][k].Trs-e[i][j][k].Tre);
        e[i][j][k].Qa=Ca*(e[i][j][k].Tas-e[i][j][k].Tae);
        Qta+=e[i][j][k].Qa;
        Qtr+=e[i][j][k].Qr;
        iant=i;jant=j;kant=k;
        end=NextElement(m,&l,&i,&j,&k);
        if ((!end)II(m!=ge.Nc)) AirUpdate(i,j,k);
        if (!end) FluidUpdate(i,j,k,iant,jant,kant);
                } //while
} //for
}


void ComputeTemperature(void)
/* Procedure for computing Heat Exchanger temperature until
   steady-steady based on tolerance TOLT.
   This routine uses the routine ComputeTemperature1 */
{
int i,j,k,m,l; double sum,sum1,TOLT=1.e-8;

/*initialization of temperature field*/
for(i=1;i<=Ne;i++) for(j=1;j<=ge.Nt;j++) for(k=1;k<=ge.Nr;k++) {
        e[i][j][k].Tae=Tae; e[i][j][k].Tre=Tre;
        e[i][j][k].Tas=Tae; e[i][j][k].Trs=Tre;
        e[i][j][k].Qa=0.;e[i][j][k].Qr=0.;
}
ComputeTemperature1();
do {
        sum=0.;
        for(i=1;i<=Ne;i++) for(j=1;j<=ge.Nt;j++) sum+=e[i][j][ge.Nr].Tas;
        ComputeTemperature1();
        sum1=0.;
        for(i=1;i<=Ne;i++) for(j=1;j<=ge.Nt;j++) sum1+=e[i][j][ge.Nr].Tas;
} while ((fabs(sum1-sum)/fabs(sum))>=TOLT);
/*compute Trs and Tas*/
Trs=0.;
for(m=1;m<=ge.Nc;m++) {
        l=ci[m].endl;
        k=ci[m].endk;
    j=l;
    if (ge.g[l][k].s==0) i=1; else i=Ne;
        Trs=Trs+e[i][j][k].Trs;
}
Trs=Trs/ge.Nc;
Tas=0.;
for(i=1;i<=Ne;i++) for(j=1;j<=ge.Nt;j++) Tas=Tas+e[i][j][ge.Nr].Tas;
Tas=Tas/(ge.Nt*Ne); // mean air heat exchanger outlet temperature
}


double ComputeE(double NUT, double CR, int CminCa)
// Compute thermal effectiveness
// Input: CR, NUT
{
double E;
if (CminCa) ComputeUACminCa(NUT,CR); else ComputeUACminCr(NUT,CR);

ComputeTemperature();
Cmin=dmin(Cat,Crt);
Cmax=dmax(Cat,Crt);
Qmax=Cmin*(Tre-Tae);
Qr=-Crt*(Trs-Tre); // Q=-Cr*(Trs-Tre) ou Q=Ca*(Tas-Tae)
Qa=Cat*(Tas-Tae);

E=Qtr/Qmax;
return(E);
}
```

```c
/* Main Program */

int main() {

/*  Read geometry of the Heat Exchanger from the input file*/
printf("Enter with the geometry file name: "); scanf("%s",name_geometry_file);
if ((file_geom=fopen(name_geometry_file,"r"))==NULL) {
    printf("Error to open the file %s\n",name_geometry_file);
    exit(EXIT_FAILURE);
}
ReadGeometry();
fclose(file_geom);

printf("Enter with NUT: "); scanf("%lf",&NUT);
printf("Enter with Cmin/Cmax: "); scanf("%lf",&CR);
printf("Enter with 1 for Cmin=Ca (air) or 0 for Cmin=Cf (fluid): "); scanf("%d",&CminCa);

Tae=35.; Tre=70.; UAt=12.; //Constant thermal input values
Ne=1000; // Used number of elements that the heat exchanger is divided

E=ComputeE(NUT,CR,CminCa);

printf("CR=%g  \n",CR);
printf("NUT=%g  \n",NUT);
if (CminCa)    printf("Cmin = Ca (air)  \n");
else          printf("Cmin = Cf (fluid)  \n");
printf("E=%g   \n",E);
return 0;
}
```

Chapter 6
Input Data Files for Simulating with the HETE Code

In Table 6.1 of the present chapter the reader can find the contents of the HETE input data files for the configurations numerically modeled in Chaps. 3 and 4 along with other configurations. All the cases shown in the table include the name of the configuration file, the corresponding geometrical sketch of the configuration, and the data to be introduced into the HETE input files. All the data input files, and others, can be found in the Springer site http://extras.springer.com.

© The Author(s) 2015
L. Cabezas-Gómez et al., *Thermal Performance Modeling of Cross-Flow Heat Exchangers*, SpringerBriefs in Applied Sciences and Technology, DOI 10.1007/978-3-319-09671-1_6

Table 6.1 HETE code input files for several configurations

Input data file name	Geometrical sketch of the input data file	Contents of the input data file
CONFIGURATION 1: "1tube1row1pass1cir.txt" • One pass cross-flow heat exchanger with one row, one tube per row and one circuit. (unmixed-mixed cross-flow heat exchanger). Figures 3.10 and 4.1(a)		1 1 1 1 1 1 1 1 1 1 1 1 1
CONFIGURATION 2: "1tube2row1pass2cir.txt" • One pass cross-flow heat exchanger with two rows, one tube per row and two circuits. (unmixed-partially mixed cross-flow heat exchanger)		2 1 1 2 1 1 1 1 1 1 1 1 1 1 2 1 2 1 1 1 1 2
CONFIGURATION 3: "1tube3row1pass3cir.txt" • One pass cross-flow heat exchanger with three rows, one tube per row and three circuits. (unmixed-partially mixed cross-flow heat exchanger)		3 1 1 3 1 1 1 1 1 1 1 1 1 1 2 1 2 1 1 1 1 2 1 3 1 3 1 1 1 1 3
CONFIGURATION 4: "1tube4row1pass4cir.txt" • One pass cross-flow heat exchanger with four rows, one tube per row and four circuits		4 1 1 4 1 1 1 1 1 1 1 1 1 1 2 1 2 1 1 1 1 2 1 3

(continued)

Table 6.1 (continued)

(unmixed-partially mixed cross-flow heat exchanger)		1 3 1 1 1 1 3 1 4 1 4 1 1 1 1 4
CONFIGURATION 11: "1tube2row2pass1cir(par).txt" • Two pass parallel-cross-flow heat exchanger with one tube per row and one circuit. Figure 4.1(b)		2 1 1 1 1 1 1 2 2 1 1 1 2 3 0 0 0
CONFIGURATION 12: "1tube3row3pass1cir(par).txt" • Three pass parallel-cross-flow heat exchanger with one tube per row and one circuit. Figure 4.1(c)		3 1 1 1 1 1 1 3 3 1 1 1 2 2 0 1 3 3 1 0 0
CONFIGURATION 13: "1tube4row4pass1cir(par).txt" Four pass parallel-cross-flow heat exchanger with one tube per row and one circuit. Figure 4.1(d)		4 1 1 1 1 1 1 4 4 1 1 1 2 2 0 1 3 2 1 1 4 3 0 0 0
CONFIGURATION 14: "1tube5row5pass1cir(par).txt" • Five pass parallel-cross-flow heat exchanger with one tube per row and one circuit		5 1 1 1 1 1 1 5 5 1 1 1 2 2 0 1 3 2 1 1 4 2 0 1 5 3 1 0 0

(continued)

Table 6.1 (continued)

CONFIGURATION 15: "1tube6row6pass1cir(par).txt" • Six pass parallel-cross-flow heat exchanger with one tube per row and one circuit	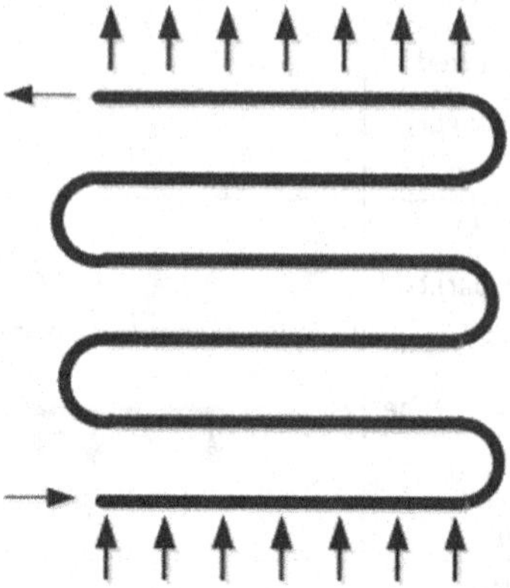	6 1 1 1 1 1 1 6 6 1 1 1 2 2 0 1 3 2 1 1 4 2 0 1 5 2 1 1 6 3 0 0 0
CONFIGURATION 20: "1tube2row2pass1cir(cc).txt" • Two pass counter-cross-flow heat exchanger with one tube per row and one circuit. Figure 4.1(e)	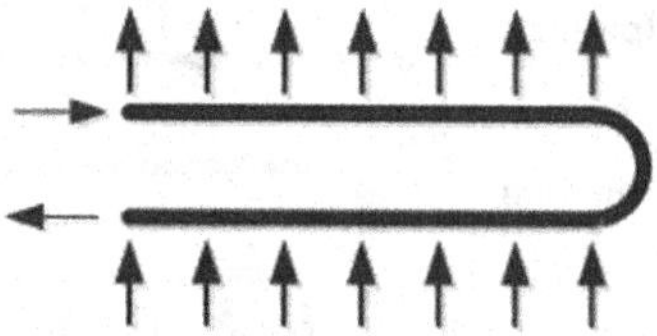	2 1 1 1 1 2 1 1 2 1 1 1 1 3 0 0 0
CONFIGURATION 21: "1tube3row3pass1cir(cc).txt" • Three pass counter-cross- flow heat exchanger with one tube per row and one circuit. Figure 4.1(f)	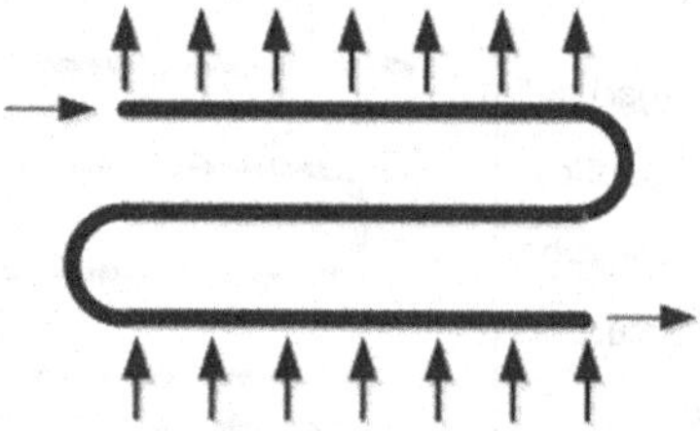	3 1 1 1 1 3 1 1 3 1 1 1 2 2 0 1 1 3 1 0 0
CONFIGURATION 22: "1tube4row4pass1cir(cc).txt" • Four pass counter-cross-flow heat exchanger with one tube per row and one circuit. Figure 4.1(g)	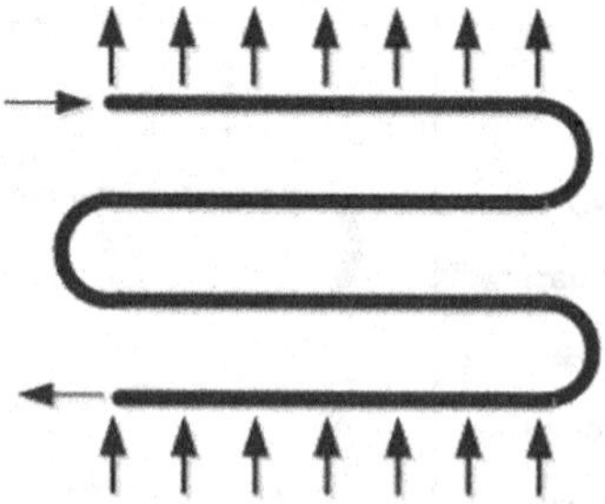	4 1 1 1 1 4 1 1 4 1 1 1 3 2 0 1 2 2 1 1 1 3 0 0 0

(continued)

Table 6.1 (continued)

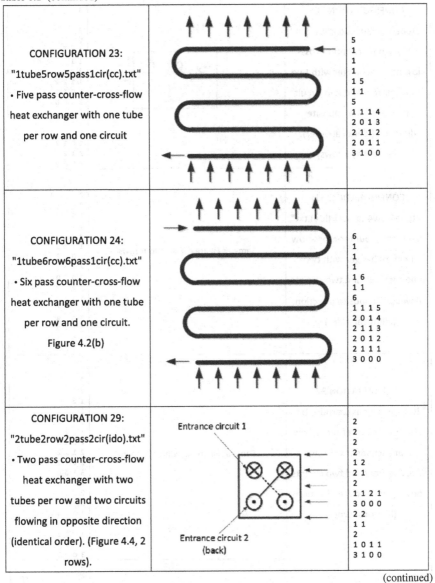

CONFIGURATION 23: "1tube5row5pass1cir(cc).txt" • Five pass counter-cross-flow heat exchanger with one tube per row and one circuit	5 1 1 1 1 5 1 1 5 1 1 1 4 2 0 1 3 2 1 1 2 2 0 1 1 3 1 0 0
CONFIGURATION 24: "1tube6row6pass1cir(cc).txt" • Six pass counter-cross-flow heat exchanger with one tube per row and one circuit. Figure 4.2(b)	6 1 1 1 1 6 1 1 6 1 1 1 5 2 0 1 4 2 1 1 3 2 0 1 2 2 1 1 1 3 0 0 0
CONFIGURATION 29: "2tube2row2pass2cir(ido).txt" • Two pass counter-cross-flow heat exchanger with two tubes per row and two circuits flowing in opposite direction (identical order). (Figure 4.4, 2 rows).	2 2 2 2 1 2 2 1 2 1 1 2 1 3 0 0 0 2 2 1 1 2 1 0 1 1 3 1 0 0

(continued)

Table 6.1 (continued)

CONFIGURATION 30: "2tube3row3pass2cir(ido).txt" • Three pass counter-cross-flow heat exchanger with two tube per row and two circuit flowing in the opposite direction (identical order). (Figure 4.4, 3 rows)	Entrance circuit 1 Entrance circuit 2 (back)	3 2 2 2 1 3 1 1 3 1 1 2 2 2 0 1 1 3 1 0 0 2 3 2 1 3 1 0 1 2 2 1 2 1 3 0 0 0
CONFIGURATION 31: "2tube4row4pass2cir(ido).txt" • Four pass counter-cross-flow heat exchanger with two tubes per row and two circuits flowing in opposite direction (identical order)	Same as Figure 4.4 with four rows	4 2 2 2 1 4 2 1 4 1 1 2 3 2 0 1 2 2 1 2 1 3 0 0 0 2 4 1 1 4 1 0 1 3 2 1 2 2 2 0 1 1 3 1 0 0
CONFIGURATION 32: "2tube5row5pass2cir(ido).txt" • Five pass counter-cross-flow heat exchanger with two tubes per row and two circuits flowing in opposite direction (identical order)	Same as Figure 4.4 with five rows	5 2 2 2 1 5 1 1 5 1 1 2 4 2 0 1 3 2 1 2 2 2 0 1 1 3 1 0 0 2 5 2 1 5 1 0 1 4 2 1 2 3 2 0 1 2 2 1 2 1 3 0 0 0
CONFIGURATION 33: "2tube6row6pass2cir(ido).txt" • Six pass counter-cross-flow		6 2 2 2 1 6 2 1 6 1 1 2 5 2 0 1 4 2 1 2 3

(continued)

Table 6.1 (continued)

heat exchanger with two tubes per row and two circuits flowing in opposite direction (identical order)	Same as Figure 4.4 with six rows	2 0 1 2 2 1 2 1 3 0 0 0 2 6 1 1 6 1 0 1 5 2 1 2 4 2 0 1 3 2 1 2 2 2 0 1 1 3 1 0 0
CONFIGURATION 34: "2tube4row2pass4cir.txt" • Two pass counter-cross-flow heat exchanger with two tubes per row, four rows and four circuits flowing in opposite direction. (Configuration of Guo et al. (2002), Figure 4.3)		4 2 2 4 1 4 2 2 2 1 1 2 2 3 0 0 0 1 3 2 1 2 1 1 2 1 3 0 0 0 2 4 1 2 2 1 0 1 2 3 1 0 0 2 3 1 1 2 1 0 1 1 3 1 0 0
CONFIGURATION 35: "1tube6row3pass2cir(cc).txt" • Three pass counter-cross-flow heat exchanger with one tube per row and two circuits flowing in the same direction (identical order). Figure 4.2(a)		6 1 1 2 1 6 1 2 3 1 1 1 4 2 0 1 2 3 1 0 0 1 5 1 1 3 1 1 1 3 2 0 1 1 3 1 0 0

Chapter 7
Heat Exchanger Effectiveness Data for Several Configurations

Heat exchanger effectiveness for 35 cross-flow configurations with important industrial applications is presented in this chapter in both plot and table forms. Both versions for the C_{min} fluid are included: (a) for the Cold (external, air); and (b) for the hot (in-tube) fluid. Effectiveness is given in terms of the heat capacity ratio, C^*, and the NTU group for the following ranges: $0 \leq C^* \leq 1.0$ and $0 \leq NTU \leq 10.0$. All the configuration input files are available in the internet site address: http://extras.springer.com. Details of most of them can be found in Chap. 6.

© The Author(s) 2015
L. Cabezas-Gómez et al., *Thermal Performance Modeling of Cross-Flow Heat Exchangers*, SpringerBriefs in Applied Sciences and Technology, DOI 10.1007/978-3-319-09671-1_7

CONFIGURATION 1

1TUBE1ROW1PASS1CIR

(a) $C_{min} = C_c$

NTU	Heat Exchanger Effectiveness										
	C^* $(C_{min} = C_c)$										
	0	0.1	0.2	0.3	0.4	0.5	0.6	0.7	0.8	0.9	1.0
0	0.0000	0.0000	0.0000	0.0000	0.0000	0.0000	0.0000	0.0000	0.0000	0.0000	0.0000
0.1	0.0952	0.0947	0.0943	0.0938	0.0934	0.0929	0.0925	0.0921	0.0916	0.0912	0.0908
0.2	0.1813	0.1796	0.1780	0.1764	0.1749	0.1733	0.1718	0.1702	0.1687	0.1673	0.1658
0.3	0.2592	0.2559	0.2526	0.2494	0.2462	0.2431	0.2400	0.2370	0.2341	0.2312	0.2283
0.4	0.3297	0.3243	0.3190	0.3139	0.3089	0.3039	0.2991	0.2944	0.2898	0.2853	0.2808
0.5	0.3935	0.3858	0.3784	0.3711	0.3641	0.3572	0.3505	0.3439	0.3376	0.3313	0.3253
0.6	0.4512	0.4412	0.4314	0.4220	0.4128	0.4039	0.3953	0.3869	0.3787	0.3708	0.3631
0.7	0.5034	0.4910	0.4789	0.4672	0.4560	0.4451	0.4345	0.4243	0.4144	0.4048	0.3955
0.8	0.5507	0.5358	0.5214	0.5076	0.4942	0.4814	0.4689	0.4570	0.4454	0.4342	0.4234
0.9	0.5934	0.5762	0.5596	0.5436	0.5283	0.5135	0.4993	0.4856	0.4724	0.4598	0.4476
1.0	0.6321	0.6126	0.5938	0.5758	0.5585	0.5420	0.5261	0.5108	0.4961	0.4821	0.4685
1.1	0.6671	0.6454	0.6245	0.6046	0.5855	0.5673	0.5498	0.5330	0.5170	0.5016	0.4868
1.2	0.6988	0.6749	0.6522	0.6304	0.6096	0.5898	0.5708	0.5527	0.5353	0.5187	0.5028
1.3	0.7275	0.7016	0.6770	0.6536	0.6312	0.6098	0.5895	0.5701	0.5515	0.5338	0.5169
1.4	0.7534	0.7257	0.6994	0.6743	0.6505	0.6278	0.6061	0.5855	0.5659	0.5471	0.5292
1.5	0.7769	0.7475	0.7195	0.6930	0.6678	0.6438	0.6209	0.5992	0.5786	0.5589	0.5402
1.6	0.7981	0.7671	0.7377	0.7097	0.6833	0.6581	0.6342	0.6115	0.5899	0.5694	0.5498
1.7	0.8173	0.7848	0.7540	0.7248	0.6972	0.6709	0.6460	0.6224	0.6000	0.5786	0.5584
1.8	0.8347	0.8008	0.7687	0.7384	0.7097	0.6824	0.6566	0.6321	0.6089	0.5869	0.5660
1.9	0.8504	0.8153	0.7820	0.7506	0.7209	0.6927	0.6661	0.6409	0.6169	0.5943	0.5728
2.0	0.8647	0.8283	0.7940	0.7616	0.7310	0.7020	0.6746	0.6487	0.6241	0.6009	0.5788
2.1	0.8775	0.8401	0.8048	0.7715	0.7401	0.7103	0.6822	0.6557	0.6305	0.6067	0.5842
2.2	0.8892	0.8508	0.8146	0.7805	0.7483	0.7178	0.6891	0.6620	0.6363	0.6120	0.5890
2.3	0.8997	0.8605	0.8234	0.7885	0.7556	0.7246	0.6953	0.6676	0.6414	0.6167	0.5933
2.4	0.9093	0.8692	0.8314	0.7958	0.7623	0.7306	0.7008	0.6727	0.6461	0.6209	0.5972
2.5	0.9179	0.8770	0.8386	0.8024	0.7683	0.7361	0.7058	0.6772	0.6502	0.6247	0.6006
2.6	0.9257	0.8842	0.8451	0.8083	0.7737	0.7410	0.7103	0.6813	0.6540	0.6281	0.6038
2.7	0.9328	0.8906	0.8510	0.8136	0.7785	0.7455	0.7143	0.6850	0.6573	0.6312	0.6065
2.8	0.9392	0.8964	0.8563	0.8185	0.7829	0.7495	0.7180	0.6883	0.6603	0.6340	0.6091
2.9	0.9450	0.9017	0.8610	0.8228	0.7869	0.7531	0.7213	0.6913	0.6631	0.6364	0.6113
3.0	0.9502	0.9065	0.8654	0.8268	0.7905	0.7564	0.7242	0.6940	0.6655	0.6387	0.6133
3.1	0.9550	0.9108	0.8693	0.8303	0.7937	0.7593	0.7269	0.6964	0.6677	0.6407	0.6152
3.2	0.9592	0.9147	0.8728	0.8336	0.7967	0.7620	0.7293	0.6986	0.6697	0.6425	0.6168
3.3	0.9631	0.9182	0.8760	0.8365	0.7993	0.7644	0.7315	0.7006	0.6715	0.6441	0.6183
3.4	0.9666	0.9214	0.8789	0.8391	0.8017	0.7665	0.7335	0.7024	0.6731	0.6456	0.6196
3.5	0.9698	0.9243	0.8815	0.8415	0.8038	0.7685	0.7353	0.7040	0.6746	0.6469	0.6208
3.6	0.9727	0.9269	0.8839	0.8436	0.8058	0.7703	0.7369	0.7055	0.6759	0.6481	0.6219
3.7	0.9753	0.9292	0.8861	0.8456	0.8075	0.7719	0.7383	0.7068	0.6771	0.6492	0.6229
3.8	0.9776	0.9314	0.8880	0.8473	0.8091	0.7733	0.7396	0.7080	0.6782	0.6502	0.6238
3.9	0.9798	0.9333	0.8897	0.8489	0.8106	0.7746	0.7408	0.7090	0.6792	0.6511	0.6246
4.0	0.9817	0.9350	0.8913	0.8503	0.8119	0.7758	0.7419	0.7100	0.6800	0.6519	0.6253
4.1	0.9834	0.9366	0.8928	0.8516	0.8131	0.7768	0.7428	0.7109	0.6808	0.6526	0.6260
4.2	0.9850	0.9380	0.8941	0.8528	0.8141	0.7778	0.7437	0.7117	0.6816	0.6532	0.6266
4.3	0.9864	0.9393	0.8952	0.8539	0.8151	0.7787	0.7445	0.7124	0.6822	0.6538	0.6271
4.4	0.9877	0.9405	0.8963	0.8548	0.8160	0.7795	0.7452	0.7130	0.6828	0.6543	0.6276
4.5	0.9889	0.9416	0.8972	0.8557	0.8167	0.7802	0.7459	0.7136	0.6833	0.6548	0.6280
4.6	0.9899	0.9425	0.8981	0.8565	0.8174	0.7808	0.7464	0.7142	0.6838	0.6553	0.6284
4.7	0.9909	0.9434	0.8989	0.8572	0.8181	0.7814	0.7470	0.7146	0.6842	0.6557	0.6288
4.8	0.9918	0.9442	0.8996	0.8578	0.8187	0.7819	0.7475	0.7151	0.6846	0.6560	0.6291
4.9	0.9926	0.9449	0.9002	0.8584	0.8192	0.7824	0.7479	0.7155	0.6850	0.6563	0.6294
5.0	0.9933	0.9455	0.9008	0.8589	0.8197	0.7828	0.7483	0.7158	0.6853	0.6566	0.6296
5.1	0.9939	0.9461	0.9014	0.8594	0.8201	0.7832	0.7486	0.7161	0.6856	0.6569	0.6299
5.2	0.9945	0.9466	0.9018	0.8598	0.8205	0.7836	0.7489	0.7164	0.6859	0.6571	0.6301
5.3	0.9950	0.9471	0.9023	0.8602	0.8209	0.7839	0.7492	0.7167	0.6861	0.6573	0.6303
5.4	0.9955	0.9475	0.9026	0.8606	0.8212	0.7842	0.7495	0.7169	0.6863	0.6575	0.6305
5.5	0.9959	0.9479	0.9030	0.8609	0.8215	0.7845	0.7497	0.7171	0.6865	0.6577	0.6306
5.6	0.9963	0.9483	0.9033	0.8612	0.8217	0.7847	0.7499	0.7173	0.6867	0.6579	0.6308
5.7	0.9967	0.9486	0.9036	0.8615	0.8220	0.7849	0.7501	0.7175	0.6868	0.6580	0.6309
5.8	0.9970	0.9489	0.9039	0.8617	0.8222	0.7851	0.7503	0.7177	0.6870	0.6581	0.6310
5.9	0.9973	0.9491	0.9041	0.8619	0.8224	0.7853	0.7505	0.7178	0.6871	0.6583	0.6311
6.0	0.9975	0.9494	0.9043	0.8621	0.8225	0.7854	0.7506	0.7179	0.6872	0.6584	0.6312

NTU	Heat Exchanger Effectiveness										
	C^* $(C_{min} = C_c)$										
	0	0.1	0.2	0.3	0.4	0.5	0.6	0.7	0.8	0.9	1.0
6.1	0.9978	0.9496	0.9045	0.8623	0.8227	0.7856	0.7507	0.7180	0.6873	0.6585	0.6313
6.2	0.9980	0.9498	0.9047	0.8624	0.8228	0.7857	0.7509	0.7182	0.6874	0.6585	0.6314
6.3	0.9982	0.9500	0.9048	0.8626	0.8230	0.7858	0.7510	0.7183	0.6875	0.6586	0.6314
6.4	0.9983	0.9501	0.9050	0.8627	0.8231	0.7859	0.7511	0.7183	0.6876	0.6587	0.6315
6.5	0.9985	0.9503	0.9051	0.8628	0.8232	0.7860	0.7512	0.7184	0.6877	0.6588	0.6316
6.6	0.9986	0.9504	0.9052	0.8629	0.8233	0.7861	0.7512	0.7185	0.6877	0.6588	0.6316
6.7	0.9988	0.9505	0.9053	0.8630	0.8234	0.7862	0.7513	0.7186	0.6878	0.6589	0.6317
6.8	0.9989	0.9506	0.9054	0.8631	0.8235	0.7863	0.7514	0.7186	0.6878	0.6589	0.6317
6.9	0.9990	0.9507	0.9055	0.8632	0.8235	0.7863	0.7514	0.7187	0.6879	0.6590	0.6318
7.0	0.9991	0.9508	0.9056	0.8633	0.8236	0.7864	0.7515	0.7187	0.6879	0.6590	0.6318
7.1	0.9992	0.9509	0.9057	0.8633	0.8236	0.7864	0.7515	0.7188	0.6880	0.6590	0.6318
7.2	0.9993	0.9509	0.9057	0.8634	0.8237	0.7865	0.7516	0.7188	0.6880	0.6591	0.6318
7.3	0.9993	0.9510	0.9058	0.8634	0.8237	0.7865	0.7516	0.7188	0.6881	0.6591	0.6319
7.4	0.9994	0.9511	0.9058	0.8635	0.8238	0.7866	0.7516	0.7189	0.6881	0.6591	0.6319
7.5	0.9994	0.9511	0.9059	0.8635	0.8238	0.7866	0.7517	0.7189	0.6881	0.6591	0.6319
7.6	0.9995	0.9512	0.9059	0.8636	0.8239	0.7866	0.7517	0.7189	0.6881	0.6592	0.6319
7.7	0.9995	0.9512	0.9060	0.8636	0.8239	0.7867	0.7517	0.7189	0.6881	0.6592	0.6320
7.8	0.9996	0.9513	0.9060	0.8636	0.8239	0.7867	0.7518	0.7190	0.6882	0.6592	0.6320
7.9	0.9996	0.9513	0.9060	0.8637	0.8240	0.7867	0.7518	0.7190	0.6882	0.6592	0.6320
8.0	0.9997	0.9513	0.9061	0.8637	0.8240	0.7867	0.7518	0.7190	0.6882	0.6592	0.6320
8.1	0.9997	0.9514	0.9061	0.8637	0.8240	0.7868	0.7518	0.7190	0.6882	0.6592	0.6320
8.2	0.9997	0.9514	0.9061	0.8637	0.8240	0.7868	0.7518	0.7190	0.6882	0.6593	0.6320
8.3	0.9998	0.9514	0.9061	0.8638	0.8240	0.7868	0.7518	0.7190	0.6882	0.6593	0.6320
8.4	0.9998	0.9514	0.9062	0.8638	0.8240	0.7868	0.7519	0.7191	0.6882	0.6593	0.6320
8.5	0.9998	0.9514	0.9062	0.8638	0.8241	0.7868	0.7519	0.7191	0.6882	0.6593	0.6320
8.6	0.9998	0.9515	0.9062	0.8638	0.8241	0.7868	0.7519	0.7191	0.6883	0.6593	0.6321
8.7	0.9998	0.9515	0.9062	0.8638	0.8241	0.7868	0.7519	0.7191	0.6883	0.6593	0.6321
8.8	0.9998	0.9515	0.9062	0.8638	0.8241	0.7868	0.7519	0.7191	0.6883	0.6593	0.6321
8.9	0.9999	0.9515	0.9062	0.8638	0.8241	0.7869	0.7519	0.7191	0.6883	0.6593	0.6321
9.0	0.9999	0.9515	0.9062	0.8638	0.8241	0.7869	0.7519	0.7191	0.6883	0.6593	0.6321
9.1	0.9999	0.9515	0.9063	0.8639	0.8241	0.7869	0.7519	0.7191	0.6883	0.6593	0.6321
9.2	0.9999	0.9515	0.9063	0.8639	0.8241	0.7869	0.7519	0.7191	0.6883	0.6593	0.6321
9.3	0.9999	0.9515	0.9063	0.8639	0.8241	0.7869	0.7519	0.7191	0.6883	0.6593	0.6321
9.4	0.9999	0.9516	0.9063	0.8639	0.8241	0.7869	0.7519	0.7191	0.6883	0.6593	0.6321
9.5	0.9999	0.9516	0.9063	0.8639	0.8242	0.7869	0.7519	0.7191	0.6883	0.6593	0.6321
9.6	0.9999	0.9516	0.9063	0.8639	0.8242	0.7869	0.7519	0.7191	0.6883	0.6593	0.6321
9.7	0.9999	0.9516	0.9063	0.8639	0.8242	0.7869	0.7519	0.7191	0.6883	0.6593	0.6321
9.8	0.9999	0.9516	0.9063	0.8639	0.8242	0.7869	0.7520	0.7191	0.6883	0.6593	0.6321
9.9	1.0000	0.9516	0.9063	0.8639	0.8242	0.7869	0.7520	0.7191	0.6883	0.6593	0.6321
10.0	1.0000	0.9516	0.9063	0.8639	0.8242	0.7869	0.7520	0.7191	0.6883	0.6593	0.6321

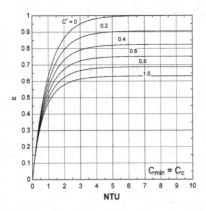

CONFIGURATION 1

1TUBE1ROW1PASS1CIR

(b) $C_{min} = C_h$

NTU	Heat Exchanger Effectiveness										
	$C^* (C_{min} = C_h)$										
	0	0.1	0.2	0.3	0.4	0.5	0.6	0.7	0.8	0.9	1.0
0	0.0000	0.0000	0.0000	0.0000	0.0000	0.0000	0.0000	0.0000	0.0000	0.0000	0.0000
0.1	0.0952	0.0947	0.0943	0.0938	0.0934	0.0929	0.0925	0.0921	0.0916	0.0912	0.0908
0.2	0.1813	0.1796	0.1780	0.1764	0.1749	0.1733	0.1718	0.1703	0.1687	0.1673	0.1658
0.3	0.2592	0.2559	0.2526	0.2494	0.2463	0.2431	0.2401	0.2371	0.2341	0.2312	0.2283
0.4	0.3297	0.3244	0.3192	0.3140	0.3090	0.3041	0.2993	0.2945	0.2899	0.2853	0.2808
0.5	0.3935	0.3860	0.3786	0.3714	0.3644	0.3575	0.3508	0.3442	0.3377	0.3314	0.3253
0.6	0.4512	0.4414	0.4319	0.4225	0.4134	0.4045	0.3958	0.3873	0.3791	0.3710	0.3631
0.7	0.5034	0.4914	0.4796	0.4681	0.4569	0.4460	0.4354	0.4250	0.4149	0.4051	0.3955
0.8	0.5507	0.5364	0.5225	0.5090	0.4957	0.4828	0.4703	0.4580	0.4462	0.4346	0.4234
0.9	0.5934	0.5771	0.5612	0.5456	0.5304	0.5156	0.5011	0.4871	0.4735	0.4603	0.4476
1.0	0.6321	0.6139	0.5960	0.5785	0.5614	0.5448	0.5286	0.5128	0.4976	0.4828	0.4685
1.1	0.6671	0.6471	0.6275	0.6082	0.5893	0.5709	0.5530	0.5357	0.5188	0.5025	0.4868
1.2	0.6988	0.6772	0.6559	0.6350	0.6144	0.5944	0.5749	0.5560	0.5376	0.5199	0.5028
1.3	0.7275	0.7045	0.6817	0.6592	0.6371	0.6155	0.5945	0.5741	0.5543	0.5353	0.5169
1.4	0.7534	0.7292	0.7051	0.6812	0.6577	0.6346	0.6122	0.5903	0.5692	0.5488	0.5292
1.5	0.7769	0.7517	0.7264	0.7012	0.6763	0.6519	0.6281	0.6049	0.5825	0.5609	0.5402
1.6	0.7981	0.7720	0.7457	0.7194	0.6933	0.6676	0.6425	0.6180	0.5944	0.5717	0.5498
1.7	0.8173	0.7906	0.7633	0.7360	0.7087	0.6818	0.6555	0.6299	0.6051	0.5813	0.5584
1.8	0.8347	0.8074	0.7794	0.7511	0.7228	0.6948	0.6674	0.6406	0.6147	0.5898	0.5660
1.9	0.8504	0.8228	0.7942	0.7650	0.7357	0.7067	0.6781	0.6503	0.6234	0.5975	0.5728
2.0	0.8647	0.8368	0.8076	0.7778	0.7476	0.7175	0.6880	0.6591	0.6312	0.6044	0.5788
2.1	0.8775	0.8496	0.8200	0.7894	0.7585	0.7275	0.6970	0.6672	0.6383	0.6106	0.5842
2.2	0.8892	0.8612	0.8313	0.8002	0.7685	0.7366	0.7052	0.6745	0.6448	0.6162	0.5890
2.3	0.8997	0.8719	0.8418	0.8101	0.7777	0.7451	0.7127	0.6811	0.6506	0.6213	0.5933
2.4	0.9093	0.8816	0.8513	0.8193	0.7862	0.7528	0.7197	0.6872	0.6559	0.6258	0.5972
2.5	0.9179	0.8905	0.8602	0.8277	0.7941	0.7600	0.7260	0.6928	0.6607	0.6299	0.6006
2.6	0.9257	0.8987	0.8683	0.8356	0.8014	0.7666	0.7319	0.6979	0.6651	0.6336	0.6038
2.7	0.9328	0.9062	0.8758	0.8428	0.8081	0.7727	0.7373	0.7026	0.6691	0.6370	0.6065
2.8	0.9392	0.9130	0.8828	0.8496	0.8144	0.7784	0.7423	0.7069	0.6727	0.6400	0.6091
2.9	0.9450	0.9193	0.8893	0.8558	0.8203	0.7836	0.7469	0.7109	0.6760	0.6428	0.6113
3.0	0.9502	0.9251	0.8952	0.8617	0.8257	0.7885	0.7512	0.7145	0.6791	0.6453	0.6133
3.1	0.9550	0.9304	0.9008	0.8671	0.8308	0.7931	0.7552	0.7179	0.6819	0.6476	0.6152
3.2	0.9592	0.9353	0.9059	0.8722	0.8355	0.7973	0.7588	0.7210	0.6844	0.6496	0.6168
3.3	0.9631	0.9398	0.9107	0.8769	0.8399	0.8013	0.7623	0.7238	0.6868	0.6515	0.6183
3.4	0.9666	0.9440	0.9152	0.8813	0.8441	0.8050	0.7654	0.7265	0.6889	0.6532	0.6196
3.5	0.9698	0.9478	0.9193	0.8855	0.8479	0.8084	0.7684	0.7289	0.6909	0.6548	0.6208
3.6	0.9727	0.9514	0.9232	0.8893	0.8516	0.8116	0.7711	0.7312	0.6927	0.6562	0.6219
3.7	0.9753	0.9546	0.9268	0.8930	0.8550	0.8147	0.7736	0.7332	0.6943	0.6575	0.6229
3.8	0.9776	0.9576	0.9302	0.8964	0.8582	0.8175	0.7760	0.7352	0.6958	0.6586	0.6238
3.9	0.9798	0.9604	0.9333	0.8996	0.8612	0.8201	0.7782	0.7369	0.6972	0.6597	0.6246
4.0	0.9817	0.9630	0.9363	0.9026	0.8640	0.8226	0.7803	0.7386	0.6985	0.6607	0.6253
4.1	0.9834	0.9654	0.9391	0.9055	0.8667	0.8249	0.7822	0.7401	0.6997	0.6615	0.6260
4.2	0.9850	0.9676	0.9417	0.9082	0.8692	0.8271	0.7840	0.7415	0.7008	0.6624	0.6266
4.3	0.9864	0.9696	0.9441	0.9107	0.8716	0.8292	0.7857	0.7429	0.7018	0.6631	0.6271
4.4	0.9877	0.9716	0.9464	0.9131	0.8738	0.8311	0.7873	0.7441	0.7027	0.6638	0.6276
4.5	0.9889	0.9733	0.9486	0.9153	0.8759	0.8329	0.7887	0.7452	0.7035	0.6644	0.6280
4.6	0.9899	0.9750	0.9506	0.9175	0.8779	0.8346	0.7901	0.7463	0.7043	0.6649	0.6284
4.7	0.9909	0.9765	0.9525	0.9195	0.8798	0.8362	0.7914	0.7472	0.7050	0.6654	0.6288
4.8	0.9918	0.9779	0.9543	0.9214	0.8816	0.8377	0.7926	0.7482	0.7057	0.6659	0.6291
4.9	0.9926	0.9792	0.9560	0.9232	0.8833	0.8392	0.7937	0.7490	0.7063	0.6663	0.6294
5.0	0.9933	0.9804	0.9576	0.9249	0.8849	0.8405	0.7948	0.7498	0.7069	0.6667	0.6296
5.1	0.9939	0.9816	0.9591	0.9266	0.8864	0.8418	0.7958	0.7505	0.7074	0.6671	0.6299
5.2	0.9945	0.9827	0.9605	0.9281	0.8878	0.8430	0.7967	0.7512	0.7079	0.6674	0.6301
5.3	0.9950	0.9837	0.9619	0.9296	0.8892	0.8441	0.7976	0.7518	0.7083	0.6677	0.6303
5.4	0.9955	0.9846	0.9632	0.9310	0.8905	0.8452	0.7984	0.7524	0.7087	0.6680	0.6305
5.5	0.9959	0.9855	0.9644	0.9323	0.8917	0.8462	0.7992	0.7530	0.7091	0.6682	0.6306
5.6	0.9963	0.9863	0.9656	0.9336	0.8929	0.8472	0.7999	0.7535	0.7094	0.6684	0.6308
5.7	0.9967	0.9870	0.9667	0.9348	0.8940	0.8481	0.8005	0.7539	0.7097	0.6686	0.6309
5.8	0.9970	0.9877	0.9677	0.9360	0.8951	0.8489	0.8012	0.7544	0.7100	0.6688	0.6310
5.9	0.9973	0.9884	0.9687	0.9371	0.8961	0.8497	0.8018	0.7548	0.7103	0.6690	0.6311
6.0	0.9975	0.9890	0.9696	0.9381	0.8970	0.8505	0.8023	0.7552	0.7105	0.6692	0.6312

NTU	Heat Exchanger Effectiveness										
	C^* ($C_{min} = C_h$)										
	0	0.1	0.2	0.3	0.4	0.5	0.6	0.7	0.8	0.9	1.0
6.1	0.9978	0.9896	0.9705	0.9391	0.8979	0.8512	0.8028	0.7555	0.7108	0.6693	0.6313
6.2	0.9980	0.9902	0.9714	0.9401	0.8988	0.8519	0.8033	0.7558	0.7110	0.6694	0.6314
6.3	0.9982	0.9907	0.9722	0.9410	0.8996	0.8526	0.8038	0.7562	0.7112	0.6695	0.6314
6.4	0.9983	0.9911	0.9729	0.9418	0.9004	0.8532	0.8042	0.7564	0.7113	0.6697	0.6315
6.5	0.9985	0.9916	0.9737	0.9427	0.9012	0.8538	0.8046	0.7567	0.7115	0.6698	0.6316
6.6	0.9986	0.9920	0.9744	0.9435	0.9019	0.8543	0.8050	0.7570	0.7117	0.6698	0.6316
6.7	0.9988	0.9924	0.9750	0.9442	0.9026	0.8548	0.8054	0.7572	0.7118	0.6699	0.6317
6.8	0.9989	0.9928	0.9757	0.9450	0.9032	0.8553	0.8057	0.7574	0.7119	0.6700	0.6317
6.9	0.9990	0.9932	0.9763	0.9457	0.9038	0.8558	0.8060	0.7576	0.7121	0.6701	0.6318
7.0	0.9991	0.9935	0.9769	0.9463	0.9044	0.8562	0.8063	0.7578	0.7122	0.6701	0.6318
7.1	0.9992	0.9938	0.9774	0.9470	0.9050	0.8567	0.8066	0.7580	0.7123	0.6702	0.6318
7.2	0.9993	0.9941	0.9780	0.9476	0.9055	0.8571	0.8069	0.7581	0.7124	0.6702	0.6318
7.3	0.9993	0.9944	0.9785	0.9482	0.9061	0.8574	0.8071	0.7583	0.7125	0.6703	0.6319
7.4	0.9994	0.9946	0.9790	0.9488	0.9066	0.8578	0.8074	0.7584	0.7125	0.6703	0.6319
7.5	0.9994	0.9949	0.9794	0.9493	0.9070	0.8581	0.8076	0.7585	0.7126	0.6704	0.6319
7.6	0.9995	0.9951	0.9799	0.9498	0.9075	0.8585	0.8078	0.7587	0.7127	0.6704	0.6319
7.7	0.9995	0.9953	0.9803	0.9503	0.9079	0.8588	0.8080	0.7588	0.7127	0.6704	0.6320
7.8	0.9996	0.9956	0.9807	0.9508	0.9083	0.8591	0.8082	0.7589	0.7128	0.6705	0.6320
7.9	0.9996	0.9958	0.9811	0.9513	0.9087	0.8594	0.8084	0.7590	0.7129	0.6705	0.6320
8.0	0.9997	0.9959	0.9815	0.9517	0.9091	0.8596	0.8085	0.7591	0.7129	0.6705	0.6320
8.1	0.9997	0.9961	0.9819	0.9522	0.9095	0.8599	0.8087	0.7592	0.7129	0.6706	0.6320
8.2	0.9997	0.9963	0.9822	0.9526	0.9098	0.8601	0.8088	0.7592	0.7130	0.6706	0.6320
8.3	0.9998	0.9964	0.9826	0.9530	0.9102	0.8603	0.8089	0.7593	0.7130	0.6706	0.6320
8.4	0.9998	0.9966	0.9829	0.9534	0.9105	0.8605	0.8091	0.7594	0.7131	0.6706	0.6320
8.5	0.9998	0.9967	0.9832	0.9537	0.9108	0.8607	0.8092	0.7595	0.7131	0.6706	0.6320
8.6	0.9998	0.9969	0.9835	0.9541	0.9111	0.8609	0.8093	0.7595	0.7131	0.6706	0.6321
8.7	0.9998	0.9970	0.9838	0.9544	0.9113	0.8611	0.8094	0.7596	0.7132	0.6707	0.6321
8.8	0.9998	0.9971	0.9841	0.9547	0.9116	0.8613	0.8095	0.7596	0.7132	0.6707	0.6321
8.9	0.9999	0.9972	0.9843	0.9551	0.9119	0.8615	0.8096	0.7597	0.7132	0.6707	0.6321
9.0	0.9999	0.9974	0.9846	0.9554	0.9121	0.8616	0.8097	0.7597	0.7132	0.6707	0.6321
9.1	0.9999	0.9975	0.9849	0.9557	0.9123	0.8618	0.8098	0.7598	0.7132	0.6707	0.6321
9.2	0.9999	0.9976	0.9851	0.9559	0.9126	0.8619	0.8099	0.7598	0.7133	0.6707	0.6321
9.3	0.9999	0.9977	0.9853	0.9562	0.9128	0.8621	0.8099	0.7598	0.7133	0.6707	0.6321
9.4	0.9999	0.9977	0.9856	0.9565	0.9130	0.8622	0.8100	0.7599	0.7133	0.6707	0.6321
9.5	0.9999	0.9978	0.9858	0.9567	0.9132	0.8623	0.8101	0.7599	0.7133	0.6707	0.6321
9.6	0.9999	0.9979	0.9860	0.9570	0.9134	0.8624	0.8101	0.7599	0.7133	0.6707	0.6321
9.7	0.9999	0.9980	0.9862	0.9572	0.9136	0.8625	0.8102	0.7600	0.7133	0.6707	0.6321
9.8	0.9999	0.9981	0.9864	0.9575	0.9137	0.8626	0.8102	0.7600	0.7134	0.6708	0.6321
9.9	1.0000	0.9981	0.9866	0.9577	0.9139	0.8627	0.8103	0.7600	0.7134	0.6708	0.6321
10.0	1.0000	0.9982	0.9867	0.9579	0.9141	0.8628	0.8103	0.7600	0.7134	0.6708	0.6321

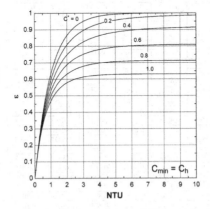

CONFIGURATION 2

1TUBE2ROW1PASS2CIR

(a) $C_{min} = C_c$

NTU	Heat Exchanger Effectiveness										
	C^* ($C_{min} = C_c$)										
	0	0.1	0.2	0.3	0.4	0.5	0.6	0.7	0.8	0.9	1.0
0	0.0000	0.0000	0.0000	0.0000	0.0000	0.0000	0.0000	0.0000	0.0000	0.0000	0.0000
0.1	0.0952	0.0947	0.0943	0.0938	0.0934	0.0929	0.0925	0.0921	0.0916	0.0912	0.0908
0.2	0.1813	0.1796	0.1780	0.1764	0.1749	0.1733	0.1718	0.1703	0.1688	0.1673	0.1658
0.3	0.2592	0.2559	0.2526	0.2494	0.2463	0.2432	0.2401	0.2371	0.2342	0.2313	0.2285
0.4	0.3297	0.3244	0.3191	0.3140	0.3091	0.3042	0.2994	0.2947	0.2901	0.2856	0.2812
0.5	0.3935	0.3859	0.3786	0.3715	0.3645	0.3577	0.3510	0.3446	0.3382	0.3321	0.3261
0.6	0.4512	0.4414	0.4318	0.4226	0.4136	0.4048	0.3963	0.3880	0.3799	0.3721	0.3645
0.7	0.5034	0.4913	0.4796	0.4682	0.4572	0.4465	0.4361	0.4261	0.4163	0.4069	0.3977
0.8	0.5507	0.5363	0.5225	0.5091	0.4961	0.4835	0.4714	0.4596	0.4483	0.4373	0.4266
0.9	0.5934	0.5770	0.5611	0.5457	0.5309	0.5166	0.5027	0.4894	0.4764	0.4640	0.4519
1.0	0.6321	0.6137	0.5959	0.5787	0.5621	0.5461	0.5307	0.5158	0.5015	0.4876	0.4743
1.1	0.6671	0.6468	0.6273	0.6084	0.5902	0.5727	0.5558	0.5395	0.5238	0.5087	0.4941
1.2	0.6988	0.6769	0.6557	0.6353	0.6156	0.5966	0.5783	0.5608	0.5438	0.5275	0.5118
1.3	0.7275	0.7040	0.6814	0.6596	0.6385	0.6183	0.5987	0.5799	0.5618	0.5444	0.5277
1.4	0.7534	0.7287	0.7047	0.6816	0.6594	0.6379	0.6172	0.5973	0.5781	0.5597	0.5420
1.5	0.7769	0.7510	0.7259	0.7017	0.6783	0.6557	0.6340	0.6131	0.5929	0.5736	0.5550
1.6	0.7981	0.7712	0.7452	0.7200	0.6956	0.6720	0.6493	0.6275	0.6064	0.5862	0.5668
1.7	0.8173	0.7897	0.7628	0.7366	0.7114	0.6869	0.6634	0.6406	0.6188	0.5977	0.5775
1.8	0.8347	0.8064	0.7788	0.7519	0.7258	0.7006	0.6762	0.6527	0.6301	0.6083	0.5874
1.9	0.8504	0.8216	0.7934	0.7659	0.7391	0.7132	0.6880	0.6638	0.6405	0.6180	0.5964
2.0	0.8647	0.8355	0.8068	0.7787	0.7513	0.7247	0.6989	0.6740	0.6500	0.6269	0.6047
2.1	0.8775	0.8481	0.8190	0.7905	0.7626	0.7354	0.7090	0.6835	0.6589	0.6352	0.6124
2.2	0.8892	0.8596	0.8302	0.8013	0.7729	0.7452	0.7183	0.6923	0.6671	0.6428	0.6194
2.3	0.8997	0.8701	0.8406	0.8113	0.7825	0.7544	0.7270	0.7004	0.6747	0.6499	0.6260
2.4	0.9093	0.8797	0.8500	0.8205	0.7914	0.7629	0.7350	0.7079	0.6817	0.6564	0.6321
2.5	0.9179	0.8885	0.8587	0.8291	0.7996	0.7707	0.7425	0.7150	0.6883	0.6626	0.6378
2.6	0.9257	0.8965	0.8668	0.8369	0.8073	0.7781	0.7494	0.7215	0.6944	0.6683	0.6431
2.7	0.9328	0.9038	0.8742	0.8443	0.8144	0.7849	0.7559	0.7276	0.7002	0.6736	0.6480
2.8	0.9392	0.9106	0.8810	0.8511	0.8210	0.7913	0.7620	0.7333	0.7055	0.6786	0.6526
2.9	0.9450	0.9167	0.8873	0.8574	0.8272	0.7972	0.7676	0.7387	0.7105	0.6832	0.6569
3.0	0.9502	0.9224	0.8932	0.8632	0.8330	0.8028	0.7730	0.7437	0.7152	0.6876	0.6609
3.1	0.9550	0.9276	0.8986	0.8687	0.8384	0.8080	0.7779	0.7484	0.7196	0.6917	0.6647
3.2	0.9592	0.9323	0.9036	0.8738	0.8434	0.8129	0.7826	0.7528	0.7238	0.6955	0.6682
3.3	0.9631	0.9367	0.9083	0.8785	0.8481	0.8175	0.7870	0.7570	0.7276	0.6991	0.6716
3.4	0.9666	0.9408	0.9126	0.8830	0.8525	0.8218	0.7911	0.7609	0.7313	0.7025	0.6747
3.5	0.9698	0.9445	0.9166	0.8871	0.8567	0.8258	0.7950	0.7646	0.7347	0.7057	0.6776
3.6	0.9727	0.9479	0.9204	0.8910	0.8605	0.8296	0.7986	0.7680	0.7380	0.7087	0.6804
3.7	0.9753	0.9511	0.9238	0.8946	0.8642	0.8332	0.8021	0.7713	0.7410	0.7116	0.6830
3.8	0.9776	0.9540	0.9271	0.8980	0.8676	0.8366	0.8053	0.7743	0.7439	0.7142	0.6855
3.9	0.9798	0.9567	0.9301	0.9012	0.8708	0.8397	0.8084	0.7772	0.7466	0.7167	0.6878
4.0	0.9817	0.9591	0.9329	0.9042	0.8739	0.8427	0.8113	0.7800	0.7492	0.7191	0.6900
4.1	0.9834	0.9614	0.9356	0.9070	0.8767	0.8455	0.8140	0.7826	0.7516	0.7214	0.6920
4.2	0.9850	0.9636	0.9380	0.9096	0.8794	0.8482	0.8166	0.7850	0.7539	0.7235	0.6940
4.3	0.9864	0.9656	0.9403	0.9121	0.8820	0.8507	0.8190	0.7873	0.7560	0.7255	0.6958
4.4	0.9877	0.9674	0.9425	0.9144	0.8844	0.8531	0.8213	0.7895	0.7581	0.7274	0.6975
4.5	0.9889	0.9691	0.9445	0.9166	0.8866	0.8553	0.8234	0.7915	0.7600	0.7291	0.6992
4.6	0.9899	0.9707	0.9464	0.9187	0.8887	0.8574	0.8255	0.7935	0.7618	0.7308	0.7007
4.7	0.9909	0.9721	0.9482	0.9207	0.8908	0.8594	0.8274	0.7953	0.7636	0.7324	0.7022
4.8	0.9918	0.9735	0.9498	0.9225	0.8927	0.8613	0.8293	0.7971	0.7652	0.7340	0.7036
4.9	0.9926	0.9747	0.9514	0.9242	0.8945	0.8631	0.8310	0.7987	0.7668	0.7354	0.7049
5.0	0.9933	0.9759	0.9529	0.9259	0.8961	0.8648	0.8327	0.8003	0.7682	0.7367	0.7061
5.1	0.9939	0.9770	0.9543	0.9274	0.8978	0.8664	0.8342	0.8018	0.7696	0.7380	0.7073
5.2	0.9945	0.9780	0.9556	0.9288	0.8993	0.8679	0.8357	0.8032	0.7709	0.7393	0.7084
5.3	0.9950	0.9790	0.9568	0.9302	0.9007	0.8694	0.8371	0.8045	0.7722	0.7404	0.7095
5.4	0.9955	0.9799	0.9579	0.9315	0.9021	0.8707	0.8384	0.8058	0.7734	0.7415	0.7105
5.5	0.9959	0.9807	0.9590	0.9327	0.9034	0.8720	0.8397	0.8070	0.7745	0.7425	0.7114
5.6	0.9963	0.9815	0.9600	0.9339	0.9046	0.8733	0.8409	0.8081	0.7756	0.7435	0.7123
5.7	0.9967	0.9822	0.9610	0.9350	0.9057	0.8744	0.8420	0.8092	0.7766	0.7445	0.7132
5.8	0.9970	0.9829	0.9619	0.9360	0.9068	0.8755	0.8431	0.8102	0.7775	0.7453	0.7140
5.9	0.9973	0.9835	0.9628	0.9370	0.9079	0.8766	0.8441	0.8112	0.7784	0.7462	0.7148
6.0	0.9975	0.9841	0.9636	0.9379	0.9088	0.8776	0.8451	0.8121	0.7793	0.7470	0.7155

NTU	Heat Exchanger Effectiveness										
	C^* ($C_{min} = C_c$)										
	0	0.1	0.2	0.3	0.4	0.5	0.6	0.7	0.8	0.9	1.0
6.1	0.9978	0.9847	0.9643	0.9388	0.9098	0.8785	0.8460	0.8130	0.7801	0.7477	0.7162
6.2	0.9980	0.9852	0.9651	0.9396	0.9107	0.8794	0.8469	0.8138	0.7809	0.7485	0.7169
6.3	0.9982	0.9857	0.9657	0.9404	0.9115	0.8802	0.8477	0.8146	0.7816	0.7491	0.7175
6.4	0.9983	0.9862	0.9664	0.9412	0.9123	0.8810	0.8485	0.8154	0.7823	0.7498	0.7181
6.5	0.9985	0.9866	0.9670	0.9419	0.9130	0.8818	0.8492	0.8161	0.7830	0.7504	0.7186
6.6	0.9986	0.9870	0.9676	0.9426	0.9138	0.8825	0.8499	0.8168	0.7836	0.7510	0.7192
6.7	0.9988	0.9874	0.9681	0.9432	0.9144	0.8832	0.8506	0.8174	0.7842	0.7515	0.7197
6.8	0.9989	0.9877	0.9686	0.9438	0.9151	0.8839	0.8512	0.8180	0.7848	0.7521	0.7201
6.9	0.9990	0.9881	0.9691	0.9444	0.9157	0.8845	0.8518	0.8186	0.7853	0.7526	0.7206
7.0	0.9991	0.9884	0.9696	0.9449	0.9163	0.8851	0.8524	0.8191	0.7859	0.7530	0.7210
7.1	0.9992	0.9887	0.9700	0.9454	0.9168	0.8856	0.8530	0.8197	0.7863	0.7535	0.7214
7.2	0.9993	0.9890	0.9704	0.9459	0.9173	0.8862	0.8535	0.8202	0.7868	0.7539	0.7218
7.3	0.9993	0.9893	0.9708	0.9464	0.9178	0.8867	0.8540	0.8206	0.7872	0.7543	0.7222
7.4	0.9994	0.9895	0.9712	0.9468	0.9183	0.8872	0.8544	0.8211	0.7877	0.7547	0.7226
7.5	0.9994	0.9897	0.9715	0.9472	0.9188	0.8876	0.8549	0.8215	0.7881	0.7551	0.7229
7.6	0.9995	0.9900	0.9719	0.9476	0.9192	0.8880	0.8553	0.8219	0.7884	0.7554	0.7232
7.7	0.9995	0.9902	0.9722	0.9480	0.9196	0.8884	0.8557	0.8223	0.7888	0.7558	0.7235
7.8	0.9996	0.9904	0.9725	0.9484	0.9200	0.8888	0.8561	0.8226	0.7891	0.7561	0.7238
7.9	0.9996	0.9906	0.9727	0.9487	0.9203	0.8892	0.8565	0.8230	0.7895	0.7564	0.7241
8.0	0.9997	0.9907	0.9730	0.9490	0.9207	0.8896	0.8568	0.8233	0.7898	0.7567	0.7243
8.1	0.9997	0.9909	0.9733	0.9493	0.9210	0.8899	0.8571	0.8236	0.7901	0.7569	0.7246
8.2	0.9997	0.9911	0.9735	0.9496	0.9213	0.8902	0.8574	0.8239	0.7903	0.7572	0.7248
8.3	0.9998	0.9912	0.9737	0.9499	0.9216	0.8905	0.8577	0.8242	0.7906	0.7574	0.7250
8.4	0.9998	0.9914	0.9740	0.9501	0.9219	0.8908	0.8580	0.8245	0.7909	0.7577	0.7252
8.5	0.9998	0.9915	0.9742	0.9504	0.9222	0.8911	0.8583	0.8247	0.7911	0.7579	0.7254
8.6	0.9998	0.9916	0.9744	0.9506	0.9224	0.8913	0.8585	0.8250	0.7913	0.7581	0.7256
8.7	0.9998	0.9917	0.9745	0.9508	0.9227	0.8916	0.8588	0.8252	0.7915	0.7583	0.7258
8.8	0.9998	0.9919	0.9747	0.9511	0.9229	0.8918	0.8590	0.8254	0.7917	0.7585	0.7260
8.9	0.9999	0.9920	0.9749	0.9513	0.9231	0.8920	0.8592	0.8256	0.7919	0.7587	0.7262
9.0	0.9999	0.9921	0.9750	0.9515	0.9233	0.8923	0.8594	0.8258	0.7921	0.7588	0.7263
9.1	0.9999	0.9922	0.9752	0.9516	0.9235	0.8925	0.8596	0.8260	0.7923	0.7590	0.7265
9.2	0.9999	0.9922	0.9753	0.9518	0.9237	0.8926	0.8598	0.8262	0.7925	0.7591	0.7266
9.3	0.9999	0.9923	0.9755	0.9520	0.9239	0.8928	0.8600	0.8264	0.7926	0.7593	0.7267
9.4	0.9999	0.9924	0.9756	0.9521	0.9241	0.8930	0.8602	0.8265	0.7928	0.7594	0.7269
9.5	0.9999	0.9925	0.9757	0.9523	0.9242	0.8932	0.8603	0.8267	0.7929	0.7596	0.7270
9.6	0.9999	0.9926	0.9758	0.9524	0.9244	0.8933	0.8605	0.8268	0.7931	0.7597	0.7271
9.7	0.9999	0.9926	0.9760	0.9526	0.9245	0.8935	0.8606	0.8270	0.7932	0.7598	0.7272
9.8	0.9999	0.9927	0.9761	0.9527	0.9247	0.8936	0.8608	0.8271	0.7933	0.7599	0.7273
9.9	1.0000	0.9928	0.9762	0.9528	0.9248	0.8937	0.8609	0.8272	0.7934	0.7600	0.7274
10.0	1.0000	0.9928	0.9763	0.9529	0.9249	0.8939	0.8610	0.8274	0.7935	0.7601	0.7275

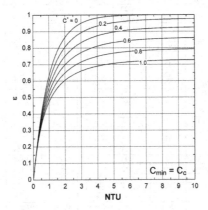

CONFIGURATION 2

1TUBE2ROW1PASS2CIR

(b) $C_{min} = C_h$

NTU	Heat Exchanger Effectiveness										
	$C^* (C_{min} = C_h)$										
	0	0.1	0.2	0.3	0.4	0.5	0.6	0.7	0.8	0.9	1.0
0	0.0000	0.0000	0.0000	0.0000	0.0000	0.0000	0.0000	0.0000	0.0000	0.0000	0.0000
0.1	0.0952	0.0947	0.0943	0.0938	0.0934	0.0929	0.0925	0.0921	0.0916	0.0912	0.0908
0.2	0.1813	0.1796	0.1780	0.1764	0.1749	0.1733	0.1718	0.1703	0.1688	0.1673	0.1658
0.3	0.2592	0.2559	0.2526	0.2494	0.2463	0.2432	0.2401	0.2372	0.2342	0.2313	0.2285
0.4	0.3297	0.3244	0.3192	0.3141	0.3091	0.3042	0.2994	0.2947	0.2901	0.2856	0.2812
0.5	0.3935	0.3860	0.3787	0.3715	0.3646	0.3577	0.3511	0.3446	0.3383	0.3321	0.3261
0.6	0.4512	0.4414	0.4319	0.4227	0.4137	0.4049	0.3964	0.3881	0.3800	0.3722	0.3645
0.7	0.5034	0.4914	0.4798	0.4684	0.4574	0.4467	0.4363	0.4263	0.4165	0.4069	0.3977
0.8	0.5507	0.5365	0.5227	0.5094	0.4964	0.4839	0.4717	0.4599	0.4484	0.4374	0.4266
0.9	0.5934	0.5772	0.5615	0.5462	0.5314	0.5171	0.5032	0.4897	0.4767	0.4641	0.4519
1.0	0.6321	0.6140	0.5964	0.5793	0.5628	0.5468	0.5313	0.5163	0.5018	0.4878	0.4743
1.1	0.6671	0.6473	0.6280	0.6093	0.5911	0.5736	0.5566	0.5401	0.5243	0.5089	0.4941
1.2	0.6988	0.6774	0.6566	0.6364	0.6167	0.5977	0.5794	0.5616	0.5444	0.5278	0.5118
1.3	0.7275	0.7047	0.6825	0.6609	0.6400	0.6197	0.6000	0.5809	0.5625	0.5448	0.5277
1.4	0.7534	0.7295	0.7061	0.6833	0.6611	0.6396	0.6187	0.5985	0.5790	0.5602	0.5420
1.5	0.7769	0.7520	0.7276	0.7037	0.6804	0.6578	0.6358	0.6145	0.5939	0.5741	0.5550
1.6	0.7981	0.7724	0.7471	0.7223	0.6980	0.6744	0.6514	0.6291	0.6076	0.5868	0.5668
1.7	0.8173	0.7910	0.7650	0.7393	0.7142	0.6896	0.6657	0.6425	0.6201	0.5984	0.5775
1.8	0.8347	0.8079	0.7813	0.7550	0.7291	0.7037	0.6789	0.6549	0.6316	0.6090	0.5874
1.9	0.8504	0.8234	0.7963	0.7693	0.7427	0.7166	0.6911	0.6662	0.6421	0.6188	0.5964
2.0	0.8647	0.8374	0.8100	0.7826	0.7554	0.7286	0.7023	0.6768	0.6519	0.6279	0.6047
2.1	0.8775	0.8503	0.8226	0.7948	0.7671	0.7397	0.7128	0.6865	0.6610	0.6362	0.6124
2.2	0.8892	0.8620	0.8342	0.8060	0.7779	0.7500	0.7225	0.6956	0.6694	0.6440	0.6194
2.3	0.8997	0.8727	0.8448	0.8165	0.7880	0.7596	0.7315	0.7040	0.6772	0.6511	0.6260
2.4	0.9093	0.8825	0.8547	0.8261	0.7973	0.7685	0.7399	0.7119	0.6844	0.6578	0.6321
2.5	0.9179	0.8915	0.8637	0.8351	0.8061	0.7768	0.7478	0.7192	0.6912	0.6641	0.6378
2.6	0.9257	0.8997	0.8721	0.8435	0.8142	0.7847	0.7552	0.7261	0.6976	0.6699	0.6431
2.7	0.9328	0.9072	0.8799	0.8513	0.8218	0.7920	0.7621	0.7326	0.7036	0.6753	0.6480
2.8	0.9392	0.9142	0.8871	0.8585	0.8289	0.7988	0.7686	0.7386	0.7091	0.6804	0.6526
2.9	0.9450	0.9205	0.8938	0.8653	0.8356	0.8053	0.7747	0.7443	0.7144	0.6852	0.6569
3.0	0.9502	0.9264	0.9000	0.8716	0.8419	0.8114	0.7805	0.7497	0.7193	0.6897	0.6609
3.1	0.9550	0.9317	0.9057	0.8775	0.8478	0.8171	0.7859	0.7547	0.7240	0.6939	0.6647
3.2	0.9592	0.9367	0.9111	0.8831	0.8534	0.8225	0.7910	0.7595	0.7283	0.6978	0.6682
3.3	0.9631	0.9412	0.9161	0.8883	0.8586	0.8276	0.7959	0.7640	0.7325	0.7016	0.6716
3.4	0.9666	0.9454	0.9207	0.8932	0.8635	0.8324	0.8005	0.7683	0.7364	0.7051	0.6747
3.5	0.9698	0.9493	0.9251	0.8978	0.8682	0.8370	0.8048	0.7723	0.7401	0.7084	0.6776
3.6	0.9727	0.9529	0.9291	0.9021	0.8726	0.8413	0.8089	0.7762	0.7435	0.7115	0.6804
3.7	0.9753	0.9561	0.9329	0.9062	0.8768	0.8454	0.8128	0.7798	0.7468	0.7145	0.6830
3.8	0.9776	0.9592	0.9365	0.9100	0.8807	0.8493	0.8165	0.7832	0.7500	0.7173	0.6855
3.9	0.9798	0.9620	0.9398	0.9137	0.8844	0.8530	0.8201	0.7865	0.7529	0.7199	0.6878
4.0	0.9817	0.9646	0.9429	0.9171	0.8880	0.8565	0.8234	0.7896	0.7558	0.7224	0.6900
4.1	0.9834	0.9670	0.9458	0.9203	0.8914	0.8598	0.8266	0.7926	0.7584	0.7248	0.6920
4.2	0.9850	0.9692	0.9485	0.9234	0.8946	0.8630	0.8296	0.7954	0.7610	0.7270	0.6940
4.3	0.9864	0.9713	0.9511	0.9263	0.8976	0.8660	0.8325	0.7981	0.7634	0.7291	0.6958
4.4	0.9877	0.9732	0.9535	0.9290	0.9005	0.8689	0.8353	0.8006	0.7657	0.7311	0.6975
4.5	0.9889	0.9750	0.9558	0.9316	0.9033	0.8717	0.8379	0.8030	0.7678	0.7330	0.6992
4.6	0.9899	0.9767	0.9579	0.9341	0.9059	0.8743	0.8404	0.8054	0.7699	0.7349	0.7007
4.7	0.9909	0.9782	0.9599	0.9364	0.9084	0.8768	0.8429	0.8076	0.7719	0.7366	0.7022
4.8	0.9918	0.9796	0.9618	0.9387	0.9108	0.8792	0.8452	0.8097	0.7738	0.7382	0.7036
4.9	0.9926	0.9809	0.9636	0.9408	0.9131	0.8815	0.8474	0.8117	0.7756	0.7398	0.7049
5.0	0.9933	0.9821	0.9653	0.9428	0.9152	0.8837	0.8495	0.8136	0.7773	0.7412	0.7061
5.1	0.9939	0.9833	0.9669	0.9447	0.9173	0.8858	0.8515	0.8155	0.7789	0.7426	0.7073
5.2	0.9945	0.9844	0.9684	0.9465	0.9193	0.8879	0.8534	0.8172	0.7804	0.7440	0.7084
5.3	0.9950	0.9853	0.9698	0.9483	0.9212	0.8898	0.8552	0.8189	0.7819	0.7452	0.7095
5.4	0.9955	0.9863	0.9712	0.9499	0.9231	0.8916	0.8570	0.8205	0.7833	0.7464	0.7105
5.5	0.9959	0.9871	0.9724	0.9515	0.9248	0.8934	0.8587	0.8221	0.7847	0.7476	0.7114
5.6	0.9963	0.9879	0.9736	0.9530	0.9265	0.8951	0.8603	0.8235	0.7860	0.7487	0.7123
5.7	0.9967	0.9887	0.9748	0.9545	0.9281	0.8968	0.8619	0.8250	0.7872	0.7497	0.7132
5.8	0.9970	0.9894	0.9759	0.9558	0.9296	0.8983	0.8634	0.8263	0.7884	0.7507	0.7140
5.9	0.9973	0.9900	0.9769	0.9571	0.9311	0.8999	0.8649	0.8276	0.7895	0.7516	0.7148
6.0	0.9975	0.9906	0.9778	0.9584	0.9325	0.9013	0.8662	0.8289	0.7906	0.7525	0.7155

NTU	Heat Exchanger Effectiveness										
	$C^* (C_{min} = C_h)$										
	0	0.1	0.2	0.3	0.4	0.5	0.6	0.7	0.8	0.9	1.0
6.1	0.9978	0.9912	0.9788	0.9596	0.9339	0.9027	0.8676	0.8301	0.7916	0.7534	0.7162
6.2	0.9980	0.9917	0.9796	0.9608	0.9352	0.9041	0.8688	0.8312	0.7926	0.7542	0.7169
6.3	0.9982	0.9922	0.9805	0.9619	0.9365	0.9054	0.8701	0.8323	0.7935	0.7550	0.7175
6.4	0.9983	0.9927	0.9813	0.9629	0.9377	0.9066	0.8713	0.8334	0.7944	0.7557	0.7181
6.5	0.9985	0.9931	0.9820	0.9639	0.9389	0.9078	0.8724	0.8344	0.7953	0.7564	0.7186
6.6	0.9986	0.9935	0.9827	0.9649	0.9400	0.9090	0.8735	0.8353	0.7961	0.7571	0.7192
6.7	0.9988	0.9939	0.9834	0.9658	0.9411	0.9101	0.8745	0.8363	0.7969	0.7577	0.7197
6.8	0.9989	0.9943	0.9840	0.9667	0.9421	0.9112	0.8756	0.8372	0.7977	0.7583	0.7201
6.9	0.9990	0.9946	0.9847	0.9676	0.9431	0.9122	0.8765	0.8380	0.7984	0.7589	0.7206
7.0	0.9991	0.9949	0.9852	0.9684	0.9441	0.9132	0.8775	0.8389	0.7991	0.7595	0.7210
7.1	0.9992	0.9952	0.9858	0.9692	0.9450	0.9142	0.8784	0.8397	0.7997	0.7600	0.7214
7.2	0.9993	0.9955	0.9863	0.9700	0.9459	0.9151	0.8793	0.8404	0.8004	0.7605	0.7218
7.3	0.9993	0.9957	0.9868	0.9707	0.9468	0.9160	0.8801	0.8412	0.8010	0.7610	0.7222
7.4	0.9994	0.9960	0.9873	0.9714	0.9476	0.9168	0.8809	0.8419	0.8016	0.7614	0.7226
7.5	0.9994	0.9962	0.9878	0.9721	0.9484	0.9177	0.8817	0.8425	0.8021	0.7619	0.7229
7.6	0.9995	0.9964	0.9882	0.9727	0.9492	0.9185	0.8825	0.8432	0.8026	0.7623	0.7232
7.7	0.9995	0.9966	0.9886	0.9733	0.9500	0.9193	0.8832	0.8438	0.8032	0.7627	0.7235
7.8	0.9996	0.9968	0.9890	0.9739	0.9507	0.9200	0.8839	0.8444	0.8036	0.7631	0.7238
7.9	0.9996	0.9970	0.9894	0.9745	0.9514	0.9208	0.8846	0.8450	0.8041	0.7635	0.7241
8.0	0.9997	0.9971	0.9898	0.9751	0.9521	0.9215	0.8852	0.8456	0.8046	0.7638	0.7243
8.1	0.9997	0.9973	0.9901	0.9756	0.9527	0.9222	0.8858	0.8461	0.8050	0.7641	0.7246
8.2	0.9997	0.9974	0.9905	0.9762	0.9534	0.9228	0.8865	0.8466	0.8054	0.7644	0.7248
8.3	0.9998	0.9976	0.9908	0.9767	0.9540	0.9235	0.8870	0.8471	0.8058	0.7647	0.7250
8.4	0.9998	0.9977	0.9911	0.9771	0.9546	0.9241	0.8876	0.8476	0.8062	0.7650	0.7252
8.5	0.9998	0.9978	0.9914	0.9776	0.9552	0.9247	0.8882	0.8481	0.8066	0.7653	0.7254
8.6	0.9998	0.9980	0.9917	0.9781	0.9557	0.9252	0.8887	0.8485	0.8069	0.7656	0.7256
8.7	0.9998	0.9981	0.9919	0.9785	0.9563	0.9258	0.8892	0.8489	0.8072	0.7658	0.7258
8.8	0.9998	0.9982	0.9922	0.9789	0.9568	0.9264	0.8897	0.8494	0.8076	0.7661	0.7260
8.9	0.9999	0.9983	0.9925	0.9793	0.9573	0.9269	0.8902	0.8497	0.8079	0.7663	0.7262
9.0	0.9999	0.9984	0.9927	0.9797	0.9578	0.9274	0.8906	0.8501	0.8082	0.7665	0.7263
9.1	0.9999	0.9984	0.9929	0.9801	0.9583	0.9279	0.8911	0.8505	0.8085	0.7667	0.7265
9.2	0.9999	0.9985	0.9931	0.9805	0.9587	0.9284	0.8915	0.8509	0.8087	0.7669	0.7266
9.3	0.9999	0.9986	0.9933	0.9808	0.9592	0.9288	0.8919	0.8512	0.8090	0.7671	0.7267
9.4	0.9999	0.9987	0.9936	0.9812	0.9596	0.9293	0.8923	0.8515	0.8092	0.7673	0.7269
9.5	0.9999	0.9987	0.9937	0.9815	0.9600	0.9297	0.8927	0.8518	0.8095	0.7675	0.7270
9.6	0.9999	0.9988	0.9939	0.9818	0.9604	0.9301	0.8931	0.8522	0.8097	0.7676	0.7271
9.7	0.9999	0.9989	0.9941	0.9822	0.9608	0.9305	0.8935	0.8524	0.8099	0.7678	0.7272
9.8	0.9999	0.9989	0.9943	0.9825	0.9612	0.9309	0.8938	0.8527	0.8102	0.7679	0.7273
9.9	1.0000	0.9990	0.9945	0.9828	0.9616	0.9313	0.8942	0.8530	0.8104	0.7681	0.7274
10.0	1.0000	0.9990	0.9946	0.9830	0.9620	0.9317	0.8945	0.8533	0.8106	0.7682	0.7275

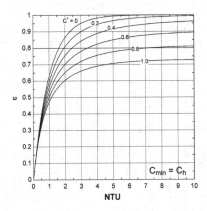

CONFIGURATION 3

1TUBE3ROW1PASS3CIR

(a) $C_{min} = C_c$

NTU	Heat Exchanger Effectiveness										
	C^* $(C_{min} = C_c)$										
	0	0.1	0.2	0.3	0.4	0.5	0.6	0.7	0.8	0.9	1.0
0	0.0000	0.0000	0.0000	0.0000	0.0000	0.0000	0.0000	0.0000	0.0000	0.0000	0.0000
0.1	0.0952	0.0947	0.0943	0.0938	0.0934	0.0929	0.0925	0.0921	0.0916	0.0912	0.0908
0.2	0.1813	0.1796	0.1780	0.1764	0.1749	0.1733	0.1718	0.1703	0.1688	0.1673	0.1658
0.3	0.2592	0.2559	0.2526	0.2494	0.2463	0.2432	0.2401	0.2372	0.2342	0.2313	0.2285
0.4	0.3297	0.3244	0.3192	0.3141	0.3091	0.3042	0.2994	0.2948	0.2902	0.2857	0.2813
0.5	0.3935	0.3860	0.3786	0.3715	0.3645	0.3578	0.3511	0.3447	0.3384	0.3322	0.3262
0.6	0.4512	0.4414	0.4319	0.4227	0.4137	0.4050	0.3965	0.3882	0.3802	0.3724	0.3648
0.7	0.5034	0.4914	0.4797	0.4684	0.4574	0.4468	0.4364	0.4264	0.4167	0.4073	0.3981
0.8	0.5507	0.5364	0.5227	0.5093	0.4964	0.4839	0.4718	0.4601	0.4488	0.4378	0.4272
0.9	0.5934	0.5771	0.5613	0.5461	0.5314	0.5171	0.5034	0.4900	0.4772	0.4648	0.4528
1.0	0.6321	0.6139	0.5962	0.5792	0.5628	0.5469	0.5315	0.5167	0.5025	0.4887	0.4754
1.1	0.6671	0.6471	0.6278	0.6091	0.5911	0.5737	0.5569	0.5407	0.5251	0.5100	0.4955
1.2	0.6988	0.6772	0.6563	0.6361	0.6167	0.5979	0.5797	0.5623	0.5454	0.5292	0.5135
1.3	0.7275	0.7045	0.6822	0.6607	0.6399	0.6198	0.6004	0.5818	0.5638	0.5465	0.5298
1.4	0.7534	0.7292	0.7057	0.6830	0.6610	0.6398	0.6193	0.5995	0.5805	0.5621	0.5445
1.5	0.7769	0.7516	0.7271	0.7033	0.6802	0.6580	0.6365	0.6157	0.5957	0.5764	0.5579
1.6	0.7981	0.7720	0.7465	0.7218	0.6978	0.6746	0.6522	0.6305	0.6096	0.5895	0.5701
1.7	0.8173	0.7905	0.7643	0.7388	0.7140	0.6899	0.6666	0.6441	0.6224	0.6014	0.5813
1.8	0.8347	0.8073	0.7805	0.7543	0.7288	0.7040	0.6799	0.6567	0.6342	0.6125	0.5916
1.9	0.8504	0.8227	0.7954	0.7686	0.7425	0.7170	0.6922	0.6683	0.6451	0.6227	0.6011
2.0	0.8647	0.8367	0.8090	0.7818	0.7551	0.7290	0.7036	0.6790	0.6552	0.6321	0.6099
2.1	0.8775	0.8494	0.8215	0.7939	0.7667	0.7401	0.7142	0.6890	0.6645	0.6409	0.6181
2.2	0.8892	0.8611	0.8330	0.8051	0.7775	0.7505	0.7240	0.6983	0.6733	0.6491	0.6258
2.3	0.8997	0.8717	0.8435	0.8154	0.7875	0.7601	0.7332	0.7070	0.6815	0.6568	0.6329
2.4	0.9093	0.8814	0.8533	0.8250	0.7968	0.7691	0.7418	0.7151	0.6891	0.6639	0.6395
2.5	0.9179	0.8903	0.8622	0.8339	0.8055	0.7774	0.7498	0.7227	0.6963	0.6706	0.6458
2.6	0.9257	0.8985	0.8705	0.8421	0.8136	0.7853	0.7573	0.7298	0.7030	0.6769	0.6517
2.7	0.9328	0.9060	0.8782	0.8498	0.8212	0.7926	0.7644	0.7365	0.7093	0.6828	0.6572
2.8	0.9392	0.9128	0.8853	0.8570	0.8283	0.7995	0.7710	0.7429	0.7153	0.6884	0.6624
2.9	0.9450	0.9191	0.8918	0.8636	0.8349	0.8060	0.7773	0.7488	0.7209	0.6937	0.6673
3.0	0.9502	0.9249	0.8979	0.8699	0.8411	0.8121	0.7832	0.7545	0.7262	0.6987	0.6719
3.1	0.9550	0.9302	0.9036	0.8757	0.8470	0.8179	0.7887	0.7598	0.7313	0.7034	0.6763
3.2	0.9592	0.9351	0.9088	0.8811	0.8525	0.8233	0.7940	0.7648	0.7361	0.7079	0.6804
3.3	0.9631	0.9396	0.9137	0.8862	0.8577	0.8284	0.7990	0.7696	0.7406	0.7121	0.6844
3.4	0.9666	0.9437	0.9183	0.8910	0.8625	0.8333	0.8037	0.7741	0.7449	0.7161	0.6881
3.5	0.9698	0.9475	0.9225	0.8955	0.8671	0.8379	0.8082	0.7784	0.7490	0.7200	0.6917
3.6	0.9727	0.9510	0.9265	0.8998	0.8715	0.8422	0.8124	0.7825	0.7528	0.7236	0.6950
3.7	0.9753	0.9543	0.9302	0.9037	0.8756	0.8463	0.8165	0.7864	0.7565	0.7271	0.6983
3.8	0.9776	0.9573	0.9337	0.9075	0.8795	0.8503	0.8203	0.7901	0.7601	0.7304	0.7013
3.9	0.9798	0.9601	0.9369	0.9110	0.8832	0.8540	0.8240	0.7937	0.7634	0.7335	0.7042
4.0	0.9817	0.9626	0.9399	0.9143	0.8867	0.8575	0.8275	0.7970	0.7666	0.7365	0.7070
4.1	0.9834	0.9650	0.9428	0.9175	0.8900	0.8609	0.8308	0.8003	0.7697	0.7394	0.7097
4.2	0.9850	0.9672	0.9454	0.9204	0.8931	0.8641	0.8340	0.8033	0.7726	0.7421	0.7122
4.3	0.9864	0.9692	0.9479	0.9233	0.8961	0.8671	0.8370	0.8063	0.7754	0.7447	0.7146
4.4	0.9877	0.9711	0.9502	0.9259	0.8989	0.8700	0.8399	0.8091	0.7781	0.7472	0.7170
4.5	0.9889	0.9729	0.9524	0.9284	0.9016	0.8728	0.8426	0.8117	0.7806	0.7496	0.7192
4.6	0.9899	0.9745	0.9545	0.9308	0.9042	0.8754	0.8453	0.8143	0.7831	0.7519	0.7213
4.7	0.9909	0.9760	0.9565	0.9330	0.9066	0.8779	0.8478	0.8168	0.7854	0.7541	0.7233
4.8	0.9918	0.9774	0.9583	0.9352	0.9089	0.8803	0.8502	0.8191	0.7876	0.7562	0.7253
4.9	0.9926	0.9787	0.9600	0.9372	0.9111	0.8826	0.8525	0.8214	0.7898	0.7583	0.7271
5.0	0.9933	0.9800	0.9617	0.9391	0.9132	0.8848	0.8547	0.8235	0.7919	0.7602	0.7289
5.1	0.9939	0.9811	0.9632	0.9410	0.9153	0.8869	0.8568	0.8256	0.7938	0.7621	0.7307
5.2	0.9945	0.9822	0.9646	0.9427	0.9172	0.8890	0.8589	0.8276	0.7957	0.7638	0.7323
5.3	0.9950	0.9831	0.9660	0.9443	0.9190	0.8909	0.8608	0.8295	0.7976	0.7656	0.7339
5.4	0.9955	0.9841	0.9673	0.9459	0.9208	0.8927	0.8627	0.8313	0.7993	0.7672	0.7354
5.5	0.9959	0.9849	0.9685	0.9474	0.9224	0.8945	0.8645	0.8331	0.8010	0.7688	0.7369
5.6	0.9963	0.9857	0.9697	0.9488	0.9240	0.8962	0.8662	0.8348	0.8026	0.7703	0.7383
5.7	0.9967	0.9865	0.9708	0.9502	0.9256	0.8978	0.8678	0.8364	0.8042	0.7718	0.7396
5.8	0.9970	0.9872	0.9718	0.9515	0.9270	0.8994	0.8694	0.8379	0.8057	0.7732	0.7409
5.9	0.9973	0.9878	0.9728	0.9527	0.9284	0.9009	0.8709	0.8394	0.8071	0.7745	0.7421
6.0	0.9975	0.9885	0.9737	0.9539	0.9298	0.9023	0.8724	0.8409	0.8085	0.7758	0.7433

NTU	Heat Exchanger Effectiveness										
	C^* ($C_{min} = C_c$)										
	0	0.1	0.2	0.3	0.4	0.5	0.6	0.7	0.8	0.9	1.0
6.1	0.9978	0.9890	0.9746	0.9550	0.9311	0.9037	0.8738	0.8423	0.8098	0.7770	0.7445
6.2	0.9980	0.9896	0.9755	0.9561	0.9323	0.9050	0.8751	0.8436	0.8111	0.7782	0.7456
6.3	0.9982	0.9901	0.9762	0.9571	0.9335	0.9063	0.8764	0.8449	0.8123	0.7794	0.7467
6.4	0.9983	0.9906	0.9770	0.9581	0.9346	0.9075	0.8777	0.8461	0.8135	0.7805	0.7477
6.5	0.9985	0.9910	0.9777	0.9590	0.9357	0.9087	0.8789	0.8473	0.8146	0.7816	0.7487
6.6	0.9986	0.9914	0.9784	0.9599	0.9367	0.9098	0.8801	0.8484	0.8157	0.7826	0.7496
6.7	0.9988	0.9918	0.9791	0.9608	0.9377	0.9109	0.8812	0.8495	0.8168	0.7836	0.7505
6.8	0.9989	0.9922	0.9797	0.9616	0.9387	0.9119	0.8822	0.8506	0.8178	0.7846	0.7514
6.9	0.9990	0.9925	0.9803	0.9624	0.9396	0.9129	0.8833	0.8516	0.8188	0.7855	0.7523
7.0	0.9991	0.9929	0.9808	0.9631	0.9405	0.9139	0.8843	0.8526	0.8198	0.7864	0.7531
7.1	0.9992	0.9932	0.9813	0.9638	0.9413	0.9148	0.8852	0.8536	0.8207	0.7872	0.7539
7.2	0.9993	0.9935	0.9819	0.9645	0.9422	0.9157	0.8861	0.8545	0.8215	0.7881	0.7546
7.3	0.9993	0.9937	0.9823	0.9652	0.9429	0.9165	0.8870	0.8554	0.8224	0.7889	0.7554
7.4	0.9994	0.9940	0.9828	0.9658	0.9437	0.9174	0.8879	0.8562	0.8232	0.7896	0.7561
7.5	0.9994	0.9943	0.9832	0.9664	0.9444	0.9182	0.8887	0.8570	0.8240	0.7904	0.7568
7.6	0.9995	0.9945	0.9837	0.9670	0.9451	0.9189	0.8895	0.8578	0.8248	0.7911	0.7574
7.7	0.9995	0.9947	0.9841	0.9676	0.9458	0.9197	0.8903	0.8586	0.8255	0.7918	0.7581
7.8	0.9996	0.9949	0.9844	0.9681	0.9464	0.9204	0.8910	0.8593	0.8262	0.7924	0.7587
7.9	0.9996	0.9951	0.9848	0.9686	0.9470	0.9211	0.8917	0.8600	0.8269	0.7931	0.7593
8.0	0.9997	0.9953	0.9852	0.9691	0.9476	0.9217	0.8924	0.8607	0.8275	0.7937	0.7598
8.1	0.9997	0.9955	0.9855	0.9696	0.9482	0.9224	0.8931	0.8614	0.8282	0.7943	0.7604
8.2	0.9997	0.9956	0.9858	0.9700	0.9488	0.9230	0.8937	0.8620	0.8288	0.7949	0.7609
8.3	0.9998	0.9958	0.9861	0.9705	0.9493	0.9236	0.8943	0.8626	0.8294	0.7954	0.7614
8.4	0.9998	0.9959	0.9864	0.9709	0.9498	0.9241	0.8949	0.8632	0.8300	0.7960	0.7619
8.5	0.9998	0.9961	0.9867	0.9713	0.9503	0.9247	0.8955	0.8638	0.8305	0.7965	0.7624
8.6	0.9998	0.9962	0.9870	0.9717	0.9508	0.9252	0.8960	0.8643	0.8310	0.7970	0.7628
8.7	0.9998	0.9963	0.9872	0.9720	0.9512	0.9257	0.8966	0.8649	0.8316	0.7975	0.7633
8.8	0.9998	0.9964	0.9875	0.9724	0.9517	0.9262	0.8971	0.8654	0.8320	0.7979	0.7637
8.9	0.9999	0.9966	0.9877	0.9727	0.9521	0.9267	0.8976	0.8659	0.8325	0.7984	0.7641
9.0	0.9999	0.9967	0.9879	0.9731	0.9525	0.9271	0.8981	0.8664	0.8330	0.7988	0.7645
9.1	0.9999	0.9968	0.9881	0.9734	0.9529	0.9276	0.8985	0.8668	0.8334	0.7992	0.7649
9.2	0.9999	0.9969	0.9883	0.9737	0.9533	0.9280	0.8990	0.8673	0.8339	0.7996	0.7653
9.3	0.9999	0.9970	0.9885	0.9740	0.9536	0.9284	0.8994	0.8677	0.8343	0.8000	0.7656
9.4	0.9999	0.9971	0.9887	0.9743	0.9540	0.9288	0.8998	0.8681	0.8347	0.8004	0.7660
9.5	0.9999	0.9971	0.9889	0.9745	0.9543	0.9292	0.9002	0.8685	0.8351	0.8008	0.7663
9.6	0.9999	0.9972	0.9891	0.9748	0.9547	0.9296	0.9006	0.8689	0.8355	0.8011	0.7666
9.7	0.9999	0.9973	0.9893	0.9750	0.9550	0.9299	0.9010	0.8693	0.8358	0.8015	0.7669
9.8	0.9999	0.9974	0.9894	0.9753	0.9553	0.9303	0.9014	0.8697	0.8362	0.8018	0.7672
9.9	1.0000	0.9974	0.9896	0.9755	0.9556	0.9306	0.9017	0.8700	0.8365	0.8021	0.7675
10.0	1.0000	0.9975	0.9897	0.9758	0.9559	0.9309	0.9020	0.8703	0.8368	0.8024	0.7678

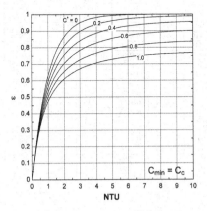

CONFIGURATION 3

1TUBE3ROW1PASS3CIR

(b) $C_{min} = C_h$

NTU	Heat Exchanger Effectiveness										
	C^* $(C_{min} = C_h)$										
	0	0.1	0.2	0.3	0.4	0.5	0.6	0.7	0.8	0.9	1.0
0	0.0000	0.0000	0.0000	0.0000	0.0000	0.0000	0.0000	0.0000	0.0000	0.0000	0.0000
0.1	0.0952	0.0947	0.0943	0.0938	0.0934	0.0929	0.0925	0.0921	0.0916	0.0912	0.0908
0.2	0.1813	0.1796	0.1780	0.1764	0.1749	0.1733	0.1718	0.1703	0.1688	0.1673	0.1658
0.3	0.2592	0.2559	0.2526	0.2494	0.2463	0.2432	0.2402	0.2372	0.2342	0.2313	0.2285
0.4	0.3297	0.3244	0.3192	0.3141	0.3091	0.3042	0.2994	0.2948	0.2902	0.2857	0.2813
0.5	0.3935	0.3860	0.3787	0.3715	0.3646	0.3578	0.3512	0.3447	0.3384	0.3322	0.3262
0.6	0.4512	0.4414	0.4320	0.4227	0.4138	0.4050	0.3965	0.3883	0.3802	0.3724	0.3648
0.7	0.5034	0.4914	0.4798	0.4685	0.4575	0.4469	0.4365	0.4265	0.4167	0.4073	0.3981
0.8	0.5507	0.5365	0.5228	0.5095	0.4966	0.4841	0.4720	0.4602	0.4489	0.4379	0.4272
0.9	0.5934	0.5772	0.5615	0.5463	0.5316	0.5173	0.5036	0.4902	0.4773	0.4648	0.4528
1.0	0.6321	0.6140	0.5965	0.5795	0.5631	0.5472	0.5318	0.5170	0.5026	0.4888	0.4754
1.1	0.6671	0.6473	0.6281	0.6095	0.5915	0.5741	0.5572	0.5410	0.5253	0.5101	0.4955
1.2	0.6988	0.6774	0.6567	0.6366	0.6172	0.5984	0.5802	0.5626	0.5457	0.5293	0.5135
1.3	0.7275	0.7048	0.6827	0.6613	0.6405	0.6204	0.6010	0.5822	0.5641	0.5466	0.5298
1.4	0.7534	0.7295	0.7063	0.6837	0.6618	0.6405	0.6199	0.6000	0.5808	0.5623	0.5445
1.5	0.7769	0.7520	0.7278	0.7041	0.6812	0.6588	0.6372	0.6163	0.5961	0.5766	0.5579
1.6	0.7981	0.7725	0.7474	0.7228	0.6989	0.6757	0.6531	0.6312	0.6101	0.5897	0.5701
1.7	0.8173	0.7911	0.7653	0.7400	0.7152	0.6911	0.6677	0.6450	0.6230	0.6017	0.5813
1.8	0.8347	0.8080	0.7816	0.7557	0.7302	0.7054	0.6811	0.6576	0.6348	0.6128	0.5916
1.9	0.8504	0.8235	0.7966	0.7701	0.7441	0.7185	0.6936	0.6693	0.6458	0.6231	0.6011
2.0	0.8647	0.8375	0.8104	0.7835	0.7569	0.7307	0.7051	0.6802	0.6560	0.6326	0.6099
2.1	0.8775	0.8504	0.8230	0.7958	0.7687	0.7420	0.7159	0.6903	0.6655	0.6414	0.6181
2.2	0.8892	0.8621	0.8347	0.8071	0.7797	0.7526	0.7259	0.6998	0.6743	0.6496	0.6258
2.3	0.8997	0.8728	0.8454	0.8177	0.7899	0.7624	0.7352	0.7086	0.6826	0.6573	0.6329
2.4	0.9093	0.8827	0.8553	0.8274	0.7995	0.7716	0.7440	0.7168	0.6903	0.6645	0.6395
2.5	0.9179	0.8916	0.8644	0.8365	0.8084	0.7802	0.7522	0.7246	0.6976	0.6713	0.6458
2.6	0.9257	0.8999	0.8728	0.8450	0.8167	0.7882	0.7599	0.7319	0.7044	0.6776	0.6517
2.7	0.9328	0.9074	0.8807	0.8529	0.8245	0.7958	0.7671	0.7388	0.7109	0.6836	0.6572
2.8	0.9392	0.9144	0.8879	0.8602	0.8318	0.8029	0.7740	0.7452	0.7169	0.6893	0.6624
2.9	0.9450	0.9207	0.8946	0.8671	0.8386	0.8096	0.7804	0.7514	0.7227	0.6946	0.6673
3.0	0.9502	0.9266	0.9009	0.8735	0.8451	0.8159	0.7865	0.7572	0.7281	0.6996	0.6719
3.1	0.9550	0.9320	0.9067	0.8796	0.8512	0.8219	0.7923	0.7626	0.7333	0.7044	0.6763
3.2	0.9592	0.9369	0.9121	0.8852	0.8569	0.8276	0.7978	0.7679	0.7382	0.7089	0.6804
3.3	0.9631	0.9415	0.9171	0.8905	0.8623	0.8329	0.8030	0.7728	0.7428	0.7132	0.6844
3.4	0.9666	0.9457	0.9218	0.8955	0.8674	0.8380	0.8079	0.7775	0.7472	0.7173	0.6881
3.5	0.9698	0.9496	0.9262	0.9002	0.8722	0.8429	0.8126	0.7820	0.7514	0.7212	0.6917
3.6	0.9727	0.9531	0.9303	0.9046	0.8768	0.8474	0.8171	0.7863	0.7554	0.7249	0.6950
3.7	0.9753	0.9564	0.9341	0.9088	0.8811	0.8518	0.8214	0.7903	0.7592	0.7284	0.6983
3.8	0.9776	0.9595	0.9377	0.9127	0.8852	0.8559	0.8254	0.7942	0.7629	0.7318	0.7013
3.9	0.9798	0.9623	0.9410	0.9164	0.8892	0.8599	0.8293	0.7979	0.7664	0.7350	0.7042
4.0	0.9817	0.9649	0.9441	0.9199	0.8929	0.8637	0.8330	0.8015	0.7697	0.7381	0.7070
4.1	0.9834	0.9673	0.9471	0.9232	0.8964	0.8673	0.8366	0.8049	0.7729	0.7410	0.7097
4.2	0.9850	0.9696	0.9499	0.9264	0.8998	0.8707	0.8399	0.8082	0.7759	0.7438	0.7122
4.3	0.9864	0.9716	0.9524	0.9293	0.9030	0.8740	0.8432	0.8113	0.7789	0.7465	0.7146
4.4	0.9877	0.9735	0.9549	0.9322	0.9060	0.8771	0.8463	0.8143	0.7817	0.7491	0.7170
4.5	0.9889	0.9753	0.9572	0.9348	0.9089	0.8801	0.8493	0.8171	0.7844	0.7515	0.7192
4.6	0.9899	0.9770	0.9593	0.9374	0.9117	0.8830	0.8522	0.8199	0.7869	0.7539	0.7213
4.7	0.9909	0.9785	0.9614	0.9398	0.9143	0.8858	0.8549	0.8225	0.7894	0.7562	0.7233
4.8	0.9918	0.9799	0.9633	0.9421	0.9169	0.8884	0.8576	0.8251	0.7918	0.7583	0.7253
4.9	0.9926	0.9812	0.9651	0.9443	0.9193	0.8910	0.8601	0.8275	0.7941	0.7604	0.7271
5.0	0.9933	0.9825	0.9668	0.9463	0.9216	0.8934	0.8625	0.8299	0.7963	0.7624	0.7289
5.1	0.9939	0.9836	0.9684	0.9483	0.9239	0.8958	0.8649	0.8321	0.7984	0.7644	0.7307
5.2	0.9945	0.9847	0.9699	0.9502	0.9260	0.8980	0.8671	0.8343	0.8004	0.7662	0.7323
5.3	0.9950	0.9857	0.9713	0.9520	0.9280	0.9002	0.8693	0.8364	0.8024	0.7680	0.7339
5.4	0.9955	0.9866	0.9727	0.9537	0.9300	0.9023	0.8714	0.8385	0.8043	0.7697	0.7354
5.5	0.9959	0.9874	0.9740	0.9553	0.9319	0.9043	0.8734	0.8404	0.8061	0.7714	0.7369
5.6	0.9963	0.9882	0.9752	0.9569	0.9337	0.9062	0.8754	0.8423	0.8079	0.7729	0.7383
5.7	0.9967	0.9890	0.9763	0.9584	0.9354	0.9080	0.8773	0.8441	0.8095	0.7745	0.7396
5.8	0.9970	0.9897	0.9774	0.9598	0.9371	0.9098	0.8791	0.8459	0.8112	0.7759	0.7409
5.9	0.9973	0.9903	0.9784	0.9612	0.9387	0.9116	0.8808	0.8476	0.8127	0.7774	0.7421
6.0	0.9975	0.9909	0.9794	0.9625	0.9402	0.9132	0.8825	0.8492	0.8143	0.7787	0.7433

NTU	Heat Exchanger Effectiveness										
	C^* ($C_{min} = C_h$)										
	0	0.1	0.2	0.3	0.4	0.5	0.6	0.7	0.8	0.9	1.0
6.1	0.9978	0.9915	0.9803	0.9637	0.9417	0.9148	0.8842	0.8508	0.8157	0.7800	0.7445
6.2	0.9980	0.9920	0.9812	0.9649	0.9431	0.9164	0.8858	0.8523	0.8171	0.7813	0.7456
6.3	0.9982	0.9925	0.9820	0.9660	0.9445	0.9179	0.8873	0.8538	0.8185	0.7825	0.7467
6.4	0.9983	0.9930	0.9828	0.9671	0.9458	0.9193	0.8888	0.8552	0.8198	0.7837	0.7477
6.5	0.9985	0.9934	0.9836	0.9682	0.9471	0.9207	0.8902	0.8566	0.8211	0.7848	0.7487
6.6	0.9986	0.9938	0.9843	0.9692	0.9483	0.9221	0.8916	0.8579	0.8223	0.7859	0.7496
6.7	0.9988	0.9942	0.9849	0.9701	0.9495	0.9234	0.8929	0.8592	0.8235	0.7870	0.7505
6.8	0.9989	0.9945	0.9856	0.9710	0.9506	0.9247	0.8942	0.8605	0.8247	0.7880	0.7514
6.9	0.9990	0.9949	0.9862	0.9719	0.9517	0.9259	0.8955	0.8617	0.8258	0.7890	0.7523
7.0	0.9991	0.9952	0.9868	0.9728	0.9528	0.9271	0.8967	0.8629	0.8269	0.7900	0.7531
7.1	0.9992	0.9954	0.9873	0.9736	0.9538	0.9282	0.8979	0.8640	0.8279	0.7909	0.7539
7.2	0.9993	0.9957	0.9879	0.9744	0.9548	0.9293	0.8990	0.8651	0.8289	0.7918	0.7546
7.3	0.9993	0.9960	0.9884	0.9751	0.9557	0.9304	0.9001	0.8662	0.8299	0.7926	0.7554
7.4	0.9994	0.9962	0.9888	0.9759	0.9567	0.9315	0.9012	0.8672	0.8309	0.7935	0.7561
7.5	0.9994	0.9964	0.9893	0.9766	0.9575	0.9325	0.9023	0.8682	0.8318	0.7943	0.7568
7.6	0.9995	0.9966	0.9897	0.9772	0.9584	0.9335	0.9033	0.8692	0.8327	0.7951	0.7574
7.7	0.9995	0.9968	0.9901	0.9779	0.9592	0.9344	0.9043	0.8701	0.8335	0.7958	0.7581
7.8	0.9996	0.9970	0.9905	0.9785	0.9601	0.9353	0.9052	0.8711	0.8344	0.7965	0.7587
7.9	0.9996	0.9972	0.9909	0.9791	0.9608	0.9362	0.9061	0.8719	0.8352	0.7972	0.7593
8.0	0.9997	0.9974	0.9913	0.9797	0.9616	0.9371	0.9070	0.8728	0.8360	0.7979	0.7598
8.1	0.9997	0.9975	0.9916	0.9802	0.9623	0.9379	0.9079	0.8737	0.8367	0.7986	0.7604
8.2	0.9997	0.9977	0.9919	0.9808	0.9630	0.9387	0.9088	0.8745	0.8375	0.7992	0.7609
8.3	0.9998	0.9978	0.9922	0.9813	0.9637	0.9395	0.9096	0.8753	0.8382	0.7998	0.7614
8.4	0.9998	0.9979	0.9925	0.9818	0.9644	0.9403	0.9104	0.8760	0.8389	0.8004	0.7619
8.5	0.9998	0.9980	0.9928	0.9822	0.9650	0.9411	0.9112	0.8768	0.8395	0.8010	0.7624
8.6	0.9998	0.9981	0.9931	0.9827	0.9657	0.9418	0.9119	0.8775	0.8402	0.8015	0.7628
8.7	0.9998	0.9982	0.9934	0.9832	0.9663	0.9425	0.9127	0.8782	0.8408	0.8021	0.7633
8.8	0.9998	0.9983	0.9936	0.9836	0.9669	0.9432	0.9134	0.8789	0.8414	0.8026	0.7637
8.9	0.9999	0.9984	0.9939	0.9840	0.9674	0.9439	0.9141	0.8795	0.8420	0.8031	0.7641
9.0	0.9999	0.9985	0.9941	0.9844	0.9680	0.9445	0.9147	0.8802	0.8426	0.8036	0.7645
9.1	0.9999	0.9986	0.9943	0.9848	0.9685	0.9451	0.9154	0.8808	0.8432	0.8041	0.7649
9.2	0.9999	0.9987	0.9945	0.9852	0.9691	0.9458	0.9160	0.8814	0.8437	0.8045	0.7653
9.3	0.9999	0.9988	0.9947	0.9855	0.9696	0.9464	0.9167	0.8820	0.8442	0.8050	0.7656
9.4	0.9999	0.9988	0.9949	0.9859	0.9701	0.9469	0.9173	0.8826	0.8447	0.8054	0.7660
9.5	0.9999	0.9989	0.9951	0.9862	0.9705	0.9475	0.9179	0.8832	0.8452	0.8058	0.7663
9.6	0.9999	0.9989	0.9953	0.9865	0.9710	0.9481	0.9184	0.8837	0.8457	0.8062	0.7666
9.7	0.9999	0.9990	0.9954	0.9869	0.9715	0.9486	0.9190	0.8842	0.8462	0.8066	0.7669
9.8	0.9999	0.9991	0.9956	0.9872	0.9719	0.9491	0.9195	0.8848	0.8466	0.8070	0.7672
9.9	1.0000	0.9991	0.9957	0.9875	0.9723	0.9496	0.9201	0.8853	0.8471	0.8073	0.7675
10.0	1.0000	0.9992	0.9959	0.9877	0.9727	0.9501	0.9206	0.8857	0.8475	0.8077	0.7678

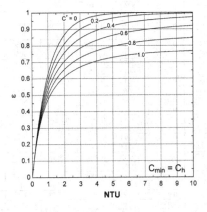

CONFIGURATION 4

1TUBE4ROW1PASS4CIR

(a) $C_{min} = C_c$

NTU	Heat Exchanger Effectiveness $C^* (C_{min} = C_c)$										
	0	0.1	0.2	0.3	0.4	0.5	0.6	0.7	0.8	0.9	1.0
0	0.0000	0.0000	0.0000	0.0000	0.0000	0.0000	0.0000	0.0000	0.0000	0.0000	0.0000
0.1	0.0952	0.0947	0.0943	0.0938	0.0934	0.0929	0.0925	0.0921	0.0916	0.0912	0.0908
0.2	0.1813	0.1796	0.1780	0.1764	0.1749	0.1733	0.1718	0.1703	0.1688	0.1673	0.1658
0.3	0.2592	0.2559	0.2526	0.2494	0.2463	0.2432	0.2402	0.2372	0.2342	0.2313	0.2285
0.4	0.3297	0.3244	0.3192	0.3141	0.3091	0.3042	0.2994	0.2948	0.2902	0.2857	0.2813
0.5	0.3935	0.3860	0.3787	0.3715	0.3646	0.3578	0.3512	0.3447	0.3384	0.3323	0.3263
0.6	0.4512	0.4414	0.4319	0.4227	0.4137	0.4050	0.3965	0.3883	0.3803	0.3724	0.3648
0.7	0.5034	0.4914	0.4797	0.4684	0.4575	0.4468	0.4365	0.4265	0.4168	0.4074	0.3982
0.8	0.5507	0.5365	0.5227	0.5094	0.4965	0.4841	0.4720	0.4603	0.4490	0.4380	0.4274
0.9	0.5934	0.5772	0.5614	0.5462	0.5315	0.5173	0.5036	0.4903	0.4774	0.4650	0.4530
1.0	0.6321	0.6139	0.5964	0.5794	0.5630	0.5471	0.5318	0.5171	0.5028	0.4890	0.4757
1.1	0.6671	0.6472	0.6279	0.6093	0.5914	0.5740	0.5573	0.5411	0.5255	0.5105	0.4960
1.2	0.6988	0.6773	0.6565	0.6364	0.6170	0.5983	0.5802	0.5628	0.5460	0.5298	0.5141
1.3	0.7275	0.7046	0.6825	0.6610	0.6403	0.6203	0.6010	0.5824	0.5645	0.5472	0.5305
1.4	0.7534	0.7293	0.7060	0.6834	0.6615	0.6404	0.6200	0.6003	0.5813	0.5630	0.5453
1.5	0.7769	0.7518	0.7275	0.7038	0.6809	0.6587	0.6373	0.6166	0.5966	0.5774	0.5589
1.6	0.7981	0.7722	0.7470	0.7225	0.6986	0.6755	0.6532	0.6316	0.6107	0.5906	0.5712
1.7	0.8173	0.7908	0.7648	0.7395	0.7149	0.6910	0.6678	0.6453	0.6237	0.6028	0.5826
1.8	0.8347	0.8077	0.7811	0.7552	0.7298	0.7052	0.6812	0.6581	0.6356	0.6140	0.5931
1.9	0.8504	0.8231	0.7961	0.7696	0.7436	0.7183	0.6937	0.6698	0.6467	0.6244	0.6028
2.0	0.8647	0.8371	0.8098	0.7828	0.7564	0.7305	0.7053	0.6807	0.6570	0.6340	0.6118
2.1	0.8775	0.8499	0.8223	0.7950	0.7682	0.7418	0.7160	0.6909	0.6666	0.6430	0.6202
2.2	0.8892	0.8616	0.8339	0.8064	0.7791	0.7523	0.7260	0.7004	0.6755	0.6514	0.6280
2.3	0.8997	0.8723	0.8445	0.8168	0.7893	0.7621	0.7354	0.7093	0.6839	0.6592	0.6354
2.4	0.9093	0.8820	0.8544	0.8265	0.7987	0.7712	0.7441	0.7176	0.6917	0.6666	0.6422
2.5	0.9179	0.8910	0.8634	0.8355	0.8076	0.7798	0.7524	0.7254	0.6991	0.6735	0.6487
2.6	0.9257	0.8992	0.8718	0.8439	0.8158	0.7878	0.7601	0.7328	0.7061	0.6800	0.6548
2.7	0.9328	0.9067	0.8795	0.8517	0.8236	0.7954	0.7674	0.7397	0.7126	0.6862	0.6605
2.8	0.9392	0.9136	0.8867	0.8590	0.8308	0.8025	0.7742	0.7463	0.7188	0.6920	0.6659
2.9	0.9450	0.9199	0.8934	0.8658	0.8376	0.8091	0.7807	0.7525	0.7247	0.6975	0.6711
3.0	0.9502	0.9257	0.8995	0.8721	0.8440	0.8154	0.7868	0.7583	0.7303	0.7027	0.6759
3.1	0.9550	0.9311	0.9053	0.8781	0.8500	0.8214	0.7926	0.7639	0.7355	0.7077	0.6806
3.2	0.9592	0.9360	0.9106	0.8837	0.8557	0.8270	0.7981	0.7692	0.7406	0.7124	0.6850
3.3	0.9631	0.9405	0.9156	0.8889	0.8610	0.8323	0.8033	0.7742	0.7453	0.7169	0.6891
3.4	0.9666	0.9447	0.9202	0.8938	0.8661	0.8374	0.8082	0.7790	0.7499	0.7212	0.6931
3.5	0.9698	0.9485	0.9245	0.8984	0.8708	0.8422	0.8129	0.7835	0.7542	0.7253	0.6969
3.6	0.9727	0.9521	0.9286	0.9028	0.8753	0.8467	0.8174	0.7879	0.7583	0.7292	0.7006
3.7	0.9753	0.9553	0.9323	0.9069	0.8796	0.8510	0.8217	0.7920	0.7623	0.7329	0.7040
3.8	0.9776	0.9584	0.9359	0.9107	0.8836	0.8552	0.8258	0.7960	0.7661	0.7365	0.7073
3.9	0.9798	0.9612	0.9392	0.9144	0.8875	0.8591	0.8297	0.7997	0.7697	0.7399	0.7105
4.0	0.9817	0.9638	0.9422	0.9178	0.8911	0.8628	0.8334	0.8034	0.7732	0.7431	0.7135
4.1	0.9834	0.9662	0.9451	0.9211	0.8946	0.8664	0.8369	0.8068	0.7765	0.7463	0.7165
4.2	0.9850	0.9684	0.9479	0.9241	0.8979	0.8698	0.8403	0.8102	0.7797	0.7493	0.7193
4.3	0.9864	0.9704	0.9504	0.9271	0.9010	0.8730	0.8436	0.8134	0.7827	0.7522	0.7219
4.4	0.9877	0.9723	0.9528	0.9298	0.9040	0.8761	0.8467	0.8164	0.7857	0.7549	0.7245
4.5	0.9889	0.9741	0.9551	0.9324	0.9069	0.8791	0.8497	0.8193	0.7885	0.7576	0.7270
4.6	0.9899	0.9757	0.9572	0.9349	0.9096	0.8819	0.8526	0.8222	0.7912	0.7602	0.7294
4.7	0.9909	0.9773	0.9592	0.9373	0.9122	0.8846	0.8553	0.8249	0.7938	0.7626	0.7317
4.8	0.9918	0.9787	0.9611	0.9395	0.9146	0.8872	0.8580	0.8275	0.7963	0.7650	0.7339
4.9	0.9926	0.9800	0.9628	0.9416	0.9170	0.8897	0.8605	0.8300	0.7988	0.7673	0.7360
5.0	0.9933	0.9812	0.9645	0.9436	0.9193	0.8921	0.8630	0.8324	0.8011	0.7695	0.7381
5.1	0.9939	0.9824	0.9661	0.9455	0.9214	0.8944	0.8653	0.8348	0.8033	0.7716	0.7401
5.2	0.9945	0.9834	0.9676	0.9474	0.9235	0.8967	0.8676	0.8370	0.8055	0.7737	0.7420
5.3	0.9950	0.9844	0.9690	0.9491	0.9255	0.8988	0.8698	0.8392	0.8076	0.7757	0.7438
5.4	0.9955	0.9853	0.9703	0.9508	0.9274	0.9008	0.8719	0.8413	0.8096	0.7776	0.7456
5.5	0.9959	0.9862	0.9716	0.9524	0.9292	0.9028	0.8739	0.8433	0.8116	0.7794	0.7473
5.6	0.9963	0.9870	0.9728	0.9539	0.9309	0.9047	0.8758	0.8452	0.8135	0.7812	0.7490
5.7	0.9967	0.9878	0.9739	0.9553	0.9326	0.9065	0.8777	0.8471	0.8153	0.7829	0.7506
5.8	0.9970	0.9885	0.9749	0.9567	0.9342	0.9082	0.8795	0.8489	0.8170	0.7846	0.7521
5.9	0.9973	0.9891	0.9760	0.9580	0.9357	0.9099	0.8813	0.8507	0.8188	0.7862	0.7536
6.0	0.9975	0.9897	0.9769	0.9592	0.9372	0.9115	0.8830	0.8523	0.8204	0.7878	0.7550

NTU	Heat Exchanger Effectiveness										
	C^* ($C_{min} = C_c$)										
	0	0.1	0.2	0.3	0.4	0.5	0.6	0.7	0.8	0.9	1.0
6.1	0.9978	0.9903	0.9778	0.9604	0.9386	0.9131	0.8846	0.8540	0.8220	0.7893	0.7565
6.2	0.9980	0.9908	0.9787	0.9616	0.9400	0.9146	0.8862	0.8556	0.8235	0.7907	0.7578
6.3	0.9982	0.9913	0.9795	0.9627	0.9413	0.9160	0.8877	0.8571	0.8250	0.7922	0.7591
6.4	0.9983	0.9918	0.9803	0.9637	0.9426	0.9174	0.8892	0.8586	0.8265	0.7935	0.7604
6.5	0.9985	0.9922	0.9810	0.9647	0.9438	0.9188	0.8906	0.8600	0.8279	0.7949	0.7616
6.6	0.9986	0.9927	0.9817	0.9657	0.9449	0.9201	0.8920	0.8614	0.8292	0.7961	0.7628
6.7	0.9988	0.9931	0.9824	0.9666	0.9461	0.9214	0.8933	0.8628	0.8305	0.7974	0.7640
6.8	0.9989	0.9934	0.9830	0.9675	0.9471	0.9226	0.8946	0.8641	0.8318	0.7986	0.7651
6.9	0.9990	0.9938	0.9836	0.9684	0.9482	0.9237	0.8958	0.8653	0.8330	0.7998	0.7662
7.0	0.9991	0.9941	0.9842	0.9692	0.9492	0.9249	0.8970	0.8665	0.8342	0.8009	0.7672
7.1	0.9992	0.9944	0.9848	0.9700	0.9502	0.9260	0.8982	0.8677	0.8354	0.8020	0.7682
7.2	0.9993	0.9947	0.9853	0.9707	0.9511	0.9270	0.8993	0.8689	0.8365	0.8031	0.7692
7.3	0.9993	0.9949	0.9858	0.9714	0.9520	0.9281	0.9004	0.8700	0.8376	0.8041	0.7702
7.4	0.9994	0.9952	0.9863	0.9721	0.9529	0.9291	0.9015	0.8711	0.8387	0.8051	0.7711
7.5	0.9994	0.9954	0.9867	0.9728	0.9537	0.9300	0.9025	0.8721	0.8397	0.8061	0.7720
7.6	0.9995	0.9956	0.9871	0.9734	0.9545	0.9309	0.9035	0.8731	0.8407	0.8070	0.7729
7.7	0.9995	0.9959	0.9876	0.9740	0.9553	0.9318	0.9045	0.8741	0.8417	0.8080	0.7738
7.8	0.9996	0.9961	0.9880	0.9746	0.9560	0.9327	0.9054	0.8751	0.8426	0.8089	0.7746
7.9	0.9996	0.9962	0.9883	0.9752	0.9568	0.9336	0.9063	0.8760	0.8435	0.8097	0.7754
8.0	0.9997	0.9964	0.9887	0.9757	0.9575	0.9344	0.9072	0.8769	0.8444	0.8106	0.7762
8.1	0.9997	0.9966	0.9890	0.9763	0.9581	0.9352	0.9081	0.8778	0.8453	0.8114	0.7769
8.2	0.9997	0.9967	0.9894	0.9768	0.9588	0.9359	0.9089	0.8786	0.8461	0.8122	0.7777
8.3	0.9998	0.9969	0.9897	0.9772	0.9594	0.9367	0.9097	0.8795	0.8469	0.8129	0.7784
8.4	0.9998	0.9970	0.9900	0.9777	0.9600	0.9374	0.9105	0.8803	0.8477	0.8137	0.7791
8.5	0.9998	0.9972	0.9903	0.9782	0.9606	0.9381	0.9112	0.8810	0.8485	0.8144	0.7798
8.6	0.9998	0.9973	0.9906	0.9786	0.9612	0.9387	0.9120	0.8818	0.8492	0.8151	0.7804
8.7	0.9998	0.9974	0.9908	0.9790	0.9618	0.9394	0.9127	0.8825	0.8499	0.8158	0.7810
8.8	0.9998	0.9975	0.9911	0.9794	0.9623	0.9400	0.9134	0.8832	0.8506	0.8165	0.7817
8.9	0.9999	0.9976	0.9913	0.9798	0.9628	0.9407	0.9140	0.8839	0.8513	0.8171	0.7823
9.0	0.9999	0.9977	0.9915	0.9802	0.9633	0.9413	0.9147	0.8846	0.8520	0.8178	0.7829
9.1	0.9999	0.9978	0.9918	0.9806	0.9638	0.9418	0.9153	0.8853	0.8526	0.8184	0.7834
9.2	0.9999	0.9979	0.9920	0.9809	0.9643	0.9424	0.9159	0.8859	0.8533	0.8190	0.7840
9.3	0.9999	0.9980	0.9922	0.9812	0.9647	0.9429	0.9165	0.8865	0.8539	0.8196	0.7845
9.4	0.9999	0.9981	0.9924	0.9816	0.9652	0.9435	0.9171	0.8871	0.8545	0.8201	0.7850
9.5	0.9999	0.9981	0.9926	0.9819	0.9656	0.9440	0.9177	0.8877	0.8550	0.8207	0.7855
9.6	0.9999	0.9982	0.9928	0.9822	0.9660	0.9445	0.9182	0.8883	0.8556	0.8212	0.7860
9.7	0.9999	0.9983	0.9929	0.9825	0.9664	0.9450	0.9188	0.8888	0.8562	0.8218	0.7865
9.8	0.9999	0.9984	0.9931	0.9828	0.9668	0.9454	0.9193	0.8894	0.8567	0.8223	0.7870
9.9	1.0000	0.9984	0.9933	0.9830	0.9672	0.9459	0.9198	0.8899	0.8572	0.8228	0.7874
10.0	1.0000	0.9985	0.9934	0.9833	0.9676	0.9463	0.9203	0.8904	0.8577	0.8232	0.7879

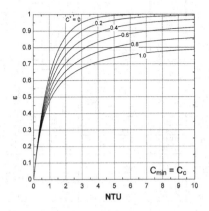

CONFIGURATION 4

1TUBE4ROW1PASS4CIR

(b) $C_{min} = C_h$

NTU	Heat Exchanger Effectiveness										
	$C^* (C_{min} = C_h)$										
	0	0.1	0.2	0.3	0.4	0.5	0.6	0.7	0.8	0.9	1.0
0	0.0000	0.0000	0.0000	0.0000	0.0000	0.0000	0.0000	0.0000	0.0000	0.0000	0.0000
0.1	0.0952	0.0947	0.0943	0.0938	0.0934	0.0929	0.0925	0.0921	0.0916	0.0912	0.0908
0.2	0.1813	0.1796	0.1780	0.1764	0.1749	0.1733	0.1718	0.1703	0.1688	0.1673	0.1658
0.3	0.2592	0.2559	0.2526	0.2494	0.2463	0.2432	0.2402	0.2372	0.2342	0.2313	0.2285
0.4	0.3297	0.3244	0.3192	0.3141	0.3091	0.3042	0.2995	0.2948	0.2902	0.2857	0.2813
0.5	0.3935	0.3860	0.3787	0.3715	0.3646	0.3578	0.3512	0.3447	0.3384	0.3323	0.3263
0.6	0.4512	0.4414	0.4320	0.4227	0.4138	0.4051	0.3966	0.3883	0.3803	0.3725	0.3648
0.7	0.5034	0.4914	0.4798	0.4685	0.4575	0.4469	0.4366	0.4266	0.4168	0.4074	0.3982
0.8	0.5507	0.5365	0.5228	0.5095	0.4966	0.4842	0.4721	0.4604	0.4490	0.4380	0.4274
0.9	0.5934	0.5772	0.5615	0.5464	0.5317	0.5174	0.5037	0.4904	0.4775	0.4651	0.4530
1.0	0.6321	0.6140	0.5965	0.5796	0.5632	0.5473	0.5320	0.5172	0.5029	0.4891	0.4757
1.1	0.6671	0.6473	0.6281	0.6096	0.5916	0.5742	0.5575	0.5413	0.5256	0.5105	0.4960
1.2	0.6988	0.6775	0.6568	0.6367	0.6173	0.5986	0.5805	0.5630	0.5461	0.5298	0.5141
1.3	0.7275	0.7048	0.6827	0.6614	0.6407	0.6207	0.6014	0.5827	0.5646	0.5473	0.5305
1.4	0.7534	0.7296	0.7064	0.6838	0.6620	0.6408	0.6204	0.6006	0.5815	0.5631	0.5453
1.5	0.7769	0.7521	0.7279	0.7043	0.6814	0.6592	0.6377	0.6170	0.5969	0.5775	0.5589
1.6	0.7981	0.7725	0.7475	0.7230	0.6992	0.6761	0.6537	0.6320	0.6110	0.5908	0.5712
1.7	0.8173	0.7911	0.7654	0.7402	0.7156	0.6916	0.6684	0.6458	0.6240	0.6029	0.5826
1.8	0.8347	0.8081	0.7818	0.7559	0.7306	0.7059	0.6819	0.6586	0.6360	0.6142	0.5931
1.9	0.8504	0.8235	0.7968	0.7704	0.7445	0.7192	0.6945	0.6704	0.6471	0.6246	0.6028
2.0	0.8647	0.8376	0.8106	0.7838	0.7574	0.7314	0.7061	0.6814	0.6575	0.6342	0.6118
2.1	0.8775	0.8504	0.8232	0.7961	0.7693	0.7428	0.7169	0.6917	0.6671	0.6432	0.6202
2.2	0.8892	0.8622	0.8349	0.8075	0.7803	0.7535	0.7271	0.7012	0.6761	0.6517	0.6280
2.3	0.8997	0.8729	0.8456	0.8181	0.7906	0.7634	0.7365	0.7102	0.6845	0.6595	0.6354
2.4	0.9093	0.8827	0.8555	0.8279	0.8002	0.7726	0.7454	0.7186	0.6924	0.6669	0.6422
2.5	0.9179	0.8917	0.8646	0.8370	0.8092	0.7813	0.7537	0.7265	0.6999	0.6739	0.6487
2.6	0.9257	0.8999	0.8731	0.8455	0.8175	0.7895	0.7615	0.7340	0.7069	0.6804	0.6548
2.7	0.9328	0.9075	0.8809	0.8534	0.8254	0.7971	0.7689	0.7410	0.7135	0.6866	0.6605
2.8	0.9392	0.9144	0.8882	0.8608	0.8328	0.8044	0.7759	0.7476	0.7198	0.6925	0.6659
2.9	0.9450	0.9208	0.8949	0.8677	0.8397	0.8112	0.7825	0.7539	0.7257	0.6980	0.6711
3.0	0.9502	0.9267	0.9012	0.8742	0.8462	0.8176	0.7887	0.7599	0.7313	0.7033	0.6759
3.1	0.9550	0.9321	0.9070	0.8803	0.8523	0.8237	0.7946	0.7655	0.7367	0.7083	0.6806
3.2	0.9592	0.9370	0.9124	0.8859	0.8581	0.8294	0.8002	0.7709	0.7417	0.7130	0.6850
3.3	0.9631	0.9416	0.9174	0.8913	0.8636	0.8349	0.8055	0.7760	0.7466	0.7176	0.6891
3.4	0.9666	0.9458	0.9222	0.8963	0.8688	0.8400	0.8106	0.7809	0.7512	0.7219	0.6931
3.5	0.9698	0.9497	0.9265	0.9010	0.8737	0.8450	0.8154	0.7855	0.7556	0.7260	0.6969
3.6	0.9727	0.9532	0.9306	0.9055	0.8783	0.8497	0.8201	0.7900	0.7598	0.7299	0.7006
3.7	0.9753	0.9565	0.9345	0.9097	0.8827	0.8541	0.8245	0.7942	0.7638	0.7337	0.7040
3.8	0.9776	0.9596	0.9381	0.9136	0.8869	0.8584	0.8287	0.7983	0.7677	0.7373	0.7073
3.9	0.9798	0.9624	0.9414	0.9174	0.8908	0.8624	0.8327	0.8022	0.7714	0.7407	0.7105
4.0	0.9817	0.9650	0.9446	0.9209	0.8946	0.8663	0.8365	0.8059	0.7749	0.7440	0.7135
4.1	0.9834	0.9674	0.9475	0.9243	0.8982	0.8700	0.8402	0.8095	0.7783	0.7472	0.7165
4.2	0.9850	0.9697	0.9503	0.9274	0.9016	0.8735	0.8437	0.8129	0.7816	0.7502	0.7193
4.3	0.9864	0.9717	0.9529	0.9304	0.9049	0.8769	0.8471	0.8162	0.7847	0.7532	0.7219
4.4	0.9877	0.9736	0.9554	0.9333	0.9080	0.8801	0.8504	0.8194	0.7877	0.7560	0.7245
4.5	0.9889	0.9754	0.9577	0.9360	0.9110	0.8832	0.8535	0.8224	0.7907	0.7587	0.7270
4.6	0.9899	0.9771	0.9598	0.9385	0.9138	0.8862	0.8565	0.8254	0.7934	0.7613	0.7294
4.7	0.9909	0.9786	0.9619	0.9410	0.9165	0.8891	0.8594	0.8282	0.7961	0.7638	0.7317
4.8	0.9918	0.9800	0.9638	0.9433	0.9191	0.8918	0.8622	0.8309	0.7987	0.7662	0.7339
4.9	0.9926	0.9813	0.9656	0.9455	0.9216	0.8944	0.8648	0.8335	0.8012	0.7686	0.7360
5.0	0.9933	0.9826	0.9673	0.9476	0.9239	0.8970	0.8674	0.8361	0.8037	0.7708	0.7381
5.1	0.9939	0.9837	0.9689	0.9496	0.9262	0.8994	0.8699	0.8385	0.8060	0.7730	0.7401
5.2	0.9945	0.9848	0.9704	0.9515	0.9284	0.9017	0.8723	0.8409	0.8082	0.7751	0.7420
5.3	0.9950	0.9858	0.9719	0.9533	0.9305	0.9040	0.8746	0.8432	0.8104	0.7771	0.7438
5.4	0.9955	0.9867	0.9732	0.9550	0.9325	0.9062	0.8769	0.8454	0.8125	0.7791	0.7456
5.5	0.9959	0.9875	0.9745	0.9567	0.9344	0.9083	0.8790	0.8475	0.8146	0.7809	0.7473
5.6	0.9963	0.9883	0.9757	0.9583	0.9363	0.9103	0.8811	0.8496	0.8165	0.7828	0.7490
5.7	0.9967	0.9891	0.9769	0.9598	0.9380	0.9122	0.8831	0.8516	0.8184	0.7845	0.7506
5.8	0.9970	0.9898	0.9779	0.9612	0.9397	0.9141	0.8851	0.8535	0.8203	0.7863	0.7521
5.9	0.9973	0.9904	0.9790	0.9626	0.9414	0.9159	0.8869	0.8554	0.8221	0.7879	0.7536
6.0	0.9975	0.9910	0.9799	0.9639	0.9430	0.9177	0.8888	0.8572	0.8238	0.7895	0.7550

NTU	Heat Exchanger Effectiveness										
	$C^* (C_{min} = C_h)$										
	0	0.1	0.2	0.3	0.4	0.5	0.6	0.7	0.8	0.9	1.0
6.1	0.9978	0.9916	0.9809	0.9652	0.9445	0.9194	0.8905	0.8589	0.8255	0.7911	0.7565
6.2	0.9980	0.9921	0.9817	0.9664	0.9460	0.9210	0.8923	0.8606	0.8271	0.7926	0.7578
6.3	0.9982	0.9926	0.9826	0.9675	0.9474	0.9226	0.8939	0.8623	0.8287	0.7940	0.7591
6.4	0.9983	0.9931	0.9834	0.9686	0.9487	0.9241	0.8955	0.8639	0.8302	0.7954	0.7604
6.5	0.9985	0.9935	0.9841	0.9697	0.9500	0.9256	0.8971	0.8654	0.8317	0.7968	0.7616
6.6	0.9986	0.9939	0.9848	0.9707	0.9513	0.9270	0.8986	0.8670	0.8331	0.7981	0.7628
6.7	0.9988	0.9943	0.9855	0.9717	0.9525	0.9284	0.9001	0.8684	0.8345	0.7994	0.7640
6.8	0.9989	0.9946	0.9861	0.9726	0.9537	0.9298	0.9015	0.8698	0.8359	0.8007	0.7651
6.9	0.9990	0.9949	0.9867	0.9735	0.9548	0.9311	0.9029	0.8712	0.8372	0.8019	0.7662
7.0	0.9991	0.9952	0.9873	0.9743	0.9559	0.9323	0.9042	0.8726	0.8385	0.8031	0.7672
7.1	0.9992	0.9955	0.9879	0.9752	0.9570	0.9335	0.9055	0.8739	0.8398	0.8042	0.7682
7.2	0.9993	0.9958	0.9884	0.9760	0.9580	0.9347	0.9068	0.8752	0.8410	0.8053	0.7692
7.3	0.9993	0.9961	0.9889	0.9767	0.9590	0.9359	0.9080	0.8764	0.8422	0.8064	0.7702
7.4	0.9994	0.9963	0.9894	0.9774	0.9600	0.9370	0.9092	0.8776	0.8433	0.8075	0.7711
7.5	0.9994	0.9965	0.9898	0.9782	0.9609	0.9381	0.9104	0.8788	0.8444	0.8085	0.7720
7.6	0.9995	0.9967	0.9903	0.9788	0.9618	0.9391	0.9115	0.8799	0.8455	0.8095	0.7729
7.7	0.9995	0.9969	0.9907	0.9795	0.9626	0.9402	0.9126	0.8810	0.8466	0.8105	0.7738
7.8	0.9996	0.9971	0.9911	0.9801	0.9635	0.9411	0.9137	0.8821	0.8476	0.8114	0.7746
7.9	0.9996	0.9973	0.9914	0.9807	0.9643	0.9421	0.9147	0.8831	0.8486	0.8123	0.7754
8.0	0.9997	0.9974	0.9918	0.9813	0.9651	0.9430	0.9157	0.8841	0.8496	0.8132	0.7762
8.1	0.9997	0.9976	0.9921	0.9818	0.9658	0.9440	0.9167	0.8851	0.8505	0.8141	0.7769
8.2	0.9997	0.9977	0.9924	0.9824	0.9666	0.9448	0.9177	0.8861	0.8514	0.8149	0.7777
8.3	0.9998	0.9979	0.9928	0.9829	0.9673	0.9457	0.9186	0.8871	0.8523	0.8157	0.7784
8.4	0.9998	0.9980	0.9930	0.9834	0.9680	0.9465	0.9195	0.8880	0.8532	0.8165	0.7791
8.5	0.9998	0.9981	0.9933	0.9839	0.9687	0.9473	0.9204	0.8889	0.8541	0.8173	0.7798
8.6	0.9998	0.9982	0.9936	0.9843	0.9693	0.9481	0.9213	0.8898	0.8549	0.8180	0.7804
8.7	0.9998	0.9983	0.9938	0.9848	0.9699	0.9489	0.9221	0.8906	0.8557	0.8188	0.7810
8.8	0.9998	0.9984	0.9941	0.9852	0.9705	0.9497	0.9229	0.8914	0.8565	0.8195	0.7817
8.9	0.9999	0.9985	0.9943	0.9856	0.9711	0.9504	0.9237	0.8923	0.8573	0.8202	0.7823
9.0	0.9999	0.9986	0.9946	0.9860	0.9717	0.9511	0.9245	0.8930	0.8580	0.8208	0.7829
9.1	0.9999	0.9987	0.9948	0.9864	0.9723	0.9518	0.9253	0.8938	0.8587	0.8215	0.7834
9.2	0.9999	0.9987	0.9950	0.9868	0.9728	0.9525	0.9260	0.8946	0.8595	0.8221	0.7840
9.3	0.9999	0.9988	0.9952	0.9872	0.9734	0.9531	0.9268	0.8953	0.8601	0.8228	0.7845
9.4	0.9999	0.9989	0.9954	0.9875	0.9739	0.9538	0.9275	0.8960	0.8608	0.8234	0.7850
9.5	0.9999	0.9989	0.9955	0.9878	0.9744	0.9544	0.9282	0.8967	0.8615	0.8240	0.7855
9.6	0.9999	0.9990	0.9957	0.9882	0.9748	0.9550	0.9288	0.8974	0.8621	0.8246	0.7860
9.7	0.9999	0.9991	0.9959	0.9885	0.9753	0.9556	0.9295	0.8981	0.8628	0.8251	0.7865
9.8	0.9999	0.9991	0.9960	0.9888	0.9758	0.9562	0.9301	0.8987	0.8634	0.8257	0.7870
9.9	1.0000	0.9992	0.9962	0.9891	0.9762	0.9567	0.9308	0.8994	0.8640	0.8262	0.7874
10.0	1.0000	0.9992	0.9963	0.9894	0.9767	0.9573	0.9314	0.9000	0.8646	0.8267	0.7879

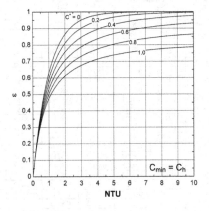

CONFIGURATION 5

1TUBE5ROW1PASS5CIR

(a) $C_{min} = C_c$

NTU	Heat Exchanger Effectiveness										
	C^* ($C_{min} = C_c$)										
	0	0.1	0.2	0.3	0.4	0.5	0.6	0.7	0.8	0.9	1.0
0	0.0000	0.0000	0.0000	0.0000	0.0000	0.0000	0.0000	0.0000	0.0000	0.0000	0.0000
0.1	0.0952	0.0947	0.0943	0.0938	0.0934	0.0929	0.0925	0.0921	0.0916	0.0912	0.0908
0.2	0.1813	0.1796	0.1780	0.1764	0.1749	0.1733	0.1718	0.1703	0.1688	0.1673	0.1658
0.3	0.2592	0.2559	0.2526	0.2494	0.2463	0.2432	0.2402	0.2372	0.2342	0.2313	0.2285
0.4	0.3297	0.3244	0.3192	0.3141	0.3091	0.3042	0.2995	0.2948	0.2902	0.2857	0.2813
0.5	0.3935	0.3860	0.3787	0.3715	0.3646	0.3578	0.3512	0.3447	0.3384	0.3323	0.3263
0.6	0.4512	0.4414	0.4320	0.4227	0.4138	0.4050	0.3966	0.3883	0.3803	0.3725	0.3649
0.7	0.5034	0.4914	0.4798	0.4685	0.4575	0.4469	0.4366	0.4266	0.4169	0.4075	0.3983
0.8	0.5507	0.5365	0.5228	0.5095	0.4966	0.4841	0.4721	0.4604	0.4491	0.4381	0.4275
0.9	0.5934	0.5772	0.5615	0.5463	0.5316	0.5174	0.5037	0.4904	0.4776	0.4652	0.4532
1.0	0.6321	0.6140	0.5964	0.5795	0.5631	0.5473	0.5320	0.5172	0.5030	0.4892	0.4759
1.1	0.6671	0.6472	0.6280	0.6094	0.5915	0.5742	0.5574	0.5413	0.5257	0.5107	0.4962
1.2	0.6988	0.6774	0.6566	0.6366	0.6172	0.5985	0.5805	0.5630	0.5462	0.5300	0.5144
1.3	0.7275	0.7047	0.6826	0.6612	0.6406	0.6206	0.6013	0.5827	0.5648	0.5475	0.5308
1.4	0.7534	0.7294	0.7062	0.6836	0.6618	0.6407	0.6203	0.6006	0.5817	0.5634	0.5457
1.5	0.7769	0.7519	0.7276	0.7041	0.6812	0.6591	0.6377	0.6170	0.5971	0.5779	0.5593
1.6	0.7981	0.7723	0.7472	0.7228	0.6990	0.6759	0.6536	0.6321	0.6112	0.5911	0.5718
1.7	0.8173	0.7909	0.7651	0.7399	0.7153	0.6914	0.6683	0.6459	0.6243	0.6034	0.5832
1.8	0.8347	0.8078	0.7814	0.7556	0.7303	0.7057	0.6818	0.6587	0.6363	0.6147	0.5938
1.9	0.8504	0.8232	0.7964	0.7700	0.7442	0.7189	0.6944	0.6705	0.6475	0.6251	0.6036
2.0	0.8647	0.8373	0.8101	0.7833	0.7570	0.7312	0.7060	0.6816	0.6578	0.6349	0.6127
2.1	0.8775	0.8501	0.8227	0.7956	0.7688	0.7425	0.7169	0.6918	0.6675	0.6439	0.6212
2.2	0.8892	0.8618	0.8343	0.8070	0.7798	0.7531	0.7270	0.7014	0.6765	0.6524	0.6291
2.3	0.8997	0.8725	0.8450	0.8175	0.7901	0.7630	0.7364	0.7104	0.6850	0.6604	0.6365
2.4	0.9093	0.8823	0.8549	0.8272	0.7996	0.7722	0.7453	0.7188	0.6930	0.6678	0.6435
2.5	0.9179	0.8913	0.8640	0.8363	0.8085	0.7809	0.7536	0.7267	0.7004	0.6749	0.6500
2.6	0.9257	0.8995	0.8724	0.8447	0.8169	0.7890	0.7614	0.7342	0.7075	0.6815	0.6562
2.7	0.9328	0.9070	0.8802	0.8526	0.8247	0.7966	0.7688	0.7412	0.7142	0.6878	0.6621
2.8	0.9392	0.9139	0.8874	0.8599	0.8320	0.8038	0.7757	0.7479	0.7205	0.6937	0.6676
2.9	0.9450	0.9203	0.8941	0.8668	0.8389	0.8106	0.7823	0.7542	0.7265	0.6993	0.6729
3.0	0.9502	0.9261	0.9003	0.8732	0.8453	0.8170	0.7885	0.7601	0.7321	0.7047	0.6778
3.1	0.9550	0.9315	0.9060	0.8792	0.8514	0.8230	0.7944	0.7658	0.7375	0.7097	0.6826
3.2	0.9592	0.9364	0.9114	0.8848	0.8571	0.8287	0.8000	0.7712	0.7427	0.7146	0.6871
3.3	0.9631	0.9409	0.9164	0.8901	0.8626	0.8342	0.8053	0.7763	0.7476	0.7192	0.6914
3.4	0.9666	0.9451	0.9211	0.8951	0.8677	0.8393	0.8104	0.7812	0.7522	0.7236	0.6955
3.5	0.9698	0.9490	0.9254	0.8998	0.8725	0.8442	0.8152	0.7859	0.7567	0.7278	0.6994
3.6	0.9727	0.9525	0.9295	0.9042	0.8771	0.8488	0.8198	0.7904	0.7609	0.7318	0.7032
3.7	0.9753	0.9558	0.9333	0.9083	0.8814	0.8532	0.8241	0.7946	0.7650	0.7357	0.7068
3.8	0.9776	0.9589	0.9369	0.9122	0.8856	0.8574	0.8283	0.7987	0.7689	0.7394	0.7102
3.9	0.9798	0.9617	0.9402	0.9159	0.8895	0.8614	0.8323	0.8026	0.7727	0.7429	0.7135
4.0	0.9817	0.9643	0.9433	0.9194	0.8932	0.8653	0.8362	0.8064	0.7763	0.7463	0.7167
4.1	0.9834	0.9667	0.9462	0.9227	0.8967	0.8689	0.8398	0.8100	0.7797	0.7496	0.7197
4.2	0.9850	0.9689	0.9490	0.9258	0.9001	0.8724	0.8433	0.8134	0.7831	0.7527	0.7227
4.3	0.9864	0.9710	0.9515	0.9288	0.9033	0.8758	0.8467	0.8167	0.7862	0.7557	0.7255
4.4	0.9877	0.9729	0.9540	0.9316	0.9064	0.8790	0.8499	0.8199	0.7893	0.7586	0.7282
4.5	0.9889	0.9746	0.9562	0.9343	0.9093	0.8820	0.8531	0.8230	0.7923	0.7614	0.7308
4.6	0.9899	0.9763	0.9584	0.9368	0.9121	0.8850	0.8560	0.8259	0.7951	0.7641	0.7333
4.7	0.9909	0.9778	0.9604	0.9392	0.9147	0.8878	0.8589	0.8288	0.7979	0.7667	0.7357
4.8	0.9918	0.9792	0.9623	0.9415	0.9173	0.8905	0.8617	0.8315	0.8005	0.7693	0.7381
4.9	0.9926	0.9806	0.9641	0.9436	0.9197	0.8931	0.8643	0.8341	0.8031	0.7717	0.7404
5.0	0.9933	0.9818	0.9658	0.9457	0.9220	0.8956	0.8669	0.8367	0.8056	0.7740	0.7425
5.1	0.9939	0.9829	0.9674	0.9476	0.9243	0.8980	0.8694	0.8391	0.8079	0.7763	0.7447
5.2	0.9945	0.9840	0.9689	0.9495	0.9264	0.9003	0.8717	0.8415	0.8103	0.7785	0.7467
5.3	0.9950	0.9850	0.9703	0.9513	0.9285	0.9025	0.8740	0.8438	0.8125	0.7806	0.7487
5.4	0.9955	0.9859	0.9716	0.9530	0.9304	0.9046	0.8762	0.8460	0.8147	0.7827	0.7506
5.5	0.9959	0.9868	0.9729	0.9546	0.9323	0.9067	0.8784	0.8482	0.8167	0.7847	0.7525
5.6	0.9963	0.9876	0.9741	0.9561	0.9341	0.9086	0.8804	0.8503	0.8188	0.7866	0.7543
5.7	0.9967	0.9883	0.9752	0.9576	0.9358	0.9105	0.8824	0.8523	0.8207	0.7885	0.7560
5.8	0.9970	0.9890	0.9763	0.9590	0.9375	0.9124	0.8844	0.8542	0.8226	0.7903	0.7577
5.9	0.9973	0.9897	0.9773	0.9604	0.9391	0.9141	0.8862	0.8561	0.8245	0.7920	0.7593
6.0	0.9975	0.9903	0.9783	0.9616	0.9406	0.9159	0.8880	0.8579	0.8263	0.7937	0.7609

NTU	Heat Exchanger Effectiveness										
	C^* $(C_{min} = C_c)$										
	0	0.1	0.2	0.3	0.4	0.5	0.6	0.7	0.8	0.9	1.0
6.1	0.9978	0.9908	0.9792	0.9629	0.9421	0.9175	0.8898	0.8597	0.8280	0.7954	0.7624
6.2	0.9980	0.9914	0.9801	0.9641	0.9435	0.9191	0.8915	0.8614	0.8297	0.7970	0.7639
6.3	0.9982	0.9919	0.9809	0.9652	0.9449	0.9206	0.8931	0.8631	0.8313	0.7985	0.7654
6.4	0.9983	0.9923	0.9817	0.9663	0.9462	0.9221	0.8947	0.8647	0.8329	0.8001	0.7668
6.5	0.9985	0.9928	0.9824	0.9673	0.9475	0.9236	0.8962	0.8662	0.8344	0.8015	0.7682
6.6	0.9986	0.9932	0.9832	0.9683	0.9487	0.9250	0.8977	0.8678	0.8359	0.8030	0.7695
6.7	0.9988	0.9936	0.9838	0.9692	0.9499	0.9263	0.8992	0.8692	0.8374	0.8043	0.7708
6.8	0.9989	0.9939	0.9845	0.9701	0.9510	0.9276	0.9006	0.8707	0.8388	0.8057	0.7721
6.9	0.9990	0.9943	0.9851	0.9710	0.9521	0.9289	0.9019	0.8721	0.8402	0.8070	0.7733
7.0	0.9991	0.9946	0.9857	0.9718	0.9532	0.9301	0.9032	0.8734	0.8415	0.8083	0.7745
7.1	0.9992	0.9949	0.9862	0.9727	0.9542	0.9313	0.9045	0.8747	0.8428	0.8095	0.7756
7.2	0.9993	0.9951	0.9867	0.9734	0.9552	0.9324	0.9057	0.8760	0.8441	0.8107	0.7768
7.3	0.9993	0.9954	0.9872	0.9742	0.9561	0.9335	0.9069	0.8773	0.8453	0.8119	0.7779
7.4	0.9994	0.9957	0.9877	0.9749	0.9571	0.9346	0.9081	0.8785	0.8465	0.8131	0.7789
7.5	0.9994	0.9959	0.9882	0.9756	0.9579	0.9356	0.9093	0.8796	0.8477	0.8142	0.7800
7.6	0.9995	0.9961	0.9886	0.9762	0.9588	0.9366	0.9104	0.8808	0.8488	0.8153	0.7810
7.7	0.9995	0.9963	0.9890	0.9769	0.9596	0.9376	0.9114	0.8819	0.8499	0.8163	0.7820
7.8	0.9996	0.9965	0.9894	0.9775	0.9604	0.9386	0.9125	0.8830	0.8510	0.8174	0.7829
7.9	0.9996	0.9967	0.9898	0.9781	0.9612	0.9395	0.9135	0.8840	0.8520	0.8184	0.7839
8.0	0.9997	0.9969	0.9901	0.9786	0.9620	0.9404	0.9145	0.8851	0.8531	0.8193	0.7848
8.1	0.9997	0.9970	0.9905	0.9792	0.9627	0.9412	0.9154	0.8861	0.8540	0.8203	0.7857
8.2	0.9997	0.9972	0.9908	0.9797	0.9634	0.9421	0.9164	0.8870	0.8550	0.8212	0.7865
8.3	0.9998	0.9973	0.9911	0.9802	0.9641	0.9429	0.9173	0.8880	0.8560	0.8221	0.7874
8.4	0.9998	0.9974	0.9914	0.9807	0.9647	0.9437	0.9182	0.8889	0.8569	0.8230	0.7882
8.5	0.9998	0.9976	0.9917	0.9811	0.9654	0.9445	0.9190	0.8898	0.8578	0.8239	0.7890
8.6	0.9998	0.9977	0.9920	0.9816	0.9660	0.9452	0.9199	0.8907	0.8587	0.8247	0.7898
8.7	0.9998	0.9978	0.9923	0.9820	0.9666	0.9459	0.9207	0.8915	0.8595	0.8256	0.7905
8.8	0.9998	0.9979	0.9925	0.9825	0.9672	0.9467	0.9215	0.8924	0.8604	0.8264	0.7913
8.9	0.9999	0.9980	0.9928	0.9829	0.9677	0.9473	0.9222	0.8932	0.8612	0.8271	0.7920
9.0	0.9999	0.9981	0.9930	0.9833	0.9683	0.9480	0.9230	0.8940	0.8620	0.8279	0.7927
9.1	0.9999	0.9982	0.9932	0.9836	0.9688	0.9487	0.9237	0.8948	0.8627	0.8286	0.7934
9.2	0.9999	0.9983	0.9934	0.9840	0.9693	0.9493	0.9244	0.8955	0.8635	0.8294	0.7941
9.3	0.9999	0.9984	0.9936	0.9844	0.9698	0.9499	0.9251	0.8962	0.8642	0.8301	0.7948
9.4	0.9999	0.9984	0.9938	0.9847	0.9703	0.9505	0.9258	0.8970	0.8649	0.8308	0.7954
9.5	0.9999	0.9985	0.9940	0.9850	0.9708	0.9511	0.9265	0.8977	0.8657	0.8315	0.7960
9.6	0.9999	0.9986	0.9942	0.9853	0.9712	0.9517	0.9271	0.8983	0.8663	0.8321	0.7966
9.7	0.9999	0.9986	0.9944	0.9856	0.9717	0.9522	0.9277	0.8990	0.8670	0.8328	0.7972
9.8	0.9999	0.9987	0.9945	0.9859	0.9721	0.9528	0.9284	0.8997	0.8677	0.8334	0.7978
9.9	1.0000	0.9988	0.9947	0.9862	0.9725	0.9533	0.9290	0.9003	0.8683	0.8340	0.7984
10.0	1.0000	0.9988	0.9948	0.9865	0.9729	0.9538	0.9295	0.9009	0.8689	0.8346	0.7990

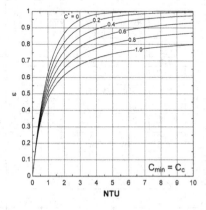

CONFIGURATION 5

1TUBE5ROW1PASS5CIR

(b) $C_{min} = C_h$

NTU	Heat Exchanger Effectiveness									
	C^* ($C_{min} = C_h$)									
	0.1	0.2	0.3	0.4	0.5	0.6	0.7	0.8	0.9	1.0
0	0.0000	0.0000	0.0000	0.0000	0.0000	0.0000	0.0000	0.0000	0.0000	0.0000	0.0000
0.1	0.0952	0.0947	0.0943	0.0938	0.0934	0.0929	0.0925	0.0921	0.0916	0.0912	0.0908
0.2	0.1813	0.1796	0.1780	0.1764	0.1749	0.1733	0.1718	0.1703	0.1688	0.1673	0.1658
0.3	0.2592	0.2559	0.2526	0.2494	0.2463	0.2432	0.2402	0.2372	0.2342	0.2313	0.2285
0.4	0.3297	0.3244	0.3192	0.3141	0.3091	0.3042	0.2995	0.2948	0.2902	0.2857	0.2813
0.5	0.3935	0.3860	0.3787	0.3716	0.3646	0.3578	0.3512	0.3447	0.3384	0.3323	0.3263
0.6	0.4512	0.4414	0.4320	0.4228	0.4138	0.4051	0.3966	0.3883	0.3803	0.3725	0.3649
0.7	0.5034	0.4914	0.4798	0.4685	0.4576	0.4469	0.4366	0.4266	0.4169	0.4075	0.3983
0.8	0.5507	0.5365	0.5228	0.5095	0.4967	0.4842	0.4721	0.4604	0.4491	0.4381	0.4275
0.9	0.5934	0.5772	0.5615	0.5464	0.5317	0.5175	0.5038	0.4905	0.4776	0.4652	0.4532
1.0	0.6321	0.6140	0.5965	0.5796	0.5632	0.5474	0.5321	0.5173	0.5030	0.4892	0.4759
1.1	0.6671	0.6473	0.6281	0.6096	0.5917	0.5743	0.5576	0.5414	0.5258	0.5107	0.4962
1.2	0.6988	0.6775	0.6568	0.6368	0.6174	0.5987	0.5806	0.5632	0.5463	0.5301	0.5144
1.3	0.7275	0.7048	0.6828	0.6614	0.6408	0.6208	0.6015	0.5829	0.5649	0.5476	0.5308
1.4	0.7534	0.7296	0.7064	0.6839	0.6621	0.6410	0.6206	0.6008	0.5818	0.5634	0.5457
1.5	0.7769	0.7521	0.7279	0.7044	0.6816	0.6594	0.6380	0.6173	0.5972	0.5779	0.5593
1.6	0.7981	0.7725	0.7475	0.7231	0.6994	0.6763	0.6540	0.6323	0.6114	0.5912	0.5718
1.7	0.8173	0.7911	0.7654	0.7403	0.7158	0.6919	0.6687	0.6462	0.6245	0.6035	0.5832
1.8	0.8347	0.8081	0.7818	0.7560	0.7308	0.7062	0.6823	0.6590	0.6365	0.6148	0.5938
1.9	0.8504	0.8235	0.7968	0.7706	0.7447	0.7195	0.6949	0.6709	0.6477	0.6253	0.6036
2.0	0.8647	0.8376	0.8106	0.7839	0.7576	0.7318	0.7066	0.6820	0.6581	0.6350	0.6127
2.1	0.8775	0.8504	0.8233	0.7963	0.7695	0.7432	0.7175	0.6923	0.6678	0.6441	0.6212
2.2	0.8892	0.8622	0.8349	0.8077	0.7806	0.7539	0.7276	0.7019	0.6769	0.6526	0.6291
2.3	0.8997	0.8729	0.8457	0.8183	0.7909	0.7638	0.7371	0.7110	0.6854	0.6606	0.6365
2.4	0.9093	0.8827	0.8556	0.8281	0.8006	0.7731	0.7461	0.7194	0.6934	0.6681	0.6435
2.5	0.9179	0.8917	0.8647	0.8373	0.8095	0.7819	0.7544	0.7274	0.7009	0.6751	0.6500
2.6	0.9257	0.9000	0.8732	0.8458	0.8180	0.7901	0.7623	0.7349	0.7080	0.6818	0.6562
2.7	0.9328	0.9075	0.8810	0.8537	0.8258	0.7978	0.7698	0.7420	0.7147	0.6880	0.6621
2.8	0.9392	0.9145	0.8883	0.8611	0.8332	0.8050	0.7768	0.7487	0.7211	0.6940	0.6676
2.9	0.9450	0.9208	0.8951	0.8680	0.8402	0.8119	0.7834	0.7551	0.7271	0.6996	0.6729
3.0	0.9502	0.9267	0.9013	0.8745	0.8467	0.8183	0.7897	0.7611	0.7328	0.7050	0.6778
3.1	0.9550	0.9321	0.9071	0.8806	0.8529	0.8245	0.7957	0.7669	0.7383	0.7101	0.6826
3.2	0.9592	0.9370	0.9126	0.8863	0.8587	0.8303	0.8014	0.7723	0.7434	0.7150	0.6871
3.3	0.9631	0.9416	0.9176	0.8916	0.8642	0.8358	0.8067	0.7775	0.7484	0.7196	0.6914
3.4	0.9666	0.9458	0.9223	0.8967	0.8694	0.8410	0.8119	0.7825	0.7531	0.7240	0.6955
3.5	0.9698	0.9497	0.9267	0.9014	0.8743	0.8460	0.8168	0.7872	0.7576	0.7282	0.6994
3.6	0.9727	0.9533	0.9308	0.9059	0.8790	0.8507	0.8214	0.7917	0.7619	0.7323	0.7032
3.7	0.9753	0.9566	0.9347	0.9101	0.8834	0.8552	0.8259	0.7961	0.7660	0.7362	0.7068
3.8	0.9776	0.9596	0.9383	0.9141	0.8876	0.8595	0.8302	0.8002	0.7700	0.7399	0.7102
3.9	0.9798	0.9625	0.9416	0.9178	0.8916	0.8636	0.8343	0.8042	0.7738	0.7434	0.7135
4.0	0.9817	0.9651	0.9448	0.9214	0.8954	0.8675	0.8382	0.8080	0.7774	0.7469	0.7167
4.1	0.9834	0.9675	0.9477	0.9247	0.8990	0.8712	0.8419	0.8117	0.7809	0.7502	0.7197
4.2	0.9850	0.9697	0.9505	0.9279	0.9025	0.8748	0.8455	0.8152	0.7843	0.7533	0.7227
4.3	0.9864	0.9718	0.9531	0.9309	0.9058	0.8782	0.8490	0.8186	0.7875	0.7564	0.7255
4.4	0.9877	0.9737	0.9556	0.9338	0.9089	0.8815	0.8523	0.8218	0.7907	0.7593	0.7282
4.5	0.9889	0.9755	0.9579	0.9365	0.9119	0.8847	0.8555	0.8250	0.7937	0.7621	0.7308
4.6	0.9899	0.9771	0.9601	0.9391	0.9148	0.8877	0.8586	0.8280	0.7966	0.7649	0.7333
4.7	0.9909	0.9787	0.9621	0.9415	0.9175	0.8906	0.8615	0.8309	0.7994	0.7675	0.7357
4.8	0.9918	0.9801	0.9640	0.9439	0.9201	0.8934	0.8644	0.8337	0.8021	0.7700	0.7381
4.9	0.9926	0.9814	0.9658	0.9461	0.9226	0.8961	0.8671	0.8364	0.8047	0.7725	0.7404
5.0	0.9933	0.9826	0.9675	0.9482	0.9250	0.8986	0.8697	0.8390	0.8072	0.7749	0.7425
5.1	0.9939	0.9838	0.9691	0.9502	0.9273	0.9011	0.8723	0.8416	0.8097	0.7772	0.7447
5.2	0.9945	0.9848	0.9707	0.9521	0.9295	0.9035	0.8748	0.8440	0.8120	0.7794	0.7467
5.3	0.9950	0.9858	0.9721	0.9539	0.9316	0.9058	0.8771	0.8464	0.8143	0.7815	0.7487
5.4	0.9955	0.9867	0.9735	0.9557	0.9337	0.9080	0.8794	0.8487	0.8165	0.7836	0.7506
5.5	0.9959	0.9876	0.9747	0.9573	0.9356	0.9102	0.8817	0.8509	0.8187	0.7856	0.7525
5.6	0.9963	0.9884	0.9760	0.9589	0.9375	0.9122	0.8838	0.8531	0.8208	0.7876	0.7543
5.7	0.9967	0.9891	0.9771	0.9604	0.9393	0.9142	0.8859	0.8552	0.8228	0.7895	0.7560
5.8	0.9970	0.9898	0.9782	0.9619	0.9410	0.9161	0.8879	0.8572	0.8247	0.7913	0.7577
5.9	0.9973	0.9905	0.9792	0.9632	0.9427	0.9180	0.8899	0.8591	0.8266	0.7931	0.7593
6.0	0.9975	0.9911	0.9802	0.9646	0.9443	0.9198	0.8918	0.8611	0.8285	0.7949	0.7609

NTU	Heat Exchanger Effectiveness										
	C^* ($C_{min} = C_h$)										
	0	0.1	0.2	0.3	0.4	0.5	0.6	0.7	0.8	0.9	1.0
6.1	0.9978	0.9916	0.9811	0.9658	0.9458	0.9215	0.8936	0.8629	0.8303	0.7965	0.7624
6.2	0.9980	0.9922	0.9820	0.9670	0.9473	0.9232	0.8954	0.8647	0.8320	0.7982	0.7639
6.3	0.9982	0.9926	0.9828	0.9682	0.9487	0.9248	0.8971	0.8664	0.8337	0.7998	0.7654
6.4	0.9983	0.9931	0.9836	0.9693	0.9501	0.9264	0.8988	0.8681	0.8354	0.8013	0.7668
6.5	0.9985	0.9935	0.9844	0.9704	0.9514	0.9279	0.9004	0.8698	0.8370	0.8028	0.7682
6.6	0.9986	0.9939	0.9851	0.9714	0.9527	0.9294	0.9020	0.8714	0.8385	0.8043	0.7695
6.7	0.9988	0.9943	0.9857	0.9724	0.9540	0.9308	0.9035	0.8729	0.8400	0.8057	0.7708
6.8	0.9989	0.9947	0.9864	0.9733	0.9552	0.9322	0.9050	0.8744	0.8415	0.8071	0.7721
6.9	0.9990	0.9950	0.9870	0.9742	0.9563	0.9335	0.9064	0.8759	0.8429	0.8084	0.7733
7.0	0.9991	0.9953	0.9876	0.9751	0.9574	0.9348	0.9078	0.8774	0.8443	0.8097	0.7745
7.1	0.9992	0.9956	0.9881	0.9759	0.9585	0.9361	0.9092	0.8787	0.8457	0.8110	0.7756
7.2	0.9993	0.9958	0.9886	0.9767	0.9595	0.9373	0.9105	0.8801	0.8470	0.8123	0.7768
7.3	0.9993	0.9961	0.9891	0.9775	0.9605	0.9385	0.9118	0.8814	0.8483	0.8135	0.7779
7.4	0.9994	0.9963	0.9896	0.9782	0.9615	0.9396	0.9131	0.8827	0.8496	0.8146	0.7789
7.5	0.9994	0.9966	0.9901	0.9789	0.9625	0.9408	0.9143	0.8840	0.8508	0.8158	0.7800
7.6	0.9995	0.9968	0.9905	0.9796	0.9634	0.9418	0.9155	0.8852	0.8520	0.8169	0.7810
7.7	0.9995	0.9970	0.9909	0.9802	0.9642	0.9429	0.9167	0.8864	0.8531	0.8180	0.7820
7.8	0.9996	0.9971	0.9913	0.9809	0.9651	0.9439	0.9178	0.8876	0.8543	0.8191	0.7829
7.9	0.9996	0.9973	0.9917	0.9815	0.9659	0.9449	0.9189	0.8887	0.8554	0.8201	0.7839
8.0	0.9997	0.9975	0.9920	0.9820	0.9667	0.9459	0.9200	0.8898	0.8565	0.8211	0.7848
8.1	0.9997	0.9976	0.9924	0.9826	0.9675	0.9468	0.9210	0.8909	0.8575	0.8221	0.7857
8.2	0.9997	0.9978	0.9927	0.9831	0.9682	0.9478	0.9220	0.8919	0.8586	0.8231	0.7865
8.3	0.9998	0.9979	0.9930	0.9837	0.9690	0.9486	0.9230	0.8930	0.8596	0.8240	0.7874
8.4	0.9998	0.9980	0.9933	0.9842	0.9697	0.9495	0.9240	0.8940	0.8606	0.8249	0.7882
8.5	0.9998	0.9981	0.9936	0.9846	0.9704	0.9504	0.9250	0.8950	0.8615	0.8258	0.7890
8.6	0.9998	0.9982	0.9938	0.9851	0.9710	0.9512	0.9259	0.8959	0.8625	0.8267	0.7898
8.7	0.9998	0.9983	0.9941	0.9855	0.9717	0.9520	0.9268	0.8969	0.8634	0.8275	0.7905
8.8	0.9998	0.9984	0.9943	0.9860	0.9723	0.9528	0.9277	0.8978	0.8643	0.8284	0.7913
8.9	0.9999	0.9985	0.9945	0.9864	0.9729	0.9535	0.9285	0.8987	0.8651	0.8292	0.7920
9.0	0.9999	0.9986	0.9948	0.9868	0.9735	0.9543	0.9294	0.8996	0.8660	0.8300	0.7927
9.1	0.9999	0.9987	0.9950	0.9872	0.9740	0.9550	0.9302	0.9004	0.8668	0.8308	0.7934
9.2	0.9999	0.9988	0.9952	0.9876	0.9746	0.9557	0.9310	0.9012	0.8677	0.8315	0.7941
9.3	0.9999	0.9988	0.9954	0.9879	0.9751	0.9564	0.9318	0.9021	0.8685	0.8323	0.7948
9.4	0.9999	0.9989	0.9956	0.9883	0.9757	0.9570	0.9325	0.9029	0.8692	0.8330	0.7954
9.5	0.9999	0.9990	0.9957	0.9886	0.9762	0.9577	0.9333	0.9036	0.8700	0.8337	0.7960
9.6	0.9999	0.9990	0.9959	0.9889	0.9767	0.9583	0.9340	0.9044	0.8708	0.8344	0.7966
9.7	0.9999	0.9991	0.9961	0.9892	0.9771	0.9590	0.9347	0.9052	0.8715	0.8351	0.7972
9.8	0.9999	0.9991	0.9962	0.9896	0.9776	0.9596	0.9354	0.9059	0.8722	0.8357	0.7978
9.9	1.0000	0.9992	0.9964	0.9898	0.9781	0.9601	0.9361	0.9066	0.8729	0.8364	0.7984
10.0	1.0000	0.9992	0.9965	0.9901	0.9785	0.9607	0.9368	0.9073	0.8736	0.8370	0.7990

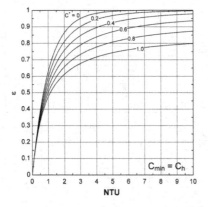

CONFIGURATION 6

1TUBE6ROW1PASS6CIR

(a) $C_{min} = C_c$

NTU	Heat Exchanger Effectiveness										
	C^* ($C_{min} = C_c$)										
	0	0.1	0.2	0.3	0.4	0.5	0.6	0.7	0.8	0.9	1.0
0	0.0000	0.0000	0.0000	0.0000	0.0000	0.0000	0.0000	0.0000	0.0000	0.0000	0.0000
0.1	0.0952	0.0947	0.0943	0.0938	0.0934	0.0929	0.0925	0.0921	0.0916	0.0912	0.0908
0.2	0.1813	0.1796	0.1780	0.1764	0.1749	0.1733	0.1718	0.1703	0.1688	0.1673	0.1658
0.3	0.2592	0.2559	0.2526	0.2494	0.2463	0.2432	0.2402	0.2372	0.2342	0.2313	0.2285
0.4	0.3297	0.3244	0.3192	0.3141	0.3091	0.3042	0.2995	0.2948	0.2902	0.2857	0.2813
0.5	0.3935	0.3860	0.3787	0.3715	0.3646	0.3578	0.3512	0.3447	0.3384	0.3323	0.3263
0.6	0.4512	0.4414	0.4320	0.4227	0.4138	0.4051	0.3966	0.3883	0.3803	0.3725	0.3649
0.7	0.5034	0.4914	0.4798	0.4685	0.4575	0.4469	0.4366	0.4266	0.4169	0.4075	0.3983
0.8	0.5507	0.5365	0.5228	0.5095	0.4966	0.4842	0.4721	0.4604	0.4491	0.4382	0.4275
0.9	0.5934	0.5772	0.5615	0.5463	0.5317	0.5175	0.5037	0.4905	0.4776	0.4652	0.4532
1.0	0.6321	0.6140	0.5965	0.5795	0.5632	0.5473	0.5321	0.5173	0.5031	0.4893	0.4760
1.1	0.6671	0.6473	0.6281	0.6095	0.5916	0.5743	0.5575	0.5414	0.5258	0.5108	0.4963
1.2	0.6988	0.6774	0.6567	0.6367	0.6173	0.5986	0.5806	0.5632	0.5464	0.5302	0.5146
1.3	0.7275	0.7047	0.6827	0.6613	0.6407	0.6207	0.6015	0.5829	0.5650	0.5477	0.5310
1.4	0.7534	0.7295	0.7063	0.6837	0.6620	0.6409	0.6205	0.6008	0.5819	0.5636	0.5460
1.5	0.7769	0.7520	0.7277	0.7042	0.6814	0.6593	0.6379	0.6173	0.5973	0.5781	0.5596
1.6	0.7981	0.7724	0.7473	0.7229	0.6992	0.6762	0.6539	0.6323	0.6115	0.5914	0.5721
1.7	0.8173	0.7910	0.7652	0.7400	0.7155	0.6917	0.6686	0.6462	0.6246	0.6037	0.5836
1.8	0.8347	0.8079	0.7816	0.7558	0.7306	0.7060	0.6822	0.6590	0.6367	0.6150	0.5942
1.9	0.8504	0.8233	0.7966	0.7702	0.7445	0.7193	0.6948	0.6709	0.6479	0.6256	0.6040
2.0	0.8647	0.8374	0.8103	0.7836	0.7573	0.7316	0.7064	0.6820	0.6583	0.6353	0.6132
2.1	0.8775	0.8502	0.8229	0.7959	0.7692	0.7430	0.7173	0.6923	0.6680	0.6445	0.6217
2.2	0.8892	0.8619	0.8346	0.8073	0.7802	0.7536	0.7275	0.7019	0.6771	0.6530	0.6297
2.3	0.8997	0.8727	0.8453	0.8178	0.7905	0.7635	0.7370	0.7110	0.6856	0.6610	0.6371
2.4	0.9093	0.8825	0.8551	0.8276	0.8001	0.7728	0.7459	0.7194	0.6936	0.6685	0.6442
2.5	0.9179	0.8914	0.8643	0.8367	0.8090	0.7815	0.7542	0.7274	0.7012	0.6756	0.6508
2.6	0.9257	0.8997	0.8727	0.8452	0.8174	0.7897	0.7621	0.7349	0.7083	0.6823	0.6570
2.7	0.9328	0.9072	0.8805	0.8531	0.8253	0.7973	0.7695	0.7420	0.7150	0.6886	0.6629
2.8	0.9392	0.9141	0.8877	0.8604	0.8326	0.8046	0.7765	0.7487	0.7214	0.6946	0.6685
2.9	0.9450	0.9205	0.8944	0.8673	0.8395	0.8114	0.7831	0.7551	0.7274	0.7003	0.6738
3.0	0.9502	0.9263	0.9007	0.8738	0.8460	0.8178	0.7894	0.7611	0.7332	0.7057	0.6789
3.1	0.9550	0.9317	0.9065	0.8798	0.8522	0.8239	0.7954	0.7669	0.7386	0.7108	0.6837
3.2	0.9592	0.9366	0.9119	0.8855	0.8579	0.8297	0.8010	0.7723	0.7438	0.7157	0.6883
3.3	0.9631	0.9412	0.9169	0.8908	0.8634	0.8351	0.8064	0.7775	0.7488	0.7204	0.6926
3.4	0.9666	0.9454	0.9216	0.8958	0.8686	0.8403	0.8115	0.7825	0.7535	0.7249	0.6968
3.5	0.9698	0.9492	0.9259	0.9005	0.8734	0.8453	0.8164	0.7872	0.7580	0.7292	0.7008
3.6	0.9727	0.9528	0.9300	0.9049	0.8781	0.8500	0.8210	0.7917	0.7624	0.7333	0.7046
3.7	0.9753	0.9561	0.9338	0.9091	0.8825	0.8544	0.8255	0.7961	0.7665	0.7372	0.7083
3.8	0.9776	0.9591	0.9374	0.9130	0.8866	0.8587	0.8297	0.8002	0.7705	0.7409	0.7118
3.9	0.9798	0.9619	0.9407	0.9168	0.8906	0.8627	0.8338	0.8042	0.7743	0.7446	0.7152
4.0	0.9817	0.9645	0.9439	0.9203	0.8943	0.8666	0.8377	0.8080	0.7780	0.7480	0.7184
4.1	0.9834	0.9669	0.9468	0.9236	0.8979	0.8703	0.8414	0.8117	0.7815	0.7514	0.7215
4.2	0.9850	0.9692	0.9496	0.9268	0.9013	0.8739	0.8450	0.8152	0.7849	0.7546	0.7245
4.3	0.9864	0.9712	0.9522	0.9297	0.9046	0.8773	0.8484	0.8186	0.7882	0.7577	0.7274
4.4	0.9877	0.9732	0.9546	0.9326	0.9077	0.8805	0.8517	0.8218	0.7913	0.7607	0.7302
4.5	0.9889	0.9749	0.9569	0.9352	0.9106	0.8836	0.8549	0.8250	0.7944	0.7636	0.7329
4.6	0.9899	0.9766	0.9590	0.9378	0.9134	0.8866	0.8579	0.8280	0.7973	0.7663	0.7355
4.7	0.9909	0.9781	0.9611	0.9402	0.9161	0.8895	0.8609	0.8309	0.8001	0.7690	0.7380
4.8	0.9918	0.9795	0.9630	0.9425	0.9187	0.8922	0.8637	0.8337	0.8028	0.7716	0.7404
4.9	0.9926	0.9808	0.9648	0.9447	0.9212	0.8949	0.8664	0.8364	0.8055	0.7741	0.7428
5.0	0.9933	0.9821	0.9665	0.9468	0.9236	0.8974	0.8690	0.8390	0.8080	0.7765	0.7450
5.1	0.9939	0.9832	0.9681	0.9488	0.9258	0.8999	0.8716	0.8416	0.8105	0.7789	0.7472
5.2	0.9945	0.9843	0.9696	0.9506	0.9280	0.9022	0.8740	0.8440	0.8129	0.7812	0.7494
5.3	0.9950	0.9853	0.9710	0.9524	0.9301	0.9045	0.8764	0.8464	0.8152	0.7834	0.7514
5.4	0.9955	0.9862	0.9723	0.9542	0.9321	0.9067	0.8786	0.8487	0.8174	0.7855	0.7534
5.5	0.9959	0.9870	0.9736	0.9558	0.9340	0.9088	0.8808	0.8509	0.8196	0.7876	0.7554
5.6	0.9963	0.9878	0.9748	0.9573	0.9358	0.9108	0.8830	0.8531	0.8217	0.7896	0.7572
5.7	0.9967	0.9886	0.9760	0.9588	0.9376	0.9128	0.8850	0.8551	0.8238	0.7916	0.7591
5.8	0.9970	0.9893	0.9770	0.9603	0.9393	0.9146	0.8870	0.8572	0.8258	0.7934	0.7608
5.9	0.9973	0.9899	0.9781	0.9616	0.9409	0.9165	0.8890	0.8591	0.8277	0.7953	0.7625
6.0	0.9975	0.9905	0.9790	0.9629	0.9425	0.9182	0.8908	0.8610	0.8295	0.7971	0.7642

NTU	Heat Exchanger Effectiveness										
	$C^* (C_{min} = C_c)$										
	0	0.1	0.2	0.3	0.4	0.5	0.6	0.7	0.8	0.9	1.0
6.1	0.9978	0.9911	0.9800	0.9642	0.9440	0.9199	0.8926	0.8629	0.8314	0.7988	0.7658
6.2	0.9980	0.9916	0.9808	0.9654	0.9454	0.9216	0.8944	0.8647	0.8331	0.8005	0.7674
6.3	0.9982	0.9921	0.9817	0.9665	0.9468	0.9231	0.8961	0.8664	0.8348	0.8022	0.7690
6.4	0.9983	0.9926	0.9824	0.9676	0.9482	0.9247	0.8977	0.8681	0.8365	0.8037	0.7705
6.5	0.9985	0.9930	0.9832	0.9686	0.9495	0.9262	0.8993	0.8697	0.8381	0.8053	0.7719
6.6	0.9986	0.9934	0.9839	0.9696	0.9507	0.9276	0.9009	0.8713	0.8397	0.8068	0.7733
6.7	0.9988	0.9938	0.9846	0.9706	0.9520	0.9290	0.9024	0.8729	0.8413	0.8083	0.7747
6.8	0.9989	0.9942	0.9852	0.9715	0.9531	0.9304	0.9039	0.8744	0.8428	0.8097	0.7760
6.9	0.9990	0.9945	0.9858	0.9724	0.9542	0.9317	0.9053	0.8759	0.8442	0.8111	0.7774
7.0	0.9991	0.9948	0.9864	0.9733	0.9553	0.9329	0.9067	0.8773	0.8456	0.8125	0.7786
7.1	0.9992	0.9951	0.9870	0.9741	0.9564	0.9341	0.9080	0.8787	0.8470	0.8138	0.7799
7.2	0.9993	0.9954	0.9875	0.9749	0.9574	0.9353	0.9093	0.8800	0.8484	0.8151	0.7811
7.3	0.9993	0.9957	0.9880	0.9756	0.9584	0.9365	0.9106	0.8813	0.8497	0.8164	0.7823
7.4	0.9994	0.9959	0.9885	0.9763	0.9593	0.9376	0.9118	0.8826	0.8510	0.8176	0.7834
7.5	0.9994	0.9961	0.9889	0.9770	0.9602	0.9387	0.9130	0.8839	0.8522	0.8188	0.7845
7.6	0.9995	0.9963	0.9893	0.9777	0.9611	0.9397	0.9142	0.8851	0.8534	0.8200	0.7856
7.7	0.9995	0.9965	0.9898	0.9783	0.9620	0.9408	0.9153	0.8863	0.8546	0.8211	0.7867
7.8	0.9996	0.9967	0.9902	0.9790	0.9628	0.9418	0.9164	0.8874	0.8558	0.8223	0.7877
7.9	0.9996	0.9969	0.9905	0.9796	0.9636	0.9427	0.9175	0.8886	0.8569	0.8234	0.7888
8.0	0.9997	0.9971	0.9909	0.9801	0.9644	0.9437	0.9185	0.8897	0.8580	0.8244	0.7898
8.1	0.9997	0.9972	0.9912	0.9807	0.9651	0.9446	0.9195	0.8908	0.8591	0.8255	0.7907
8.2	0.9997	0.9974	0.9916	0.9812	0.9658	0.9455	0.9205	0.8918	0.8601	0.8265	0.7917
8.3	0.9998	0.9975	0.9919	0.9817	0.9665	0.9463	0.9215	0.8928	0.8612	0.8275	0.7926
8.4	0.9998	0.9976	0.9922	0.9822	0.9672	0.9472	0.9224	0.8938	0.8622	0.8284	0.7935
8.5	0.9998	0.9978	0.9925	0.9827	0.9679	0.9480	0.9234	0.8948	0.8632	0.8294	0.7944
8.6	0.9998	0.9979	0.9927	0.9832	0.9685	0.9488	0.9243	0.8957	0.8641	0.8303	0.7953
8.7	0.9998	0.9980	0.9930	0.9836	0.9691	0.9495	0.9251	0.8967	0.8651	0.8312	0.7961
8.8	0.9998	0.9981	0.9932	0.9840	0.9697	0.9503	0.9260	0.8976	0.8660	0.8321	0.7970
8.9	0.9999	0.9982	0.9935	0.9844	0.9703	0.9510	0.9268	0.8985	0.8669	0.8330	0.7978
9.0	0.9999	0.9983	0.9937	0.9848	0.9709	0.9517	0.9276	0.8993	0.8677	0.8338	0.7986
9.1	0.9999	0.9984	0.9939	0.9852	0.9714	0.9524	0.9284	0.9002	0.8686	0.8347	0.7993
9.2	0.9999	0.9985	0.9941	0.9856	0.9720	0.9531	0.9292	0.9010	0.8694	0.8355	0.8001
9.3	0.9999	0.9985	0.9943	0.9859	0.9725	0.9537	0.9300	0.9018	0.8703	0.8363	0.8008
9.4	0.9999	0.9986	0.9945	0.9863	0.9730	0.9544	0.9307	0.9026	0.8711	0.8370	0.8016
9.5	0.9999	0.9987	0.9947	0.9866	0.9735	0.9550	0.9314	0.9034	0.8718	0.8378	0.8023
9.6	0.9999	0.9987	0.9949	0.9869	0.9740	0.9556	0.9321	0.9041	0.8726	0.8386	0.8030
9.7	0.9999	0.9988	0.9951	0.9873	0.9744	0.9562	0.9328	0.9049	0.8734	0.8393	0.8036
9.8	0.9999	0.9989	0.9952	0.9876	0.9749	0.9568	0.9335	0.9056	0.8741	0.8400	0.8043
9.9	1.0000	0.9989	0.9954	0.9878	0.9753	0.9573	0.9341	0.9063	0.8748	0.8407	0.8050
10.0	1.0000	0.9990	0.9955	0.9881	0.9757	0.9579	0.9348	0.9070	0.8755	0.8414	0.8056

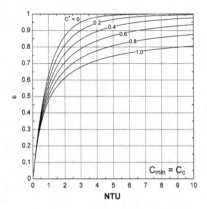

CONFIGURATION 6

1TUBE6ROW1PASS6CIR

(b) $C_{min} = C_h$

NTU	Heat Exchanger Effectiveness										
	$C^* (C_{min} = C_h)$										
	0	0.1	0.2	0.3	0.4	0.5	0.6	0.7	0.8	0.9	1.0
0	0.0000	0.0000	0.0000	0.0000	0.0000	0.0000	0.0000	0.0000	0.0000	0.0000	0.0000
0.1	0.0952	0.0947	0.0943	0.0938	0.0934	0.0929	0.0925	0.0921	0.0916	0.0912	0.0908
0.2	0.1813	0.1796	0.1780	0.1764	0.1749	0.1733	0.1718	0.1703	0.1688	0.1673	0.1658
0.3	0.2592	0.2559	0.2526	0.2494	0.2463	0.2432	0.2402	0.2372	0.2342	0.2313	0.2285
0.4	0.3297	0.3244	0.3192	0.3141	0.3091	0.3042	0.2995	0.2948	0.2902	0.2857	0.2813
0.5	0.3935	0.3860	0.3787	0.3716	0.3646	0.3578	0.3512	0.3447	0.3384	0.3323	0.3263
0.6	0.4512	0.4414	0.4320	0.4228	0.4138	0.4051	0.3966	0.3883	0.3803	0.3725	0.3649
0.7	0.5034	0.4914	0.4798	0.4685	0.4576	0.4469	0.4366	0.4266	0.4169	0.4075	0.3983
0.8	0.5507	0.5365	0.5228	0.5095	0.4967	0.4842	0.4721	0.4604	0.4491	0.4382	0.4275
0.9	0.5934	0.5772	0.5616	0.5464	0.5317	0.5175	0.5038	0.4905	0.4777	0.4652	0.4532
1.0	0.6321	0.6140	0.5965	0.5796	0.5632	0.5474	0.5321	0.5174	0.5031	0.4893	0.4760
1.1	0.6671	0.6473	0.6281	0.6096	0.5917	0.5744	0.5576	0.5415	0.5259	0.5108	0.4963
1.2	0.6988	0.6775	0.6568	0.6368	0.6174	0.5987	0.5807	0.5633	0.5464	0.5302	0.5146
1.3	0.7275	0.7048	0.6828	0.6615	0.6408	0.6209	0.6016	0.5830	0.5650	0.5477	0.5310
1.4	0.7534	0.7296	0.7064	0.6839	0.6621	0.6411	0.6207	0.6010	0.5820	0.5636	0.5460
1.5	0.7769	0.7521	0.7279	0.7044	0.6816	0.6595	0.6381	0.6174	0.5974	0.5782	0.5596
1.6	0.7981	0.7725	0.7475	0.7232	0.6995	0.6764	0.6541	0.6325	0.6116	0.5915	0.5721
1.7	0.8173	0.7911	0.7655	0.7403	0.7158	0.6920	0.6689	0.6464	0.6247	0.6038	0.5836
1.8	0.8347	0.8081	0.7819	0.7561	0.7309	0.7064	0.6825	0.6593	0.6368	0.6151	0.5942
1.9	0.8504	0.8235	0.7969	0.7706	0.7449	0.7197	0.6951	0.6712	0.6481	0.6257	0.6040
2.0	0.8647	0.8376	0.8107	0.7840	0.7577	0.7320	0.7068	0.6823	0.6585	0.6354	0.6132
2.1	0.8775	0.8505	0.8233	0.7964	0.7697	0.7434	0.7177	0.6926	0.6682	0.6446	0.6217
2.2	0.8892	0.8622	0.8350	0.8078	0.7808	0.7541	0.7279	0.7023	0.6774	0.6531	0.6297
2.3	0.8997	0.8729	0.8457	0.8184	0.7911	0.7641	0.7375	0.7114	0.6859	0.6611	0.6371
2.4	0.9093	0.8828	0.8556	0.8282	0.8008	0.7734	0.7464	0.7199	0.6939	0.6687	0.6442
2.5	0.9179	0.8918	0.8648	0.8374	0.8098	0.7822	0.7548	0.7279	0.7015	0.6758	0.6508
2.6	0.9257	0.9000	0.8733	0.8459	0.8182	0.7904	0.7627	0.7354	0.7086	0.6825	0.6570
2.7	0.9328	0.9076	0.8811	0.8538	0.8261	0.7981	0.7702	0.7426	0.7154	0.6888	0.6629
2.8	0.9392	0.9145	0.8884	0.8613	0.8335	0.8054	0.7773	0.7493	0.7218	0.6948	0.6685
2.9	0.9450	0.9209	0.8951	0.8682	0.8405	0.8123	0.7839	0.7557	0.7279	0.7005	0.6738
3.0	0.9502	0.9267	0.9014	0.8747	0.8470	0.8188	0.7903	0.7618	0.7336	0.7059	0.6789
3.1	0.9550	0.9321	0.9072	0.8808	0.8532	0.8249	0.7963	0.7676	0.7391	0.7111	0.6837
3.2	0.9592	0.9371	0.9126	0.8865	0.8590	0.8307	0.8020	0.7731	0.7444	0.7160	0.6883
3.3	0.9631	0.9416	0.9177	0.8918	0.8645	0.8363	0.8074	0.7783	0.7493	0.7207	0.6926
3.4	0.9666	0.9458	0.9224	0.8969	0.8698	0.8415	0.8126	0.7833	0.7541	0.7252	0.6968
3.5	0.9698	0.9497	0.9268	0.9016	0.8747	0.8465	0.8175	0.7881	0.7587	0.7295	0.7008
3.6	0.9727	0.9533	0.9309	0.9061	0.8794	0.8513	0.8222	0.7927	0.7630	0.7336	0.7046
3.7	0.9753	0.9566	0.9348	0.9103	0.8838	0.8558	0.8267	0.7971	0.7672	0.7375	0.7083
3.8	0.9776	0.9597	0.9384	0.9143	0.8880	0.8601	0.8310	0.8012	0.7712	0.7413	0.7118
3.9	0.9798	0.9625	0.9417	0.9181	0.8921	0.8642	0.8351	0.8053	0.7751	0.7449	0.7152
4.0	0.9817	0.9651	0.9449	0.9216	0.8959	0.8682	0.8391	0.8091	0.7788	0.7484	0.7184
4.1	0.9834	0.9675	0.9479	0.9250	0.8995	0.8719	0.8429	0.8128	0.7823	0.7518	0.7215
4.2	0.9850	0.9697	0.9506	0.9282	0.9030	0.8755	0.8465	0.8164	0.7858	0.7550	0.7245
4.3	0.9864	0.9718	0.9533	0.9312	0.9063	0.8790	0.8500	0.8198	0.7891	0.7581	0.7274
4.4	0.9877	0.9737	0.9557	0.9341	0.9094	0.8823	0.8533	0.8232	0.7923	0.7611	0.7302
4.5	0.9889	0.9755	0.9580	0.9368	0.9124	0.8855	0.8566	0.8263	0.7953	0.7640	0.7329
4.6	0.9899	0.9772	0.9602	0.9394	0.9153	0.8885	0.8597	0.8294	0.7983	0.7668	0.7355
4.7	0.9909	0.9787	0.9622	0.9418	0.9180	0.8914	0.8627	0.8324	0.8012	0.7695	0.7380
4.8	0.9918	0.9801	0.9641	0.9442	0.9207	0.8943	0.8656	0.8352	0.8039	0.7722	0.7404
4.9	0.9926	0.9814	0.9660	0.9464	0.9232	0.8970	0.8683	0.8380	0.8066	0.7747	0.7428
5.0	0.9933	0.9827	0.9677	0.9485	0.9256	0.8996	0.8710	0.8407	0.8092	0.7771	0.7450
5.1	0.9939	0.9838	0.9693	0.9505	0.9279	0.9021	0.8736	0.8433	0.8117	0.7795	0.7472
5.2	0.9945	0.9849	0.9708	0.9524	0.9301	0.9045	0.8761	0.8458	0.8141	0.7818	0.7494
5.3	0.9950	0.9858	0.9722	0.9543	0.9323	0.9068	0.8785	0.8482	0.8165	0.7840	0.7514
5.4	0.9955	0.9868	0.9736	0.9560	0.9343	0.9090	0.8809	0.8505	0.8188	0.7862	0.7534
5.5	0.9959	0.9876	0.9749	0.9577	0.9363	0.9112	0.8831	0.8528	0.8210	0.7883	0.7554
5.6	0.9963	0.9884	0.9761	0.9593	0.9382	0.9133	0.8853	0.8550	0.8231	0.7903	0.7572
5.7	0.9967	0.9892	0.9772	0.9608	0.9400	0.9153	0.8874	0.8572	0.8252	0.7923	0.7591
5.8	0.9970	0.9899	0.9783	0.9622	0.9417	0.9172	0.8895	0.8592	0.8272	0.7942	0.7608
5.9	0.9973	0.9905	0.9794	0.9636	0.9434	0.9191	0.8915	0.8612	0.8292	0.7961	0.7625
6.0	0.9975	0.9911	0.9803	0.9649	0.9450	0.9209	0.8934	0.8632	0.8311	0.7979	0.7642

NTU	Heat Exchanger Effectiveness										
	C^* ($C_{min} = C_h$)										
	0	0.1	0.2	0.3	0.4	0.5	0.6	0.7	0.8	0.9	1.0
6.1	0.9978	0.9917	0.9813	0.9662	0.9465	0.9227	0.8953	0.8651	0.8330	0.7996	0.7658
6.2	0.9980	0.9922	0.9821	0.9674	0.9480	0.9244	0.8971	0.8670	0.8348	0.8013	0.7674
6.3	0.9982	0.9927	0.9830	0.9686	0.9495	0.9260	0.8989	0.8687	0.8365	0.8030	0.7690
6.4	0.9983	0.9931	0.9838	0.9697	0.9509	0.9276	0.9006	0.8705	0.8382	0.8046	0.7705
6.5	0.9985	0.9936	0.9845	0.9708	0.9522	0.9292	0.9022	0.8722	0.8399	0.8062	0.7719
6.6	0.9986	0.9940	0.9852	0.9718	0.9535	0.9307	0.9039	0.8738	0.8415	0.8078	0.7733
6.7	0.9988	0.9943	0.9859	0.9728	0.9548	0.9321	0.9054	0.8755	0.8431	0.8092	0.7747
6.8	0.9989	0.9947	0.9865	0.9737	0.9560	0.9335	0.9069	0.8770	0.8446	0.8107	0.7760
6.9	0.9990	0.9950	0.9871	0.9746	0.9571	0.9349	0.9084	0.8785	0.8461	0.8121	0.7774
7.0	0.9991	0.9953	0.9877	0.9755	0.9582	0.9362	0.9099	0.8800	0.8476	0.8135	0.7786
7.1	0.9992	0.9956	0.9883	0.9763	0.9593	0.9375	0.9113	0.8815	0.8490	0.8149	0.7799
7.2	0.9993	0.9959	0.9888	0.9771	0.9604	0.9387	0.9126	0.8829	0.8504	0.8162	0.7811
7.3	0.9993	0.9961	0.9893	0.9779	0.9614	0.9399	0.9140	0.8843	0.8518	0.8175	0.7823
7.4	0.9994	0.9964	0.9897	0.9786	0.9624	0.9411	0.9153	0.8856	0.8531	0.8187	0.7834
7.5	0.9994	0.9966	0.9902	0.9793	0.9633	0.9422	0.9165	0.8869	0.8544	0.8200	0.7845
7.6	0.9995	0.9968	0.9906	0.9800	0.9642	0.9433	0.9177	0.8882	0.8557	0.8212	0.7856
7.7	0.9995	0.9970	0.9910	0.9806	0.9651	0.9444	0.9189	0.8894	0.8569	0.8223	0.7867
7.8	0.9996	0.9972	0.9914	0.9813	0.9660	0.9455	0.9201	0.8907	0.8581	0.8235	0.7877
7.9	0.9996	0.9973	0.9918	0.9819	0.9668	0.9465	0.9212	0.8919	0.8593	0.8246	0.7888
8.0	0.9997	0.9975	0.9921	0.9824	0.9676	0.9475	0.9224	0.8930	0.8604	0.8257	0.7898
8.1	0.9997	0.9976	0.9925	0.9830	0.9684	0.9484	0.9234	0.8941	0.8616	0.8267	0.7907
8.2	0.9997	0.9978	0.9928	0.9835	0.9692	0.9494	0.9245	0.8953	0.8627	0.8278	0.7917
8.3	0.9998	0.9979	0.9931	0.9841	0.9699	0.9503	0.9255	0.8963	0.8637	0.8288	0.7926
8.4	0.9998	0.9980	0.9934	0.9846	0.9706	0.9512	0.9265	0.8974	0.8648	0.8298	0.7935
8.5	0.9998	0.9981	0.9937	0.9850	0.9713	0.9520	0.9275	0.8984	0.8658	0.8308	0.7944
8.6	0.9998	0.9982	0.9939	0.9855	0.9720	0.9529	0.9285	0.8994	0.8668	0.8317	0.7953
8.7	0.9998	0.9984	0.9942	0.9860	0.9726	0.9537	0.9294	0.9004	0.8678	0.8326	0.7961
8.8	0.9998	0.9984	0.9944	0.9864	0.9732	0.9545	0.9303	0.9014	0.8688	0.8336	0.7970
8.9	0.9999	0.9985	0.9947	0.9868	0.9738	0.9553	0.9312	0.9023	0.8697	0.8345	0.7978
9.0	0.9999	0.9986	0.9949	0.9872	0.9744	0.9560	0.9321	0.9033	0.8706	0.8353	0.7986
9.1	0.9999	0.9987	0.9951	0.9876	0.9750	0.9568	0.9329	0.9042	0.8715	0.8362	0.7993
9.2	0.9999	0.9988	0.9953	0.9880	0.9756	0.9575	0.9338	0.9051	0.8724	0.8370	0.8001
9.3	0.9999	0.9988	0.9955	0.9883	0.9761	0.9582	0.9346	0.9059	0.8733	0.8378	0.8008
9.4	0.9999	0.9989	0.9957	0.9887	0.9766	0.9589	0.9354	0.9068	0.8741	0.8386	0.8016
9.5	0.9999	0.9990	0.9958	0.9890	0.9771	0.9595	0.9361	0.9076	0.8749	0.8394	0.8023
9.6	0.9999	0.9990	0.9960	0.9893	0.9776	0.9602	0.9369	0.9084	0.8758	0.8402	0.8030
9.7	0.9999	0.9991	0.9962	0.9896	0.9781	0.9608	0.9377	0.9092	0.8766	0.8409	0.8036
9.8	0.9999	0.9991	0.9963	0.9899	0.9786	0.9614	0.9384	0.9100	0.8773	0.8417	0.8043
9.9	1.0000	0.9992	0.9965	0.9902	0.9791	0.9620	0.9391	0.9108	0.8781	0.8424	0.8050
10.0	1.0000	0.9992	0.9966	0.9905	0.9795	0.9626	0.9398	0.9115	0.8789	0.8431	0.8056

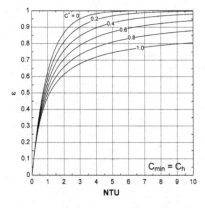

CONFIGURATION 7

1TUBE10ROW1PASS10CIR

(a) $C_{min} = C_c$

NTU	Heat Exchanger Effectiveness										
	$C^* (C_{min} = C_c)$										
	0	0.1	0.2	0.3	0.4	0.5	0.6	0.7	0.8	0.9	1.0
0	0.0000	0.0000	0.0000	0.0000	0.0000	0.0000	0.0000	0.0000	0.0000	0.0000	0.0000
0.1	0.0952	0.0947	0.0943	0.0938	0.0934	0.0929	0.0925	0.0921	0.0916	0.0912	0.0908
0.2	0.1813	0.1796	0.1780	0.1764	0.1749	0.1733	0.1718	0.1703	0.1688	0.1673	0.1658
0.3	0.2592	0.2559	0.2526	0.2494	0.2463	0.2432	0.2402	0.2372	0.2342	0.2313	0.2285
0.4	0.3297	0.3244	0.3192	0.3141	0.3091	0.3042	0.2995	0.2948	0.2902	0.2857	0.2813
0.5	0.3935	0.3860	0.3787	0.3716	0.3646	0.3578	0.3512	0.3448	0.3385	0.3323	0.3263
0.6	0.4512	0.4414	0.4320	0.4228	0.4138	0.4051	0.3966	0.3884	0.3803	0.3725	0.3649
0.7	0.5034	0.4914	0.4798	0.4685	0.4576	0.4469	0.4366	0.4266	0.4169	0.4075	0.3984
0.8	0.5507	0.5365	0.5228	0.5095	0.4967	0.4842	0.4722	0.4605	0.4492	0.4382	0.4276
0.9	0.5934	0.5772	0.5615	0.5464	0.5317	0.5175	0.5038	0.4906	0.4777	0.4653	0.4533
1.0	0.6321	0.6140	0.5965	0.5796	0.5632	0.5474	0.5322	0.5174	0.5032	0.4894	0.4761
1.1	0.6671	0.6473	0.6281	0.6096	0.5917	0.5744	0.5577	0.5416	0.5260	0.5110	0.4965
1.2	0.6988	0.6774	0.6568	0.6368	0.6174	0.5988	0.5808	0.5634	0.5466	0.5304	0.5148
1.3	0.7275	0.7048	0.6828	0.6615	0.6408	0.6209	0.6017	0.5831	0.5652	0.5479	0.5313
1.4	0.7534	0.7295	0.7064	0.6839	0.6622	0.6411	0.6208	0.6011	0.5822	0.5639	0.5463
1.5	0.7769	0.7520	0.7279	0.7044	0.6816	0.6596	0.6382	0.6176	0.5977	0.5785	0.5600
1.6	0.7981	0.7725	0.7475	0.7231	0.6995	0.6765	0.6543	0.6327	0.6119	0.5919	0.5725
1.7	0.8173	0.7911	0.7654	0.7403	0.7159	0.6921	0.6690	0.6467	0.6251	0.6042	0.5841
1.8	0.8347	0.8080	0.7818	0.7561	0.7310	0.7065	0.6827	0.6596	0.6372	0.6156	0.5947
1.9	0.8504	0.8235	0.7968	0.7706	0.7449	0.7198	0.6953	0.6715	0.6485	0.6262	0.6046
2.0	0.8647	0.8375	0.8106	0.7840	0.7578	0.7321	0.7070	0.6826	0.6590	0.6360	0.6139
2.1	0.8775	0.8504	0.8233	0.7963	0.7697	0.7436	0.7180	0.6930	0.6688	0.6452	0.6225
2.2	0.8892	0.8621	0.8349	0.8077	0.7808	0.7543	0.7282	0.7027	0.6779	0.6538	0.6305
2.3	0.8997	0.8729	0.8456	0.8183	0.7912	0.7642	0.7378	0.7118	0.6865	0.6619	0.6381
2.4	0.9093	0.8827	0.8555	0.8282	0.8008	0.7736	0.7467	0.7204	0.6946	0.6695	0.6452
2.5	0.9179	0.8917	0.8647	0.8373	0.8098	0.7824	0.7552	0.7284	0.7022	0.6767	0.6519
2.6	0.9257	0.8999	0.8732	0.8458	0.8182	0.7906	0.7631	0.7360	0.7094	0.6835	0.6582
2.7	0.9328	0.9075	0.8810	0.8538	0.8261	0.7983	0.7706	0.7432	0.7162	0.6899	0.6642
2.8	0.9392	0.9144	0.8883	0.8612	0.8335	0.8056	0.7777	0.7500	0.7227	0.6960	0.6699
2.9	0.9450	0.9208	0.8950	0.8681	0.8405	0.8125	0.7844	0.7565	0.7288	0.7017	0.6753
3.0	0.9502	0.9266	0.9012	0.8746	0.8471	0.8190	0.7908	0.7626	0.7347	0.7072	0.6804
3.1	0.9550	0.9320	0.9071	0.8807	0.8533	0.8252	0.7968	0.7684	0.7402	0.7125	0.6853
3.2	0.9592	0.9369	0.9125	0.8864	0.8591	0.8310	0.8025	0.7739	0.7455	0.7175	0.6900
3.3	0.9631	0.9415	0.9175	0.8918	0.8646	0.8366	0.8080	0.7792	0.7506	0.7222	0.6945
3.4	0.9666	0.9457	0.9222	0.8968	0.8698	0.8418	0.8132	0.7843	0.7554	0.7268	0.6987
3.5	0.9698	0.9496	0.9266	0.9015	0.8748	0.8468	0.8182	0.7891	0.7600	0.7312	0.7028
3.6	0.9727	0.9532	0.9307	0.9060	0.8795	0.8516	0.8229	0.7937	0.7645	0.7354	0.7067
3.7	0.9753	0.9565	0.9346	0.9102	0.8839	0.8562	0.8274	0.7982	0.7687	0.7394	0.7105
3.8	0.9776	0.9595	0.9382	0.9142	0.8881	0.8605	0.8318	0.8024	0.7728	0.7433	0.7141
3.9	0.9798	0.9623	0.9415	0.9180	0.8921	0.8646	0.8359	0.8065	0.7767	0.7470	0.7176
4.0	0.9817	0.9649	0.9447	0.9215	0.8960	0.8686	0.8399	0.8104	0.7805	0.7506	0.7210
4.1	0.9834	0.9673	0.9477	0.9249	0.8996	0.8724	0.8437	0.8142	0.7841	0.7540	0.7242
4.2	0.9850	0.9696	0.9504	0.9281	0.9031	0.8760	0.8474	0.8178	0.7877	0.7574	0.7273
4.3	0.9864	0.9716	0.9530	0.9311	0.9064	0.8795	0.8509	0.8213	0.7910	0.7606	0.7303
4.4	0.9877	0.9736	0.9555	0.9340	0.9095	0.8828	0.8543	0.8246	0.7943	0.7637	0.7332
4.5	0.9889	0.9753	0.9578	0.9367	0.9125	0.8860	0.8576	0.8279	0.7974	0.7667	0.7360
4.6	0.9899	0.9770	0.9599	0.9392	0.9154	0.8890	0.8607	0.8310	0.8005	0.7696	0.7387
4.7	0.9909	0.9785	0.9620	0.9417	0.9182	0.8920	0.8637	0.8340	0.8034	0.7724	0.7414
4.8	0.9918	0.9799	0.9639	0.9440	0.9208	0.8948	0.8667	0.8370	0.8063	0.7751	0.7439
4.9	0.9926	0.9813	0.9657	0.9462	0.9233	0.8975	0.8695	0.8398	0.8090	0.7777	0.7464
5.0	0.9933	0.9825	0.9674	0.9484	0.9257	0.9001	0.8722	0.8425	0.8117	0.7803	0.7488
5.1	0.9939	0.9836	0.9690	0.9504	0.9280	0.9027	0.8748	0.8452	0.8143	0.7828	0.7511
5.2	0.9945	0.9847	0.9706	0.9523	0.9303	0.9051	0.8774	0.8477	0.8168	0.7852	0.7533
5.3	0.9950	0.9857	0.9720	0.9541	0.9324	0.9074	0.8798	0.8502	0.8192	0.7875	0.7555
5.4	0.9955	0.9866	0.9734	0.9558	0.9344	0.9097	0.8822	0.8526	0.8216	0.7898	0.7576
5.5	0.9959	0.9875	0.9746	0.9575	0.9364	0.9119	0.8845	0.8549	0.8239	0.7920	0.7597
5.6	0.9963	0.9883	0.9758	0.9591	0.9383	0.9140	0.8867	0.8572	0.8261	0.7941	0.7617
5.7	0.9967	0.9890	0.9770	0.9606	0.9401	0.9160	0.8889	0.8594	0.8283	0.7962	0.7637
5.8	0.9970	0.9897	0.9781	0.9621	0.9419	0.9179	0.8909	0.8615	0.8304	0.7982	0.7656
5.9	0.9973	0.9903	0.9791	0.9634	0.9435	0.9198	0.8930	0.8636	0.8325	0.8002	0.7674
6.0	0.9975	0.9909	0.9801	0.9648	0.9451	0.9217	0.8949	0.8656	0.8344	0.8021	0.7692

NTU	Heat Exchanger Effectiveness										
	C^* ($C_{min} = C_c$)										
	0	0.1	0.2	0.3	0.4	0.5	0.6	0.7	0.8	0.9	1.0
6.1	0.9978	0.9915	0.9810	0.9660	0.9467	0.9234	0.8968	0.8676	0.8364	0.8040	0.7710
6.2	0.9980	0.9920	0.9819	0.9672	0.9482	0.9252	0.8987	0.8695	0.8383	0.8058	0.7727
6.3	0.9982	0.9925	0.9827	0.9684	0.9496	0.9268	0.9005	0.8713	0.8401	0.8076	0.7743
6.4	0.9983	0.9930	0.9835	0.9695	0.9510	0.9284	0.9022	0.8731	0.8419	0.8093	0.7760
6.5	0.9985	0.9934	0.9842	0.9706	0.9524	0.9300	0.9039	0.8749	0.8437	0.8110	0.7776
6.6	0.9986	0.9938	0.9850	0.9716	0.9537	0.9315	0.9056	0.8766	0.8454	0.8127	0.7791
6.7	0.9988	0.9942	0.9856	0.9726	0.9549	0.9329	0.9072	0.8783	0.8471	0.8143	0.7806
6.8	0.9989	0.9945	0.9863	0.9735	0.9561	0.9344	0.9087	0.8799	0.8487	0.8158	0.7821
6.9	0.9990	0.9949	0.9869	0.9744	0.9573	0.9357	0.9102	0.8815	0.8503	0.8174	0.7835
7.0	0.9991	0.9952	0.9875	0.9753	0.9584	0.9371	0.9117	0.8830	0.8518	0.8189	0.7850
7.1	0.9992	0.9955	0.9880	0.9761	0.9595	0.9384	0.9131	0.8845	0.8533	0.8204	0.7863
7.2	0.9993	0.9957	0.9885	0.9769	0.9605	0.9396	0.9145	0.8860	0.8548	0.8218	0.7877
7.3	0.9993	0.9960	0.9890	0.9777	0.9616	0.9408	0.9159	0.8875	0.8563	0.8232	0.7890
7.4	0.9994	0.9962	0.9895	0.9784	0.9625	0.9420	0.9172	0.8889	0.8577	0.8246	0.7903
7.5	0.9994	0.9964	0.9899	0.9791	0.9635	0.9432	0.9185	0.8902	0.8591	0.8259	0.7916
7.6	0.9995	0.9967	0.9904	0.9798	0.9644	0.9443	0.9198	0.8916	0.8604	0.8272	0.7928
7.7	0.9995	0.9969	0.9908	0.9804	0.9653	0.9454	0.9210	0.8929	0.8618	0.8285	0.7940
7.8	0.9996	0.9970	0.9912	0.9811	0.9662	0.9464	0.9222	0.8942	0.8630	0.8298	0.7952
7.9	0.9996	0.9972	0.9915	0.9817	0.9670	0.9475	0.9234	0.8954	0.8643	0.8310	0.7964
8.0	0.9997	0.9974	0.9919	0.9822	0.9678	0.9485	0.9245	0.8966	0.8656	0.8322	0.7975
8.1	0.9997	0.9975	0.9922	0.9828	0.9686	0.9494	0.9256	0.8978	0.8668	0.8334	0.7986
8.2	0.9997	0.9977	0.9926	0.9833	0.9693	0.9504	0.9267	0.8990	0.8680	0.8346	0.7997
8.3	0.9998	0.9978	0.9929	0.9839	0.9701	0.9513	0.9278	0.9001	0.8691	0.8357	0.8008
8.4	0.9998	0.9979	0.9932	0.9844	0.9708	0.9522	0.9288	0.9012	0.8703	0.8368	0.8018
8.5	0.9998	0.9980	0.9934	0.9848	0.9715	0.9531	0.9298	0.9023	0.8714	0.8379	0.8028
8.6	0.9998	0.9982	0.9937	0.9853	0.9721	0.9539	0.9308	0.9034	0.8725	0.8390	0.8039
8.7	0.9998	0.9983	0.9940	0.9857	0.9728	0.9547	0.9318	0.9045	0.8736	0.8400	0.8048
8.8	0.9998	0.9984	0.9942	0.9862	0.9734	0.9555	0.9327	0.9055	0.8746	0.8411	0.8058
8.9	0.9999	0.9984	0.9944	0.9866	0.9740	0.9563	0.9336	0.9065	0.8756	0.8421	0.8068
9.0	0.9999	0.9985	0.9947	0.9870	0.9746	0.9571	0.9346	0.9075	0.8767	0.8431	0.8077
9.1	0.9999	0.9986	0.9949	0.9874	0.9752	0.9578	0.9354	0.9084	0.8776	0.8440	0.8086
9.2	0.9999	0.9987	0.9951	0.9878	0.9757	0.9586	0.9363	0.9094	0.8786	0.8450	0.8095
9.3	0.9999	0.9988	0.9953	0.9881	0.9763	0.9593	0.9371	0.9103	0.8796	0.8459	0.8104
9.4	0.9999	0.9988	0.9955	0.9885	0.9768	0.9600	0.9380	0.9112	0.8805	0.8469	0.8113
9.5	0.9999	0.9989	0.9956	0.9888	0.9773	0.9607	0.9388	0.9121	0.8814	0.8478	0.8121
9.6	0.9999	0.9990	0.9958	0.9891	0.9778	0.9613	0.9396	0.9130	0.8823	0.8487	0.8130
9.7	0.9999	0.9990	0.9960	0.9894	0.9783	0.9620	0.9403	0.9138	0.8832	0.8495	0.8138
9.8	0.9999	0.9991	0.9961	0.9897	0.9788	0.9626	0.9411	0.9147	0.8841	0.8504	0.8146
9.9	1.0000	0.9991	0.9963	0.9900	0.9792	0.9632	0.9418	0.9155	0.8850	0.8512	0.8154
10.0	1.0000	0.9992	0.9964	0.9903	0.9797	0.9638	0.9426	0.9163	0.8858	0.8521	0.8162

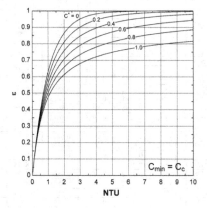

CONFIGURATION 7

1TUBE10ROW1PASS10CIR

(b) $C_{min} = C_h$

NTU	Heat Exchanger Effectiveness										
	C^* ($C_{min} = C_h$)										
	0	0.1	0.2	0.3	0.4	0.5	0.6	0.7	0.8	0.9	1.0
0	0.0000	0.0000	0.0000	0.0000	0.0000	0.0000	0.0000	0.0000	0.0000	0.0000	0.0000
0.1	0.0952	0.0947	0.0943	0.0938	0.0934	0.0929	0.0925	0.0921	0.0916	0.0912	0.0908
0.2	0.1813	0.1796	0.1780	0.1764	0.1749	0.1733	0.1718	0.1703	0.1688	0.1673	0.1658
0.3	0.2592	0.2559	0.2526	0.2494	0.2463	0.2432	0.2402	0.2372	0.2342	0.2313	0.2285
0.4	0.3297	0.3244	0.3192	0.3141	0.3091	0.3042	0.2995	0.2948	0.2902	0.2857	0.2813
0.5	0.3935	0.3860	0.3787	0.3716	0.3646	0.3578	0.3512	0.3448	0.3385	0.3323	0.3263
0.6	0.4512	0.4414	0.4320	0.4228	0.4138	0.4051	0.3966	0.3884	0.3803	0.3725	0.3649
0.7	0.5034	0.4914	0.4798	0.4685	0.4576	0.4470	0.4367	0.4267	0.4170	0.4075	0.3984
0.8	0.5507	0.5365	0.5228	0.5095	0.4967	0.4842	0.4722	0.4605	0.4492	0.4382	0.4276
0.9	0.5934	0.5772	0.5616	0.5464	0.5317	0.5175	0.5038	0.4906	0.4777	0.4653	0.4533
1.0	0.6321	0.6140	0.5965	0.5796	0.5633	0.5475	0.5322	0.5174	0.5032	0.4894	0.4761
1.1	0.6671	0.6473	0.6282	0.6096	0.5917	0.5744	0.5577	0.5416	0.5260	0.5110	0.4965
1.2	0.6988	0.6775	0.6568	0.6368	0.6175	0.5988	0.5808	0.5634	0.5466	0.5304	0.5148
1.3	0.7275	0.7048	0.6828	0.6615	0.6409	0.6210	0.6017	0.5832	0.5652	0.5479	0.5313
1.4	0.7534	0.7296	0.7064	0.6840	0.6622	0.6412	0.6208	0.6012	0.5822	0.5639	0.5463
1.5	0.7769	0.7521	0.7279	0.7045	0.6817	0.6597	0.6383	0.6177	0.5977	0.5785	0.5600
1.6	0.7981	0.7725	0.7476	0.7232	0.6996	0.6766	0.6543	0.6328	0.6120	0.5919	0.5725
1.7	0.8173	0.7912	0.7655	0.7404	0.7160	0.6922	0.6691	0.6467	0.6251	0.6042	0.5841
1.8	0.8347	0.8081	0.7819	0.7562	0.7311	0.7066	0.6828	0.6596	0.6373	0.6156	0.5947
1.9	0.8504	0.8235	0.7969	0.7707	0.7450	0.7199	0.6954	0.6716	0.6485	0.6262	0.6046
2.0	0.8647	0.8376	0.8107	0.7841	0.7579	0.7323	0.7072	0.6828	0.6590	0.6361	0.6139
2.1	0.8775	0.8505	0.8234	0.7965	0.7699	0.7437	0.7181	0.6931	0.6688	0.6453	0.6225
2.2	0.8892	0.8622	0.8351	0.8079	0.7810	0.7545	0.7284	0.7029	0.6780	0.6539	0.6305
2.3	0.8997	0.8730	0.8458	0.8185	0.7914	0.7645	0.7380	0.7120	0.6866	0.6620	0.6381
2.4	0.9093	0.8828	0.8557	0.8284	0.8010	0.7738	0.7469	0.7205	0.6947	0.6696	0.6452
2.5	0.9179	0.8918	0.8649	0.8376	0.8101	0.7826	0.7554	0.7286	0.7024	0.6767	0.6519
2.6	0.9257	0.9000	0.8734	0.8461	0.8185	0.7908	0.7634	0.7362	0.7096	0.6835	0.6582
2.7	0.9328	0.9076	0.8812	0.8540	0.8264	0.7986	0.7709	0.7434	0.7164	0.6899	0.6642
2.8	0.9392	0.9145	0.8885	0.8615	0.8339	0.8059	0.7780	0.7502	0.7229	0.6960	0.6699
2.9	0.9450	0.9209	0.8952	0.8684	0.8408	0.8128	0.7847	0.7567	0.7290	0.7018	0.6753
3.0	0.9502	0.9268	0.9015	0.8749	0.8474	0.8194	0.7911	0.7628	0.7348	0.7073	0.6804
3.1	0.9550	0.9321	0.9073	0.8810	0.8536	0.8255	0.7971	0.7687	0.7404	0.7126	0.6853
3.2	0.9592	0.9371	0.9128	0.8868	0.8595	0.8314	0.8029	0.7742	0.7457	0.7176	0.6900
3.3	0.9631	0.9417	0.9178	0.8921	0.8650	0.8370	0.8084	0.7795	0.7508	0.7224	0.6945
3.4	0.9666	0.9459	0.9226	0.8972	0.8703	0.8423	0.8136	0.7846	0.7556	0.7269	0.6987
3.5	0.9698	0.9498	0.9270	0.9019	0.8752	0.8473	0.8186	0.7894	0.7603	0.7313	0.7028
3.6	0.9727	0.9533	0.9311	0.9064	0.8799	0.8521	0.8233	0.7941	0.7647	0.7355	0.7067
3.7	0.9753	0.9566	0.9349	0.9107	0.8844	0.8566	0.8279	0.7985	0.7690	0.7395	0.7105
3.8	0.9776	0.9597	0.9385	0.9147	0.8886	0.8610	0.8322	0.8028	0.7731	0.7434	0.7141
3.9	0.9798	0.9625	0.9419	0.9184	0.8927	0.8652	0.8364	0.8069	0.7770	0.7472	0.7176
4.0	0.9817	0.9651	0.9451	0.9220	0.8965	0.8691	0.8404	0.8108	0.7808	0.7507	0.7210
4.1	0.9834	0.9675	0.9480	0.9254	0.9002	0.8729	0.8442	0.8146	0.7844	0.7542	0.7242
4.2	0.9850	0.9698	0.9508	0.9286	0.9037	0.8766	0.8479	0.8182	0.7880	0.7575	0.7273
4.3	0.9864	0.9718	0.9534	0.9316	0.9070	0.8801	0.8515	0.8217	0.7914	0.7608	0.7303
4.4	0.9877	0.9738	0.9559	0.9345	0.9102	0.8834	0.8549	0.8251	0.7946	0.7639	0.7332
4.5	0.9889	0.9755	0.9582	0.9372	0.9132	0.8866	0.8582	0.8284	0.7978	0.7669	0.7360
4.6	0.9899	0.9772	0.9604	0.9398	0.9161	0.8897	0.8613	0.8315	0.8008	0.7698	0.7387
4.7	0.9909	0.9787	0.9624	0.9423	0.9188	0.8927	0.8644	0.8346	0.8038	0.7726	0.7414
4.8	0.9918	0.9801	0.9643	0.9446	0.9215	0.8955	0.8673	0.8375	0.8067	0.7753	0.7439
4.9	0.9926	0.9815	0.9661	0.9469	0.9240	0.8983	0.8702	0.8404	0.8094	0.7779	0.7464
5.0	0.9933	0.9827	0.9679	0.9490	0.9265	0.9009	0.8729	0.8431	0.8121	0.7805	0.7488
5.1	0.9939	0.9838	0.9695	0.9510	0.9288	0.9034	0.8756	0.8458	0.8147	0.7830	0.7511
5.2	0.9945	0.9849	0.9710	0.9529	0.9310	0.9059	0.8781	0.8483	0.8172	0.7854	0.7533
5.3	0.9950	0.9859	0.9724	0.9548	0.9332	0.9083	0.8806	0.8508	0.8197	0.7877	0.7555
5.4	0.9955	0.9868	0.9738	0.9565	0.9352	0.9105	0.8830	0.8533	0.8221	0.7900	0.7576
5.5	0.9959	0.9877	0.9751	0.9582	0.9372	0.9127	0.8853	0.8556	0.8244	0.7922	0.7597
5.6	0.9963	0.9885	0.9763	0.9598	0.9391	0.9148	0.8875	0.8579	0.8266	0.7944	0.7617
5.7	0.9967	0.9892	0.9774	0.9613	0.9410	0.9169	0.8897	0.8601	0.8288	0.7965	0.7637
5.8	0.9970	0.9899	0.9785	0.9627	0.9427	0.9189	0.8918	0.8623	0.8309	0.7985	0.7656
5.9	0.9973	0.9905	0.9796	0.9641	0.9444	0.9208	0.8939	0.8644	0.8330	0.8005	0.7674
6.0	0.9975	0.9911	0.9805	0.9655	0.9460	0.9226	0.8959	0.8664	0.8350	0.8024	0.7692

NTU	Heat Exchanger Effectiveness										
	C^* ($C_{min} = C_h$)										
	0	0.1	0.2	0.3	0.4	0.5	0.6	0.7	0.8	0.9	1.0
6.1	0.9978	0.9917	0.9815	0.9667	0.9476	0.9244	0.8978	0.8684	0.8370	0.8043	0.7710
6.2	0.9980	0.9922	0.9823	0.9680	0.9491	0.9262	0.8997	0.8703	0.8389	0.8061	0.7727
6.3	0.9982	0.9927	0.9832	0.9691	0.9506	0.9278	0.9015	0.8722	0.8407	0.8079	0.7743
6.4	0.9983	0.9932	0.9840	0.9702	0.9520	0.9295	0.9032	0.8740	0.8426	0.8096	0.7760
6.5	0.9985	0.9936	0.9847	0.9713	0.9533	0.9310	0.9050	0.8758	0.8443	0.8113	0.7776
6.6	0.9986	0.9940	0.9854	0.9723	0.9546	0.9326	0.9066	0.8775	0.8461	0.8130	0.7791
6.7	0.9988	0.9944	0.9861	0.9733	0.9559	0.9341	0.9083	0.8792	0.8477	0.8146	0.7806
6.8	0.9989	0.9947	0.9867	0.9743	0.9571	0.9355	0.9098	0.8809	0.8494	0.8162	0.7821
6.9	0.9990	0.9950	0.9873	0.9752	0.9583	0.9369	0.9114	0.8825	0.8510	0.8177	0.7835
7.0	0.9991	0.9953	0.9879	0.9760	0.9594	0.9382	0.9129	0.8840	0.8526	0.8193	0.7850
7.1	0.9992	0.9956	0.9885	0.9769	0.9605	0.9395	0.9143	0.8856	0.8541	0.8207	0.7863
7.2	0.9993	0.9959	0.9890	0.9777	0.9616	0.9408	0.9157	0.8871	0.8556	0.8222	0.7877
7.3	0.9993	0.9961	0.9895	0.9784	0.9626	0.9421	0.9171	0.8885	0.8570	0.8236	0.7890
7.4	0.9994	0.9964	0.9899	0.9792	0.9636	0.9433	0.9185	0.8899	0.8585	0.8250	0.7903
7.5	0.9994	0.9966	0.9904	0.9799	0.9646	0.9444	0.9198	0.8913	0.8599	0.8263	0.7916
7.6	0.9995	0.9968	0.9908	0.9806	0.9655	0.9456	0.9211	0.8927	0.8612	0.8277	0.7928
7.7	0.9995	0.9970	0.9912	0.9812	0.9664	0.9467	0.9223	0.8940	0.8626	0.8290	0.7940
7.8	0.9996	0.9972	0.9916	0.9818	0.9673	0.9477	0.9236	0.8953	0.8639	0.8302	0.7952
7.9	0.9996	0.9973	0.9920	0.9825	0.9681	0.9488	0.9247	0.8966	0.8652	0.8315	0.7964
8.0	0.9997	0.9975	0.9923	0.9830	0.9689	0.9498	0.9259	0.8978	0.8664	0.8327	0.7975
8.1	0.9997	0.9977	0.9927	0.9836	0.9697	0.9508	0.9270	0.8990	0.8677	0.8339	0.7986
8.2	0.9997	0.9978	0.9930	0.9841	0.9705	0.9518	0.9281	0.9002	0.8689	0.8350	0.7997
8.3	0.9998	0.9979	0.9933	0.9847	0.9712	0.9527	0.9292	0.9014	0.8701	0.8362	0.8008
8.4	0.9998	0.9980	0.9936	0.9852	0.9720	0.9536	0.9303	0.9025	0.8712	0.8373	0.8018
8.5	0.9998	0.9982	0.9939	0.9856	0.9727	0.9545	0.9313	0.9037	0.8724	0.8384	0.8028
8.6	0.9998	0.9983	0.9941	0.9861	0.9733	0.9554	0.9323	0.9047	0.8735	0.8395	0.8039
8.7	0.9998	0.9984	0.9944	0.9866	0.9740	0.9562	0.9333	0.9058	0.8746	0.8406	0.8048
8.8	0.9998	0.9985	0.9946	0.9870	0.9746	0.9570	0.9343	0.9069	0.8756	0.8416	0.8058
8.9	0.9999	0.9986	0.9948	0.9874	0.9752	0.9578	0.9352	0.9079	0.8767	0.8426	0.8068
9.0	0.9999	0.9986	0.9951	0.9878	0.9758	0.9586	0.9361	0.9089	0.8777	0.8436	0.8077
9.1	0.9999	0.9987	0.9953	0.9882	0.9764	0.9594	0.9370	0.9099	0.8787	0.8446	0.8086
9.2	0.9999	0.9988	0.9955	0.9886	0.9770	0.9601	0.9379	0.9109	0.8797	0.8456	0.8095
9.3	0.9999	0.9989	0.9957	0.9889	0.9775	0.9609	0.9388	0.9118	0.8807	0.8465	0.8104
9.4	0.9999	0.9989	0.9958	0.9893	0.9781	0.9616	0.9396	0.9127	0.8817	0.8475	0.8113
9.5	0.9999	0.9990	0.9960	0.9896	0.9786	0.9623	0.9405	0.9136	0.8826	0.8484	0.8121
9.6	0.9999	0.9991	0.9962	0.9899	0.9791	0.9629	0.9413	0.9145	0.8835	0.8493	0.8130
9.7	0.9999	0.9991	0.9963	0.9902	0.9796	0.9636	0.9421	0.9154	0.8844	0.8502	0.8138
9.8	0.9999	0.9992	0.9965	0.9905	0.9801	0.9642	0.9429	0.9163	0.8853	0.8510	0.8146
9.9	1.0000	0.9992	0.9966	0.9908	0.9805	0.9649	0.9436	0.9171	0.8862	0.8519	0.8154
10.0	1.0000	0.9993	0.9968	0.9911	0.9810	0.9655	0.9444	0.9180	0.8870	0.8527	0.8162

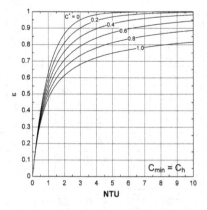

CONFIGURATION 8

1TUBE20ROW1PASS20CIR

(a) $C_{min} = C_c$

NTU	Heat Exchanger Effectiveness										
	$C^* (C_{min} = C_c)$										
	0	0.1	0.2	0.3	0.4	0.5	0.6	0.7	0.8	0.9	1.0
0	0.0000	0.0000	0.0000	0.0000	0.0000	0.0000	0.0000	0.0000	0.0000	0.0000	0.0000
0.1	0.0952	0.0947	0.0943	0.0938	0.0934	0.0929	0.0925	0.0921	0.0916	0.0912	0.0908
0.2	0.1813	0.1796	0.1780	0.1764	0.1749	0.1733	0.1718	0.1703	0.1688	0.1673	0.1658
0.3	0.2592	0.2559	0.2526	0.2494	0.2463	0.2432	0.2402	0.2372	0.2342	0.2313	0.2285
0.4	0.3297	0.3244	0.3192	0.3141	0.3091	0.3042	0.2995	0.2948	0.2902	0.2857	0.2814
0.5	0.3935	0.3860	0.3787	0.3716	0.3646	0.3578	0.3512	0.3448	0.3385	0.3323	0.3263
0.6	0.4512	0.4414	0.4320	0.4228	0.4138	0.4051	0.3966	0.3884	0.3803	0.3725	0.3650
0.7	0.5034	0.4914	0.4798	0.4685	0.4576	0.4470	0.4367	0.4267	0.4170	0.4076	0.3984
0.8	0.5507	0.5365	0.5228	0.5095	0.4967	0.4842	0.4722	0.4605	0.4492	0.4383	0.4277
0.9	0.5934	0.5772	0.5616	0.5464	0.5317	0.5176	0.5038	0.4906	0.4778	0.4654	0.4534
1.0	0.6321	0.6140	0.5965	0.5796	0.5633	0.5475	0.5322	0.5175	0.5032	0.4895	0.4762
1.1	0.6671	0.6473	0.6282	0.6096	0.5917	0.5744	0.5577	0.5416	0.5261	0.5111	0.4966
1.2	0.6988	0.6775	0.6568	0.6368	0.6175	0.5988	0.5808	0.5634	0.5467	0.5305	0.5149
1.3	0.7275	0.7048	0.6828	0.6615	0.6409	0.6210	0.6018	0.5832	0.5653	0.5480	0.5314
1.4	0.7534	0.7296	0.7064	0.6840	0.6622	0.6412	0.6209	0.6012	0.5823	0.5640	0.5464
1.5	0.7769	0.7521	0.7279	0.7045	0.6817	0.6597	0.6384	0.6177	0.5978	0.5786	0.5601
1.6	0.7981	0.7725	0.7476	0.7232	0.6996	0.6766	0.6544	0.6329	0.6121	0.5920	0.5727
1.7	0.8173	0.7911	0.7655	0.7404	0.7160	0.6922	0.6692	0.6469	0.6253	0.6044	0.5843
1.8	0.8347	0.8081	0.7819	0.7562	0.7311	0.7066	0.6829	0.6598	0.6374	0.6158	0.5950
1.9	0.8504	0.8235	0.7969	0.7707	0.7451	0.7200	0.6955	0.6718	0.6487	0.6264	0.6049
2.0	0.8647	0.8376	0.8107	0.7841	0.7580	0.7323	0.7073	0.6829	0.6592	0.6363	0.6141
2.1	0.8775	0.8505	0.8234	0.7965	0.7699	0.7438	0.7183	0.6933	0.6691	0.6455	0.6228
2.2	0.8892	0.8622	0.8351	0.8079	0.7811	0.7545	0.7285	0.7031	0.6783	0.6542	0.6309
2.3	0.8997	0.8729	0.8458	0.8186	0.7914	0.7646	0.7381	0.7122	0.6869	0.6623	0.6385
2.4	0.9093	0.8828	0.8557	0.8284	0.8011	0.7739	0.7471	0.7208	0.6950	0.6699	0.6456
2.5	0.9179	0.8918	0.8649	0.8376	0.8101	0.7827	0.7556	0.7289	0.7027	0.6771	0.6523
2.6	0.9257	0.9000	0.8733	0.8461	0.8186	0.7910	0.7636	0.7365	0.7099	0.6840	0.6587
2.7	0.9328	0.9076	0.8812	0.8541	0.8265	0.7988	0.7711	0.7437	0.7168	0.6904	0.6647
2.8	0.9392	0.9145	0.8885	0.8615	0.8339	0.8061	0.7782	0.7505	0.7233	0.6965	0.6705
2.9	0.9450	0.9209	0.8952	0.8684	0.8409	0.8130	0.7849	0.7570	0.7294	0.7023	0.6759
3.0	0.9502	0.9267	0.9015	0.8750	0.8475	0.8195	0.7913	0.7632	0.7353	0.7079	0.6811
3.1	0.9550	0.9321	0.9073	0.8811	0.8537	0.8257	0.7974	0.7690	0.7409	0.7132	0.6860
3.2	0.9592	0.9371	0.9128	0.8868	0.8596	0.8316	0.8032	0.7746	0.7462	0.7182	0.6907
3.3	0.9631	0.9416	0.9178	0.8922	0.8651	0.8372	0.8087	0.7800	0.7513	0.7230	0.6952
3.4	0.9666	0.9458	0.9225	0.8972	0.8704	0.8425	0.8139	0.7851	0.7562	0.7276	0.6995
3.5	0.9698	0.9497	0.9269	0.9020	0.8753	0.8475	0.8189	0.7899	0.7609	0.7321	0.7037
3.6	0.9727	0.9533	0.9311	0.9065	0.8801	0.8523	0.8237	0.7946	0.7654	0.7363	0.7077
3.7	0.9753	0.9566	0.9349	0.9107	0.8845	0.8569	0.8283	0.7991	0.7697	0.7404	0.7115
3.8	0.9776	0.9597	0.9385	0.9147	0.8888	0.8613	0.8326	0.8033	0.7738	0.7443	0.7151
3.9	0.9798	0.9625	0.9419	0.9185	0.8928	0.8654	0.8368	0.8075	0.7778	0.7481	0.7187
4.0	0.9817	0.9651	0.9450	0.9220	0.8967	0.8694	0.8408	0.8114	0.7816	0.7517	0.7221
4.1	0.9834	0.9675	0.9480	0.9254	0.9003	0.8732	0.8447	0.8152	0.7853	0.7552	0.7253
4.2	0.9850	0.9697	0.9508	0.9286	0.9038	0.8769	0.8484	0.8189	0.7888	0.7586	0.7285
4.3	0.9864	0.9718	0.9534	0.9317	0.9071	0.8804	0.8520	0.8224	0.7922	0.7618	0.7316
4.4	0.9877	0.9737	0.9559	0.9345	0.9103	0.8837	0.8554	0.8258	0.7956	0.7650	0.7345
4.5	0.9889	0.9755	0.9582	0.9373	0.9133	0.8870	0.8587	0.8291	0.7988	0.7680	0.7374
4.6	0.9899	0.9772	0.9603	0.9399	0.9162	0.8901	0.8619	0.8323	0.8018	0.7710	0.7401
4.7	0.9909	0.9787	0.9624	0.9423	0.9190	0.8930	0.8650	0.8354	0.8048	0.7738	0.7428
4.8	0.9918	0.9801	0.9643	0.9447	0.9217	0.8959	0.8679	0.8383	0.8077	0.7766	0.7454
4.9	0.9926	0.9814	0.9661	0.9469	0.9242	0.8986	0.8708	0.8412	0.8105	0.7793	0.7479
5.0	0.9933	0.9827	0.9678	0.9490	0.9267	0.9013	0.8735	0.8440	0.8133	0.7819	0.7504
5.1	0.9939	0.9838	0.9694	0.9510	0.9290	0.9038	0.8762	0.8467	0.8159	0.7844	0.7527
5.2	0.9945	0.9849	0.9710	0.9530	0.9312	0.9063	0.8788	0.8493	0.8185	0.7869	0.7550
5.3	0.9950	0.9859	0.9724	0.9548	0.9334	0.9087	0.8813	0.8518	0.8210	0.7893	0.7573
5.4	0.9955	0.9868	0.9738	0.9566	0.9354	0.9110	0.8837	0.8543	0.8234	0.7916	0.7595
5.5	0.9959	0.9876	0.9751	0.9582	0.9374	0.9132	0.8860	0.8567	0.8257	0.7938	0.7616
5.6	0.9963	0.9884	0.9763	0.9598	0.9393	0.9153	0.8883	0.8590	0.8280	0.7960	0.7636
5.7	0.9967	0.9892	0.9774	0.9613	0.9412	0.9173	0.8905	0.8612	0.8302	0.7982	0.7657
5.8	0.9970	0.9899	0.9785	0.9628	0.9429	0.9193	0.8926	0.8634	0.8324	0.8003	0.7676
5.9	0.9973	0.9905	0.9795	0.9642	0.9446	0.9213	0.8947	0.8655	0.8345	0.8023	0.7695
6.0	0.9975	0.9911	0.9805	0.9655	0.9463	0.9231	0.8967	0.8676	0.8366	0.8043	0.7714

NTU	Heat Exchanger Effectiveness										
	C^* ($C_{min} = C_{air}$)										
	0	0.1	0.2	0.3	0.4	0.5	0.6	0.7	0.8	0.9	1.0
6.1	0.9978	0.9917	0.9814	0.9668	0.9478	0.9249	0.8986	0.8696	0.8386	0.8062	0.7732
6.2	0.9980	0.9922	0.9823	0.9680	0.9494	0.9267	0.9005	0.8716	0.8405	0.8081	0.7750
6.3	0.9982	0.9927	0.9831	0.9692	0.9508	0.9284	0.9023	0.8735	0.8424	0.8099	0.7767
6.4	0.9983	0.9931	0.9839	0.9703	0.9522	0.9300	0.9041	0.8753	0.8443	0.8117	0.7784
6.5	0.9985	0.9936	0.9847	0.9714	0.9536	0.9316	0.9059	0.8771	0.8461	0.8135	0.7800
6.6	0.9986	0.9940	0.9854	0.9724	0.9549	0.9331	0.9076	0.8789	0.8478	0.8152	0.7816
6.7	0.9988	0.9943	0.9861	0.9734	0.9562	0.9346	0.9092	0.8806	0.8496	0.8169	0.7832
6.8	0.9989	0.9947	0.9867	0.9743	0.9574	0.9360	0.9108	0.8823	0.8513	0.8185	0.7848
6.9	0.9990	0.9950	0.9873	0.9752	0.9586	0.9374	0.9124	0.8839	0.8529	0.8201	0.7863
7.0	0.9991	0.9953	0.9879	0.9761	0.9597	0.9388	0.9139	0.8855	0.8545	0.8217	0.7877
7.1	0.9992	0.9956	0.9884	0.9769	0.9608	0.9401	0.9153	0.8871	0.8561	0.8232	0.7892
7.2	0.9993	0.9959	0.9889	0.9777	0.9619	0.9414	0.9168	0.8886	0.8576	0.8247	0.7906
7.3	0.9993	0.9961	0.9894	0.9785	0.9629	0.9427	0.9182	0.8901	0.8592	0.8262	0.7920
7.4	0.9994	0.9964	0.9899	0.9792	0.9639	0.9439	0.9196	0.8916	0.8606	0.8276	0.7933
7.5	0.9994	0.9966	0.9904	0.9799	0.9648	0.9451	0.9209	0.8930	0.8621	0.8290	0.7947
7.6	0.9995	0.9968	0.9908	0.9806	0.9658	0.9462	0.9222	0.8944	0.8635	0.8304	0.7960
7.7	0.9995	0.9970	0.9912	0.9813	0.9667	0.9473	0.9235	0.8957	0.8649	0.8318	0.7973
7.8	0.9996	0.9972	0.9916	0.9819	0.9675	0.9484	0.9247	0.8971	0.8662	0.8331	0.7985
7.9	0.9996	0.9973	0.9920	0.9825	0.9684	0.9494	0.9259	0.8984	0.8676	0.8344	0.7997
8.0	0.9997	0.9975	0.9923	0.9831	0.9692	0.9505	0.9271	0.8996	0.8689	0.8357	0.8009
8.1	0.9997	0.9976	0.9926	0.9837	0.9700	0.9515	0.9282	0.9009	0.8701	0.8369	0.8021
8.2	0.9997	0.9978	0.9930	0.9842	0.9708	0.9524	0.9294	0.9021	0.8714	0.8381	0.8033
8.3	0.9998	0.9979	0.9933	0.9847	0.9715	0.9534	0.9305	0.9033	0.8726	0.8394	0.8044
8.4	0.9998	0.9980	0.9936	0.9852	0.9722	0.9543	0.9315	0.9045	0.8738	0.8405	0.8055
8.5	0.9998	0.9981	0.9938	0.9857	0.9729	0.9552	0.9326	0.9056	0.8750	0.8417	0.8066
8.6	0.9998	0.9983	0.9941	0.9862	0.9736	0.9561	0.9336	0.9067	0.8762	0.8428	0.8077
8.7	0.9998	0.9984	0.9943	0.9866	0.9743	0.9569	0.9346	0.9078	0.8773	0.8440	0.8088
8.8	0.9998	0.9984	0.9946	0.9870	0.9749	0.9578	0.9356	0.9089	0.8784	0.8451	0.8098
8.9	0.9999	0.9985	0.9948	0.9875	0.9755	0.9586	0.9366	0.9100	0.8795	0.8461	0.8108
9.0	0.9999	0.9986	0.9950	0.9879	0.9761	0.9594	0.9375	0.9110	0.8806	0.8472	0.8118
9.1	0.9999	0.9987	0.9952	0.9882	0.9767	0.9601	0.9384	0.9120	0.8817	0.8483	0.8128
9.2	0.9999	0.9988	0.9954	0.9886	0.9773	0.9609	0.9393	0.9130	0.8827	0.8493	0.8138
9.3	0.9999	0.9989	0.9956	0.9890	0.9778	0.9616	0.9402	0.9140	0.8837	0.8503	0.8148
9.4	0.9999	0.9989	0.9958	0.9893	0.9784	0.9623	0.9411	0.9150	0.8847	0.8513	0.8157
9.5	0.9999	0.9990	0.9960	0.9897	0.9789	0.9630	0.9419	0.9159	0.8857	0.8523	0.8166
9.6	0.9999	0.9990	0.9962	0.9900	0.9794	0.9637	0.9428	0.9168	0.8867	0.8532	0.8175
9.7	0.9999	0.9991	0.9963	0.9903	0.9799	0.9644	0.9436	0.9177	0.8876	0.8542	0.8184
9.8	0.9999	0.9991	0.9965	0.9906	0.9804	0.9650	0.9444	0.9186	0.8886	0.8551	0.8193
9.9	1.0000	0.9992	0.9966	0.9909	0.9808	0.9657	0.9451	0.9195	0.8895	0.8560	0.8202
10.0	1.0000	0.9992	0.9967	0.9912	0.9813	0.9663	0.9459	0.9204	0.8904	0.8569	0.8210

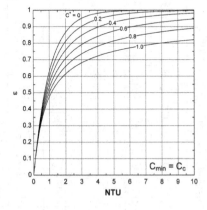

CONFIGURATION 8

1TUBE20ROW1PASS20CIR

(b) $C_{min} = C_h$

NTU	Heat Exchanger Effectiveness										
	C^* ($C_{min} = C_h$)										
	0	0.1	0.2	0.3	0.4	0.5	0.6	0.7	0.8	0.9	1.0
0	0.0000	0.0000	0.0000	0.0000	0.0000	0.0000	0.0000	0.0000	0.0000	0.0000	0.0000
0.1	0.0952	0.0947	0.0943	0.0938	0.0934	0.0929	0.0925	0.0921	0.0916	0.0912	0.0908
0.2	0.1813	0.1796	0.1780	0.1764	0.1749	0.1733	0.1718	0.1703	0.1688	0.1673	0.1658
0.3	0.2592	0.2559	0.2526	0.2494	0.2463	0.2432	0.2402	0.2372	0.2342	0.2313	0.2285
0.4	0.3297	0.3244	0.3192	0.3141	0.3091	0.3042	0.2995	0.2948	0.2902	0.2857	0.2814
0.5	0.3935	0.3860	0.3787	0.3716	0.3646	0.3578	0.3512	0.3448	0.3385	0.3323	0.3263
0.6	0.4512	0.4414	0.4320	0.4228	0.4138	0.4051	0.3966	0.3884	0.3803	0.3725	0.3650
0.7	0.5034	0.4914	0.4798	0.4685	0.4576	0.4470	0.4367	0.4267	0.4170	0.4076	0.3984
0.8	0.5507	0.5365	0.5228	0.5095	0.4967	0.4842	0.4722	0.4605	0.4492	0.4383	0.4277
0.9	0.5934	0.5772	0.5616	0.5464	0.5317	0.5176	0.5039	0.4906	0.4778	0.4654	0.4534
1.0	0.6321	0.6140	0.5965	0.5796	0.5633	0.5475	0.5322	0.5175	0.5032	0.4895	0.4762
1.1	0.6671	0.6473	0.6282	0.6096	0.5917	0.5745	0.5578	0.5416	0.5261	0.5111	0.4966
1.2	0.6988	0.6775	0.6568	0.6368	0.6175	0.5989	0.5808	0.5634	0.5467	0.5305	0.5149
1.3	0.7275	0.7048	0.6828	0.6615	0.6409	0.6210	0.6018	0.5832	0.5653	0.5480	0.5314
1.4	0.7534	0.7296	0.7064	0.6840	0.6623	0.6412	0.6209	0.6013	0.5823	0.5640	0.5464
1.5	0.7769	0.7521	0.7280	0.7045	0.6818	0.6597	0.6384	0.6178	0.5978	0.5786	0.5601
1.6	0.7981	0.7725	0.7476	0.7233	0.6996	0.6767	0.6544	0.6329	0.6121	0.5920	0.5727
1.7	0.8173	0.7912	0.7655	0.7405	0.7160	0.6923	0.6692	0.6469	0.6253	0.6044	0.5843
1.8	0.8347	0.8081	0.7819	0.7562	0.7311	0.7067	0.6829	0.6598	0.6374	0.6158	0.5950
1.9	0.8504	0.8235	0.7970	0.7708	0.7451	0.7200	0.6956	0.6718	0.6487	0.6264	0.6049
2.0	0.8647	0.8376	0.8107	0.7842	0.7580	0.7324	0.7073	0.6829	0.6593	0.6363	0.6141
2.1	0.8775	0.8505	0.8234	0.7965	0.7700	0.7439	0.7183	0.6934	0.6691	0.6456	0.6228
2.2	0.8892	0.8622	0.8351	0.8080	0.7811	0.7546	0.7286	0.7031	0.6783	0.6542	0.6309
2.3	0.8997	0.8730	0.8458	0.8186	0.7915	0.7646	0.7382	0.7122	0.6869	0.6623	0.6385
2.4	0.9093	0.8828	0.8558	0.8285	0.8011	0.7740	0.7472	0.7208	0.6950	0.6700	0.6456
2.5	0.9179	0.8918	0.8649	0.8376	0.8102	0.7828	0.7556	0.7289	0.7027	0.6772	0.6523
2.6	0.9257	0.9000	0.8734	0.8462	0.8186	0.7910	0.7636	0.7365	0.7099	0.6840	0.6587
2.7	0.9328	0.9076	0.8812	0.8541	0.8266	0.7988	0.7712	0.7438	0.7168	0.6904	0.6647
2.8	0.9392	0.9145	0.8885	0.8616	0.8340	0.8062	0.7783	0.7506	0.7233	0.6965	0.6705
2.9	0.9450	0.9209	0.8953	0.8685	0.8410	0.8131	0.7850	0.7571	0.7295	0.7024	0.6759
3.0	0.9502	0.9268	0.9016	0.8750	0.8476	0.8196	0.7914	0.7632	0.7353	0.7079	0.6811
3.1	0.9550	0.9322	0.9074	0.8811	0.8538	0.8258	0.7975	0.7691	0.7409	0.7132	0.6860
3.2	0.9592	0.9371	0.9128	0.8869	0.8597	0.8317	0.8033	0.7747	0.7463	0.7182	0.6907
3.3	0.9631	0.9417	0.9179	0.8922	0.8652	0.8373	0.8088	0.7800	0.7514	0.7230	0.6952
3.4	0.9666	0.9459	0.9226	0.8973	0.8705	0.8426	0.8140	0.7851	0.7563	0.7277	0.6995
3.5	0.9698	0.9498	0.9270	0.9021	0.8755	0.8476	0.8190	0.7900	0.7609	0.7321	0.7037
3.6	0.9727	0.9534	0.9311	0.9066	0.8802	0.8524	0.8238	0.7947	0.7654	0.7363	0.7077
3.7	0.9753	0.9567	0.9350	0.9108	0.8846	0.8570	0.8284	0.7991	0.7697	0.7404	0.7115
3.8	0.9776	0.9597	0.9386	0.9148	0.8889	0.8614	0.8327	0.8034	0.7739	0.7443	0.7151
3.9	0.9798	0.9625	0.9420	0.9186	0.8929	0.8656	0.8369	0.8076	0.7778	0.7481	0.7187
4.0	0.9817	0.9651	0.9451	0.9222	0.8968	0.8695	0.8410	0.8115	0.7817	0.7517	0.7221
4.1	0.9834	0.9676	0.9481	0.9255	0.9005	0.8734	0.8448	0.8153	0.7853	0.7552	0.7253
4.2	0.9850	0.9698	0.9509	0.9288	0.9039	0.8770	0.8485	0.8190	0.7889	0.7586	0.7285
4.3	0.9864	0.9719	0.9535	0.9318	0.9073	0.8805	0.8521	0.8225	0.7923	0.7619	0.7316
4.4	0.9877	0.9738	0.9560	0.9347	0.9105	0.8839	0.8555	0.8260	0.7956	0.7650	0.7345
4.5	0.9889	0.9756	0.9583	0.9374	0.9135	0.8871	0.8589	0.8293	0.7988	0.7681	0.7374
4.6	0.9899	0.9772	0.9604	0.9400	0.9164	0.8902	0.8620	0.8324	0.8019	0.7710	0.7401
4.7	0.9909	0.9787	0.9625	0.9425	0.9192	0.8932	0.8651	0.8355	0.8049	0.7739	0.7428
4.8	0.9918	0.9802	0.9644	0.9448	0.9218	0.8961	0.8681	0.8385	0.8078	0.7767	0.7454
4.9	0.9926	0.9815	0.9662	0.9470	0.9244	0.8988	0.8709	0.8414	0.8106	0.7793	0.7479
5.0	0.9933	0.9827	0.9679	0.9492	0.9268	0.9015	0.8737	0.8441	0.8134	0.7820	0.7504
5.1	0.9939	0.9839	0.9696	0.9512	0.9292	0.9040	0.8764	0.8468	0.8160	0.7845	0.7527
5.2	0.9945	0.9849	0.9711	0.9531	0.9314	0.9065	0.8790	0.8494	0.8186	0.7869	0.7550
5.3	0.9950	0.9859	0.9725	0.9550	0.9336	0.9089	0.8815	0.8520	0.8211	0.7893	0.7573
5.4	0.9955	0.9868	0.9739	0.9567	0.9356	0.9112	0.8839	0.8544	0.8235	0.7917	0.7595
5.5	0.9959	0.9877	0.9752	0.9584	0.9376	0.9134	0.8862	0.8568	0.8258	0.7939	0.7616
5.6	0.9963	0.9885	0.9764	0.9600	0.9395	0.9155	0.8885	0.8591	0.8281	0.7961	0.7636
5.7	0.9967	0.9892	0.9775	0.9615	0.9414	0.9176	0.8907	0.8614	0.8304	0.7983	0.7657
5.8	0.9970	0.9899	0.9786	0.9630	0.9431	0.9196	0.8928	0.8636	0.8325	0.8003	0.7676
5.9	0.9973	0.9906	0.9796	0.9644	0.9448	0.9215	0.8949	0.8657	0.8346	0.8024	0.7695
6.0	0.9975	0.9912	0.9806	0.9657	0.9465	0.9234	0.8969	0.8678	0.8367	0.8044	0.7714

NTU	Heat Exchanger Effectiveness										
	$C^* (C_{min} = C_h)$										
	0	0.1	0.2	0.3	0.4	0.5	0.6	0.7	0.8	0.9	1.0
6.1	0.9978	0.9917	0.9815	0.9670	0.9481	0.9252	0.8989	0.8698	0.8387	0.8063	0.7732
6.2	0.9980	0.9922	0.9824	0.9682	0.9496	0.9269	0.9008	0.8718	0.8407	0.8082	0.7750
6.3	0.9982	0.9927	0.9833	0.9694	0.9510	0.9286	0.9026	0.8737	0.8426	0.8100	0.7767
6.4	0.9983	0.9932	0.9840	0.9705	0.9525	0.9303	0.9044	0.8755	0.8444	0.8118	0.7784
6.5	0.9985	0.9936	0.9848	0.9715	0.9538	0.9318	0.9061	0.8773	0.8462	0.8136	0.7800
6.6	0.9986	0.9940	0.9855	0.9726	0.9551	0.9334	0.9078	0.8791	0.8480	0.8153	0.7816
6.7	0.9988	0.9944	0.9862	0.9736	0.9564	0.9349	0.9095	0.8808	0.8497	0.8170	0.7832
6.8	0.9989	0.9947	0.9868	0.9745	0.9576	0.9363	0.9111	0.8825	0.8514	0.8186	0.7848
6.9	0.9990	0.9951	0.9874	0.9754	0.9588	0.9377	0.9126	0.8842	0.8531	0.8202	0.7863
7.0	0.9991	0.9954	0.9880	0.9763	0.9599	0.9391	0.9142	0.8858	0.8547	0.8218	0.7877
7.1	0.9992	0.9956	0.9885	0.9771	0.9611	0.9404	0.9156	0.8873	0.8563	0.8233	0.7892
7.2	0.9993	0.9959	0.9891	0.9779	0.9621	0.9417	0.9171	0.8889	0.8578	0.8248	0.7906
7.3	0.9993	0.9962	0.9896	0.9787	0.9631	0.9430	0.9185	0.8904	0.8593	0.8263	0.7920
7.4	0.9994	0.9964	0.9900	0.9794	0.9641	0.9442	0.9199	0.8918	0.8608	0.8277	0.7933
7.5	0.9994	0.9966	0.9905	0.9801	0.9651	0.9454	0.9212	0.8932	0.8623	0.8291	0.7947
7.6	0.9995	0.9968	0.9909	0.9808	0.9660	0.9465	0.9225	0.8946	0.8637	0.8305	0.7960
7.7	0.9995	0.9970	0.9913	0.9815	0.9669	0.9476	0.9238	0.8960	0.8651	0.8319	0.7973
7.8	0.9996	0.9972	0.9917	0.9821	0.9678	0.9487	0.9250	0.8973	0.8664	0.8332	0.7985
7.9	0.9996	0.9974	0.9921	0.9827	0.9687	0.9498	0.9262	0.8987	0.8678	0.8345	0.7997
8.0	0.9997	0.9975	0.9924	0.9833	0.9695	0.9508	0.9274	0.8999	0.8691	0.8358	0.8009
8.1	0.9997	0.9977	0.9927	0.9838	0.9703	0.9518	0.9286	0.9012	0.8704	0.8370	0.8021
8.2	0.9997	0.9978	0.9931	0.9844	0.9711	0.9528	0.9297	0.9024	0.8716	0.8383	0.8033
8.3	0.9998	0.9979	0.9934	0.9849	0.9718	0.9537	0.9308	0.9036	0.8729	0.8395	0.8044
8.4	0.9998	0.9981	0.9937	0.9854	0.9725	0.9547	0.9319	0.9048	0.8741	0.8407	0.8055
8.5	0.9998	0.9982	0.9939	0.9859	0.9732	0.9556	0.9330	0.9059	0.8753	0.8418	0.8066
8.6	0.9998	0.9983	0.9942	0.9864	0.9739	0.9564	0.9340	0.9071	0.8764	0.8430	0.8077
8.7	0.9998	0.9984	0.9944	0.9868	0.9746	0.9573	0.9350	0.9082	0.8776	0.8441	0.8088
8.8	0.9998	0.9985	0.9947	0.9872	0.9752	0.9581	0.9360	0.9093	0.8787	0.8452	0.8098
8.9	0.9999	0.9986	0.9949	0.9876	0.9758	0.9589	0.9370	0.9103	0.8798	0.8463	0.8108
9.0	0.9999	0.9987	0.9951	0.9880	0.9764	0.9597	0.9379	0.9114	0.8809	0.8474	0.8118
9.1	0.9999	0.9987	0.9953	0.9884	0.9770	0.9605	0.9388	0.9124	0.8819	0.8484	0.8128
9.2	0.9999	0.9988	0.9955	0.9888	0.9776	0.9613	0.9397	0.9134	0.8830	0.8494	0.8138
9.3	0.9999	0.9989	0.9957	0.9892	0.9781	0.9620	0.9406	0.9144	0.8840	0.8504	0.8148
9.4	0.9999	0.9989	0.9959	0.9895	0.9787	0.9627	0.9415	0.9154	0.8850	0.8514	0.8157
9.5	0.9999	0.9990	0.9961	0.9898	0.9792	0.9634	0.9423	0.9163	0.8860	0.8524	0.8166
9.6	0.9999	0.9991	0.9962	0.9902	0.9797	0.9641	0.9432	0.9172	0.8870	0.8534	0.8175
9.7	0.9999	0.9991	0.9964	0.9905	0.9802	0.9648	0.9440	0.9181	0.8879	0.8543	0.8184
9.8	0.9999	0.9992	0.9965	0.9908	0.9807	0.9654	0.9448	0.9190	0.8889	0.8553	0.8193
9.9	1.0000	0.9992	0.9967	0.9911	0.9811	0.9661	0.9456	0.9199	0.8898	0.8562	0.8202
10.0	1.0000	0.9993	0.9968	0.9913	0.9816	0.9667	0.9463	0.9208	0.8907	0.8571	0.8210

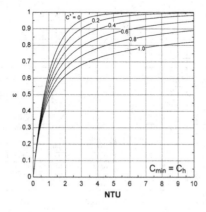

CONFIGURATION 9

1TUBE50ROW1PASS50CIR

(a) $C_{min} = C_c$

NTU	Heat Exchanger Effectiveness										
	C^* ($C_{min} = C_c$)										
	0	0.1	0.2	0.3	0.4	0.5	0.6	0.7	0.8	0.9	1.0
0	0.0000	0.0000	0.0000	0.0000	0.0000	0.0000	0.0000	0.0000	0.0000	0.0000	0.0000
0.1	0.0952	0.0947	0.0943	0.0938	0.0934	0.0929	0.0925	0.0921	0.0916	0.0912	0.0908
0.2	0.1813	0.1796	0.1780	0.1764	0.1749	0.1733	0.1718	0.1703	0.1688	0.1673	0.1658
0.3	0.2592	0.2559	0.2526	0.2494	0.2463	0.2432	0.2402	0.2372	0.2342	0.2313	0.2285
0.4	0.3297	0.3244	0.3192	0.3141	0.3091	0.3042	0.2995	0.2948	0.2902	0.2857	0.2814
0.5	0.3935	0.3860	0.3787	0.3716	0.3646	0.3578	0.3512	0.3448	0.3385	0.3323	0.3263
0.6	0.4512	0.4414	0.4320	0.4228	0.4138	0.4051	0.3966	0.3884	0.3804	0.3725	0.3650
0.7	0.5034	0.4914	0.4798	0.4685	0.4576	0.4470	0.4367	0.4267	0.4170	0.4076	0.3984
0.8	0.5507	0.5365	0.5228	0.5095	0.4967	0.4842	0.4722	0.4605	0.4492	0.4383	0.4277
0.9	0.5934	0.5772	0.5616	0.5464	0.5317	0.5176	0.5039	0.4906	0.4778	0.4654	0.4534
1.0	0.6321	0.6140	0.5965	0.5796	0.5633	0.5475	0.5322	0.5175	0.5032	0.4895	0.4762
1.1	0.6671	0.6473	0.6282	0.6096	0.5917	0.5745	0.5578	0.5416	0.5261	0.5111	0.4966
1.2	0.6988	0.6775	0.6568	0.6368	0.6175	0.5989	0.5808	0.5635	0.5467	0.5305	0.5149
1.3	0.7275	0.7048	0.6828	0.6615	0.6409	0.6210	0.6018	0.5833	0.5653	0.5481	0.5314
1.4	0.7534	0.7296	0.7064	0.6840	0.6623	0.6412	0.6209	0.6013	0.5823	0.5641	0.5465
1.5	0.7769	0.7521	0.7280	0.7045	0.6818	0.6597	0.6384	0.6178	0.5979	0.5787	0.5602
1.6	0.7981	0.7725	0.7476	0.7233	0.6996	0.6767	0.6545	0.6329	0.6122	0.5921	0.5727
1.7	0.8173	0.7912	0.7655	0.7405	0.7160	0.6923	0.6692	0.6469	0.6253	0.6045	0.5843
1.8	0.8347	0.8081	0.7819	0.7562	0.7312	0.7067	0.6829	0.6598	0.6375	0.6159	0.5950
1.9	0.8504	0.8235	0.7970	0.7708	0.7451	0.7200	0.6956	0.6718	0.6488	0.6265	0.6050
2.0	0.8647	0.8376	0.8107	0.7842	0.7580	0.7324	0.7074	0.6830	0.6593	0.6364	0.6142
2.1	0.8775	0.8505	0.8234	0.7966	0.7700	0.7439	0.7183	0.6934	0.6692	0.6456	0.6229
2.2	0.8892	0.8622	0.8351	0.8080	0.7811	0.7546	0.7286	0.7032	0.6784	0.6543	0.6310
2.3	0.8997	0.8730	0.8458	0.8186	0.7915	0.7646	0.7382	0.7123	0.6870	0.6624	0.6386
2.4	0.9093	0.8828	0.8558	0.8285	0.8012	0.7740	0.7472	0.7209	0.6951	0.6701	0.6457
2.5	0.9179	0.8918	0.8649	0.8376	0.8102	0.7828	0.7557	0.7290	0.7028	0.6773	0.6525
2.6	0.9257	0.9000	0.8734	0.8462	0.8187	0.7911	0.7637	0.7366	0.7101	0.6841	0.6588
2.7	0.9328	0.9076	0.8813	0.8541	0.8266	0.7989	0.7712	0.7438	0.7169	0.6906	0.6649
2.8	0.9392	0.9145	0.8885	0.8616	0.8340	0.8062	0.7783	0.7507	0.7234	0.6967	0.6706
2.9	0.9450	0.9209	0.8953	0.8685	0.8410	0.8131	0.7851	0.7572	0.7296	0.7025	0.6761
3.0	0.9502	0.9268	0.9016	0.8750	0.8476	0.8197	0.7915	0.7634	0.7355	0.7081	0.6813
3.1	0.9550	0.9322	0.9074	0.8812	0.8539	0.8259	0.7976	0.7692	0.7411	0.7134	0.6862
3.2	0.9592	0.9371	0.9128	0.8869	0.8597	0.8318	0.8034	0.7748	0.7464	0.7184	0.6909
3.3	0.9631	0.9417	0.9179	0.8923	0.8653	0.8373	0.8089	0.7802	0.7515	0.7232	0.6955
3.4	0.9666	0.9459	0.9226	0.8973	0.8705	0.8427	0.8141	0.7853	0.7564	0.7279	0.6998
3.5	0.9698	0.9498	0.9270	0.9021	0.8755	0.8477	0.8191	0.7902	0.7611	0.7323	0.7039
3.6	0.9727	0.9533	0.9311	0.9066	0.8802	0.8525	0.8239	0.7948	0.7656	0.7366	0.7079
3.7	0.9753	0.9567	0.9350	0.9108	0.8847	0.8571	0.8285	0.7993	0.7699	0.7406	0.7117
3.8	0.9776	0.9597	0.9386	0.9148	0.8889	0.8615	0.8329	0.8036	0.7741	0.7446	0.7154
3.9	0.9798	0.9625	0.9420	0.9186	0.8930	0.8656	0.8371	0.8077	0.7780	0.7484	0.7190
4.0	0.9817	0.9651	0.9451	0.9222	0.8968	0.8696	0.8411	0.8117	0.7819	0.7520	0.7224
4.1	0.9834	0.9676	0.9481	0.9256	0.9005	0.8735	0.8450	0.8155	0.7856	0.7555	0.7257
4.2	0.9850	0.9698	0.9509	0.9288	0.9040	0.8771	0.8487	0.8192	0.7891	0.7589	0.7288
4.3	0.9864	0.9719	0.9535	0.9318	0.9073	0.8806	0.8523	0.8228	0.7926	0.7622	0.7319
4.4	0.9877	0.9738	0.9560	0.9347	0.9105	0.8840	0.8557	0.8262	0.7959	0.7653	0.7349
4.5	0.9889	0.9756	0.9583	0.9374	0.9136	0.8872	0.8590	0.8295	0.7991	0.7684	0.7377
4.6	0.9899	0.9772	0.9604	0.9400	0.9165	0.8903	0.8622	0.8327	0.8022	0.7714	0.7405
4.7	0.9909	0.9787	0.9625	0.9425	0.9193	0.8933	0.8653	0.8357	0.8052	0.7742	0.7432
4.8	0.9918	0.9802	0.9644	0.9448	0.9219	0.8962	0.8683	0.8387	0.8081	0.7770	0.7458
4.9	0.9926	0.9815	0.9662	0.9471	0.9245	0.8989	0.8711	0.8416	0.8110	0.7797	0.7484
5.0	0.9933	0.9827	0.9679	0.9492	0.9269	0.9016	0.8739	0.8444	0.8137	0.7824	0.7508
5.1	0.9939	0.9839	0.9696	0.9512	0.9293	0.9042	0.8766	0.8471	0.8164	0.7849	0.7532
5.2	0.9945	0.9849	0.9711	0.9532	0.9315	0.9066	0.8792	0.8497	0.8189	0.7874	0.7555
5.3	0.9950	0.9859	0.9725	0.9550	0.9337	0.9090	0.8817	0.8523	0.8214	0.7898	0.7578
5.4	0.9955	0.9868	0.9739	0.9567	0.9357	0.9113	0.8841	0.8547	0.8239	0.7921	0.7600
5.5	0.9959	0.9877	0.9752	0.9584	0.9377	0.9135	0.8864	0.8571	0.8262	0.7944	0.7621
5.6	0.9963	0.9885	0.9764	0.9600	0.9396	0.9157	0.8887	0.8595	0.8285	0.7966	0.7642
5.7	0.9967	0.9892	0.9775	0.9615	0.9415	0.9177	0.8909	0.8617	0.8308	0.7987	0.7662
5.8	0.9970	0.9899	0.9786	0.9630	0.9432	0.9197	0.8931	0.8639	0.8330	0.8009	0.7682
5.9	0.9973	0.9905	0.9797	0.9644	0.9449	0.9217	0.8951	0.8661	0.8351	0.8029	0.7701
6.0	0.9975	0.9911	0.9806	0.9657	0.9466	0.9235	0.8972	0.8681	0.8372	0.8049	0.7720

NTU	Heat Exchanger Effectiveness										
	C^* ($C_{min} = C_c$)										
	0	0.1	0.2	0.3	0.4	0.5	0.6	0.7	0.8	0.9	1.0
6.1	0.9978	0.9917	0.9816	0.9670	0.9482	0.9253	0.8991	0.8702	0.8392	0.8068	0.7738
6.2	0.9980	0.9922	0.9824	0.9682	0.9497	0.9271	0.9010	0.8721	0.8411	0.8087	0.7756
6.3	0.9982	0.9927	0.9833	0.9694	0.9511	0.9288	0.9029	0.8741	0.8431	0.8106	0.7774
6.4	0.9983	0.9932	0.9840	0.9705	0.9526	0.9304	0.9047	0.8759	0.8449	0.8124	0.7791
6.5	0.9985	0.9936	0.9848	0.9716	0.9539	0.9320	0.9064	0.8778	0.8468	0.8142	0.7807
6.6	0.9986	0.9940	0.9855	0.9726	0.9552	0.9336	0.9081	0.8795	0.8485	0.8159	0.7824
6.7	0.9988	0.9944	0.9862	0.9736	0.9565	0.9351	0.9098	0.8813	0.8503	0.8176	0.7840
6.8	0.9989	0.9947	0.9868	0.9745	0.9577	0.9365	0.9114	0.8830	0.8520	0.8193	0.7855
6.9	0.9990	0.9951	0.9874	0.9754	0.9589	0.9379	0.9130	0.8846	0.8537	0.8209	0.7870
7.0	0.9991	0.9954	0.9880	0.9763	0.9601	0.9393	0.9145	0.8862	0.8553	0.8225	0.7885
7.1	0.9992	0.9956	0.9885	0.9772	0.9612	0.9406	0.9160	0.8878	0.8569	0.8240	0.7900
7.2	0.9993	0.9959	0.9891	0.9780	0.9622	0.9419	0.9174	0.8893	0.8584	0.8255	0.7914
7.3	0.9993	0.9962	0.9896	0.9787	0.9633	0.9432	0.9188	0.8908	0.8600	0.8270	0.7928
7.4	0.9994	0.9964	0.9900	0.9795	0.9643	0.9444	0.9202	0.8923	0.8615	0.8285	0.7942
7.5	0.9994	0.9966	0.9905	0.9802	0.9652	0.9456	0.9216	0.8937	0.8629	0.8299	0.7956
7.6	0.9995	0.9968	0.9909	0.9809	0.9662	0.9467	0.9229	0.8952	0.8644	0.8313	0.7969
7.7	0.9995	0.9970	0.9913	0.9815	0.9671	0.9478	0.9241	0.8965	0.8658	0.8327	0.7982
7.8	0.9996	0.9972	0.9917	0.9821	0.9679	0.9489	0.9254	0.8979	0.8671	0.8340	0.7994
7.9	0.9996	0.9974	0.9921	0.9827	0.9688	0.9500	0.9266	0.8992	0.8685	0.8354	0.8007
8.0	0.9997	0.9975	0.9924	0.9833	0.9696	0.9510	0.9278	0.9005	0.8698	0.8366	0.8019
8.1	0.9997	0.9977	0.9927	0.9839	0.9704	0.9520	0.9290	0.9017	0.8711	0.8379	0.8031
8.2	0.9997	0.9978	0.9931	0.9844	0.9712	0.9530	0.9301	0.9030	0.8724	0.8392	0.8043
8.3	0.9998	0.9979	0.9934	0.9849	0.9719	0.9540	0.9312	0.9042	0.8736	0.8404	0.8055
8.4	0.9998	0.9981	0.9937	0.9854	0.9726	0.9549	0.9323	0.9054	0.8748	0.8416	0.8066
8.5	0.9998	0.9982	0.9939	0.9859	0.9733	0.9558	0.9334	0.9065	0.8760	0.8428	0.8077
8.6	0.9998	0.9983	0.9942	0.9864	0.9740	0.9567	0.9344	0.9077	0.8772	0.8439	0.8088
8.7	0.9998	0.9984	0.9944	0.9868	0.9747	0.9575	0.9354	0.9088	0.8784	0.8451	0.8099
8.8	0.9998	0.9985	0.9947	0.9873	0.9753	0.9584	0.9364	0.9099	0.8795	0.8462	0.8110
8.9	0.9999	0.9986	0.9949	0.9877	0.9759	0.9592	0.9374	0.9110	0.8806	0.8473	0.8120
9.0	0.9999	0.9987	0.9951	0.9881	0.9766	0.9600	0.9383	0.9120	0.8817	0.8484	0.8130
9.1	0.9999	0.9987	0.9953	0.9885	0.9771	0.9608	0.9393	0.9130	0.8828	0.8495	0.8140
9.2	0.9999	0.9988	0.9955	0.9888	0.9777	0.9615	0.9402	0.9141	0.8839	0.8505	0.8150
9.3	0.9999	0.9989	0.9957	0.9892	0.9783	0.9623	0.9411	0.9151	0.8849	0.8515	0.8160
9.4	0.9999	0.9989	0.9959	0.9896	0.9788	0.9630	0.9420	0.9160	0.8859	0.8526	0.8170
9.5	0.9999	0.9990	0.9961	0.9899	0.9793	0.9637	0.9428	0.9170	0.8869	0.8536	0.8179
9.6	0.9999	0.9991	0.9962	0.9902	0.9798	0.9644	0.9436	0.9179	0.8879	0.8545	0.8189
9.7	0.9999	0.9991	0.9964	0.9905	0.9803	0.9650	0.9445	0.9189	0.8889	0.8555	0.8198
9.8	0.9999	0.9992	0.9966	0.9908	0.9808	0.9657	0.9453	0.9198	0.8899	0.8565	0.8207
9.9	1.0000	0.9992	0.9967	0.9911	0.9813	0.9663	0.9461	0.9207	0.8908	0.8574	0.8216
10.0	1.0000	0.9993	0.9968	0.9914	0.9817	0.9670	0.9468	0.9215	0.8917	0.8583	0.8224

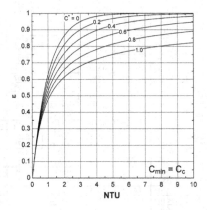

CONFIGURATION 9

1TUBE50ROW1PASS50CIR

(b) $C_{min} = C_h$

NTU	Heat Exchanger Effectiveness										
	$C^* (C_{min} = C_h)$										
	0	0.1	0.2	0.3	0.4	0.5	0.6	0.7	0.8	0.9	1.0
0	0.0000	0.0000	0.0000	0.0000	0.0000	0.0000	0.0000	0.0000	0.0000	0.0000	0.0000
0.1	0.0952	0.0947	0.0943	0.0938	0.0934	0.0929	0.0925	0.0921	0.0916	0.0912	0.0908
0.2	0.1813	0.1796	0.1780	0.1764	0.1749	0.1733	0.1718	0.1703	0.1688	0.1673	0.1658
0.3	0.2592	0.2559	0.2526	0.2494	0.2463	0.2432	0.2402	0.2372	0.2342	0.2313	0.2285
0.4	0.3297	0.3244	0.3192	0.3141	0.3091	0.3042	0.2995	0.2948	0.2902	0.2857	0.2814
0.5	0.3935	0.3860	0.3787	0.3716	0.3646	0.3578	0.3512	0.3448	0.3385	0.3323	0.3263
0.6	0.4512	0.4414	0.4320	0.4228	0.4138	0.4051	0.3966	0.3884	0.3804	0.3725	0.3650
0.7	0.5034	0.4914	0.4798	0.4685	0.4576	0.4470	0.4367	0.4267	0.4170	0.4076	0.3984
0.8	0.5507	0.5365	0.5228	0.5095	0.4967	0.4842	0.4722	0.4605	0.4492	0.4383	0.4277
0.9	0.5934	0.5772	0.5616	0.5464	0.5317	0.5176	0.5039	0.4906	0.4778	0.4654	0.4534
1.0	0.6321	0.6140	0.5965	0.5796	0.5633	0.5475	0.5322	0.5175	0.5032	0.4895	0.4762
1.1	0.6671	0.6473	0.6282	0.6096	0.5917	0.5745	0.5578	0.5416	0.5261	0.5111	0.4966
1.2	0.6988	0.6775	0.6568	0.6368	0.6175	0.5989	0.5809	0.5635	0.5467	0.5305	0.5149
1.3	0.7275	0.7048	0.6828	0.6615	0.6409	0.6210	0.6018	0.5833	0.5653	0.5481	0.5314
1.4	0.7534	0.7296	0.7064	0.6840	0.6623	0.6412	0.6209	0.6013	0.5823	0.5641	0.5465
1.5	0.7769	0.7521	0.7280	0.7045	0.6818	0.6597	0.6384	0.6178	0.5979	0.5787	0.5602
1.6	0.7981	0.7725	0.7476	0.7233	0.6996	0.6767	0.6545	0.6329	0.6122	0.5921	0.5727
1.7	0.8173	0.7912	0.7655	0.7405	0.7160	0.6923	0.6692	0.6469	0.6253	0.6045	0.5843
1.8	0.8347	0.8081	0.7819	0.7563	0.7312	0.7067	0.6829	0.6598	0.6375	0.6159	0.5950
1.9	0.8504	0.8235	0.7970	0.7708	0.7451	0.7200	0.6956	0.6718	0.6488	0.6265	0.6050
2.0	0.8647	0.8376	0.8108	0.7842	0.7580	0.7324	0.7074	0.6830	0.6593	0.6364	0.6142
2.1	0.8775	0.8505	0.8234	0.7966	0.7700	0.7439	0.7183	0.6934	0.6692	0.6456	0.6229
2.2	0.8892	0.8622	0.8351	0.8080	0.7811	0.7546	0.7286	0.7032	0.6784	0.6543	0.6310
2.3	0.8997	0.8730	0.8458	0.8186	0.7915	0.7647	0.7382	0.7123	0.6870	0.6624	0.6386
2.4	0.9093	0.8828	0.8558	0.8285	0.8012	0.7740	0.7472	0.7209	0.6951	0.6701	0.6457
2.5	0.9179	0.8918	0.8649	0.8377	0.8102	0.7828	0.7557	0.7290	0.7028	0.6773	0.6525
2.6	0.9257	0.9000	0.8734	0.8462	0.8187	0.7911	0.7637	0.7366	0.7101	0.6841	0.6588
2.7	0.9328	0.9076	0.8813	0.8542	0.8266	0.7989	0.7712	0.7439	0.7169	0.6906	0.6649
2.8	0.9392	0.9145	0.8885	0.8616	0.8340	0.8062	0.7784	0.7507	0.7234	0.6967	0.6706
2.9	0.9450	0.9209	0.8953	0.8685	0.8411	0.8131	0.7851	0.7572	0.7296	0.7025	0.6761
3.0	0.9502	0.9268	0.9016	0.8751	0.8477	0.8197	0.7915	0.7634	0.7355	0.7081	0.6813
3.1	0.9550	0.9322	0.9074	0.8812	0.8539	0.8259	0.7976	0.7692	0.7411	0.7134	0.6862
3.2	0.9592	0.9371	0.9128	0.8869	0.8597	0.8318	0.8034	0.7748	0.7464	0.7184	0.6909
3.3	0.9631	0.9417	0.9179	0.8923	0.8653	0.8374	0.8089	0.7802	0.7516	0.7232	0.6955
3.4	0.9666	0.9459	0.9226	0.8973	0.8705	0.8427	0.8141	0.7853	0.7564	0.7279	0.6998
3.5	0.9698	0.9498	0.9270	0.9021	0.8755	0.8477	0.8191	0.7902	0.7611	0.7323	0.7039
3.6	0.9727	0.9534	0.9312	0.9066	0.8802	0.8525	0.8239	0.7948	0.7656	0.7366	0.7079
3.7	0.9753	0.9567	0.9350	0.9108	0.8847	0.8571	0.8285	0.7993	0.7699	0.7406	0.7117
3.8	0.9776	0.9597	0.9386	0.9149	0.8890	0.8615	0.8329	0.8036	0.7741	0.7446	0.7154
3.9	0.9798	0.9625	0.9420	0.9186	0.8930	0.8657	0.8371	0.8078	0.7781	0.7484	0.7190
4.0	0.9817	0.9651	0.9452	0.9222	0.8969	0.8697	0.8411	0.8117	0.7819	0.7520	0.7224
4.1	0.9834	0.9676	0.9481	0.9256	0.9005	0.8735	0.8450	0.8155	0.7856	0.7555	0.7257
4.2	0.9850	0.9698	0.9509	0.9288	0.9040	0.8772	0.8487	0.8192	0.7892	0.7589	0.7288
4.3	0.9864	0.9719	0.9535	0.9318	0.9074	0.8807	0.8523	0.8228	0.7926	0.7622	0.7319
4.4	0.9877	0.9738	0.9560	0.9347	0.9105	0.8840	0.8557	0.8262	0.7959	0.7654	0.7349
4.5	0.9889	0.9756	0.9583	0.9375	0.9136	0.8873	0.8590	0.8295	0.7991	0.7684	0.7377
4.6	0.9899	0.9772	0.9605	0.9401	0.9165	0.8904	0.8622	0.8327	0.8022	0.7714	0.7405
4.7	0.9909	0.9787	0.9625	0.9425	0.9193	0.8934	0.8653	0.8358	0.8053	0.7743	0.7432
4.8	0.9918	0.9802	0.9644	0.9449	0.9219	0.8962	0.8683	0.8387	0.8082	0.7770	0.7458
4.9	0.9926	0.9815	0.9662	0.9471	0.9245	0.8990	0.8712	0.8416	0.8110	0.7797	0.7484
5.0	0.9933	0.9827	0.9680	0.9492	0.9269	0.9016	0.8739	0.8444	0.8137	0.7824	0.7508
5.1	0.9939	0.9839	0.9696	0.9513	0.9293	0.9042	0.8766	0.8471	0.8164	0.7849	0.7532
5.2	0.9945	0.9849	0.9711	0.9532	0.9315	0.9067	0.8792	0.8498	0.8190	0.7874	0.7555
5.3	0.9950	0.9859	0.9725	0.9550	0.9337	0.9090	0.8817	0.8523	0.8215	0.7898	0.7578
5.4	0.9955	0.9868	0.9739	0.9568	0.9358	0.9113	0.8841	0.8548	0.8239	0.7921	0.7600
5.5	0.9959	0.9877	0.9752	0.9584	0.9377	0.9136	0.8865	0.8572	0.8263	0.7944	0.7621
5.6	0.9963	0.9885	0.9764	0.9600	0.9397	0.9157	0.8888	0.8595	0.8286	0.7966	0.7642
5.7	0.9967	0.9892	0.9776	0.9616	0.9415	0.9178	0.8910	0.8618	0.8308	0.7988	0.7662
5.8	0.9970	0.9899	0.9786	0.9630	0.9433	0.9198	0.8931	0.8640	0.8330	0.8009	0.7682
5.9	0.9973	0.9906	0.9797	0.9644	0.9450	0.9217	0.8952	0.8661	0.8351	0.8029	0.7701
6.0	0.9975	0.9912	0.9806	0.9658	0.9466	0.9236	0.8972	0.8682	0.8372	0.8049	0.7720

NTU	Heat Exchanger Effectiveness C^* ($C_{min} = C_h$)										
	0	0.1	0.2	0.3	0.4	0.5	0.6	0.7	0.8	0.9	1.0
6.1	0.9978	0.9917	0.9816	0.9670	0.9482	0.9254	0.8992	0.8702	0.8392	0.8069	0.7738
6.2	0.9980	0.9922	0.9824	0.9683	0.9497	0.9271	0.9011	0.8722	0.8412	0.8088	0.7756
6.3	0.9982	0.9927	0.9833	0.9694	0.9512	0.9288	0.9029	0.8741	0.8431	0.8106	0.7774
6.4	0.9983	0.9932	0.9841	0.9705	0.9526	0.9305	0.9047	0.8760	0.8450	0.8124	0.7791
6.5	0.9985	0.9936	0.9848	0.9716	0.9540	0.9321	0.9065	0.8778	0.8468	0.8142	0.7807
6.6	0.9986	0.9940	0.9855	0.9726	0.9553	0.9336	0.9082	0.8796	0.8486	0.8159	0.7824
6.7	0.9988	0.9944	0.9862	0.9736	0.9565	0.9351	0.9098	0.8813	0.8503	0.8176	0.7840
6.8	0.9989	0.9947	0.9868	0.9746	0.9578	0.9366	0.9114	0.8830	0.8520	0.8193	0.7855
6.9	0.9990	0.9951	0.9874	0.9755	0.9589	0.9380	0.9130	0.8846	0.8537	0.8209	0.7870
7.0	0.9991	0.9954	0.9880	0.9763	0.9601	0.9393	0.9145	0.8863	0.8553	0.8225	0.7885
7.1	0.9992	0.9956	0.9886	0.9772	0.9612	0.9407	0.9160	0.8878	0.8569	0.8240	0.7900
7.2	0.9993	0.9959	0.9891	0.9780	0.9623	0.9420	0.9175	0.8894	0.8585	0.8256	0.7914
7.3	0.9993	0.9962	0.9896	0.9788	0.9633	0.9432	0.9189	0.8909	0.8600	0.8270	0.7928
7.4	0.9994	0.9964	0.9900	0.9795	0.9643	0.9444	0.9203	0.8924	0.8615	0.8285	0.7942
7.5	0.9994	0.9966	0.9905	0.9802	0.9653	0.9456	0.9216	0.8938	0.8630	0.8299	0.7956
7.6	0.9995	0.9968	0.9909	0.9809	0.9662	0.9468	0.9229	0.8952	0.8644	0.8313	0.7969
7.7	0.9995	0.9970	0.9913	0.9815	0.9671	0.9479	0.9242	0.8966	0.8658	0.8327	0.7982
7.8	0.9996	0.9972	0.9917	0.9822	0.9680	0.9490	0.9255	0.8979	0.8672	0.8340	0.7994
7.9	0.9996	0.9974	0.9921	0.9828	0.9688	0.9501	0.9267	0.8992	0.8685	0.8354	0.8007
8.0	0.9997	0.9975	0.9924	0.9834	0.9696	0.9511	0.9279	0.9005	0.8698	0.8367	0.8019
8.1	0.9997	0.9977	0.9928	0.9839	0.9704	0.9521	0.9290	0.9018	0.8711	0.8379	0.8031
8.2	0.9997	0.9978	0.9931	0.9845	0.9712	0.9531	0.9302	0.9030	0.8724	0.8392	0.8043
8.3	0.9998	0.9979	0.9934	0.9850	0.9720	0.9540	0.9313	0.9042	0.8737	0.8404	0.8055
8.4	0.9998	0.9981	0.9937	0.9855	0.9727	0.9550	0.9324	0.9054	0.8749	0.8416	0.8066
8.5	0.9998	0.9982	0.9939	0.9860	0.9734	0.9559	0.9334	0.9066	0.8761	0.8428	0.8077
8.6	0.9998	0.9983	0.9942	0.9864	0.9741	0.9567	0.9345	0.9077	0.8773	0.8440	0.8088
8.7	0.9998	0.9984	0.9945	0.9869	0.9747	0.9576	0.9355	0.9088	0.8784	0.8451	0.8099
8.8	0.9998	0.9985	0.9947	0.9873	0.9754	0.9584	0.9365	0.9099	0.8796	0.8462	0.8110
8.9	0.9999	0.9986	0.9949	0.9877	0.9760	0.9593	0.9375	0.9110	0.8807	0.8473	0.8120
9.0	0.9999	0.9987	0.9951	0.9881	0.9766	0.9600	0.9384	0.9121	0.8818	0.8484	0.8130
9.1	0.9999	0.9987	0.9954	0.9885	0.9772	0.9608	0.9393	0.9131	0.8828	0.8495	0.8140
9.2	0.9999	0.9988	0.9956	0.9889	0.9778	0.9616	0.9403	0.9141	0.8839	0.8505	0.8150
9.3	0.9999	0.9989	0.9957	0.9892	0.9783	0.9623	0.9411	0.9151	0.8849	0.8516	0.8160
9.4	0.9999	0.9989	0.9959	0.9896	0.9788	0.9630	0.9420	0.9161	0.8860	0.8526	0.8170
9.5	0.9999	0.9990	0.9961	0.9899	0.9794	0.9637	0.9429	0.9171	0.8870	0.8536	0.8179
9.6	0.9999	0.9991	0.9963	0.9902	0.9799	0.9644	0.9437	0.9180	0.8880	0.8546	0.8189
9.7	0.9999	0.9991	0.9964	0.9905	0.9804	0.9651	0.9445	0.9189	0.8889	0.8555	0.8198
9.8	0.9999	0.9992	0.9966	0.9908	0.9808	0.9658	0.9453	0.9198	0.8899	0.8565	0.8207
9.9	1.0000	0.9992	0.9967	0.9911	0.9813	0.9664	0.9461	0.9207	0.8908	0.8574	0.8216
10.0	1.0000	0.9993	0.9968	0.9914	0.9818	0.9670	0.9469	0.9216	0.8918	0.8584	0.8224

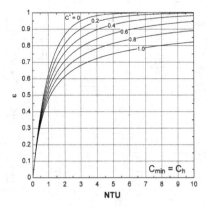

CONFIGURATION 10

1TUBE100ROW1PASS100CIR

(a) $C_{min} = C_c$

NTU	Heat Exchanger Effectiveness										
	$C^* (C_{min} = C_c)$										
	0	0.1	0.2	0.3	0.4	0.5	0.6	0.7	0.8	0.9	1.0
0	0.0000	0.0000	0.0000	0.0000	0.0000	0.0000	0.0000	0.0000	0.0000	0.0000	0.0000
0.1	0.0952	0.0947	0.0943	0.0938	0.0934	0.0929	0.0925	0.0921	0.0916	0.0912	0.0908
0.2	0.1813	0.1796	0.1780	0.1764	0.1749	0.1733	0.1718	0.1703	0.1688	0.1673	0.1658
0.3	0.2592	0.2559	0.2526	0.2494	0.2463	0.2432	0.2402	0.2372	0.2342	0.2313	0.2285
0.4	0.3297	0.3244	0.3192	0.3141	0.3091	0.3042	0.2995	0.2948	0.2902	0.2857	0.2814
0.5	0.3935	0.3860	0.3787	0.3716	0.3646	0.3578	0.3512	0.3448	0.3385	0.3323	0.3263
0.6	0.4512	0.4414	0.4320	0.4228	0.4138	0.4051	0.3966	0.3884	0.3804	0.3726	0.3650
0.7	0.5034	0.4914	0.4798	0.4685	0.4576	0.4470	0.4367	0.4267	0.4170	0.4076	0.3984
0.8	0.5507	0.5365	0.5228	0.5095	0.4967	0.4842	0.4722	0.4605	0.4492	0.4383	0.4277
0.9	0.5934	0.5772	0.5616	0.5464	0.5317	0.5176	0.5039	0.4906	0.4778	0.4654	0.4534
1.0	0.6321	0.6140	0.5965	0.5796	0.5633	0.5475	0.5322	0.5175	0.5033	0.4895	0.4762
1.1	0.6671	0.6473	0.6282	0.6096	0.5917	0.5745	0.5578	0.5416	0.5261	0.5111	0.4966
1.2	0.6988	0.6775	0.6568	0.6368	0.6175	0.5989	0.5809	0.5635	0.5467	0.5305	0.5149
1.3	0.7275	0.7048	0.6828	0.6615	0.6409	0.6210	0.6018	0.5833	0.5653	0.5481	0.5314
1.4	0.7534	0.7296	0.7064	0.6840	0.6623	0.6412	0.6209	0.6013	0.5823	0.5641	0.5465
1.5	0.7769	0.7521	0.7280	0.7045	0.6818	0.6597	0.6384	0.6178	0.5979	0.5787	0.5602
1.6	0.7981	0.7725	0.7476	0.7233	0.6996	0.6767	0.6545	0.6329	0.6122	0.5921	0.5728
1.7	0.8173	0.7912	0.7655	0.7405	0.7160	0.6923	0.6692	0.6469	0.6253	0.6045	0.5843
1.8	0.8347	0.8081	0.7819	0.7563	0.7312	0.7067	0.6829	0.6598	0.6375	0.6159	0.5950
1.9	0.8504	0.8235	0.7970	0.7708	0.7451	0.7200	0.6956	0.6718	0.6488	0.6265	0.6050
2.0	0.8647	0.8376	0.8108	0.7842	0.7580	0.7324	0.7074	0.6830	0.6593	0.6364	0.6142
2.1	0.8775	0.8505	0.8234	0.7966	0.7700	0.7439	0.7184	0.6934	0.6692	0.6456	0.6229
2.2	0.8892	0.8622	0.8351	0.8080	0.7811	0.7546	0.7286	0.7032	0.6784	0.6543	0.6310
2.3	0.8997	0.8730	0.8458	0.8186	0.7915	0.7647	0.7382	0.7123	0.6870	0.6624	0.6386
2.4	0.9093	0.8828	0.8558	0.8285	0.8012	0.7740	0.7472	0.7209	0.6952	0.6701	0.6457
2.5	0.9179	0.8918	0.8649	0.8377	0.8102	0.7828	0.7557	0.7290	0.7028	0.6773	0.6525
2.6	0.9257	0.9000	0.8734	0.8462	0.8187	0.7911	0.7637	0.7366	0.7101	0.6841	0.6589
2.7	0.9328	0.9076	0.8813	0.8542	0.8266	0.7989	0.7712	0.7439	0.7169	0.6906	0.6649
2.8	0.9392	0.9145	0.8885	0.8616	0.8341	0.8062	0.7784	0.7507	0.7234	0.6967	0.6706
2.9	0.9450	0.9209	0.8953	0.8686	0.8411	0.8132	0.7851	0.7572	0.7296	0.7025	0.6761
3.0	0.9502	0.9268	0.9016	0.8751	0.8477	0.8197	0.7915	0.7634	0.7355	0.7081	0.6813
3.1	0.9550	0.9322	0.9074	0.8812	0.8539	0.8259	0.7976	0.7693	0.7411	0.7134	0.6862
3.2	0.9592	0.9371	0.9128	0.8869	0.8597	0.8318	0.8034	0.7749	0.7465	0.7184	0.6910
3.3	0.9631	0.9417	0.9179	0.8923	0.8653	0.8374	0.8089	0.7802	0.7516	0.7233	0.6955
3.4	0.9666	0.9459	0.9226	0.8973	0.8705	0.8427	0.8141	0.7853	0.7565	0.7279	0.6998
3.5	0.9698	0.9498	0.9270	0.9021	0.8755	0.8477	0.8192	0.7902	0.7612	0.7323	0.7040
3.6	0.9727	0.9534	0.9312	0.9066	0.8802	0.8525	0.8239	0.7949	0.7656	0.7366	0.7079
3.7	0.9753	0.9567	0.9350	0.9109	0.8847	0.8571	0.8285	0.7993	0.7700	0.7407	0.7118
3.8	0.9776	0.9597	0.9386	0.9149	0.8890	0.8615	0.8329	0.8036	0.7741	0.7446	0.7155
3.9	0.9798	0.9625	0.9420	0.9186	0.8930	0.8657	0.8371	0.8078	0.7781	0.7484	0.7190
4.0	0.9817	0.9651	0.9452	0.9222	0.8969	0.8697	0.8411	0.8118	0.7819	0.7520	0.7224
4.1	0.9834	0.9676	0.9481	0.9256	0.9005	0.8735	0.8450	0.8156	0.7856	0.7556	0.7257
4.2	0.9850	0.9698	0.9509	0.9288	0.9040	0.8772	0.8487	0.8193	0.7892	0.7590	0.7289
4.3	0.9864	0.9719	0.9535	0.9318	0.9074	0.8807	0.8523	0.8228	0.7926	0.7622	0.7320
4.4	0.9877	0.9738	0.9560	0.9347	0.9106	0.8840	0.8558	0.8262	0.7960	0.7654	0.7349
4.5	0.9889	0.9756	0.9583	0.9375	0.9136	0.8873	0.8591	0.8295	0.7992	0.7685	0.7378
4.6	0.9899	0.9772	0.9605	0.9401	0.9165	0.8904	0.8623	0.8327	0.8023	0.7714	0.7406
4.7	0.9909	0.9787	0.9625	0.9425	0.9193	0.8934	0.8653	0.8358	0.8053	0.7743	0.7433
4.8	0.9918	0.9802	0.9644	0.9449	0.9220	0.8962	0.8683	0.8388	0.8082	0.7771	0.7459
4.9	0.9926	0.9815	0.9662	0.9471	0.9245	0.8990	0.8712	0.8417	0.8110	0.7798	0.7484
5.0	0.9933	0.9827	0.9680	0.9492	0.9269	0.9017	0.8740	0.8445	0.8138	0.7824	0.7509
5.1	0.9939	0.9839	0.9696	0.9513	0.9293	0.9042	0.8766	0.8472	0.8164	0.7850	0.7533
5.2	0.9945	0.9849	0.9711	0.9532	0.9315	0.9067	0.8792	0.8498	0.8190	0.7874	0.7556
5.3	0.9950	0.9859	0.9725	0.9550	0.9337	0.9091	0.8817	0.8523	0.8215	0.7898	0.7579
5.4	0.9955	0.9868	0.9739	0.9568	0.9358	0.9114	0.8842	0.8548	0.8239	0.7922	0.7600
5.5	0.9959	0.9877	0.9752	0.9584	0.9378	0.9136	0.8865	0.8572	0.8263	0.7945	0.7622
5.6	0.9963	0.9885	0.9764	0.9600	0.9397	0.9157	0.8888	0.8595	0.8286	0.7967	0.7643
5.7	0.9967	0.9892	0.9776	0.9616	0.9415	0.9178	0.8910	0.8618	0.8309	0.7988	0.7663
5.8	0.9970	0.9899	0.9786	0.9630	0.9433	0.9198	0.8931	0.8640	0.8330	0.8009	0.7683
5.9	0.9973	0.9906	0.9797	0.9644	0.9450	0.9217	0.8952	0.8661	0.8352	0.8030	0.7702
6.0	0.9975	0.9912	0.9806	0.9658	0.9466	0.9236	0.8972	0.8682	0.8372	0.8050	0.7721

NTU	Heat Exchanger Effectiveness										
	C^* ($C_{min} = C_c$)										
	0	0.1	0.2	0.3	0.4	0.5	0.6	0.7	0.8	0.9	1.0
6.1	0.9978	0.9917	0.9816	0.9670	0.9482	0.9254	0.8992	0.8702	0.8393	0.8069	0.7739
6.2	0.9980	0.9922	0.9824	0.9683	0.9497	0.9272	0.9011	0.8722	0.8412	0.8088	0.7757
6.3	0.9982	0.9927	0.9833	0.9694	0.9512	0.9289	0.9030	0.8741	0.8431	0.8107	0.7775
6.4	0.9983	0.9932	0.9841	0.9705	0.9526	0.9305	0.9047	0.8760	0.8450	0.8125	0.7792
6.5	0.9985	0.9936	0.9848	0.9716	0.9540	0.9321	0.9065	0.8778	0.8469	0.8143	0.7808
6.6	0.9986	0.9940	0.9855	0.9726	0.9553	0.9336	0.9082	0.8796	0.8486	0.8160	0.7825
6.7	0.9988	0.9944	0.9862	0.9736	0.9565	0.9351	0.9099	0.8814	0.8504	0.8177	0.7841
6.8	0.9989	0.9947	0.9868	0.9746	0.9578	0.9366	0.9115	0.8831	0.8521	0.8194	0.7856
6.9	0.9990	0.9951	0.9874	0.9755	0.9590	0.9380	0.9130	0.8847	0.8538	0.8210	0.7872
7.0	0.9991	0.9954	0.9880	0.9764	0.9601	0.9394	0.9146	0.8863	0.8554	0.8226	0.7886
7.1	0.9992	0.9956	0.9886	0.9772	0.9612	0.9407	0.9161	0.8879	0.8570	0.8241	0.7901
7.2	0.9993	0.9959	0.9891	0.9780	0.9623	0.9420	0.9175	0.8894	0.8586	0.8257	0.7915
7.3	0.9993	0.9962	0.9896	0.9788	0.9633	0.9432	0.9189	0.8909	0.8601	0.8271	0.7930
7.4	0.9994	0.9964	0.9900	0.9795	0.9643	0.9445	0.9203	0.8924	0.8616	0.8286	0.7943
7.5	0.9994	0.9966	0.9905	0.9802	0.9653	0.9457	0.9217	0.8939	0.8630	0.8300	0.7957
7.6	0.9995	0.9968	0.9909	0.9809	0.9662	0.9468	0.9230	0.8953	0.8645	0.8314	0.7970
7.7	0.9995	0.9970	0.9913	0.9815	0.9671	0.9479	0.9242	0.8966	0.8659	0.8328	0.7983
7.8	0.9996	0.9972	0.9917	0.9822	0.9680	0.9490	0.9255	0.8980	0.8673	0.8342	0.7996
7.9	0.9996	0.9974	0.9921	0.9828	0.9688	0.9501	0.9267	0.8993	0.8686	0.8355	0.8008
8.0	0.9997	0.9975	0.9924	0.9834	0.9697	0.9511	0.9279	0.9006	0.8699	0.8368	0.8021
8.1	0.9997	0.9977	0.9928	0.9839	0.9705	0.9521	0.9291	0.9019	0.8712	0.8381	0.8033
8.2	0.9997	0.9978	0.9931	0.9845	0.9712	0.9531	0.9302	0.9031	0.8725	0.8393	0.8044
8.3	0.9998	0.9979	0.9934	0.9850	0.9720	0.9541	0.9313	0.9043	0.8738	0.8405	0.8056
8.4	0.9998	0.9981	0.9937	0.9855	0.9727	0.9550	0.9324	0.9055	0.8750	0.8418	0.8067
8.5	0.9998	0.9982	0.9939	0.9860	0.9734	0.9559	0.9335	0.9067	0.8762	0.8429	0.8079
8.6	0.9998	0.9983	0.9942	0.9864	0.9741	0.9568	0.9345	0.9078	0.8774	0.8441	0.8090
8.7	0.9998	0.9984	0.9945	0.9869	0.9747	0.9576	0.9355	0.9089	0.8785	0.8452	0.8101
8.8	0.9998	0.9985	0.9947	0.9873	0.9754	0.9585	0.9365	0.9100	0.8797	0.8464	0.8111
8.9	0.9999	0.9986	0.9949	0.9877	0.9760	0.9593	0.9375	0.9111	0.8808	0.8475	0.8122
9.0	0.9999	0.9987	0.9951	0.9881	0.9766	0.9601	0.9385	0.9122	0.8819	0.8486	0.8132
9.1	0.9999	0.9987	0.9954	0.9885	0.9772	0.9609	0.9394	0.9132	0.8830	0.8496	0.8142
9.2	0.9999	0.9988	0.9956	0.9889	0.9778	0.9616	0.9403	0.9142	0.8840	0.8507	0.8152
9.3	0.9999	0.9989	0.9957	0.9892	0.9783	0.9624	0.9412	0.9152	0.8851	0.8517	0.8162
9.4	0.9999	0.9989	0.9959	0.9896	0.9789	0.9631	0.9421	0.9162	0.8861	0.8527	0.8172
9.5	0.9999	0.9990	0.9961	0.9899	0.9794	0.9638	0.9429	0.9172	0.8871	0.8537	0.8181
9.6	0.9999	0.9991	0.9963	0.9902	0.9799	0.9645	0.9438	0.9181	0.8881	0.8547	0.8190
9.7	0.9999	0.9991	0.9964	0.9906	0.9804	0.9651	0.9446	0.9190	0.8891	0.8557	0.8200
9.8	0.9999	0.9992	0.9966	0.9909	0.9809	0.9658	0.9454	0.9199	0.8900	0.8567	0.8209
9.9	1.0000	0.9992	0.9967	0.9911	0.9813	0.9664	0.9462	0.9208	0.8910	0.8576	0.8218
10.0	1.0000	0.9993	0.9968	0.9914	0.9818	0.9671	0.9470	0.9217	0.8919	0.8585	0.8226

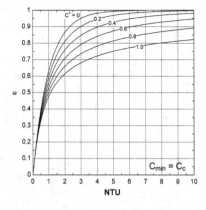

CONFIGURATION 10

1TUBE100ROW1PASS100CIR

(b) $C_{min} = C_h$

NTU	Heat Exchanger Effectiveness										
	C^* ($C_{min} = C_h$)										
	0	0.1	0.2	0.3	0.4	0.5	0.6	0.7	0.8	0.9	1.0
0	0.0000	0.0000	0.0000	0.0000	0.0000	0.0000	0.0000	0.0000	0.0000	0.0000	0.0000
0.1	0.0952	0.0947	0.0943	0.0938	0.0934	0.0929	0.0925	0.0921	0.0916	0.0912	0.0908
0.2	0.1813	0.1796	0.1780	0.1764	0.1749	0.1733	0.1718	0.1703	0.1688	0.1673	0.1658
0.3	0.2592	0.2559	0.2526	0.2494	0.2463	0.2432	0.2402	0.2372	0.2342	0.2313	0.2285
0.4	0.3297	0.3244	0.3192	0.3141	0.3091	0.3042	0.2995	0.2948	0.2902	0.2857	0.2814
0.5	0.3935	0.3860	0.3787	0.3716	0.3646	0.3578	0.3512	0.3448	0.3385	0.3323	0.3263
0.6	0.4512	0.4414	0.4320	0.4228	0.4138	0.4051	0.3966	0.3884	0.3804	0.3726	0.3650
0.7	0.5034	0.4914	0.4798	0.4685	0.4576	0.4470	0.4367	0.4267	0.4170	0.4076	0.3984
0.8	0.5507	0.5365	0.5228	0.5095	0.4967	0.4842	0.4722	0.4605	0.4492	0.4383	0.4277
0.9	0.5934	0.5772	0.5616	0.5464	0.5317	0.5176	0.5039	0.4906	0.4778	0.4654	0.4534
1.0	0.6321	0.6140	0.5965	0.5796	0.5633	0.5475	0.5322	0.5175	0.5033	0.4895	0.4762
1.1	0.6671	0.6473	0.6282	0.6096	0.5917	0.5745	0.5578	0.5416	0.5261	0.5111	0.4966
1.2	0.6988	0.6775	0.6568	0.6368	0.6175	0.5989	0.5809	0.5635	0.5467	0.5305	0.5149
1.3	0.7275	0.7048	0.6828	0.6615	0.6409	0.6210	0.6018	0.5833	0.5653	0.5481	0.5314
1.4	0.7534	0.7296	0.7064	0.6840	0.6623	0.6412	0.6209	0.6013	0.5823	0.5641	0.5465
1.5	0.7769	0.7521	0.7280	0.7045	0.6818	0.6597	0.6384	0.6178	0.5979	0.5787	0.5602
1.6	0.7981	0.7725	0.7476	0.7233	0.6996	0.6767	0.6545	0.6329	0.6122	0.5921	0.5728
1.7	0.8173	0.7912	0.7655	0.7405	0.7160	0.6923	0.6693	0.6469	0.6253	0.6045	0.5843
1.8	0.8347	0.8081	0.7819	0.7563	0.7312	0.7067	0.6829	0.6598	0.6375	0.6159	0.5950
1.9	0.8504	0.8235	0.7970	0.7708	0.7451	0.7200	0.6956	0.6718	0.6488	0.6265	0.6050
2.0	0.8647	0.8376	0.8108	0.7842	0.7580	0.7324	0.7074	0.6830	0.6593	0.6364	0.6142
2.1	0.8775	0.8505	0.8234	0.7966	0.7700	0.7439	0.7184	0.6934	0.6692	0.6456	0.6229
2.2	0.8892	0.8622	0.8351	0.8080	0.7811	0.7546	0.7286	0.7032	0.6784	0.6543	0.6310
2.3	0.8997	0.8730	0.8458	0.8186	0.7915	0.7647	0.7382	0.7123	0.6870	0.6624	0.6386
2.4	0.9093	0.8828	0.8558	0.8285	0.8012	0.7740	0.7472	0.7209	0.6952	0.6701	0.6457
2.5	0.9179	0.8918	0.8649	0.8377	0.8102	0.7828	0.7557	0.7290	0.7028	0.6773	0.6525
2.6	0.9257	0.9000	0.8734	0.8462	0.8187	0.7911	0.7637	0.7366	0.7101	0.6841	0.6589
2.7	0.9328	0.9076	0.8813	0.8542	0.8266	0.7989	0.7712	0.7439	0.7169	0.6906	0.6649
2.8	0.9392	0.9145	0.8885	0.8616	0.8341	0.8062	0.7784	0.7507	0.7234	0.6967	0.6706
2.9	0.9450	0.9209	0.8953	0.8686	0.8411	0.8132	0.7851	0.7572	0.7296	0.7025	0.6761
3.0	0.9502	0.9268	0.9016	0.8751	0.8477	0.8197	0.7915	0.7634	0.7355	0.7081	0.6813
3.1	0.9550	0.9322	0.9074	0.8812	0.8539	0.8259	0.7976	0.7693	0.7411	0.7134	0.6862
3.2	0.9592	0.9371	0.9128	0.8869	0.8597	0.8318	0.8034	0.7749	0.7465	0.7184	0.6910
3.3	0.9631	0.9417	0.9179	0.8923	0.8653	0.8374	0.8089	0.7802	0.7516	0.7233	0.6955
3.4	0.9666	0.9459	0.9226	0.8974	0.8706	0.8427	0.8141	0.7853	0.7565	0.7279	0.6998
3.5	0.9698	0.9498	0.9270	0.9021	0.8755	0.8477	0.8192	0.7902	0.7612	0.7323	0.7040
3.6	0.9727	0.9534	0.9312	0.9066	0.8802	0.8525	0.8239	0.7949	0.7656	0.7366	0.7079
3.7	0.9753	0.9567	0.9350	0.9109	0.8847	0.8571	0.8285	0.7993	0.7700	0.7407	0.7118
3.8	0.9776	0.9597	0.9386	0.9149	0.8890	0.8615	0.8329	0.8037	0.7741	0.7446	0.7155
3.9	0.9798	0.9625	0.9420	0.9186	0.8930	0.8657	0.8371	0.8078	0.7781	0.7484	0.7190
4.0	0.9817	0.9651	0.9452	0.9222	0.8969	0.8697	0.8411	0.8118	0.7819	0.7520	0.7224
4.1	0.9834	0.9676	0.9481	0.9256	0.9005	0.8735	0.8450	0.8156	0.7856	0.7556	0.7257
4.2	0.9850	0.9698	0.9509	0.9288	0.9040	0.8772	0.8487	0.8193	0.7892	0.7590	0.7289
4.3	0.9864	0.9719	0.9535	0.9318	0.9074	0.8807	0.8523	0.8228	0.7926	0.7622	0.7320
4.4	0.9877	0.9738	0.9560	0.9347	0.9106	0.8841	0.8558	0.8262	0.7960	0.7654	0.7349
4.5	0.9889	0.9756	0.9583	0.9375	0.9136	0.8873	0.8591	0.8295	0.7992	0.7685	0.7378
4.6	0.9899	0.9772	0.9605	0.9401	0.9165	0.8904	0.8623	0.8327	0.8023	0.7714	0.7406
4.7	0.9909	0.9787	0.9625	0.9425	0.9193	0.8934	0.8653	0.8358	0.8053	0.7743	0.7433
4.8	0.9918	0.9802	0.9644	0.9449	0.9220	0.8962	0.8683	0.8388	0.8082	0.7771	0.7459
4.9	0.9926	0.9815	0.9663	0.9471	0.9245	0.8990	0.8712	0.8417	0.8110	0.7798	0.7484
5.0	0.9933	0.9827	0.9680	0.9492	0.9270	0.9017	0.8740	0.8445	0.8138	0.7824	0.7509
5.1	0.9939	0.9839	0.9696	0.9513	0.9293	0.9042	0.8766	0.8472	0.8164	0.7850	0.7533
5.2	0.9945	0.9849	0.9711	0.9532	0.9315	0.9067	0.8792	0.8498	0.8190	0.7874	0.7556
5.3	0.9950	0.9859	0.9725	0.9550	0.9337	0.9091	0.8817	0.8523	0.8215	0.7898	0.7579
5.4	0.9955	0.9868	0.9739	0.9568	0.9358	0.9114	0.8842	0.8548	0.8240	0.7922	0.7600
5.5	0.9959	0.9877	0.9752	0.9585	0.9378	0.9136	0.8865	0.8572	0.8263	0.7945	0.7622
5.6	0.9963	0.9885	0.9764	0.9601	0.9397	0.9157	0.8888	0.8595	0.8286	0.7967	0.7643
5.7	0.9967	0.9892	0.9776	0.9616	0.9415	0.9178	0.8910	0.8618	0.8309	0.7988	0.7663
5.8	0.9970	0.9899	0.9786	0.9630	0.9433	0.9198	0.8931	0.8640	0.8330	0.8009	0.7683
5.9	0.9973	0.9906	0.9797	0.9644	0.9450	0.9217	0.8952	0.8661	0.8352	0.8030	0.7702
6.0	0.9975	0.9912	0.9806	0.9658	0.9466	0.9236	0.8972	0.8682	0.8372	0.8050	0.7721

NTU	Heat Exchanger Effectiveness										
	$C^*\ (C_{min} = C_h)$										
	0	0.1	0.2	0.3	0.4	0.5	0.6	0.7	0.8	0.9	1.0
6.1	0.9978	0.9917	0.9816	0.9670	0.9482	0.9254	0.8992	0.8703	0.8393	0.8069	0.7739
6.2	0.9980	0.9922	0.9824	0.9683	0.9497	0.9272	0.9011	0.8722	0.8412	0.8088	0.7757
6.3	0.9982	0.9927	0.9833	0.9694	0.9512	0.9289	0.9030	0.8742	0.8432	0.8107	0.7775
6.4	0.9983	0.9932	0.9841	0.9706	0.9526	0.9305	0.9048	0.8760	0.8450	0.8125	0.7792
6.5	0.9985	0.9936	0.9848	0.9716	0.9540	0.9321	0.9065	0.8779	0.8469	0.8143	0.7808
6.6	0.9986	0.9940	0.9855	0.9727	0.9553	0.9336	0.9082	0.8796	0.8486	0.8160	0.7825
6.7	0.9988	0.9944	0.9862	0.9736	0.9566	0.9351	0.9099	0.8814	0.8504	0.8177	0.7841
6.8	0.9989	0.9947	0.9868	0.9746	0.9578	0.9366	0.9115	0.8831	0.8521	0.8194	0.7856
6.9	0.9990	0.9951	0.9874	0.9755	0.9590	0.9380	0.9130	0.8847	0.8538	0.8210	0.7872
7.0	0.9991	0.9954	0.9880	0.9764	0.9601	0.9394	0.9146	0.8863	0.8554	0.8226	0.7886
7.1	0.9992	0.9956	0.9886	0.9772	0.9612	0.9407	0.9161	0.8879	0.8570	0.8241	0.7901
7.2	0.9993	0.9959	0.9891	0.9780	0.9623	0.9420	0.9175	0.8895	0.8586	0.8257	0.7915
7.3	0.9993	0.9962	0.9896	0.9788	0.9633	0.9433	0.9189	0.8910	0.8601	0.8272	0.7930
7.4	0.9994	0.9964	0.9900	0.9795	0.9643	0.9445	0.9203	0.8924	0.8616	0.8286	0.7943
7.5	0.9994	0.9966	0.9905	0.9802	0.9653	0.9457	0.9217	0.8939	0.8631	0.8300	0.7957
7.6	0.9995	0.9968	0.9909	0.9809	0.9662	0.9468	0.9230	0.8953	0.8645	0.8314	0.7970
7.7	0.9995	0.9970	0.9913	0.9815	0.9671	0.9479	0.9243	0.8967	0.8659	0.8328	0.7983
7.8	0.9996	0.9972	0.9917	0.9822	0.9680	0.9490	0.9255	0.8980	0.8673	0.8342	0.7996
7.9	0.9996	0.9974	0.9921	0.9828	0.9689	0.9501	0.9267	0.8993	0.8686	0.8355	0.8008
8.0	0.9997	0.9975	0.9924	0.9834	0.9697	0.9511	0.9279	0.9006	0.8699	0.8368	0.8021
8.1	0.9997	0.9977	0.9928	0.9839	0.9705	0.9521	0.9291	0.9019	0.8712	0.8381	0.8033
8.2	0.9997	0.9978	0.9931	0.9845	0.9712	0.9531	0.9302	0.9031	0.8725	0.8393	0.8044
8.3	0.9998	0.9979	0.9934	0.9850	0.9720	0.9541	0.9313	0.9043	0.8738	0.8406	0.8056
8.4	0.9998	0.9981	0.9937	0.9855	0.9727	0.9550	0.9324	0.9055	0.8750	0.8418	0.8067
8.5	0.9998	0.9982	0.9940	0.9860	0.9734	0.9559	0.9335	0.9067	0.8762	0.8429	0.8079
8.6	0.9998	0.9983	0.9942	0.9864	0.9741	0.9568	0.9345	0.9078	0.8774	0.8441	0.8090
8.7	0.9998	0.9984	0.9945	0.9869	0.9748	0.9576	0.9356	0.9089	0.8785	0.8453	0.8101
8.8	0.9998	0.9985	0.9947	0.9873	0.9754	0.9585	0.9366	0.9100	0.8797	0.8464	0.8111
8.9	0.9999	0.9986	0.9949	0.9877	0.9760	0.9593	0.9375	0.9111	0.8808	0.8475	0.8122
9.0	0.9999	0.9987	0.9951	0.9881	0.9766	0.9601	0.9385	0.9122	0.8819	0.8486	0.8132
9.1	0.9999	0.9987	0.9954	0.9885	0.9772	0.9609	0.9394	0.9132	0.8830	0.8496	0.8142
9.2	0.9999	0.9988	0.9956	0.9889	0.9778	0.9616	0.9403	0.9142	0.8840	0.8507	0.8152
9.3	0.9999	0.9989	0.9957	0.9892	0.9783	0.9624	0.9412	0.9152	0.8851	0.8517	0.8162
9.4	0.9999	0.9989	0.9959	0.9896	0.9789	0.9631	0.9421	0.9162	0.8861	0.8527	0.8172
9.5	0.9999	0.9990	0.9961	0.9899	0.9794	0.9638	0.9430	0.9172	0.8871	0.8538	0.8181
9.6	0.9999	0.9991	0.9963	0.9902	0.9799	0.9645	0.9438	0.9181	0.8881	0.8547	0.8190
9.7	0.9999	0.9991	0.9964	0.9906	0.9804	0.9652	0.9446	0.9190	0.8891	0.8557	0.8200
9.8	0.9999	0.9992	0.9966	0.9909	0.9809	0.9658	0.9454	0.9199	0.8900	0.8567	0.8209
9.9	1.0000	0.9992	0.9967	0.9911	0.9813	0.9665	0.9462	0.9208	0.8910	0.8576	0.8218
10.0	1.0000	0.9993	0.9969	0.9914	0.9818	0.9671	0.9470	0.9217	0.8919	0.8585	0.8226

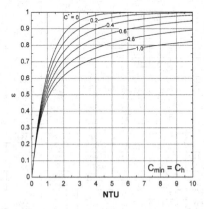

CONFIGURATION 11

1TUBE2ROW2PASS1CIR(PAR)

(a) $C_{min} = C_c$

NTU	Heat Exchanger Effectiveness										
	$C^* (C_{min} = C_c)$										
	0	0.1	0.2	0.3	0.4	0.5	0.6	0.7	0.8	0.9	1.0
0	0.0000	0.0000	0.0000	0.0000	0.0000	0.0000	0.0000	0.0000	0.0000	0.0000	0.0000
0.1	0.0952	0.0947	0.0942	0.0938	0.0933	0.0929	0.0924	0.0920	0.0915	0.0911	0.0907
0.2	0.1813	0.1796	0.1779	0.1762	0.1745	0.1729	0.1713	0.1697	0.1681	0.1666	0.1651
0.3	0.2592	0.2556	0.2521	0.2486	0.2453	0.2419	0.2387	0.2355	0.2323	0.2292	0.2262
0.4	0.3297	0.3238	0.3180	0.3124	0.3069	0.3015	0.2963	0.2912	0.2813	0.2813	0.2765
0.5	0.3935	0.3849	0.3766	0.3685	0.3606	0.3530	0.3456	0.3384	0.3314	0.3246	0.3180
0.6	0.4512	0.4397	0.4286	0.4179	0.4075	0.3975	0.3878	0.3784	0.3693	0.3606	0.3521
0.7	0.5034	0.4888	0.4748	0.4613	0.4484	0.4359	0.4239	0.4123	0.4012	0.3905	0.3802
0.8	0.5507	0.5329	0.5159	0.4996	0.4840	0.4691	0.4548	0.4411	0.4279	0.4153	0.4032
0.9	0.5934	0.5724	0.5524	0.5334	0.5152	0.4978	0.4812	0.4654	0.4503	0.4359	0.4221
1.0	0.6321	0.6079	0.5849	0.5630	0.5423	0.5226	0.5038	0.4860	0.4691	0.4529	0.4376
1.1	0.6671	0.6397	0.6137	0.5892	0.5660	0.5440	0.5232	0.5034	0.4847	0.4670	0.4502
1.2	0.6988	0.6682	0.6394	0.6122	0.5866	0.5624	0.5396	0.5181	0.4977	0.4785	0.4603
1.3	0.7275	0.6938	0.6622	0.6325	0.6045	0.5783	0.5536	0.5304	0.5085	0.4879	0.4684
1.4	0.7534	0.7167	0.6824	0.6503	0.6202	0.5920	0.5655	0.5407	0.5174	0.4955	0.4749
1.5	0.7769	0.7373	0.7004	0.6659	0.6337	0.6037	0.5756	0.5493	0.5246	0.5016	0.4799
1.6	0.7981	0.7557	0.7163	0.6797	0.6455	0.6137	0.5840	0.5564	0.5305	0.5064	0.4838
1.7	0.8173	0.7723	0.7305	0.6917	0.6557	0.6223	0.5911	0.5622	0.5352	0.5101	0.4866
1.8	0.8347	0.7871	0.7430	0.7023	0.6645	0.6295	0.5971	0.5669	0.5389	0.5129	0.4887
1.9	0.8504	0.8003	0.7542	0.7115	0.6721	0.6357	0.6020	0.5708	0.5418	0.5150	0.4900
2.0	0.8647	0.8122	0.7640	0.7196	0.6787	0.6409	0.6060	0.5738	0.5440	0.5164	0.4908
2.1	0.8775	0.8229	0.7727	0.7266	0.6843	0.6453	0.6093	0.5761	0.5455	0.5172	0.4910
2.2	0.8892	0.8324	0.7804	0.7328	0.6890	0.6489	0.6119	0.5779	0.5466	0.5176	0.4909
2.3	0.8997	0.8410	0.7873	0.7381	0.6931	0.6518	0.6140	0.5792	0.5472	0.5177	0.4905
2.4	0.9093	0.8486	0.7933	0.7427	0.6965	0.6543	0.6155	0.5800	0.5474	0.5174	0.4898
2.5	0.9179	0.8554	0.7986	0.7467	0.6994	0.6562	0.6167	0.5805	0.5473	0.5169	0.4889
2.6	0.9257	0.8615	0.8032	0.7501	0.7018	0.6577	0.6175	0.5807	0.5470	0.5161	0.4878
2.7	0.9328	0.8670	0.8073	0.7531	0.7038	0.6589	0.6180	0.5806	0.5464	0.5152	0.4865
2.8	0.9392	0.8719	0.8109	0.7556	0.7054	0.6597	0.6182	0.5803	0.5457	0.5141	0.4852
2.9	0.9450	0.8762	0.8140	0.7577	0.7067	0.6603	0.6182	0.5798	0.5449	0.5129	0.4837
3.0	0.9502	0.8801	0.8167	0.7595	0.7077	0.6607	0.6180	0.5792	0.5439	0.5117	0.4822
3.1	0.9550	0.8835	0.8191	0.7610	0.7084	0.6608	0.6177	0.5785	0.5428	0.5103	0.4807
3.2	0.9592	0.8866	0.8212	0.7622	0.7090	0.6608	0.6172	0.5776	0.5417	0.5090	0.4792
3.3	0.9631	0.8893	0.8230	0.7632	0.7093	0.6607	0.6166	0.5767	0.5405	0.5076	0.4776
3.4	0.9666	0.8918	0.8245	0.7640	0.7096	0.6604	0.6160	0.5757	0.5393	0.5062	0.4760
3.5	0.9698	0.8939	0.8258	0.7647	0.7096	0.6600	0.6152	0.5747	0.5380	0.5047	0.4745
3.6	0.9727	0.8958	0.8270	0.7651	0.7096	0.6596	0.6144	0.5737	0.5368	0.5033	0.4729
3.7	0.9753	0.8975	0.8279	0.7655	0.7094	0.6590	0.6136	0.5726	0.5355	0.5019	0.4714
3.8	0.9776	0.8990	0.8287	0.7657	0.7092	0.6584	0.6127	0.5715	0.5342	0.5005	0.4699
3.9	0.9798	0.9003	0.8294	0.7659	0.7089	0.6578	0.6118	0.5704	0.5329	0.4991	0.4684
4.0	0.9817	0.9015	0.8299	0.7659	0.7086	0.6571	0.6109	0.5693	0.5317	0.4977	0.4670
4.1	0.9834	0.9025	0.8303	0.7659	0.7082	0.6564	0.6100	0.5682	0.5305	0.4964	0.4656
4.2	0.9850	0.9034	0.8307	0.7658	0.7077	0.6557	0.6091	0.5671	0.5292	0.4951	0.4642
4.3	0.9864	0.9042	0.8310	0.7656	0.7073	0.6550	0.6081	0.5660	0.5281	0.4938	0.4629
4.4	0.9877	0.9049	0.8312	0.7655	0.7068	0.6543	0.6072	0.5649	0.5269	0.4926	0.4616
4.5	0.9889	0.9055	0.8313	0.7652	0.7063	0.6535	0.6063	0.5639	0.5258	0.4914	0.4603
4.6	0.9899	0.9060	0.8314	0.7650	0.7057	0.6528	0.6054	0.5629	0.5247	0.4902	0.4591
4.7	0.9909	0.9064	0.8314	0.7647	0.7052	0.6521	0.6045	0.5619	0.5236	0.4891	0.4580
4.8	0.9918	0.9068	0.8314	0.7644	0.7047	0.6514	0.6037	0.5609	0.5226	0.4880	0.4569
4.9	0.9926	0.9071	0.8314	0.7641	0.7041	0.6507	0.6028	0.5600	0.5216	0.4870	0.4558
5.0	0.9933	0.9074	0.8313	0.7637	0.7036	0.6500	0.6020	0.5591	0.5206	0.4860	0.4548
5.1	0.9939	0.9077	0.8312	0.7634	0.7031	0.6493	0.6012	0.5582	0.5197	0.4850	0.4538
5.2	0.9945	0.9079	0.8311	0.7631	0.7025	0.6486	0.6005	0.5574	0.5188	0.4841	0.4528
5.3	0.9950	0.9080	0.8310	0.7627	0.7020	0.6479	0.5997	0.5566	0.5179	0.4832	0.4519
5.4	0.9955	0.9081	0.8309	0.7624	0.7015	0.6473	0.5990	0.5558	0.5171	0.4823	0.4510
5.5	0.9959	0.9082	0.8307	0.7620	0.7010	0.6467	0.5983	0.5550	0.5162	0.4815	0.4502
5.6	0.9963	0.9083	0.8305	0.7617	0.7005	0.6461	0.5976	0.5543	0.5155	0.4807	0.4493
5.7	0.9967	0.9084	0.8304	0.7613	0.7000	0.6455	0.5969	0.5536	0.5147	0.4799	0.4486
5.8	0.9970	0.9084	0.8302	0.7610	0.6995	0.6449	0.5963	0.5529	0.5140	0.4792	0.4478
5.9	0.9973	0.9085	0.8300	0.7606	0.6991	0.6444	0.5957	0.5522	0.5133	0.4784	0.4471
6.0	0.9975	0.9085	0.8298	0.7603	0.6987	0.6439	0.5951	0.5516	0.5127	0.4778	0.4464

NTU	Heat Exchanger Effectiveness										
	C^* ($C_{min} = C_c$)										
	0	0.1	0.2	0.3	0.4	0.5	0.6	0.7	0.8	0.9	1.0
6.1	0.9978	0.9085	0.8297	0.7600	0.6982	0.6434	0.5945	0.5510	0.5120	0.4771	0.4458
6.2	0.9980	0.9085	0.8295	0.7597	0.6978	0.6429	0.5940	0.5504	0.5114	0.4765	0.4451
6.3	0.9982	0.9084	0.8293	0.7594	0.6974	0.6424	0.5935	0.5498	0.5109	0.4759	0.4445
6.4	0.9983	0.9084	0.8291	0.7591	0.6970	0.6420	0.5930	0.5493	0.5103	0.4753	0.4440
6.5	0.9985	0.9084	0.8289	0.7588	0.6966	0.6415	0.5925	0.5488	0.5098	0.4748	0.4434
6.6	0.9986	0.9083	0.8288	0.7585	0.6963	0.6411	0.5920	0.5483	0.5093	0.4743	0.4429
6.7	0.9988	0.9083	0.8286	0.7582	0.6959	0.6407	0.5916	0.5478	0.5088	0.4738	0.4424
6.8	0.9989	0.9083	0.8284	0.7580	0.6956	0.6403	0.5912	0.5474	0.5083	0.4733	0.4419
6.9	0.9990	0.9082	0.8283	0.7577	0.6953	0.6399	0.5908	0.5470	0.5079	0.4729	0.4415
7.0	0.9991	0.9082	0.8281	0.7575	0.6950	0.6396	0.5904	0.5466	0.5074	0.4724	0.4410
7.1	0.9992	0.9081	0.8280	0.7572	0.6947	0.6393	0.5900	0.5462	0.5070	0.4720	0.4406
7.2	0.9993	0.9081	0.8278	0.7570	0.6944	0.6389	0.5897	0.5458	0.5067	0.4716	0.4402
7.3	0.9993	0.9080	0.8277	0.7568	0.6941	0.6386	0.5893	0.5454	0.5063	0.4713	0.4399
7.4	0.9994	0.9080	0.8275	0.7566	0.6939	0.6383	0.5890	0.5451	0.5059	0.4709	0.4395
7.5	0.9994	0.9079	0.8274	0.7564	0.6936	0.6380	0.5887	0.5448	0.5056	0.4706	0.4392
7.6	0.9995	0.9079	0.8272	0.7562	0.6934	0.6378	0.5884	0.5445	0.5053	0.4702	0.4388
7.7	0.9995	0.9078	0.8271	0.7560	0.6931	0.6375	0.5881	0.5442	0.5050	0.4699	0.4385
7.8	0.9996	0.9077	0.8270	0.7558	0.6929	0.6372	0.5878	0.5439	0.5047	0.4696	0.4382
7.9	0.9996	0.9077	0.8269	0.7556	0.6927	0.6370	0.5876	0.5436	0.5044	0.4694	0.4379
8.0	0.9997	0.9076	0.8267	0.7555	0.6925	0.6368	0.5873	0.5433	0.5041	0.4691	0.4377
8.1	0.9997	0.9076	0.8266	0.7553	0.6923	0.6366	0.5871	0.5431	0.5039	0.4688	0.4374
8.2	0.9997	0.9076	0.8265	0.7552	0.6921	0.6363	0.5869	0.5429	0.5036	0.4686	0.4372
8.3	0.9998	0.9075	0.8264	0.7550	0.6919	0.6361	0.5866	0.5426	0.5034	0.4683	0.4369
8.4	0.9998	0.9075	0.8263	0.7549	0.6918	0.6360	0.5864	0.5424	0.5032	0.4681	0.4367
8.5	0.9998	0.9074	0.8262	0.7547	0.6916	0.6358	0.5863	0.5422	0.5030	0.4679	0.4365
8.6	0.9998	0.9074	0.8261	0.7546	0.6915	0.6356	0.5861	0.5420	0.5028	0.4677	0.4363
8.7	0.9998	0.9073	0.8261	0.7545	0.6913	0.6354	0.5859	0.5418	0.5026	0.4675	0.4361
8.8	0.9998	0.9073	0.8260	0.7544	0.6912	0.6353	0.5857	0.5417	0.5024	0.4673	0.4359
8.9	0.9999	0.9072	0.8259	0.7543	0.6910	0.6351	0.5856	0.5415	0.5022	0.4672	0.4358
9.0	0.9999	0.9072	0.8258	0.7541	0.6909	0.6350	0.5854	0.5413	0.5021	0.4670	0.4356
9.1	0.9999	0.9072	0.8257	0.7540	0.6908	0.6349	0.5853	0.5412	0.5019	0.4668	0.4354
9.2	0.9999	0.9071	0.8257	0.7540	0.6907	0.6347	0.5851	0.5410	0.5018	0.4667	0.4353
9.3	0.9999	0.9071	0.8256	0.7539	0.6906	0.6346	0.5850	0.5409	0.5016	0.4665	0.4351
9.4	0.9999	0.9071	0.8255	0.7538	0.6905	0.6345	0.5849	0.5408	0.5015	0.4664	0.4350
9.5	0.9999	0.9070	0.8255	0.7537	0.6904	0.6344	0.5847	0.5406	0.5014	0.4663	0.4349
9.6	0.9999	0.9070	0.8254	0.7536	0.6903	0.6343	0.5846	0.5405	0.5012	0.4662	0.4348
9.7	0.9999	0.9070	0.8254	0.7535	0.6902	0.6342	0.5845	0.5404	0.5011	0.4660	0.4346
9.8	0.9999	0.9069	0.8253	0.7535	0.6901	0.6341	0.5844	0.5403	0.5010	0.4659	0.4345
9.9	1.0000	0.9069	0.8252	0.7534	0.6900	0.6340	0.5843	0.5402	0.5009	0.4658	0.4344
10.0	1.0000	0.9069	0.8252	0.7533	0.6899	0.6339	0.5842	0.5401	0.5008	0.4657	0.4343

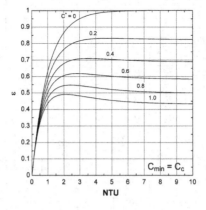

CONFIGURATION 11

1TUBE2ROW2PASS1CIR(PAR)

(b) $C_{min} = C_h$

NTU	Heat Exchanger Effectiveness										
	C^* ($C_{min} = C_h$)										
	0	0.1	0.2	0.3	0.4	0.5	0.6	0.7	0.8	0.9	1.0
0	0.0000	0.0000	0.0000	0.0000	0.0000	0.0000	0.0000	0.0000	0.0000	0.0000	0.0000
0.1	0.0952	0.0947	0.0942	0.0938	0.0933	0.0929	0.0924	0.0920	0.0915	0.0911	0.0907
0.2	0.1813	0.1796	0.1779	0.1762	0.1745	0.1729	0.1713	0.1697	0.1681	0.1666	0.1651
0.3	0.2592	0.2556	0.2521	0.2486	0.2453	0.2419	0.2387	0.2355	0.2323	0.2292	0.2262
0.4	0.3297	0.3238	0.3180	0.3124	0.3069	0.3015	0.2963	0.2912	0.2862	0.2813	0.2765
0.5	0.3935	0.3849	0.3766	0.3685	0.3606	0.3530	0.3456	0.3384	0.3314	0.3246	0.3180
0.6	0.4512	0.4397	0.4286	0.4178	0.4074	0.3974	0.3877	0.3784	0.3693	0.3606	0.3521
0.7	0.5034	0.4888	0.4748	0.4613	0.4483	0.4358	0.4238	0.4123	0.4011	0.3905	0.3802
0.8	0.5507	0.5329	0.5158	0.4995	0.4839	0.4690	0.4547	0.4410	0.4278	0.4153	0.4032
0.9	0.5934	0.5724	0.5523	0.5332	0.5150	0.4976	0.4810	0.4653	0.4502	0.4358	0.4221
1.0	0.6321	0.6078	0.5847	0.5628	0.5420	0.5223	0.5035	0.4858	0.4689	0.4528	0.4376
1.1	0.6671	0.6395	0.6134	0.5888	0.5655	0.5435	0.5227	0.5030	0.4844	0.4668	0.4502
1.2	0.6988	0.6680	0.6389	0.6116	0.5859	0.5618	0.5390	0.5175	0.4973	0.4783	0.4603
1.3	0.7275	0.6934	0.6615	0.6317	0.6037	0.5774	0.5528	0.5297	0.5080	0.4876	0.4684
1.4	0.7534	0.7163	0.6816	0.6492	0.6190	0.5908	0.5644	0.5397	0.5167	0.4951	0.4749
1.5	0.7769	0.7367	0.6993	0.6645	0.6322	0.6021	0.5741	0.5480	0.5237	0.5011	0.4799
1.6	0.7981	0.7549	0.7150	0.6779	0.6436	0.6118	0.5823	0.5549	0.5294	0.5058	0.4838
1.7	0.8173	0.7713	0.7288	0.6896	0.6533	0.6199	0.5890	0.5604	0.5339	0.5094	0.4866
1.8	0.8347	0.7858	0.7409	0.6996	0.6616	0.6266	0.5944	0.5647	0.5373	0.5120	0.4887
1.9	0.8504	0.7988	0.7516	0.7083	0.6686	0.6322	0.5988	0.5681	0.5399	0.5139	0.4900
2.0	0.8647	0.8104	0.7609	0.7158	0.6745	0.6368	0.6023	0.5707	0.5417	0.5152	0.4908
2.1	0.8775	0.8207	0.7691	0.7221	0.6794	0.6404	0.6050	0.5726	0.5429	0.5159	0.4910
2.2	0.8892	0.8299	0.7762	0.7275	0.6834	0.6433	0.6069	0.5738	0.5436	0.5161	0.4909
2.3	0.8997	0.8380	0.7823	0.7320	0.6866	0.6455	0.6082	0.5745	0.5438	0.5159	0.4905
2.4	0.9093	0.8452	0.7875	0.7357	0.6891	0.6470	0.6090	0.5747	0.5436	0.5154	0.4898
2.5	0.9179	0.8515	0.7920	0.7387	0.6909	0.6480	0.6094	0.5746	0.5431	0.5147	0.4889
2.6	0.9257	0.8570	0.7958	0.7411	0.6923	0.6485	0.6093	0.5741	0.5423	0.5137	0.4878
2.7	0.9328	0.8619	0.7990	0.7430	0.6931	0.6486	0.6089	0.5733	0.5413	0.5125	0.4865
2.8	0.9392	0.8661	0.8015	0.7443	0.6935	0.6484	0.6082	0.5722	0.5401	0.5112	0.4852
2.9	0.9450	0.8698	0.8036	0.7452	0.6936	0.6478	0.6072	0.5710	0.5387	0.5098	0.4837
3.0	0.9502	0.8730	0.8052	0.7457	0.6933	0.6470	0.6060	0.5696	0.5372	0.5082	0.4822
3.1	0.9550	0.8757	0.8064	0.7458	0.6927	0.6459	0.6046	0.5681	0.5356	0.5066	0.4807
3.2	0.9592	0.8779	0.8073	0.7457	0.6918	0.6446	0.6031	0.5664	0.5339	0.5050	0.4792
3.3	0.9631	0.8798	0.8078	0.7452	0.6907	0.6431	0.6014	0.5646	0.5322	0.5033	0.4776
3.4	0.9666	0.8814	0.8080	0.7445	0.6894	0.6415	0.5996	0.5628	0.5304	0.5016	0.4760
3.5	0.9698	0.8826	0.8079	0.7435	0.6879	0.6397	0.5977	0.5609	0.5285	0.4999	0.4745
3.6	0.9727	0.8836	0.8075	0.7424	0.6863	0.6378	0.5957	0.5589	0.5267	0.4982	0.4729
3.7	0.9753	0.8843	0.8070	0.7410	0.6845	0.6358	0.5936	0.5569	0.5248	0.4965	0.4714
3.8	0.9776	0.8848	0.8063	0.7395	0.6826	0.6337	0.5915	0.5549	0.5229	0.4948	0.4699
3.9	0.9798	0.8851	0.8053	0.7379	0.6806	0.6316	0.5894	0.5529	0.5211	0.4931	0.4684
4.0	0.9817	0.8851	0.8043	0.7362	0.6785	0.6294	0.5873	0.5509	0.5192	0.4915	0.4670
4.1	0.9834	0.8850	0.8030	0.7343	0.6763	0.6271	0.5851	0.5489	0.5174	0.4899	0.4656
4.2	0.9850	0.8848	0.8017	0.7323	0.6741	0.6248	0.5829	0.5468	0.5156	0.4883	0.4642
4.3	0.9864	0.8844	0.8002	0.7303	0.6718	0.6225	0.5807	0.5448	0.5138	0.4867	0.4629
4.4	0.9877	0.8839	0.7986	0.7282	0.6695	0.6202	0.5785	0.5428	0.5121	0.4852	0.4616
4.5	0.9889	0.8833	0.7970	0.7260	0.6671	0.6179	0.5763	0.5409	0.5103	0.4838	0.4603
4.6	0.9899	0.8825	0.7952	0.7237	0.6647	0.6155	0.5742	0.5389	0.5087	0.4823	0.4591
4.7	0.9909	0.8817	0.7934	0.7214	0.6623	0.6132	0.5720	0.5370	0.5070	0.4809	0.4580
4.8	0.9918	0.8808	0.7915	0.7191	0.6598	0.6108	0.5699	0.5352	0.5054	0.4796	0.4569
4.9	0.9926	0.8798	0.7896	0.7167	0.6574	0.6085	0.5678	0.5333	0.5038	0.4783	0.4558
5.0	0.9933	0.8787	0.7876	0.7143	0.6549	0.6062	0.5657	0.5315	0.5023	0.4770	0.4548
5.1	0.9939	0.8776	0.7855	0.7119	0.6525	0.6039	0.5636	0.5297	0.5008	0.4758	0.4538
5.2	0.9945	0.8764	0.7834	0.7095	0.6500	0.6016	0.5616	0.5280	0.4994	0.4746	0.4528
5.3	0.9950	0.8752	0.7813	0.7070	0.6476	0.5993	0.5596	0.5263	0.4980	0.4734	0.4519
5.4	0.9955	0.8739	0.7792	0.7046	0.6451	0.5971	0.5576	0.5246	0.4966	0.4723	0.4510
5.5	0.9959	0.8726	0.7770	0.7021	0.6427	0.5949	0.5557	0.5230	0.4953	0.4713	0.4502
5.6	0.9963	0.8713	0.7748	0.6997	0.6403	0.5927	0.5538	0.5214	0.4940	0.4702	0.4493
5.7	0.9967	0.8699	0.7726	0.6972	0.6379	0.5905	0.5519	0.5199	0.4927	0.4692	0.4486
5.8	0.9970	0.8684	0.7704	0.6947	0.6355	0.5884	0.5501	0.5184	0.4915	0.4683	0.4478
5.9	0.9973	0.8670	0.7681	0.6923	0.6332	0.5863	0.5483	0.5169	0.4903	0.4674	0.4471
6.0	0.9975	0.8655	0.7659	0.6898	0.6308	0.5842	0.5466	0.5155	0.4892	0.4665	0.4464

NTU	Heat Exchanger Effectiveness										
	C^* ($C_{min} = C_h$)										
	0	0.1	0.2	0.3	0.4	0.5	0.6	0.7	0.8	0.9	1.0
6.1	0.9978	0.8640	0.7637	0.6874	0.6285	0.5822	0.5449	0.5141	0.4881	0.4656	0.4458
6.2	0.9980	0.8625	0.7614	0.6850	0.6262	0.5802	0.5432	0.5128	0.4870	0.4648	0.4451
6.3	0.9982	0.8610	0.7591	0.6825	0.6240	0.5782	0.5416	0.5114	0.4860	0.4640	0.4445
6.4	0.9983	0.8595	0.7569	0.6801	0.6217	0.5763	0.5400	0.5102	0.4850	0.4632	0.4440
6.5	0.9985	0.8579	0.7546	0.6778	0.6195	0.5744	0.5384	0.5089	0.4840	0.4625	0.4434
6.6	0.9986	0.8564	0.7524	0.6754	0.6173	0.5725	0.5369	0.5077	0.4831	0.4618	0.4429
6.7	0.9988	0.8548	0.7501	0.6730	0.6152	0.5707	0.5354	0.5065	0.4822	0.4611	0.4424
6.8	0.9989	0.8532	0.7479	0.6707	0.6131	0.5688	0.5339	0.5054	0.4814	0.4605	0.4419
6.9	0.9990	0.8516	0.7456	0.6684	0.6110	0.5671	0.5325	0.5043	0.4805	0.4599	0.4415
7.0	0.9991	0.8500	0.7434	0.6661	0.6089	0.5653	0.5311	0.5032	0.4797	0.4593	0.4410
7.1	0.9992	0.8484	0.7412	0.6638	0.6068	0.5636	0.5298	0.5022	0.4789	0.4587	0.4406
7.2	0.9993	0.8468	0.7390	0.6616	0.6048	0.5620	0.5284	0.5012	0.4782	0.4582	0.4402
7.3	0.9993	0.8452	0.7368	0.6594	0.6029	0.5603	0.5272	0.5002	0.4775	0.4576	0.4399
7.4	0.9994	0.8436	0.7346	0.6572	0.6009	0.5587	0.5259	0.4993	0.4768	0.4571	0.4395
7.5	0.9994	0.8420	0.7324	0.6550	0.5990	0.5572	0.5247	0.4983	0.4761	0.4566	0.4392
7.6	0.9995	0.8404	0.7303	0.6528	0.5971	0.5556	0.5235	0.4975	0.4754	0.4562	0.4388
7.7	0.9995	0.8388	0.7281	0.6507	0.5952	0.5541	0.5223	0.4966	0.4748	0.4557	0.4385
7.8	0.9996	0.8372	0.7260	0.6486	0.5934	0.5527	0.5212	0.4958	0.4742	0.4553	0.4382
7.9	0.9996	0.8356	0.7238	0.6465	0.5916	0.5512	0.5201	0.4949	0.4736	0.4549	0.4379
8.0	0.9997	0.8340	0.7217	0.6444	0.5898	0.5498	0.5190	0.4942	0.4731	0.4545	0.4377
8.1	0.9997	0.8324	0.7196	0.6424	0.5881	0.5484	0.5180	0.4934	0.4725	0.4541	0.4374
8.2	0.9997	0.8308	0.7175	0.6404	0.5864	0.5471	0.5170	0.4927	0.4720	0.4538	0.4372
8.3	0.9998	0.8292	0.7155	0.6384	0.5847	0.5457	0.5160	0.4920	0.4715	0.4534	0.4369
8.4	0.9998	0.8277	0.7134	0.6364	0.5830	0.5445	0.5150	0.4913	0.4710	0.4531	0.4367
8.5	0.9998	0.8261	0.7114	0.6345	0.5814	0.5432	0.5141	0.4906	0.4706	0.4528	0.4365
8.6	0.9998	0.8245	0.7094	0.6326	0.5798	0.5419	0.5132	0.4900	0.4701	0.4525	0.4363
8.7	0.9998	0.8229	0.7074	0.6307	0.5782	0.5407	0.5123	0.4893	0.4697	0.4522	0.4361
8.8	0.9998	0.8214	0.7054	0.6288	0.5767	0.5396	0.5114	0.4887	0.4693	0.4519	0.4359
8.9	0.9999	0.8198	0.7034	0.6269	0.5752	0.5384	0.5106	0.4881	0.4689	0.4517	0.4358
9.0	0.9999	0.8182	0.7014	0.6251	0.5737	0.5373	0.5098	0.4876	0.4685	0.4514	0.4356
9.1	0.9999	0.8167	0.6995	0.6233	0.5722	0.5362	0.5090	0.4870	0.4681	0.4512	0.4354
9.2	0.9999	0.8151	0.6976	0.6215	0.5708	0.5351	0.5082	0.4865	0.4678	0.4509	0.4353
9.3	0.9999	0.8136	0.6956	0.6198	0.5694	0.5340	0.5075	0.4860	0.4675	0.4507	0.4351
9.4	0.9999	0.8121	0.6937	0.6180	0.5680	0.5330	0.5067	0.4855	0.4671	0.4505	0.4350
9.5	0.9999	0.8105	0.6919	0.6163	0.5666	0.5320	0.5060	0.4850	0.4668	0.4503	0.4349
9.6	0.9999	0.8090	0.6900	0.6146	0.5653	0.5310	0.5054	0.4846	0.4665	0.4501	0.4348
9.7	0.9999	0.8075	0.6882	0.6130	0.5639	0.5300	0.5047	0.4841	0.4662	0.4499	0.4346
9.8	0.9999	0.8060	0.6863	0.6113	0.5627	0.5291	0.5040	0.4837	0.4659	0.4497	0.4345
9.9	1.0000	0.8045	0.6845	0.6097	0.5614	0.5282	0.5034	0.4833	0.4657	0.4496	0.4344
10.0	1.0000	0.8030	0.6827	0.6081	0.5601	0.5273	0.5028	0.4828	0.4654	0.4494	0.4343

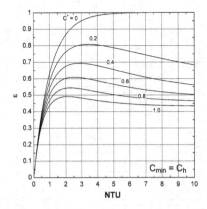

CONFIGURATION 12

1TUBE3ROW3PASS1CIR(PAR)

(a) $C_{min} = C_c$

NTU	Heat Exchanger Effectiveness $C^*(C_{min} = C_c)$										
	0	0.1	0.2	0.3	0.4	0.5	0.6	0.7	0.8	0.9	1.0
0	0.0000	0.0000	0.0000	0.0000	0.0000	0.0000	0.0000	0.0000	0.0000	0.0000	0.0000
0.1	0.0952	0.0947	0.0942	0.0938	0.0933	0.0929	0.0924	0.0920	0.0915	0.0911	0.0906
0.2	0.1813	0.1795	0.1778	0.1761	0.1745	0.1728	0.1712	0.1696	0.1680	0.1665	0.1649
0.3	0.2592	0.2556	0.2520	0.2485	0.2451	0.2417	0.2384	0.2352	0.2320	0.2289	0.2259
0.4	0.3297	0.3237	0.3178	0.3121	0.3066	0.3011	0.2958	0.2907	0.2856	0.2807	0.2759
0.5	0.3935	0.3847	0.3762	0.3680	0.3600	0.3523	0.3448	0.3375	0.3304	0.3236	0.3169
0.6	0.4512	0.4394	0.4281	0.4172	0.4066	0.3964	0.3866	0.3771	0.3680	0.3592	0.3506
0.7	0.5034	0.4885	0.4741	0.4604	0.4472	0.4345	0.4223	0.4107	0.3994	0.3887	0.3783
0.8	0.5507	0.5324	0.5150	0.4984	0.4825	0.4673	0.4528	0.4390	0.4258	0.4131	0.4010
0.9	0.5934	0.5718	0.5513	0.5318	0.5132	0.4956	0.4789	0.4630	0.4478	0.4334	0.4197
1.0	0.6321	0.6071	0.5835	0.5611	0.5400	0.5200	0.5011	0.4832	0.4663	0.4502	0.4350
1.1	0.6671	0.6387	0.6121	0.5870	0.5633	0.5411	0.5201	0.5003	0.4817	0.4641	0.4475
1.2	0.6988	0.6671	0.6374	0.6097	0.5836	0.5592	0.5363	0.5148	0.4946	0.4756	0.4578
1.3	0.7275	0.6925	0.6600	0.6296	0.6013	0.5748	0.5501	0.5270	0.5054	0.4851	0.4661
1.4	0.7534	0.7153	0.6800	0.6471	0.6166	0.5883	0.5619	0.5373	0.5143	0.4929	0.4730
1.5	0.7769	0.7357	0.6977	0.6625	0.6300	0.5998	0.5719	0.5459	0.5218	0.4994	0.4785
1.6	0.7981	0.7540	0.7134	0.6761	0.6416	0.6098	0.5804	0.5532	0.5280	0.5047	0.4831
1.7	0.8173	0.7704	0.7274	0.6879	0.6517	0.6183	0.5876	0.5593	0.5332	0.5091	0.4868
1.8	0.8347	0.7851	0.7398	0.6984	0.6604	0.6257	0.5938	0.5645	0.5375	0.5127	0.4897
1.9	0.8504	0.7983	0.7508	0.7075	0.6680	0.6320	0.5990	0.5688	0.5411	0.5156	0.4922
2.0	0.8647	0.8101	0.7605	0.7155	0.6746	0.6374	0.6034	0.5724	0.5440	0.5180	0.4941
2.1	0.8775	0.8206	0.7692	0.7226	0.6803	0.6420	0.6071	0.5754	0.5464	0.5199	0.4957
2.2	0.8892	0.8301	0.7768	0.7287	0.6853	0.6460	0.6103	0.5779	0.5484	0.5215	0.4970
2.3	0.8997	0.8385	0.7836	0.7341	0.6896	0.6493	0.6129	0.5800	0.5501	0.5228	0.4980
2.4	0.9093	0.8461	0.7896	0.7388	0.6932	0.6522	0.6152	0.5817	0.5514	0.5239	0.4989
2.5	0.9179	0.8529	0.7949	0.7429	0.6964	0.6546	0.6170	0.5831	0.5525	0.5248	0.4996
2.6	0.9257	0.8590	0.7996	0.7465	0.6992	0.6567	0.6186	0.5843	0.5534	0.5255	0.5001
2.7	0.9328	0.8644	0.8037	0.7497	0.7015	0.6585	0.6199	0.5853	0.5542	0.5260	0.5006
2.8	0.9392	0.8693	0.8074	0.7524	0.7035	0.6600	0.6210	0.5861	0.5548	0.5265	0.5010
2.9	0.9450	0.8736	0.8106	0.7548	0.7053	0.6612	0.6219	0.5868	0.5553	0.5269	0.5014
3.0	0.9502	0.8775	0.8134	0.7569	0.7068	0.6623	0.6227	0.5874	0.5557	0.5273	0.5017
3.1	0.9550	0.8810	0.8160	0.7587	0.7080	0.6632	0.6233	0.5878	0.5561	0.5276	0.5020
3.2	0.9592	0.8841	0.8182	0.7602	0.7091	0.6639	0.6239	0.5882	0.5564	0.5279	0.5022
3.3	0.9631	0.8869	0.8201	0.7616	0.7101	0.6646	0.6243	0.5885	0.5566	0.5281	0.5025
3.4	0.9666	0.8893	0.8218	0.7627	0.7108	0.6651	0.6247	0.5888	0.5569	0.5283	0.5028
3.5	0.9698	0.8915	0.8234	0.7638	0.7115	0.6655	0.6250	0.5890	0.5571	0.5286	0.5030
3.6	0.9727	0.8935	0.8247	0.7646	0.7121	0.6659	0.6252	0.5892	0.5573	0.5288	0.5033
3.7	0.9753	0.8953	0.8258	0.7654	0.7125	0.6662	0.6254	0.5894	0.5575	0.5290	0.5036
3.8	0.9776	0.8968	0.8269	0.7660	0.7129	0.6665	0.6256	0.5896	0.5576	0.5292	0.5038
3.9	0.9798	0.8982	0.8278	0.7666	0.7133	0.6667	0.6258	0.5897	0.5578	0.5295	0.5041
4.0	0.9817	0.8995	0.8285	0.7670	0.7136	0.6669	0.6259	0.5899	0.5580	0.5297	0.5044
4.1	0.9834	0.9006	0.8292	0.7674	0.7138	0.6670	0.6261	0.5900	0.5582	0.5299	0.5047
4.2	0.9850	0.9016	0.8298	0.7678	0.7140	0.6672	0.6262	0.5902	0.5584	0.5302	0.5050
4.3	0.9864	0.9025	0.8303	0.7681	0.7142	0.6673	0.6263	0.5903	0.5586	0.5304	0.5054
4.4	0.9877	0.9033	0.8308	0.7683	0.7143	0.6674	0.6264	0.5905	0.5588	0.5307	0.5057
4.5	0.9889	0.9039	0.8312	0.7685	0.7144	0.6675	0.6265	0.5906	0.5590	0.5310	0.5061
4.6	0.9899	0.9046	0.8315	0.7687	0.7145	0.6676	0.6266	0.5908	0.5592	0.5313	0.5064
4.7	0.9909	0.9051	0.8318	0.7689	0.7146	0.6676	0.6267	0.5909	0.5594	0.5316	0.5068
4.8	0.9918	0.9056	0.8320	0.7690	0.7147	0.6677	0.6268	0.5911	0.5596	0.5318	0.5071
4.9	0.9926	0.9060	0.8323	0.7691	0.7148	0.6678	0.6269	0.5913	0.5599	0.5322	0.5075
5.0	0.9933	0.9064	0.8324	0.7692	0.7148	0.6679	0.6271	0.5914	0.5601	0.5325	0.5079
5.1	0.9939	0.9067	0.8326	0.7693	0.7149	0.6679	0.6272	0.5916	0.5604	0.5328	0.5082
5.2	0.9945	0.9070	0.8327	0.7693	0.7149	0.6680	0.6273	0.5918	0.5606	0.5331	0.5086
5.3	0.9950	0.9073	0.8329	0.7694	0.7150	0.6681	0.6274	0.5920	0.5609	0.5334	0.5090
5.4	0.9955	0.9075	0.8330	0.7694	0.7150	0.6682	0.6276	0.5922	0.5611	0.5337	0.5094
5.5	0.9959	0.9077	0.8330	0.7695	0.7151	0.6682	0.6277	0.5924	0.5614	0.5341	0.5098
5.6	0.9963	0.9079	0.8331	0.7695	0.7151	0.6683	0.6278	0.5926	0.5617	0.5344	0.5102
5.7	0.9967	0.9081	0.8332	0.7695	0.7152	0.6684	0.6280	0.5928	0.5619	0.5347	0.5106
5.8	0.9970	0.9082	0.8332	0.7696	0.7152	0.6685	0.6281	0.5930	0.5622	0.5350	0.5109
5.9	0.9973	0.9083	0.8333	0.7696	0.7152	0.6686	0.6283	0.5932	0.5625	0.5354	0.5113
6.0	0.9975	0.9084	0.8333	0.7696	0.7153	0.6687	0.6284	0.5934	0.5627	0.5357	0.5117

NTU	Heat Exchanger Effectiveness										
	C^* ($C_{min} = C_c$)										
	0	0.1	0.2	0.3	0.4	0.5	0.6	0.7	0.8	0.9	1.0
6.1	0.9978	0.9085	0.8333	0.7696	0.7153	0.6688	0.6286	0.5936	0.5630	0.5360	0.5121
6.2	0.9980	0.9086	0.8334	0.7697	0.7154	0.6689	0.6287	0.5938	0.5633	0.5364	0.5124
6.3	0.9982	0.9087	0.8334	0.7697	0.7154	0.6690	0.6289	0.5940	0.5635	0.5367	0.5128
6.4	0.9983	0.9087	0.8334	0.7697	0.7155	0.6691	0.6290	0.5942	0.5638	0.5370	0.5132
6.5	0.9985	0.9088	0.8334	0.7697	0.7156	0.6692	0.6292	0.5945	0.5641	0.5373	0.5135
6.6	0.9986	0.9088	0.8334	0.7698	0.7156	0.6693	0.6293	0.5947	0.5644	0.5376	0.5139
6.7	0.9988	0.9089	0.8335	0.7698	0.7157	0.6694	0.6295	0.5949	0.5646	0.5379	0.5142
6.8	0.9989	0.9089	0.8335	0.7698	0.7157	0.6695	0.6297	0.5951	0.5649	0.5382	0.5146
6.9	0.9990	0.9089	0.8335	0.7698	0.7158	0.6696	0.6298	0.5953	0.5651	0.5385	0.5149
7.0	0.9991	0.9090	0.8335	0.7698	0.7159	0.6697	0.6300	0.5955	0.5654	0.5388	0.5153
7.1	0.9992	0.9090	0.8335	0.7699	0.7159	0.6698	0.6301	0.5957	0.5656	0.5391	0.5156
7.2	0.9993	0.9090	0.8335	0.7699	0.7160	0.6699	0.6303	0.5959	0.5659	0.5394	0.5159
7.3	0.9993	0.9090	0.8335	0.7699	0.7160	0.6700	0.6305	0.5961	0.5661	0.5397	0.5162
7.4	0.9994	0.9090	0.8335	0.7700	0.7161	0.6701	0.6306	0.5963	0.5664	0.5400	0.5166
7.5	0.9994	0.9091	0.8335	0.7700	0.7162	0.6703	0.6308	0.5965	0.5666	0.5403	0.5169
7.6	0.9995	0.9091	0.8335	0.7700	0.7162	0.6704	0.6309	0.5967	0.5669	0.5405	0.5172
7.7	0.9995	0.9091	0.8335	0.7701	0.7163	0.6705	0.6311	0.5969	0.5671	0.5408	0.5174
7.8	0.9996	0.9091	0.8335	0.7701	0.7164	0.6706	0.6312	0.5971	0.5673	0.5411	0.5177
7.9	0.9996	0.9091	0.8336	0.7701	0.7164	0.6707	0.6314	0.5973	0.5676	0.5413	0.5180
8.0	0.9997	0.9091	0.8336	0.7702	0.7165	0.6708	0.6315	0.5975	0.5678	0.5416	0.5183
8.1	0.9997	0.9091	0.8336	0.7702	0.7166	0.6709	0.6317	0.5977	0.5680	0.5418	0.5186
8.2	0.9997	0.9091	0.8336	0.7702	0.7167	0.6710	0.6318	0.5979	0.5682	0.5421	0.5188
8.3	0.9998	0.9091	0.8336	0.7703	0.7167	0.6711	0.6320	0.5980	0.5684	0.5423	0.5191
8.4	0.9998	0.9091	0.8336	0.7703	0.7168	0.6712	0.6321	0.5982	0.5686	0.5425	0.5193
8.5	0.9998	0.9091	0.8336	0.7703	0.7169	0.6713	0.6322	0.5984	0.5688	0.5427	0.5196
8.6	0.9998	0.9091	0.8336	0.7704	0.7169	0.6714	0.6324	0.5986	0.5690	0.5430	0.5198
8.7	0.9998	0.9091	0.8336	0.7704	0.7170	0.6715	0.6325	0.5987	0.5692	0.5432	0.5200
8.8	0.9998	0.9091	0.8336	0.7704	0.7171	0.6716	0.6326	0.5989	0.5694	0.5434	0.5203
8.9	0.9999	0.9091	0.8337	0.7705	0.7171	0.6717	0.6328	0.5990	0.5696	0.5436	0.5205
9.0	0.9999	0.9091	0.8337	0.7705	0.7172	0.6718	0.6329	0.5992	0.5697	0.5438	0.5207
9.1	0.9999	0.9091	0.8337	0.7705	0.7172	0.6719	0.6330	0.5993	0.5699	0.5440	0.5209
9.2	0.9999	0.9091	0.8337	0.7706	0.7173	0.6720	0.6331	0.5995	0.5701	0.5442	0.5211
9.3	0.9999	0.9091	0.8337	0.7706	0.7174	0.6721	0.6333	0.5996	0.5702	0.5443	0.5213
9.4	0.9999	0.9091	0.8337	0.7706	0.7174	0.6722	0.6334	0.5998	0.5704	0.5445	0.5215
9.5	0.9999	0.9091	0.8337	0.7707	0.7175	0.6723	0.6335	0.5999	0.5706	0.5447	0.5217
9.6	0.9999	0.9091	0.8337	0.7707	0.7175	0.6724	0.6336	0.6000	0.5707	0.5449	0.5219
9.7	0.9999	0.9091	0.8338	0.7707	0.7176	0.6724	0.6337	0.6002	0.5709	0.5450	0.5220
9.8	0.9999	0.9091	0.8338	0.7708	0.7177	0.6725	0.6338	0.6003	0.5710	0.5452	0.5222
9.9	1.0000	0.9091	0.8338	0.7708	0.7177	0.6726	0.6339	0.6004	0.5711	0.5453	0.5224
10.0	1.0000	0.9091	0.8338	0.7708	0.7178	0.6727	0.6340	0.6005	0.5713	0.5455	0.5225

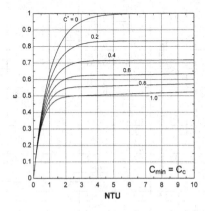

CONFIGURATION 12

1TUBE3ROW3PASS1CIR(PAR)

(b) $C_{min} = C_h$

NTU	Heat Exchanger Effectiveness										
	C^* ($C_{min} = C_h$)										
	0	0.1	0.2	0.3	0.4	0.5	0.6	0.7	0.8	0.9	1.0
0	0.0000	0.0000	0.0000	0.0000	0.0000	0.0000	0.0000	0.0000	0.0000	0.0000	0.0000
0.1	0.0952	0.0947	0.0942	0.0938	0.0933	0.0929	0.0924	0.0920	0.0915	0.0911	0.0906
0.2	0.1813	0.1795	0.1778	0.1761	0.1745	0.1728	0.1712	0.1696	0.1680	0.1665	0.1649
0.3	0.2592	0.2556	0.2520	0.2485	0.2451	0.2417	0.2384	0.2352	0.2320	0.2289	0.2259
0.4	0.3297	0.3237	0.3178	0.3121	0.3066	0.3011	0.2958	0.2907	0.2856	0.2807	0.2759
0.5	0.3935	0.3847	0.3762	0.3680	0.3601	0.3523	0.3448	0.3375	0.3305	0.3236	0.3169
0.6	0.4512	0.4394	0.4281	0.4172	0.4066	0.3965	0.3866	0.3772	0.3680	0.3592	0.3506
0.7	0.5034	0.4885	0.4741	0.4604	0.4472	0.4345	0.4224	0.4107	0.3995	0.3887	0.3783
0.8	0.5507	0.5324	0.5150	0.4984	0.4825	0.4673	0.4529	0.4390	0.4258	0.4131	0.4010
0.9	0.5934	0.5718	0.5513	0.5318	0.5133	0.4957	0.4789	0.4630	0.4478	0.4334	0.4197
1.0	0.6321	0.6071	0.5835	0.5612	0.5401	0.5201	0.5012	0.4833	0.4663	0.4502	0.4350
1.1	0.6671	0.6388	0.6121	0.5870	0.5634	0.5411	0.5202	0.5004	0.4817	0.4641	0.4475
1.2	0.6988	0.6671	0.6375	0.6097	0.5837	0.5593	0.5364	0.5149	0.4946	0.4756	0.4578
1.3	0.7275	0.6926	0.6601	0.6297	0.6014	0.5749	0.5502	0.5271	0.5054	0.4851	0.4661
1.4	0.7534	0.7154	0.6801	0.6473	0.6168	0.5884	0.5620	0.5374	0.5144	0.4930	0.4730
1.5	0.7769	0.7358	0.6978	0.6627	0.6302	0.6000	0.5720	0.5461	0.5219	0.4995	0.4785
1.6	0.7981	0.7541	0.7136	0.6763	0.6418	0.6100	0.5806	0.5534	0.5282	0.5048	0.4831
1.7	0.8173	0.7705	0.7276	0.6882	0.6520	0.6186	0.5879	0.5596	0.5334	0.5092	0.4868
1.8	0.8347	0.7853	0.7401	0.6987	0.6608	0.6260	0.5941	0.5647	0.5377	0.5128	0.4897
1.9	0.8504	0.7985	0.7511	0.7079	0.6685	0.6324	0.5994	0.5691	0.5413	0.5157	0.4922
2.0	0.8647	0.8103	0.7609	0.7160	0.6752	0.6379	0.6039	0.5728	0.5443	0.5181	0.4941
2.1	0.8775	0.8209	0.7696	0.7231	0.6810	0.6426	0.6077	0.5758	0.5468	0.5201	0.4957
2.2	0.8892	0.8304	0.7774	0.7294	0.6860	0.6467	0.6109	0.5784	0.5488	0.5218	0.4970
2.3	0.8997	0.8389	0.7842	0.7349	0.6904	0.6502	0.6137	0.5806	0.5505	0.5231	0.4980
2.4	0.9093	0.8466	0.7903	0.7398	0.6942	0.6532	0.6161	0.5824	0.5519	0.5242	0.4989
2.5	0.9179	0.8535	0.7958	0.7440	0.6976	0.6558	0.6181	0.5840	0.5531	0.5251	0.4996
2.6	0.9257	0.8596	0.8006	0.7478	0.7005	0.6580	0.6198	0.5853	0.5541	0.5258	0.5001
2.7	0.9328	0.8652	0.8049	0.7511	0.7030	0.6599	0.6213	0.5864	0.5550	0.5265	0.5006
2.8	0.9392	0.8701	0.8087	0.7540	0.7053	0.6616	0.6225	0.5874	0.5557	0.5270	0.5010
2.9	0.9450	0.8746	0.8121	0.7566	0.7072	0.6631	0.6236	0.5882	0.5563	0.5275	0.5014
3.0	0.9502	0.8786	0.8152	0.7590	0.7090	0.6644	0.6246	0.5889	0.5568	0.5279	0.5017
3.1	0.9550	0.8822	0.8179	0.7610	0.7105	0.6655	0.6254	0.5895	0.5573	0.5283	0.5020
3.2	0.9592	0.8855	0.8204	0.7628	0.7118	0.6666	0.6262	0.5901	0.5577	0.5286	0.5022
3.3	0.9631	0.8884	0.8226	0.7645	0.7131	0.6675	0.6269	0.5906	0.5581	0.5289	0.5025
3.4	0.9666	0.8911	0.8246	0.7660	0.7142	0.6683	0.6275	0.5911	0.5585	0.5292	0.5028
3.5	0.9698	0.8935	0.8264	0.7673	0.7152	0.6690	0.6281	0.5916	0.5589	0.5295	0.5030
3.6	0.9727	0.8957	0.8280	0.7685	0.7161	0.6697	0.6286	0.5920	0.5592	0.5298	0.5033
3.7	0.9753	0.8976	0.8295	0.7697	0.7169	0.6704	0.6291	0.5924	0.5596	0.5301	0.5036
3.8	0.9776	0.8994	0.8309	0.7707	0.7177	0.6710	0.6296	0.5928	0.5599	0.5304	0.5038
3.9	0.9798	0.9011	0.8321	0.7717	0.7185	0.6716	0.6301	0.5932	0.5603	0.5308	0.5041
4.0	0.9817	0.9026	0.8333	0.7726	0.7192	0.6722	0.6306	0.5936	0.5607	0.5311	0.5044
4.1	0.9834	0.9040	0.8344	0.7734	0.7199	0.6728	0.6311	0.5941	0.5610	0.5314	0.5047
4.2	0.9850	0.9052	0.8354	0.7742	0.7206	0.6733	0.6316	0.5945	0.5614	0.5318	0.5050
4.3	0.9864	0.9064	0.8364	0.7750	0.7212	0.6739	0.6321	0.5949	0.5618	0.5321	0.5054
4.4	0.9877	0.9075	0.8373	0.7758	0.7219	0.6744	0.6326	0.5954	0.5622	0.5325	0.5057
4.5	0.9889	0.9085	0.8381	0.7765	0.7225	0.6750	0.6331	0.5958	0.5626	0.5329	0.5061
4.6	0.9899	0.9095	0.8390	0.7773	0.7232	0.6756	0.6336	0.5963	0.5631	0.5333	0.5064
4.7	0.9909	0.9104	0.8398	0.7780	0.7238	0.6762	0.6341	0.5968	0.5635	0.5337	0.5068
4.8	0.9918	0.9112	0.8406	0.7787	0.7245	0.6768	0.6346	0.5973	0.5639	0.5341	0.5071
4.9	0.9926	0.9120	0.8414	0.7794	0.7251	0.6774	0.6352	0.5978	0.5644	0.5345	0.5075
5.0	0.9933	0.9128	0.8421	0.7801	0.7258	0.6780	0.6358	0.5983	0.5649	0.5349	0.5079
5.1	0.9939	0.9135	0.8429	0.7809	0.7265	0.6786	0.6363	0.5988	0.5653	0.5353	0.5082
5.2	0.9945	0.9143	0.8436	0.7816	0.7271	0.6792	0.6369	0.5993	0.5658	0.5357	0.5086
5.3	0.9950	0.9149	0.8443	0.7823	0.7278	0.6799	0.6375	0.5999	0.5663	0.5362	0.5090
5.4	0.9955	0.9156	0.8451	0.7830	0.7285	0.6805	0.6381	0.6004	0.5668	0.5366	0.5094
5.5	0.9959	0.9163	0.8458	0.7838	0.7293	0.6812	0.6387	0.6010	0.5673	0.5371	0.5098
5.6	0.9963	0.9169	0.8465	0.7845	0.7300	0.6819	0.6394	0.6016	0.5678	0.5375	0.5102
5.7	0.9967	0.9175	0.8473	0.7853	0.7307	0.6826	0.6400	0.6021	0.5683	0.5380	0.5106
5.8	0.9970	0.9182	0.8480	0.7861	0.7315	0.6833	0.6407	0.6027	0.5688	0.5384	0.5109
5.9	0.9973	0.9188	0.8488	0.7869	0.7322	0.6840	0.6413	0.6033	0.5694	0.5388	0.5113
6.0	0.9975	0.9194	0.8495	0.7877	0.7330	0.6847	0.6420	0.6039	0.5699	0.5393	0.5117

NTU	Heat Exchanger Effectiveness										
	C^* ($C_{min} = C_h$)										
	0	0.1	0.2	0.3	0.4	0.5	0.6	0.7	0.8	0.9	1.0
6.1	0.9978	0.9200	0.8503	0.7885	0.7338	0.6855	0.6426	0.6045	0.5704	0.5397	0.5121
6.2	0.9980	0.9206	0.8511	0.7893	0.7346	0.6862	0.6433	0.6051	0.5709	0.5402	0.5124
6.3	0.9982	0.9212	0.8518	0.7901	0.7354	0.6870	0.6440	0.6057	0.5714	0.5406	0.5128
6.4	0.9983	0.9218	0.8526	0.7909	0.7362	0.6877	0.6447	0.6063	0.5719	0.5411	0.5132
6.5	0.9985	0.9224	0.8534	0.7918	0.7371	0.6885	0.6454	0.6069	0.5725	0.5415	0.5135
6.6	0.9986	0.9230	0.8542	0.7926	0.7379	0.6893	0.6461	0.6075	0.5730	0.5419	0.5139
6.7	0.9988	0.9236	0.8550	0.7935	0.7387	0.6901	0.6467	0.6081	0.5735	0.5423	0.5142
6.8	0.9989	0.9242	0.8558	0.7944	0.7396	0.6908	0.6474	0.6087	0.5740	0.5428	0.5146
6.9	0.9990	0.9248	0.8567	0.7953	0.7404	0.6916	0.6481	0.6093	0.5745	0.5432	0.5149
7.0	0.9991	0.9254	0.8575	0.7961	0.7413	0.6924	0.6488	0.6099	0.5750	0.5436	0.5153
7.1	0.9992	0.9260	0.8583	0.7970	0.7421	0.6932	0.6495	0.6105	0.5755	0.5440	0.5156
7.2	0.9993	0.9266	0.8591	0.7979	0.7430	0.6940	0.6502	0.6111	0.5760	0.5444	0.5159
7.3	0.9993	0.9272	0.8600	0.7988	0.7439	0.6948	0.6509	0.6117	0.5765	0.5448	0.5162
7.4	0.9994	0.9278	0.8608	0.7997	0.7448	0.6956	0.6516	0.6122	0.5770	0.5452	0.5166
7.5	0.9994	0.9284	0.8617	0.8007	0.7456	0.6964	0.6523	0.6128	0.5774	0.5456	0.5169
7.6	0.9995	0.9291	0.8625	0.8016	0.7465	0.6972	0.6530	0.6134	0.5779	0.5460	0.5172
7.7	0.9995	0.9297	0.8634	0.8025	0.7474	0.6979	0.6537	0.6140	0.5784	0.5463	0.5174
7.8	0.9996	0.9303	0.8643	0.8034	0.7483	0.6987	0.6543	0.6145	0.5788	0.5467	0.5177
7.9	0.9996	0.9309	0.8651	0.8043	0.7492	0.6995	0.6550	0.6151	0.5793	0.5471	0.5180
8.0	0.9997	0.9315	0.8660	0.8053	0.7501	0.7003	0.6557	0.6157	0.5797	0.5474	0.5183
8.1	0.9997	0.9321	0.8669	0.8062	0.7509	0.7011	0.6564	0.6162	0.5802	0.5478	0.5186
8.2	0.9997	0.9328	0.8678	0.8071	0.7518	0.7019	0.6570	0.6168	0.5806	0.5481	0.5188
8.3	0.9998	0.9334	0.8686	0.8081	0.7527	0.7027	0.6577	0.6173	0.5810	0.5484	0.5191
8.4	0.9998	0.9340	0.8695	0.8090	0.7536	0.7034	0.6583	0.6178	0.5815	0.5488	0.5193
8.5	0.9998	0.9346	0.8704	0.8099	0.7545	0.7042	0.6590	0.6183	0.5819	0.5491	0.5196
8.6	0.9998	0.9353	0.8713	0.8109	0.7553	0.7050	0.6596	0.6189	0.5823	0.5494	0.5198
8.7	0.9998	0.9359	0.8721	0.8118	0.7562	0.7057	0.6603	0.6194	0.5827	0.5497	0.5200
8.8	0.9998	0.9365	0.8730	0.8127	0.7571	0.7065	0.6609	0.6199	0.5831	0.5500	0.5203
8.9	0.9999	0.9371	0.8739	0.8137	0.7579	0.7073	0.6615	0.6204	0.5835	0.5503	0.5205
9.0	0.9999	0.9377	0.8748	0.8146	0.7588	0.7080	0.6621	0.6209	0.5838	0.5506	0.5207
9.1	0.9999	0.9384	0.8757	0.8155	0.7597	0.7087	0.6627	0.6214	0.5842	0.5509	0.5209
9.2	0.9999	0.9390	0.8765	0.8164	0.7605	0.7095	0.6633	0.6218	0.5846	0.5511	0.5211
9.3	0.9999	0.9396	0.8774	0.8173	0.7614	0.7102	0.6639	0.6223	0.5849	0.5514	0.5213
9.4	0.9999	0.9402	0.8783	0.8183	0.7622	0.7109	0.6645	0.6228	0.5853	0.5517	0.5215
9.5	0.9999	0.9408	0.8792	0.8192	0.7631	0.7117	0.6651	0.6232	0.5856	0.5519	0.5217
9.6	0.9999	0.9414	0.8800	0.8201	0.7639	0.7124	0.6657	0.6237	0.5860	0.5522	0.5219
9.7	0.9999	0.9421	0.8809	0.8210	0.7647	0.7131	0.6663	0.6241	0.5863	0.5524	0.5220
9.8	0.9999	0.9427	0.8817	0.8219	0.7655	0.7138	0.6668	0.6245	0.5866	0.5527	0.5222
9.9	1.0000	0.9433	0.8826	0.8228	0.7664	0.7145	0.6674	0.6250	0.5870	0.5529	0.5224
10.0	1.0000	0.9439	0.8835	0.8237	0.7672	0.7152	0.6679	0.6254	0.5873	0.5531	0.5225

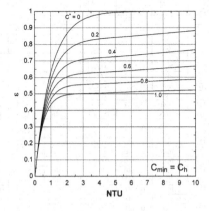

CONFIGURATION 13

1TUBE4ROW4PASS1CIR(PAR)

(a) $C_{min} = C_c$

NTU	Heat Exchanger Effectiveness										
	C^* ($C_{min} = C_c$)										
	0	0.1	0.2	0.3	0.4	0.5	0.6	0.7	0.8	0.9	1.0
0	0.0000	0.0000	0.0000	0.0000	0.0000	0.0000	0.0000	0.0000	0.0000	0.0000	0.0000
0.1	0.0952	0.0947	0.0942	0.0938	0.0933	0.0929	0.0924	0.0920	0.0915	0.0911	0.0906
0.2	0.1813	0.1795	0.1778	0.1761	0.1745	0.1728	0.1712	0.1696	0.1680	0.1664	0.1649
0.3	0.2592	0.2555	0.2520	0.2485	0.2450	0.2417	0.2384	0.2351	0.2319	0.2288	0.2257
0.4	0.3297	0.3236	0.3178	0.3120	0.3064	0.3010	0.2957	0.2905	0.2854	0.2805	0.2756
0.5	0.3935	0.3847	0.3761	0.3679	0.3598	0.3521	0.3445	0.3372	0.3301	0.3232	0.3165
0.6	0.4512	0.4393	0.4279	0.4169	0.4063	0.3961	0.3862	0.3767	0.3675	0.3586	0.3501
0.7	0.5034	0.4883	0.4739	0.4600	0.4467	0.4340	0.4218	0.4100	0.3988	0.3880	0.3776
0.8	0.5507	0.5322	0.5147	0.4979	0.4819	0.4667	0.4521	0.4382	0.4249	0.4123	0.4001
0.9	0.5934	0.5716	0.5509	0.5312	0.5125	0.4948	0.4780	0.4620	0.4468	0.4324	0.4186
1.0	0.6321	0.6068	0.5830	0.5604	0.5392	0.5191	0.5001	0.4821	0.4651	0.4490	0.4337
1.1	0.6671	0.6384	0.6115	0.5861	0.5623	0.5400	0.5189	0.4991	0.4804	0.4627	0.4461
1.2	0.6988	0.6667	0.6368	0.6087	0.5825	0.5580	0.5350	0.5134	0.4931	0.4741	0.4562
1.3	0.7275	0.6921	0.6592	0.6286	0.6000	0.5734	0.5486	0.5254	0.5038	0.4835	0.4645
1.4	0.7534	0.7148	0.6791	0.6460	0.6153	0.5868	0.5603	0.5356	0.5127	0.4912	0.4713
1.5	0.7769	0.7352	0.6968	0.6613	0.6285	0.5982	0.5702	0.5442	0.5201	0.4976	0.4768
1.6	0.7981	0.7534	0.7124	0.6747	0.6401	0.6081	0.5786	0.5514	0.5262	0.5029	0.4812
1.7	0.8173	0.7698	0.7263	0.6865	0.6501	0.6166	0.5858	0.5575	0.5313	0.5072	0.4849
1.8	0.8347	0.7844	0.7386	0.6969	0.6588	0.6239	0.5919	0.5626	0.5356	0.5107	0.4878
1.9	0.8504	0.7975	0.7496	0.7060	0.6663	0.6301	0.5971	0.5669	0.5391	0.5136	0.4902
2.0	0.8647	0.8093	0.7593	0.7140	0.6729	0.6355	0.6015	0.5704	0.5420	0.5160	0.4921
2.1	0.8775	0.8198	0.7679	0.7210	0.6786	0.6401	0.6052	0.5735	0.5445	0.5180	0.4937
2.2	0.8892	0.8293	0.7755	0.7271	0.6835	0.6441	0.6084	0.5760	0.5465	0.5195	0.4949
2.3	0.8997	0.8377	0.7822	0.7325	0.6878	0.6475	0.6110	0.5781	0.5481	0.5208	0.4959
2.4	0.9093	0.8453	0.7882	0.7372	0.6914	0.6503	0.6133	0.5798	0.5494	0.5218	0.4967
2.5	0.9179	0.8520	0.7935	0.7413	0.6946	0.6528	0.6152	0.5813	0.5505	0.5227	0.4973
2.6	0.9257	0.8581	0.7982	0.7449	0.6974	0.6549	0.6168	0.5825	0.5515	0.5233	0.4978
2.7	0.9328	0.8635	0.8023	0.7481	0.6998	0.6567	0.6181	0.5835	0.5522	0.5239	0.4981
2.8	0.9392	0.8684	0.8060	0.7508	0.7019	0.6583	0.6193	0.5843	0.5528	0.5243	0.4984
2.9	0.9450	0.8727	0.8092	0.7532	0.7037	0.6596	0.6202	0.5850	0.5533	0.5246	0.4987
3.0	0.9502	0.8766	0.8121	0.7553	0.7052	0.6607	0.6210	0.5855	0.5536	0.5249	0.4988
3.1	0.9550	0.8801	0.8147	0.7572	0.7065	0.6616	0.6217	0.5860	0.5540	0.5251	0.4989
3.2	0.9592	0.8832	0.8169	0.7588	0.7076	0.6624	0.6222	0.5864	0.5542	0.5252	0.4990
3.3	0.9631	0.8860	0.8189	0.7602	0.7086	0.6631	0.6227	0.5867	0.5544	0.5253	0.4991
3.4	0.9666	0.8885	0.8206	0.7614	0.7095	0.6637	0.6231	0.5869	0.5545	0.5254	0.4991
3.5	0.9698	0.8907	0.8222	0.7625	0.7102	0.6641	0.6234	0.5871	0.5547	0.5255	0.4991
3.6	0.9727	0.8927	0.8235	0.7634	0.7108	0.6645	0.6236	0.5873	0.5547	0.5255	0.4991
3.7	0.9753	0.8945	0.8247	0.7642	0.7113	0.6649	0.6238	0.5874	0.5548	0.5255	0.4991
3.8	0.9776	0.8961	0.8258	0.7649	0.7118	0.6652	0.6240	0.5875	0.5549	0.5256	0.4991
3.9	0.9798	0.8975	0.8267	0.7655	0.7121	0.6654	0.6241	0.5876	0.5549	0.5256	0.4991
4.0	0.9817	0.8988	0.8276	0.7660	0.7125	0.6656	0.6243	0.5876	0.5549	0.5255	0.4990
4.1	0.9834	0.8999	0.8283	0.7665	0.7127	0.6657	0.6243	0.5877	0.5549	0.5255	0.4990
4.2	0.9850	0.9009	0.8289	0.7669	0.7130	0.6659	0.6244	0.5877	0.5549	0.5255	0.4990
4.3	0.9864	0.9018	0.8295	0.7672	0.7132	0.6660	0.6245	0.5877	0.5549	0.5255	0.4989
4.4	0.9877	0.9026	0.8300	0.7675	0.7133	0.6661	0.6245	0.5877	0.5549	0.5254	0.4989
4.5	0.9889	0.9033	0.8304	0.7677	0.7135	0.6662	0.6246	0.5877	0.5549	0.5254	0.4988
4.6	0.9899	0.9040	0.8308	0.7680	0.7136	0.6662	0.6246	0.5877	0.5549	0.5254	0.4988
4.7	0.9909	0.9046	0.8311	0.7681	0.7137	0.6663	0.6246	0.5877	0.5548	0.5253	0.4987
4.8	0.9918	0.9051	0.8314	0.7683	0.7138	0.6663	0.6246	0.5877	0.5548	0.5253	0.4986
4.9	0.9926	0.9055	0.8317	0.7684	0.7138	0.6663	0.6246	0.5877	0.5548	0.5253	0.4986
5.0	0.9933	0.9059	0.8319	0.7685	0.7139	0.6664	0.6246	0.5877	0.5548	0.5252	0.4985
5.1	0.9939	0.9063	0.8321	0.7686	0.7140	0.6664	0.6246	0.5877	0.5547	0.5252	0.4984
5.2	0.9945	0.9066	0.8322	0.7687	0.7140	0.6664	0.6246	0.5877	0.5547	0.5251	0.4984
5.3	0.9950	0.9069	0.8324	0.7688	0.7140	0.6664	0.6246	0.5876	0.5547	0.5251	0.4983
5.4	0.9955	0.9071	0.8325	0.7689	0.7140	0.6664	0.6246	0.5876	0.5546	0.5250	0.4982
5.5	0.9959	0.9074	0.8326	0.7689	0.7141	0.6664	0.6246	0.5876	0.5546	0.5250	0.4981
5.6	0.9963	0.9076	0.8327	0.7690	0.7141	0.6664	0.6246	0.5876	0.5546	0.5249	0.4980
5.7	0.9967	0.9077	0.8328	0.7690	0.7141	0.6664	0.6246	0.5876	0.5545	0.5248	0.4980
5.8	0.9970	0.9079	0.8329	0.7690	0.7141	0.6664	0.6246	0.5875	0.5545	0.5248	0.4979
5.9	0.9973	0.9080	0.8329	0.7690	0.7141	0.6664	0.6246	0.5875	0.5544	0.5247	0.4978
6.0	0.9975	0.9082	0.8330	0.7691	0.7141	0.6664	0.6245	0.5875	0.5544	0.5246	0.4977

NTU	Heat Exchanger Effectiveness										
	C^* ($C_{min} = C_c$)										
	0	0.1	0.2	0.3	0.4	0.5	0.6	0.7	0.8	0.9	1.0
6.1	0.9978	0.9083	0.8330	0.7691	0.7141	0.6664	0.6245	0.5875	0.5544	0.5246	0.4976
6.2	0.9980	0.9084	0.8331	0.7691	0.7141	0.6664	0.6245	0.5874	0.5543	0.5245	0.4975
6.3	0.9982	0.9085	0.8331	0.7691	0.7141	0.6664	0.6245	0.5874	0.5543	0.5244	0.4974
6.4	0.9983	0.9085	0.8331	0.7691	0.7141	0.6664	0.6245	0.5874	0.5542	0.5244	0.4973
6.5	0.9985	0.9086	0.8332	0.7691	0.7141	0.6664	0.6245	0.5873	0.5542	0.5243	0.4972
6.6	0.9986	0.9087	0.8332	0.7691	0.7141	0.6664	0.6244	0.5873	0.5541	0.5242	0.4971
6.7	0.9988	0.9087	0.8332	0.7691	0.7141	0.6664	0.6244	0.5873	0.5540	0.5241	0.4970
6.8	0.9989	0.9088	0.8332	0.7691	0.7141	0.6663	0.6244	0.5872	0.5540	0.5240	0.4969
6.9	0.9990	0.9088	0.8332	0.7692	0.7141	0.6663	0.6244	0.5872	0.5539	0.5240	0.4968
7.0	0.9991	0.9088	0.8333	0.7692	0.7141	0.6663	0.6244	0.5872	0.5539	0.5239	0.4966
7.1	0.9992	0.9089	0.8333	0.7692	0.7141	0.6663	0.6243	0.5871	0.5538	0.5238	0.4965
7.2	0.9993	0.9089	0.8333	0.7692	0.7141	0.6663	0.6243	0.5871	0.5537	0.5237	0.4964
7.3	0.9993	0.9089	0.8333	0.7692	0.7141	0.6663	0.6243	0.5870	0.5537	0.5236	0.4963
7.4	0.9994	0.9089	0.8333	0.7692	0.7141	0.6663	0.6243	0.5870	0.5536	0.5235	0.4962
7.5	0.9994	0.9090	0.8333	0.7692	0.7141	0.6663	0.6242	0.5869	0.5536	0.5234	0.4961
7.6	0.9995	0.9090	0.8333	0.7692	0.7141	0.6663	0.6242	0.5869	0.5535	0.5233	0.4959
7.7	0.9995	0.9090	0.8333	0.7692	0.7141	0.6662	0.6242	0.5869	0.5534	0.5232	0.4958
7.8	0.9996	0.9090	0.8333	0.7692	0.7141	0.6662	0.6242	0.5868	0.5534	0.5231	0.4957
7.9	0.9996	0.9090	0.8333	0.7692	0.7141	0.6662	0.6241	0.5868	0.5533	0.5231	0.4956
8.0	0.9997	0.9090	0.8333	0.7692	0.7141	0.6662	0.6241	0.5867	0.5532	0.5230	0.4955
8.1	0.9997	0.9090	0.8333	0.7692	0.7141	0.6662	0.6241	0.5867	0.5531	0.5229	0.4953
8.2	0.9997	0.9090	0.8333	0.7692	0.7141	0.6662	0.6241	0.5866	0.5531	0.5228	0.4952
8.3	0.9998	0.9090	0.8333	0.7692	0.7141	0.6662	0.6240	0.5866	0.5530	0.5227	0.4951
8.4	0.9998	0.9091	0.8333	0.7692	0.7141	0.6661	0.6240	0.5865	0.5529	0.5226	0.4950
8.5	0.9998	0.9091	0.8333	0.7692	0.7141	0.6661	0.6240	0.5865	0.5529	0.5225	0.4948
8.6	0.9998	0.9091	0.8333	0.7692	0.7140	0.6661	0.6239	0.5864	0.5528	0.5224	0.4947
8.7	0.9998	0.9091	0.8333	0.7691	0.7140	0.6661	0.6239	0.5864	0.5527	0.5223	0.4946
8.8	0.9998	0.9091	0.8333	0.7691	0.7140	0.6661	0.6239	0.5863	0.5526	0.5222	0.4945
8.9	0.9999	0.9091	0.8333	0.7691	0.7140	0.6661	0.6238	0.5863	0.5526	0.5221	0.4944
9.0	0.9999	0.9091	0.8333	0.7691	0.7140	0.6660	0.6238	0.5862	0.5525	0.5220	0.4942
9.1	0.9999	0.9091	0.8333	0.7691	0.7140	0.6660	0.6238	0.5862	0.5524	0.5219	0.4941
9.2	0.9999	0.9091	0.8333	0.7691	0.7140	0.6660	0.6237	0.5861	0.5523	0.5218	0.4940
9.3	0.9999	0.9091	0.8333	0.7691	0.7140	0.6660	0.6237	0.5861	0.5523	0.5217	0.4939
9.4	0.9999	0.9091	0.8333	0.7691	0.7140	0.6660	0.6237	0.5860	0.5522	0.5216	0.4938
9.5	0.9999	0.9091	0.8333	0.7691	0.7140	0.6659	0.6236	0.5860	0.5521	0.5215	0.4937
9.6	0.9999	0.9091	0.8333	0.7691	0.7140	0.6659	0.6236	0.5859	0.5521	0.5214	0.4936
9.7	0.9999	0.9091	0.8333	0.7691	0.7140	0.6659	0.6236	0.5859	0.5520	0.5213	0.4934
9.8	0.9999	0.9091	0.8333	0.7691	0.7140	0.6659	0.6235	0.5858	0.5519	0.5212	0.4933
9.9	1.0000	0.9091	0.8333	0.7691	0.7139	0.6659	0.6235	0.5857	0.5518	0.5212	0.4932
10.0	1.0000	0.9091	0.8333	0.7691	0.7139	0.6658	0.6235	0.5857	0.5518	0.5211	0.4931

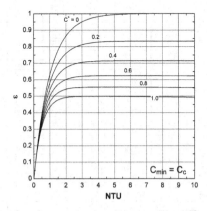

CONFIGURATION 13

1TUBE4ROW4PASS1CIR(PAR)

(b) $C_{min} = C_h$

NTU	Heat Exchanger Effectiveness $C^* (C_{min} = C_h)$										
	0	0.1	0.2	0.3	0.4	0.5	0.6	0.7	0.8	0.9	1.0
0	0.0000	0.0000	0.0000	0.0000	0.0000	0.0000	0.0000	0.0000	0.0000	0.0000	0.0000
0.1	0.0952	0.0947	0.0942	0.0938	0.0933	0.0929	0.0924	0.0920	0.0915	0.0911	0.0906
0.2	0.1813	0.1795	0.1778	0.1761	0.1745	0.1728	0.1712	0.1696	0.1680	0.1664	0.1649
0.3	0.2592	0.2555	0.2520	0.2485	0.2450	0.2417	0.2384	0.2351	0.2319	0.2288	0.2257
0.4	0.3297	0.3236	0.3178	0.3120	0.3064	0.3010	0.2957	0.2905	0.2854	0.2805	0.2756
0.5	0.3935	0.3847	0.3761	0.3679	0.3598	0.3521	0.3445	0.3372	0.3301	0.3232	0.3165
0.6	0.4512	0.4393	0.4279	0.4169	0.4063	0.3961	0.3862	0.3767	0.3675	0.3586	0.3501
0.7	0.5034	0.4883	0.4739	0.4600	0.4467	0.4340	0.4218	0.4100	0.3988	0.3880	0.3776
0.8	0.5507	0.5322	0.5147	0.4979	0.4819	0.4667	0.4521	0.4382	0.4249	0.4122	0.4001
0.9	0.5934	0.5716	0.5508	0.5312	0.5125	0.4948	0.4780	0.4620	0.4468	0.4323	0.4186
1.0	0.6321	0.6068	0.5830	0.5604	0.5392	0.5191	0.5001	0.4821	0.4651	0.4490	0.4337
1.1	0.6671	0.6384	0.6114	0.5861	0.5623	0.5399	0.5189	0.4990	0.4804	0.4627	0.4461
1.2	0.6988	0.6667	0.6367	0.6087	0.5825	0.5579	0.5349	0.5133	0.4931	0.4741	0.4562
1.3	0.7275	0.6921	0.6591	0.6285	0.6000	0.5734	0.5486	0.5254	0.5037	0.4835	0.4645
1.4	0.7534	0.7148	0.6790	0.6459	0.6152	0.5867	0.5602	0.5356	0.5126	0.4912	0.4713
1.5	0.7769	0.7351	0.6967	0.6612	0.6285	0.5982	0.5701	0.5441	0.5200	0.4976	0.4768
1.6	0.7981	0.7534	0.7123	0.6746	0.6399	0.6080	0.5785	0.5513	0.5262	0.5028	0.4812
1.7	0.8173	0.7697	0.7262	0.6864	0.6499	0.6165	0.5857	0.5574	0.5313	0.5072	0.4849
1.8	0.8347	0.7843	0.7385	0.6967	0.6586	0.6237	0.5918	0.5624	0.5355	0.5107	0.4878
1.9	0.8504	0.7974	0.7494	0.7058	0.6661	0.6299	0.5969	0.5667	0.5390	0.5136	0.4902
2.0	0.8647	0.8092	0.7591	0.7137	0.6726	0.6353	0.6013	0.5703	0.5419	0.5160	0.4921
2.1	0.8775	0.8197	0.7676	0.7207	0.6783	0.6398	0.6050	0.5732	0.5443	0.5179	0.4937
2.2	0.8892	0.8291	0.7752	0.7268	0.6831	0.6437	0.6081	0.5757	0.5463	0.5194	0.4949
2.3	0.8997	0.8375	0.7819	0.7321	0.6873	0.6471	0.6107	0.5778	0.5479	0.5207	0.4959
2.4	0.9093	0.8450	0.7878	0.7367	0.6910	0.6499	0.6129	0.5795	0.5492	0.5217	0.4967
2.5	0.9179	0.8518	0.7930	0.7408	0.6941	0.6523	0.6148	0.5809	0.5503	0.5225	0.4973
2.6	0.9257	0.8578	0.7977	0.7443	0.6968	0.6544	0.6163	0.5821	0.5512	0.5232	0.4978
2.7	0.9328	0.8632	0.8018	0.7474	0.6991	0.6561	0.6176	0.5831	0.5519	0.5237	0.4981
2.8	0.9392	0.8680	0.8054	0.7501	0.7011	0.6576	0.6187	0.5838	0.5525	0.5241	0.4984
2.9	0.9450	0.8723	0.8085	0.7524	0.7028	0.6588	0.6196	0.5845	0.5529	0.5244	0.4987
3.0	0.9502	0.8761	0.8113	0.7544	0.7043	0.6598	0.6203	0.5850	0.5533	0.5247	0.4988
3.1	0.9550	0.8795	0.8138	0.7562	0.7055	0.6607	0.6209	0.5854	0.5535	0.5249	0.4989
3.2	0.9592	0.8826	0.8159	0.7577	0.7065	0.6614	0.6214	0.5857	0.5538	0.5250	0.4990
3.3	0.9631	0.8853	0.8178	0.7590	0.7074	0.6620	0.6218	0.5860	0.5539	0.5251	0.4991
3.4	0.9666	0.8877	0.8194	0.7601	0.7082	0.6625	0.6221	0.5862	0.5540	0.5252	0.4991
3.5	0.9698	0.8898	0.8209	0.7610	0.7088	0.6629	0.6223	0.5863	0.5541	0.5252	0.4991
3.6	0.9727	0.8917	0.8221	0.7618	0.7093	0.6632	0.6225	0.5864	0.5542	0.5252	0.4991
3.7	0.9753	0.8934	0.8232	0.7625	0.7097	0.6634	0.6226	0.5865	0.5542	0.5252	0.4991
3.8	0.9776	0.8949	0.8241	0.7630	0.7100	0.6636	0.6227	0.5865	0.5542	0.5252	0.4991
3.9	0.9798	0.8962	0.8249	0.7635	0.7102	0.6637	0.6228	0.5865	0.5542	0.5252	0.4991
4.0	0.9817	0.8974	0.8256	0.7639	0.7104	0.6638	0.6228	0.5865	0.5541	0.5252	0.4990
4.1	0.9834	0.8984	0.8261	0.7641	0.7106	0.6638	0.6228	0.5865	0.5541	0.5251	0.4990
4.2	0.9850	0.8993	0.8266	0.7644	0.7106	0.6638	0.6228	0.5864	0.5541	0.5251	0.4990
4.3	0.9864	0.9000	0.8270	0.7645	0.7107	0.6638	0.6227	0.5864	0.5540	0.5250	0.4989
4.4	0.9877	0.9007	0.8273	0.7647	0.7107	0.6638	0.6226	0.5863	0.5539	0.5250	0.4989
4.5	0.9889	0.9013	0.8275	0.7647	0.7107	0.6637	0.6226	0.5862	0.5539	0.5249	0.4988
4.6	0.9899	0.9018	0.8277	0.7647	0.7106	0.6636	0.6225	0.5861	0.5538	0.5248	0.4988
4.7	0.9909	0.9022	0.8278	0.7647	0.7105	0.6635	0.6224	0.5860	0.5537	0.5248	0.4987
4.8	0.9918	0.9025	0.8279	0.7647	0.7104	0.6634	0.6223	0.5859	0.5536	0.5247	0.4986
4.9	0.9926	0.9028	0.8279	0.7646	0.7103	0.6633	0.6222	0.5858	0.5535	0.5246	0.4986
5.0	0.9933	0.9030	0.8279	0.7645	0.7102	0.6632	0.6220	0.5857	0.5535	0.5245	0.4985
5.1	0.9939	0.9032	0.8279	0.7644	0.7100	0.6630	0.6219	0.5856	0.5534	0.5245	0.4984
5.2	0.9945	0.9033	0.8278	0.7642	0.7099	0.6629	0.6218	0.5855	0.5533	0.5244	0.4984
5.3	0.9950	0.9034	0.8277	0.7640	0.7097	0.6627	0.6216	0.5854	0.5532	0.5243	0.4983
5.4	0.9955	0.9035	0.8276	0.7639	0.7095	0.6625	0.6215	0.5853	0.5531	0.5242	0.4982
5.5	0.9959	0.9035	0.8274	0.7636	0.7093	0.6624	0.6213	0.5851	0.5529	0.5241	0.4981
5.6	0.9963	0.9035	0.8272	0.7634	0.7091	0.6622	0.6212	0.5850	0.5528	0.5240	0.4980
5.7	0.9967	0.9034	0.8270	0.7632	0.7089	0.6620	0.6210	0.5849	0.5527	0.5239	0.4980
5.8	0.9970	0.9033	0.8268	0.7629	0.7086	0.6618	0.6208	0.5847	0.5526	0.5238	0.4979
5.9	0.9973	0.9032	0.8266	0.7627	0.7084	0.6616	0.6207	0.5846	0.5525	0.5237	0.4978
6.0	0.9975	0.9031	0.8263	0.7624	0.7082	0.6614	0.6205	0.5844	0.5524	0.5236	0.4977

NTU	Heat Exchanger Effectiveness										
	C^* ($C_{min} = C_h$)										
	0	0.1	0.2	0.3	0.4	0.5	0.6	0.7	0.8	0.9	1.0
6.1	0.9978	0.9029	0.8261	0.7621	0.7079	0.6611	0.6203	0.5843	0.5522	0.5235	0.4976
6.2	0.9980	0.9028	0.8258	0.7618	0.7076	0.6609	0.6201	0.5841	0.5521	0.5234	0.4975
6.3	0.9982	0.9026	0.8255	0.7615	0.7074	0.6607	0.6199	0.5840	0.5520	0.5233	0.4974
6.4	0.9983	0.9024	0.8252	0.7612	0.7071	0.6605	0.6197	0.5838	0.5518	0.5231	0.4973
6.5	0.9985	0.9022	0.8248	0.7609	0.7068	0.6602	0.6195	0.5836	0.5517	0.5230	0.4972
6.6	0.9986	0.9019	0.8245	0.7606	0.7065	0.6600	0.6193	0.5835	0.5515	0.5229	0.4971
6.7	0.9988	0.9017	0.8242	0.7602	0.7062	0.6597	0.6191	0.5833	0.5514	0.5228	0.4970
6.8	0.9989	0.9014	0.8238	0.7599	0.7059	0.6595	0.6189	0.5831	0.5512	0.5226	0.4969
6.9	0.9990	0.9011	0.8234	0.7595	0.7056	0.6592	0.6187	0.5829	0.5511	0.5225	0.4968
7.0	0.9991	0.9009	0.8231	0.7592	0.7053	0.6590	0.6185	0.5827	0.5509	0.5224	0.4966
7.1	0.9992	0.9006	0.8227	0.7588	0.7050	0.6587	0.6183	0.5826	0.5508	0.5222	0.4965
7.2	0.9993	0.9002	0.8223	0.7584	0.7047	0.6584	0.6180	0.5824	0.5506	0.5221	0.4964
7.3	0.9993	0.8999	0.8219	0.7581	0.7044	0.6582	0.6178	0.5822	0.5504	0.5220	0.4963
7.4	0.9994	0.8996	0.8215	0.7577	0.7040	0.6579	0.6176	0.5820	0.5503	0.5218	0.4962
7.5	0.9994	0.8993	0.8210	0.7573	0.7037	0.6576	0.6173	0.5818	0.5501	0.5217	0.4961
7.6	0.9995	0.8989	0.8206	0.7569	0.7033	0.6573	0.6171	0.5816	0.5499	0.5215	0.4959
7.7	0.9995	0.8986	0.8202	0.7565	0.7030	0.6570	0.6168	0.5814	0.5497	0.5214	0.4958
7.8	0.9996	0.8982	0.8197	0.7561	0.7026	0.6567	0.6166	0.5811	0.5496	0.5212	0.4957
7.9	0.9996	0.8978	0.8193	0.7556	0.7023	0.6564	0.6163	0.5809	0.5494	0.5211	0.4956
8.0	0.9997	0.8974	0.8188	0.7552	0.7019	0.6561	0.6161	0.5807	0.5492	0.5210	0.4955
8.1	0.9997	0.8970	0.8184	0.7548	0.7015	0.6558	0.6158	0.5805	0.5490	0.5208	0.4953
8.2	0.9997	0.8967	0.8179	0.7544	0.7012	0.6555	0.6155	0.5803	0.5487	0.5207	0.4952
8.3	0.9998	0.8962	0.8174	0.7539	0.7008	0.6551	0.6153	0.5801	0.5487	0.5205	0.4951
8.4	0.9998	0.8958	0.8169	0.7535	0.7004	0.6548	0.6150	0.5798	0.5485	0.5204	0.4950
8.5	0.9998	0.8954	0.8164	0.7530	0.7000	0.6545	0.6147	0.5796	0.5483	0.5202	0.4948
8.6	0.9998	0.8950	0.8159	0.7526	0.6996	0.6542	0.6145	0.5794	0.5481	0.5201	0.4947
8.7	0.9998	0.8946	0.8154	0.7521	0.6992	0.6538	0.6142	0.5792	0.5479	0.5199	0.4946
8.8	0.9998	0.8941	0.8149	0.7516	0.6988	0.6535	0.6139	0.5789	0.5477	0.5198	0.4945
8.9	0.9999	0.8937	0.8144	0.7512	0.6984	0.6532	0.6136	0.5787	0.5476	0.5196	0.4944
9.0	0.9999	0.8933	0.8139	0.7507	0.6980	0.6528	0.6133	0.5785	0.5474	0.5195	0.4942
9.1	0.9999	0.8928	0.8134	0.7502	0.6976	0.6525	0.6131	0.5782	0.5472	0.5193	0.4941
9.2	0.9999	0.8924	0.8128	0.7497	0.6972	0.6521	0.6128	0.5780	0.5470	0.5192	0.4940
9.3	0.9999	0.8919	0.8123	0.7492	0.6968	0.6518	0.6125	0.5778	0.5468	0.5190	0.4939
9.4	0.9999	0.8914	0.8118	0.7487	0.6963	0.6514	0.6122	0.5775	0.5466	0.5189	0.4938
9.5	0.9999	0.8910	0.8112	0.7482	0.6959	0.6511	0.6119	0.5773	0.5464	0.5187	0.4937
9.6	0.9999	0.8905	0.8107	0.7477	0.6955	0.6507	0.6116	0.5770	0.5462	0.5186	0.4936
9.7	0.9999	0.8900	0.8101	0.7472	0.6950	0.6503	0.6113	0.5768	0.5460	0.5184	0.4934
9.8	0.9999	0.8895	0.8096	0.7467	0.6946	0.6500	0.6110	0.5766	0.5459	0.5183	0.4933
9.9	1.0000	0.8890	0.8090	0.7462	0.6942	0.6496	0.6107	0.5763	0.5457	0.5181	0.4932
10.0	1.0000	0.8885	0.8084	0.7457	0.6937	0.6492	0.6104	0.5761	0.5455	0.5180	0.4931

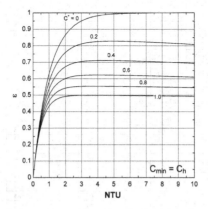

CONFIGURATION 14

1TUBE5ROW5PASS1CIR(PAR)

(b) $C_{min} = C_c$

NTU	Heat Exchanger Effectiveness										
	$C^* (C_{min} = C_c)$										
	0	0.1	0.2	0.3	0.4	0.5	0.6	0.7	0.8	0.9	1.0
0	0.0000	0.0000	0.0000	0.0000	0.0000	0.0000	0.0000	0.0000	0.0000	0.0000	0.0000
0.1	0.0952	0.0947	0.0942	0.0938	0.0933	0.0929	0.0924	0.0920	0.0915	0.0911	0.0906
0.2	0.1813	0.1795	0.1778	0.1761	0.1745	0.1728	0.1712	0.1696	0.1680	0.1664	0.1649
0.3	0.2592	0.2555	0.2520	0.2485	0.2450	0.2416	0.2383	0.2351	0.2319	0.2288	0.2257
0.4	0.3297	0.3236	0.3177	0.3120	0.3064	0.3009	0.2956	0.2904	0.2853	0.2804	0.2755
0.5	0.3935	0.3846	0.3761	0.3678	0.3597	0.3520	0.3444	0.3371	0.3300	0.3231	0.3164
0.6	0.4512	0.4393	0.4278	0.4168	0.4062	0.3959	0.3860	0.3765	0.3673	0.3584	0.3498
0.7	0.5034	0.4883	0.4738	0.4599	0.4466	0.4338	0.4215	0.4098	0.3985	0.3877	0.3773
0.8	0.5507	0.5322	0.5145	0.4977	0.4817	0.4664	0.4518	0.4379	0.4246	0.4119	0.3998
0.9	0.5934	0.5715	0.5507	0.5309	0.5122	0.4945	0.4776	0.4616	0.4464	0.4319	0.4182
1.0	0.6321	0.6067	0.5827	0.5601	0.5388	0.5187	0.4996	0.4816	0.4646	0.4485	0.4333
1.1	0.6671	0.6383	0.6112	0.5858	0.5619	0.5395	0.5184	0.4985	0.4798	0.4622	0.4456
1.2	0.6988	0.6666	0.6364	0.6083	0.5820	0.5574	0.5344	0.5128	0.4926	0.4736	0.4557
1.3	0.7275	0.6919	0.6588	0.6281	0.5995	0.5729	0.5480	0.5249	0.5032	0.4829	0.4640
1.4	0.7534	0.7146	0.6787	0.6455	0.6147	0.5862	0.5597	0.5350	0.5121	0.4907	0.4707
1.5	0.7769	0.7349	0.6963	0.6608	0.6279	0.5976	0.5695	0.5436	0.5195	0.4971	0.4763
1.6	0.7981	0.7532	0.7120	0.6742	0.6394	0.6074	0.5780	0.5508	0.5256	0.5024	0.4808
1.7	0.8173	0.7695	0.7258	0.6859	0.6494	0.6159	0.5851	0.5568	0.5308	0.5067	0.4845
1.8	0.8347	0.7841	0.7381	0.6963	0.6581	0.6232	0.5913	0.5620	0.5351	0.5103	0.4875
1.9	0.8504	0.7972	0.7490	0.7053	0.6656	0.6295	0.5964	0.5663	0.5386	0.5132	0.4899
2.0	0.8647	0.8090	0.7587	0.7133	0.6722	0.6348	0.6009	0.5699	0.5416	0.5157	0.4919
2.1	0.8775	0.8195	0.7673	0.7203	0.6779	0.6394	0.6046	0.5729	0.5440	0.5177	0.4935
2.2	0.8892	0.8289	0.7749	0.7264	0.6828	0.6434	0.6078	0.5755	0.5461	0.5193	0.4948
2.3	0.8997	0.8373	0.7816	0.7318	0.6871	0.6468	0.6105	0.5776	0.5478	0.5206	0.4959
2.4	0.9093	0.8449	0.7876	0.7365	0.6908	0.6497	0.6128	0.5794	0.5492	0.5217	0.4967
2.5	0.9179	0.8517	0.7929	0.7406	0.6940	0.6522	0.6147	0.5809	0.5504	0.5226	0.4974
2.6	0.9257	0.8577	0.7976	0.7443	0.6968	0.6544	0.6164	0.5822	0.5513	0.5234	0.4980
2.7	0.9328	0.8631	0.8017	0.7474	0.6992	0.6562	0.6178	0.5832	0.5521	0.5240	0.4984
2.8	0.9392	0.8680	0.8054	0.7502	0.7013	0.6578	0.6189	0.5841	0.5528	0.5245	0.4988
2.9	0.9450	0.8723	0.8087	0.7526	0.7031	0.6591	0.6199	0.5849	0.5533	0.5249	0.4991
3.0	0.9502	0.8762	0.8116	0.7548	0.7047	0.6603	0.6208	0.5855	0.5538	0.5252	0.4993
3.1	0.9550	0.8797	0.8141	0.7566	0.7060	0.6613	0.6215	0.5860	0.5541	0.5254	0.4995
3.2	0.9592	0.8828	0.8164	0.7583	0.7072	0.6621	0.6221	0.5864	0.5544	0.5257	0.4997
3.3	0.9631	0.8856	0.8184	0.7597	0.7082	0.6628	0.6226	0.5867	0.5547	0.5258	0.4998
3.4	0.9666	0.8881	0.8201	0.7609	0.7091	0.6634	0.6230	0.5870	0.5549	0.5260	0.4999
3.5	0.9698	0.8904	0.8217	0.7620	0.7098	0.6639	0.6234	0.5873	0.5550	0.5261	0.5000
3.6	0.9727	0.8924	0.8231	0.7630	0.7105	0.6644	0.6236	0.5875	0.5552	0.5262	0.5000
3.7	0.9753	0.8942	0.8243	0.7638	0.7110	0.6647	0.6239	0.5876	0.5553	0.5263	0.5001
3.8	0.9776	0.8958	0.8254	0.7645	0.7115	0.6651	0.6241	0.5878	0.5554	0.5263	0.5001
3.9	0.9798	0.8972	0.8263	0.7652	0.7119	0.6653	0.6243	0.5879	0.5554	0.5264	0.5002
4.0	0.9817	0.8985	0.8272	0.7657	0.7123	0.6656	0.6244	0.5880	0.5555	0.5264	0.5002
4.1	0.9834	0.8996	0.8279	0.7662	0.7126	0.6657	0.6245	0.5880	0.5555	0.5264	0.5002
4.2	0.9850	0.9006	0.8286	0.7666	0.7128	0.6659	0.6246	0.5881	0.5556	0.5265	0.5002
4.3	0.9864	0.9015	0.8292	0.7670	0.7131	0.6660	0.6247	0.5882	0.5556	0.5265	0.5003
4.4	0.9877	0.9024	0.8297	0.7673	0.7133	0.6662	0.6248	0.5882	0.5556	0.5265	0.5003
4.5	0.9889	0.9031	0.8301	0.7675	0.7134	0.6663	0.6248	0.5882	0.5557	0.5265	0.5003
4.6	0.9899	0.9037	0.8305	0.7678	0.7136	0.6663	0.6249	0.5883	0.5557	0.5265	0.5003
4.7	0.9909	0.9043	0.8309	0.7680	0.7137	0.6664	0.6249	0.5883	0.5557	0.5265	0.5003
4.8	0.9918	0.9048	0.8312	0.7682	0.7138	0.6665	0.6250	0.5883	0.5557	0.5266	0.5003
4.9	0.9926	0.9053	0.8314	0.7683	0.7139	0.6665	0.6250	0.5883	0.5557	0.5266	0.5003
5.0	0.9933	0.9057	0.8317	0.7684	0.7139	0.6665	0.6250	0.5883	0.5557	0.5266	0.5004
5.1	0.9939	0.9061	0.8319	0.7686	0.7140	0.6666	0.6250	0.5883	0.5557	0.5266	0.5004
5.2	0.9945	0.9064	0.8321	0.7687	0.7141	0.6666	0.6250	0.5884	0.5557	0.5266	0.5004
5.3	0.9950	0.9067	0.8322	0.7687	0.7141	0.6666	0.6251	0.5884	0.5558	0.5266	0.5004
5.4	0.9955	0.9070	0.8324	0.7688	0.7141	0.6666	0.6251	0.5884	0.5558	0.5266	0.5004
5.5	0.9959	0.9072	0.8325	0.7689	0.7142	0.6667	0.6251	0.5884	0.5558	0.5266	0.5004
5.6	0.9963	0.9074	0.8326	0.7689	0.7142	0.6667	0.6251	0.5884	0.5558	0.5266	0.5004
5.7	0.9967	0.9076	0.8327	0.7690	0.7142	0.6667	0.6251	0.5884	0.5558	0.5266	0.5005
5.8	0.9970	0.9078	0.8328	0.7690	0.7142	0.6667	0.6251	0.5884	0.5558	0.5266	0.5005
5.9	0.9973	0.9079	0.8329	0.7691	0.7142	0.6667	0.6251	0.5884	0.5558	0.5267	0.5005
6.0	0.9975	0.9081	0.8329	0.7691	0.7143	0.6667	0.6251	0.5884	0.5558	0.5267	0.5005

| NTU | Heat Exchanger Effectiveness | | | | | | | | | | |
| | C^* ($C_{min} = C_c$) | | | | | | | | | | |
	0	0.1	0.2	0.3	0.4	0.5	0.6	0.7	0.8	0.9	1.0
6.1	0.9978	0.9082	0.8330	0.7691	0.7143	0.6667	0.6251	0.5884	0.5558	0.5267	0.5005
6.2	0.9980	0.9083	0.8330	0.7691	0.7143	0.6667	0.6251	0.5884	0.5558	0.5267	0.5005
6.3	0.9982	0.9084	0.8331	0.7691	0.7143	0.6667	0.6251	0.5884	0.5558	0.5267	0.5006
6.4	0.9983	0.9085	0.8331	0.7692	0.7143	0.6667	0.6251	0.5884	0.5558	0.5267	0.5006
6.5	0.9985	0.9085	0.8331	0.7692	0.7143	0.6667	0.6251	0.5884	0.5558	0.5267	0.5006
6.6	0.9986	0.9086	0.8332	0.7692	0.7143	0.6667	0.6251	0.5884	0.5558	0.5267	0.5006
6.7	0.9988	0.9086	0.8332	0.7692	0.7143	0.6667	0.6251	0.5884	0.5558	0.5267	0.5006
6.8	0.9989	0.9087	0.8332	0.7692	0.7143	0.6667	0.6251	0.5884	0.5558	0.5268	0.5006
6.9	0.9990	0.9087	0.8332	0.7692	0.7143	0.6667	0.6251	0.5884	0.5559	0.5268	0.5007
7.0	0.9991	0.9088	0.8332	0.7692	0.7143	0.6667	0.6251	0.5884	0.5559	0.5268	0.5007
7.1	0.9992	0.9088	0.8332	0.7692	0.7143	0.6667	0.6251	0.5884	0.5559	0.5268	0.5007
7.2	0.9993	0.9089	0.8333	0.7692	0.7143	0.6667	0.6251	0.5884	0.5559	0.5268	0.5007
7.3	0.9993	0.9089	0.8333	0.7692	0.7143	0.6667	0.6251	0.5884	0.5559	0.5268	0.5008
7.4	0.9994	0.9089	0.8333	0.7692	0.7143	0.6667	0.6251	0.5884	0.5559	0.5269	0.5008
7.5	0.9994	0.9089	0.8333	0.7692	0.7143	0.6667	0.6251	0.5885	0.5559	0.5269	0.5008
7.6	0.9995	0.9089	0.8333	0.7692	0.7143	0.6667	0.6251	0.5885	0.5559	0.5269	0.5008
7.7	0.9995	0.9090	0.8333	0.7692	0.7143	0.6667	0.6251	0.5885	0.5559	0.5269	0.5009
7.8	0.9996	0.9090	0.8333	0.7692	0.7143	0.6667	0.6251	0.5885	0.5559	0.5269	0.5009
7.9	0.9996	0.9090	0.8333	0.7692	0.7143	0.6667	0.6251	0.5885	0.5560	0.5269	0.5009
8.0	0.9997	0.9090	0.8333	0.7692	0.7143	0.6667	0.6251	0.5885	0.5560	0.5270	0.5010
8.1	0.9997	0.9090	0.8333	0.7692	0.7143	0.6667	0.6251	0.5885	0.5560	0.5270	0.5010
8.2	0.9997	0.9090	0.8333	0.7692	0.7143	0.6667	0.6251	0.5885	0.5560	0.5270	0.5010
8.3	0.9998	0.9090	0.8333	0.7692	0.7143	0.6667	0.6251	0.5885	0.5560	0.5270	0.5011
8.4	0.9998	0.9090	0.8333	0.7692	0.7143	0.6667	0.6251	0.5885	0.5560	0.5271	0.5011
8.5	0.9998	0.9090	0.8333	0.7692	0.7143	0.6667	0.6252	0.5885	0.5560	0.5271	0.5011
8.6	0.9998	0.9091	0.8333	0.7692	0.7143	0.6667	0.6252	0.5885	0.5560	0.5271	0.5012
8.7	0.9998	0.9091	0.8333	0.7692	0.7143	0.6667	0.6252	0.5885	0.5561	0.5271	0.5012
8.8	0.9998	0.9091	0.8333	0.7692	0.7143	0.6667	0.6252	0.5885	0.5561	0.5271	0.5012
8.9	0.9999	0.9091	0.8333	0.7692	0.7143	0.6667	0.6252	0.5886	0.5561	0.5272	0.5013
9.0	0.9999	0.9091	0.8333	0.7692	0.7143	0.6667	0.6252	0.5886	0.5561	0.5272	0.5013
9.1	0.9999	0.9091	0.8333	0.7692	0.7143	0.6667	0.6252	0.5886	0.5561	0.5272	0.5013
9.2	0.9999	0.9091	0.8333	0.7692	0.7143	0.6668	0.6252	0.5886	0.5561	0.5272	0.5014
9.3	0.9999	0.9091	0.8333	0.7692	0.7143	0.6668	0.6252	0.5886	0.5562	0.5273	0.5014
9.4	0.9999	0.9091	0.8333	0.7692	0.7143	0.6668	0.6252	0.5886	0.5562	0.5273	0.5014
9.5	0.9999	0.9091	0.8333	0.7692	0.7143	0.6668	0.6252	0.5886	0.5562	0.5273	0.5015
9.6	0.9999	0.9091	0.8333	0.7692	0.7143	0.6668	0.6252	0.5886	0.5562	0.5274	0.5015
9.7	0.9999	0.9091	0.8333	0.7692	0.7143	0.6668	0.6252	0.5886	0.5562	0.5274	0.5016
9.8	0.9999	0.9091	0.8333	0.7692	0.7143	0.6668	0.6252	0.5886	0.5562	0.5274	0.5016
9.9	1.0000	0.9091	0.8333	0.7692	0.7143	0.6668	0.6252	0.5886	0.5563	0.5274	0.5016
10.0	1.0000	0.9091	0.8333	0.7692	0.7143	0.6668	0.6252	0.5887	0.5563	0.5275	0.5017

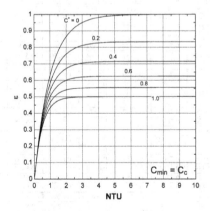

CONFIGURATION 14

1TUBE5ROW5PASS1CIR(PAR)

(b) $C_{min} = C_h$

NTU	Heat Exchanger Effectiveness										
	C^* ($C_{min} = C_h$)										
	0	0.1	0.2	0.3	0.4	0.5	0.6	0.7	0.8	0.9	1.0
0	0.0000	0.0000	0.0000	0.0000	0.0000	0.0000	0.0000	0.0000	0.0000	0.0000	0.0000
0.1	0.0952	0.0947	0.0942	0.0938	0.0933	0.0929	0.0924	0.0920	0.0915	0.0911	0.0906
0.2	0.1813	0.1795	0.1778	0.1761	0.1745	0.1728	0.1712	0.1696	0.1680	0.1664	0.1649
0.3	0.2592	0.2555	0.2520	0.2485	0.2450	0.2416	0.2383	0.2351	0.2319	0.2288	0.2257
0.4	0.3297	0.3236	0.3177	0.3120	0.3064	0.3009	0.2956	0.2904	0.2853	0.2804	0.2755
0.5	0.3935	0.3846	0.3761	0.3678	0.3597	0.3520	0.3444	0.3371	0.3300	0.3231	0.3164
0.6	0.4512	0.4393	0.4278	0.4168	0.4062	0.3959	0.3860	0.3765	0.3673	0.3584	0.3498
0.7	0.5034	0.4883	0.4738	0.4599	0.4466	0.4338	0.4215	0.4098	0.3985	0.3877	0.3773
0.8	0.5507	0.5322	0.5145	0.4977	0.4817	0.4664	0.4518	0.4379	0.4246	0.4119	0.3998
0.9	0.5934	0.5715	0.5507	0.5309	0.5122	0.4945	0.4776	0.4616	0.4464	0.4319	0.4182
1.0	0.6321	0.6067	0.5827	0.5601	0.5388	0.5187	0.4996	0.4816	0.4646	0.4485	0.4333
1.1	0.6671	0.6383	0.6112	0.5858	0.5619	0.5395	0.5184	0.4986	0.4799	0.4622	0.4456
1.2	0.6988	0.6666	0.6365	0.6083	0.5820	0.5574	0.5344	0.5128	0.4926	0.4736	0.4557
1.3	0.7275	0.6919	0.6589	0.6281	0.5995	0.5729	0.5480	0.5249	0.5032	0.4829	0.4640
1.4	0.7534	0.7146	0.6787	0.6455	0.6147	0.5862	0.5597	0.5350	0.5121	0.4907	0.4707
1.5	0.7769	0.7349	0.6964	0.6608	0.6280	0.5976	0.5696	0.5436	0.5195	0.4971	0.4763
1.6	0.7981	0.7532	0.7120	0.6742	0.6395	0.6075	0.5780	0.5508	0.5256	0.5024	0.4808
1.7	0.8173	0.7695	0.7259	0.6860	0.6494	0.6160	0.5852	0.5569	0.5308	0.5067	0.4845
1.8	0.8347	0.7841	0.7382	0.6963	0.6581	0.6232	0.5913	0.5620	0.5351	0.5103	0.4875
1.9	0.8504	0.7972	0.7491	0.7054	0.6657	0.6295	0.5965	0.5663	0.5386	0.5133	0.4899
2.0	0.8647	0.8090	0.7588	0.7134	0.6722	0.6349	0.6009	0.5699	0.5416	0.5157	0.4919
2.1	0.8775	0.8195	0.7674	0.7204	0.6779	0.6395	0.6047	0.5730	0.5441	0.5177	0.4935
2.2	0.8892	0.8289	0.7750	0.7265	0.6829	0.6435	0.6079	0.5755	0.5461	0.5193	0.4948
2.3	0.8997	0.8374	0.7817	0.7319	0.6872	0.6469	0.6106	0.5777	0.5478	0.5207	0.4959
2.4	0.9093	0.8450	0.7877	0.7366	0.6909	0.6499	0.6129	0.5795	0.5493	0.5218	0.4967
2.5	0.9179	0.8517	0.7930	0.7408	0.6941	0.6524	0.6149	0.5810	0.5504	0.5227	0.4974
2.6	0.9257	0.8578	0.7977	0.7444	0.6970	0.6545	0.6165	0.5823	0.5514	0.5234	0.4980
2.7	0.9328	0.8632	0.8019	0.7476	0.6994	0.6564	0.6179	0.5834	0.5522	0.5240	0.4984
2.8	0.9392	0.8681	0.8056	0.7504	0.7015	0.6580	0.6191	0.5843	0.5529	0.5245	0.4988
2.9	0.9450	0.8725	0.8089	0.7529	0.7033	0.6594	0.6201	0.5850	0.5534	0.5249	0.4991
3.0	0.9502	0.8764	0.8118	0.7550	0.7049	0.6605	0.6210	0.5856	0.5539	0.5252	0.4993
3.1	0.9550	0.8799	0.8144	0.7569	0.7063	0.6615	0.6217	0.5862	0.5542	0.5255	0.4995
3.2	0.9592	0.8830	0.8167	0.7586	0.7075	0.6624	0.6223	0.5866	0.5546	0.5257	0.4997
3.3	0.9631	0.8859	0.8187	0.7601	0.7086	0.6631	0.6229	0.5870	0.5548	0.5259	0.4998
3.4	0.9666	0.8884	0.8205	0.7613	0.7095	0.6638	0.6233	0.5873	0.5550	0.5261	0.4999
3.5	0.9698	0.8907	0.8221	0.7625	0.7103	0.6643	0.6237	0.5875	0.5552	0.5262	0.5000
3.6	0.9727	0.8927	0.8236	0.7635	0.7110	0.6648	0.6240	0.5877	0.5554	0.5263	0.5000
3.7	0.9753	0.8945	0.8248	0.7643	0.7116	0.6652	0.6243	0.5879	0.5555	0.5264	0.5001
3.8	0.9776	0.8961	0.8259	0.7651	0.7121	0.6656	0.6245	0.5881	0.5556	0.5264	0.5001
3.9	0.9798	0.8976	0.8270	0.7658	0.7125	0.6659	0.6247	0.5882	0.5557	0.5265	0.5002
4.0	0.9817	0.8989	0.8278	0.7664	0.7129	0.6661	0.6249	0.5883	0.5557	0.5265	0.5002
4.1	0.9834	0.9001	0.8287	0.7669	0.7133	0.6664	0.6250	0.5884	0.5558	0.5266	0.5002
4.2	0.9850	0.9012	0.8294	0.7674	0.7136	0.6666	0.6252	0.5885	0.5558	0.5266	0.5002
4.3	0.9864	0.9022	0.8300	0.7678	0.7139	0.6667	0.6253	0.5886	0.5559	0.5266	0.5003
4.4	0.9877	0.9030	0.8306	0.7682	0.7141	0.6669	0.6254	0.5886	0.5559	0.5266	0.5003
4.5	0.9889	0.9038	0.8311	0.7685	0.7143	0.6670	0.6255	0.5887	0.5560	0.5267	0.5003
4.6	0.9899	0.9045	0.8316	0.7689	0.7145	0.6672	0.6256	0.5888	0.5560	0.5267	0.5003
4.7	0.9909	0.9052	0.8320	0.7691	0.7147	0.6673	0.6256	0.5888	0.5560	0.5267	0.5003
4.8	0.9918	0.9057	0.8324	0.7694	0.7149	0.6674	0.6257	0.5888	0.5561	0.5267	0.5003
4.9	0.9926	0.9063	0.8327	0.7696	0.7150	0.6675	0.6258	0.5889	0.5561	0.5267	0.5003
5.0	0.9933	0.9068	0.8331	0.7698	0.7152	0.6676	0.6258	0.5889	0.5561	0.5268	0.5004
5.1	0.9939	0.9072	0.8334	0.7700	0.7153	0.6677	0.6259	0.5890	0.5561	0.5268	0.5004
5.2	0.9945	0.9076	0.8336	0.7702	0.7154	0.6677	0.6259	0.5890	0.5562	0.5268	0.5004
5.3	0.9950	0.9080	0.8339	0.7704	0.7155	0.6678	0.6260	0.5890	0.5562	0.5268	0.5004
5.4	0.9955	0.9083	0.8341	0.7705	0.7156	0.6679	0.6260	0.5891	0.5562	0.5268	0.5004
5.5	0.9959	0.9086	0.8343	0.7707	0.7157	0.6680	0.6261	0.5891	0.5562	0.5269	0.5004
5.6	0.9963	0.9089	0.8346	0.7708	0.7158	0.6680	0.6261	0.5891	0.5563	0.5269	0.5004
5.7	0.9967	0.9092	0.8348	0.7710	0.7159	0.6681	0.6262	0.5892	0.5563	0.5269	0.5005
5.8	0.9970	0.9095	0.8350	0.7711	0.7160	0.6682	0.6262	0.5892	0.5563	0.5269	0.5005
5.9	0.9973	0.9097	0.8351	0.7713	0.7161	0.6682	0.6263	0.5892	0.5563	0.5269	0.5005
6.0	0.9975	0.9099	0.8353	0.7714	0.7162	0.6683	0.6263	0.5893	0.5564	0.5269	0.5005

NTU	Heat Exchanger Effectiveness										
	C^* ($C_{min} = C_h$)										
	0	0.1	0.2	0.3	0.4	0.5	0.6	0.7	0.8	0.9	1.0
6.1	0.9978	0.9102	0.8355	0.7715	0.7163	0.6684	0.6264	0.5893	0.5564	0.5270	0.5005
6.2	0.9980	0.9104	0.8357	0.7717	0.7164	0.6684	0.6264	0.5893	0.5564	0.5270	0.5005
6.3	0.9982	0.9106	0.8358	0.7718	0.7165	0.6685	0.6265	0.5894	0.5564	0.5270	0.5006
6.4	0.9983	0.9108	0.8360	0.7719	0.7166	0.6686	0.6265	0.5894	0.5565	0.5270	0.5006
6.5	0.9985	0.9110	0.8362	0.7720	0.7167	0.6686	0.6266	0.5895	0.5565	0.5271	0.5006
6.6	0.9986	0.9111	0.8363	0.7722	0.7168	0.6687	0.6266	0.5895	0.5565	0.5271	0.5006
6.7	0.9988	0.9113	0.8365	0.7723	0.7169	0.6688	0.6267	0.5895	0.5566	0.5271	0.5006
6.8	0.9989	0.9115	0.8367	0.7724	0.7170	0.6689	0.6267	0.5896	0.5566	0.5271	0.5006
6.9	0.9990	0.9117	0.8368	0.7726	0.7171	0.6689	0.6268	0.5896	0.5566	0.5272	0.5007
7.0	0.9991	0.9119	0.8370	0.7727	0.7172	0.6690	0.6268	0.5897	0.5567	0.5272	0.5007
7.1	0.9992	0.9120	0.8372	0.7728	0.7173	0.6691	0.6269	0.5897	0.5567	0.5272	0.5007
7.2	0.9993	0.9122	0.8373	0.7730	0.7174	0.6692	0.6269	0.5898	0.5567	0.5272	0.5007
7.3	0.9993	0.9124	0.8375	0.7731	0.7175	0.6692	0.6270	0.5898	0.5568	0.5273	0.5008
7.4	0.9994	0.9125	0.8377	0.7732	0.7176	0.6693	0.6271	0.5898	0.5568	0.5273	0.5008
7.5	0.9994	0.9127	0.8378	0.7734	0.7177	0.6694	0.6271	0.5899	0.5569	0.5273	0.5008
7.6	0.9995	0.9129	0.8380	0.7735	0.7178	0.6695	0.6272	0.5899	0.5569	0.5274	0.5008
7.7	0.9995	0.9131	0.8382	0.7737	0.7180	0.6696	0.6273	0.5900	0.5569	0.5274	0.5009
7.8	0.9996	0.9132	0.8384	0.7738	0.7181	0.6697	0.6273	0.5901	0.5570	0.5274	0.5009
7.9	0.9996	0.9134	0.8385	0.7740	0.7182	0.6698	0.6274	0.5901	0.5570	0.5275	0.5009
8.0	0.9997	0.9136	0.8387	0.7741	0.7183	0.6698	0.6275	0.5902	0.5571	0.5275	0.5010
8.1	0.9997	0.9138	0.8389	0.7743	0.7184	0.6699	0.6275	0.5902	0.5571	0.5276	0.5010
8.2	0.9997	0.9139	0.8391	0.7744	0.7186	0.6700	0.6276	0.5903	0.5572	0.5276	0.5010
8.3	0.9998	0.9141	0.8393	0.7746	0.7187	0.6701	0.6277	0.5903	0.5572	0.5276	0.5011
8.4	0.9998	0.9143	0.8395	0.7748	0.7188	0.6702	0.6278	0.5904	0.5573	0.5277	0.5011
8.5	0.9998	0.9145	0.8397	0.7749	0.7189	0.6703	0.6279	0.5905	0.5573	0.5277	0.5011
8.6	0.9998	0.9147	0.8399	0.7751	0.7191	0.6704	0.6279	0.5905	0.5574	0.5278	0.5012
8.7	0.9998	0.9149	0.8401	0.7753	0.7192	0.6705	0.6280	0.5906	0.5574	0.5278	0.5012
8.8	0.9998	0.9151	0.8403	0.7754	0.7193	0.6706	0.6281	0.5907	0.5575	0.5278	0.5012
8.9	0.9999	0.9153	0.8405	0.7756	0.7195	0.6707	0.6282	0.5907	0.5575	0.5279	0.5013
9.0	0.9999	0.9155	0.8407	0.7758	0.7196	0.6709	0.6283	0.5908	0.5576	0.5279	0.5013
9.1	0.9999	0.9157	0.8409	0.7760	0.7198	0.6710	0.6284	0.5909	0.5576	0.5280	0.5013
9.2	0.9999	0.9159	0.8411	0.7762	0.7199	0.6711	0.6284	0.5909	0.5577	0.5280	0.5014
9.3	0.9999	0.9161	0.8414	0.7763	0.7201	0.6712	0.6285	0.5910	0.5577	0.5281	0.5014
9.4	0.9999	0.9163	0.8416	0.7765	0.7202	0.6713	0.6286	0.5911	0.5578	0.5281	0.5014
9.5	0.9999	0.9165	0.8418	0.7767	0.7204	0.6714	0.6287	0.5912	0.5579	0.5282	0.5015
9.6	0.9999	0.9167	0.8420	0.7769	0.7205	0.6716	0.6288	0.5912	0.5579	0.5282	0.5015
9.7	0.9999	0.9169	0.8423	0.7771	0.7207	0.6717	0.6289	0.5913	0.5580	0.5283	0.5016
9.8	0.9999	0.9171	0.8425	0.7773	0.7208	0.6718	0.6290	0.5914	0.5580	0.5283	0.5016
9.9	1.0000	0.9173	0.8427	0.7775	0.7210	0.6719	0.6291	0.5915	0.5581	0.5284	0.5016
10.0	1.0000	0.9176	0.8430	0.7777	0.7211	0.6721	0.6292	0.5915	0.5582	0.5284	0.5017

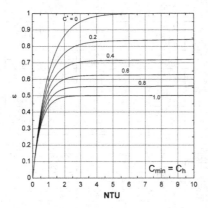

CONFIGURATION 15

1TUBE6ROW6PASS1CIR(PAR)

(a) $C_{min} = C_c$

NTU	Heat Exchanger Effectiveness										
	C^* ($C_{min} = C_c$)										
	0	0.1	0.2	0.3	0.4	0.5	0.6	0.7	0.8	0.9	1.0
0	0.0000	0.0000	0.0000	0.0000	0.0000	0.0000	0.0000	0.0000	0.0000	0.0000	0.0000
0.1	0.0952	0.0947	0.0942	0.0938	0.0933	0.0929	0.0924	0.0920	0.0915	0.0911	0.0906
0.2	0.1813	0.1795	0.1778	0.1761	0.1745	0.1728	0.1712	0.1696	0.1680	0.1664	0.1649
0.3	0.2592	0.2555	0.2520	0.2484	0.2450	0.2416	0.2383	0.2351	0.2319	0.2287	0.2257
0.4	0.3297	0.3236	0.3177	0.3120	0.3063	0.3009	0.2955	0.2903	0.2853	0.2803	0.2755
0.5	0.3935	0.3846	0.3761	0.3677	0.3597	0.3519	0.3443	0.3370	0.3299	0.3230	0.3163
0.6	0.4512	0.4393	0.4278	0.4167	0.4061	0.3958	0.3859	0.3764	0.3672	0.3583	0.3497
0.7	0.5034	0.4882	0.4737	0.4598	0.4464	0.4337	0.4214	0.4096	0.3983	0.3875	0.3771
0.8	0.5507	0.5321	0.5144	0.4976	0.4815	0.4662	0.4516	0.4377	0.4244	0.4117	0.3995
0.9	0.5934	0.5714	0.5506	0.5308	0.5121	0.4943	0.4774	0.4614	0.4461	0.4317	0.4179
1.0	0.6321	0.6066	0.5826	0.5600	0.5386	0.5184	0.4994	0.4814	0.4643	0.4482	0.4330
1.1	0.6671	0.6382	0.6111	0.5856	0.5617	0.5392	0.5181	0.4982	0.4795	0.4619	0.4453
1.2	0.6988	0.6665	0.6363	0.6081	0.5818	0.5571	0.5341	0.5125	0.4922	0.4732	0.4554
1.3	0.7275	0.6918	0.6586	0.6279	0.5992	0.5725	0.5477	0.5245	0.5028	0.4826	0.4636
1.4	0.7534	0.7145	0.6785	0.6452	0.6144	0.5858	0.5593	0.5346	0.5117	0.4903	0.4704
1.5	0.7769	0.7348	0.6961	0.6605	0.6276	0.5972	0.5691	0.5432	0.5190	0.4967	0.4759
1.6	0.7981	0.7530	0.7117	0.6738	0.6391	0.6070	0.5776	0.5504	0.5252	0.5019	0.4804
1.7	0.8173	0.7693	0.7256	0.6856	0.6490	0.6155	0.5847	0.5564	0.5303	0.5063	0.4840
1.8	0.8347	0.7839	0.7379	0.6959	0.6577	0.6228	0.5908	0.5615	0.5346	0.5099	0.4870
1.9	0.8504	0.7970	0.7487	0.7050	0.6652	0.6290	0.5960	0.5658	0.5382	0.5128	0.4895
2.0	0.8647	0.8088	0.7584	0.7129	0.6718	0.6344	0.6004	0.5694	0.5412	0.5153	0.4915
2.1	0.8775	0.8193	0.7670	0.7199	0.6774	0.6390	0.6042	0.5725	0.5436	0.5173	0.4931
2.2	0.8892	0.8287	0.7746	0.7260	0.6824	0.6430	0.6074	0.5751	0.5457	0.5189	0.4944
2.3	0.8997	0.8371	0.7813	0.7314	0.6866	0.6464	0.6101	0.5772	0.5474	0.5202	0.4955
2.4	0.9093	0.8447	0.7873	0.7361	0.6904	0.6493	0.6124	0.5790	0.5488	0.5214	0.4963
2.5	0.9179	0.8514	0.7926	0.7402	0.6936	0.6518	0.6143	0.5805	0.5500	0.5223	0.4970
2.6	0.9257	0.8575	0.7973	0.7439	0.6964	0.6540	0.6160	0.5818	0.5510	0.5230	0.4976
2.7	0.9328	0.8629	0.8014	0.7471	0.6988	0.6558	0.6174	0.5829	0.5518	0.5236	0.4981
2.8	0.9392	0.8678	0.8051	0.7498	0.7009	0.6574	0.6186	0.5838	0.5524	0.5241	0.4984
2.9	0.9450	0.8721	0.8084	0.7523	0.7027	0.6588	0.6196	0.5845	0.5530	0.5245	0.4987
3.0	0.9502	0.8760	0.8113	0.7544	0.7043	0.6599	0.6204	0.5851	0.5534	0.5248	0.4990
3.1	0.9550	0.8795	0.8138	0.7563	0.7057	0.6609	0.6212	0.5856	0.5538	0.5251	0.4992
3.2	0.9592	0.8826	0.8161	0.7579	0.7069	0.6618	0.6218	0.5861	0.5541	0.5253	0.4993
3.3	0.9631	0.8854	0.8181	0.7594	0.7079	0.6625	0.6223	0.5864	0.5544	0.5255	0.4994
3.4	0.9666	0.8879	0.8199	0.7606	0.7088	0.6631	0.6227	0.5867	0.5546	0.5256	0.4995
3.5	0.9698	0.8902	0.8214	0.7617	0.7095	0.6637	0.6231	0.5870	0.5547	0.5258	0.4996
3.6	0.9727	0.8922	0.8228	0.7627	0.7102	0.6641	0.6234	0.5872	0.5549	0.5259	0.4997
3.7	0.9753	0.8940	0.8241	0.7635	0.7108	0.6645	0.6236	0.5874	0.5550	0.5259	0.4997
3.8	0.9776	0.8956	0.8251	0.7643	0.7113	0.6648	0.6239	0.5875	0.5551	0.5260	0.4997
3.9	0.9798	0.8970	0.8261	0.7649	0.7117	0.6651	0.6240	0.5876	0.5552	0.5260	0.4998
4.0	0.9817	0.8983	0.8270	0.7655	0.7121	0.6653	0.6242	0.5877	0.5552	0.5261	0.4998
4.1	0.9834	0.8994	0.8277	0.7660	0.7124	0.6655	0.6243	0.5878	0.5553	0.5261	0.4998
4.2	0.9850	0.9005	0.8284	0.7664	0.7126	0.6657	0.6244	0.5879	0.5553	0.5261	0.4998
4.3	0.9864	0.9014	0.8290	0.7668	0.7129	0.6659	0.6245	0.5879	0.5553	0.5261	0.4998
4.4	0.9877	0.9022	0.8295	0.7671	0.7131	0.6660	0.6246	0.5880	0.5554	0.5262	0.4998
4.5	0.9889	0.9030	0.8299	0.7674	0.7133	0.6661	0.6247	0.5880	0.5554	0.5262	0.4998
4.6	0.9899	0.9036	0.8304	0.7676	0.7134	0.6662	0.6247	0.5881	0.5554	0.5262	0.4998
4.7	0.9909	0.9042	0.8307	0.7678	0.7135	0.6663	0.6248	0.5881	0.5554	0.5262	0.4999
4.8	0.9918	0.9047	0.8310	0.7680	0.7136	0.6663	0.6248	0.5881	0.5554	0.5262	0.4999
4.9	0.9926	0.9052	0.8313	0.7682	0.7137	0.6664	0.6248	0.5881	0.5554	0.5262	0.4999
5.0	0.9933	0.9056	0.8315	0.7683	0.7138	0.6664	0.6248	0.5881	0.5555	0.5262	0.4999
5.1	0.9939	0.9060	0.8318	0.7684	0.7139	0.6665	0.6249	0.5881	0.5555	0.5262	0.4999
5.2	0.9945	0.9063	0.8320	0.7685	0.7139	0.6665	0.6249	0.5881	0.5555	0.5262	0.4999
5.3	0.9950	0.9066	0.8321	0.7686	0.7140	0.6665	0.6249	0.5881	0.5555	0.5262	0.4999
5.4	0.9955	0.9069	0.8323	0.7687	0.7140	0.6665	0.6249	0.5882	0.5555	0.5262	0.4998
5.5	0.9959	0.9071	0.8324	0.7688	0.7141	0.6666	0.6249	0.5882	0.5555	0.5262	0.4998
5.6	0.9963	0.9073	0.8325	0.7689	0.7141	0.6666	0.6249	0.5882	0.5555	0.5262	0.4998
5.7	0.9967	0.9075	0.8326	0.7689	0.7141	0.6666	0.6249	0.5882	0.5555	0.5262	0.4998
5.8	0.9970	0.9077	0.8327	0.7689	0.7142	0.6666	0.6249	0.5882	0.5555	0.5262	0.4998
5.9	0.9973	0.9079	0.8328	0.7690	0.7142	0.6666	0.6249	0.5882	0.5555	0.5262	0.4998
6.0	0.9975	0.9080	0.8329	0.7690	0.7142	0.6666	0.6249	0.5882	0.5555	0.5262	0.4998

NTU	Heat Exchanger Effectiveness										
	C^* ($C_{min} = C_c$)										
	0	0.1	0.2	0.3	0.4	0.5	0.6	0.7	0.8	0.9	1.0
6.1	0.9978	0.9081	0.8329	0.7690	0.7142	0.6666	0.6250	0.5882	0.5555	0.5262	0.4998
6.2	0.9980	0.9082	0.8330	0.7691	0.7142	0.6666	0.6250	0.5882	0.5555	0.5262	0.4998
6.3	0.9982	0.9083	0.8330	0.7691	0.7142	0.6666	0.6250	0.5882	0.5555	0.5262	0.4998
6.4	0.9983	0.9084	0.8331	0.7691	0.7142	0.6666	0.6250	0.5882	0.5555	0.5262	0.4998
6.5	0.9985	0.9085	0.8331	0.7691	0.7142	0.6666	0.6250	0.5882	0.5555	0.5262	0.4998
6.6	0.9986	0.9085	0.8331	0.7691	0.7142	0.6666	0.6250	0.5882	0.5555	0.5262	0.4998
6.7	0.9988	0.9086	0.8331	0.7692	0.7142	0.6666	0.6250	0.5882	0.5555	0.5262	0.4998
6.8	0.9989	0.9087	0.8332	0.7692	0.7143	0.6666	0.6250	0.5882	0.5555	0.5262	0.4998
6.9	0.9990	0.9087	0.8332	0.7692	0.7143	0.6666	0.6250	0.5882	0.5555	0.5262	0.4998
7.0	0.9991	0.9088	0.8332	0.7692	0.7143	0.6666	0.6250	0.5882	0.5555	0.5262	0.4998
7.1	0.9992	0.9088	0.8332	0.7692	0.7143	0.6666	0.6250	0.5882	0.5555	0.5262	0.4998
7.2	0.9993	0.9088	0.8332	0.7692	0.7143	0.6666	0.6250	0.5882	0.5555	0.5262	0.4998
7.3	0.9993	0.9089	0.8333	0.7692	0.7143	0.6666	0.6250	0.5882	0.5555	0.5262	0.4998
7.4	0.9994	0.9089	0.8333	0.7692	0.7143	0.6666	0.6250	0.5882	0.5555	0.5262	0.4998
7.5	0.9994	0.9089	0.8333	0.7692	0.7143	0.6666	0.6250	0.5882	0.5555	0.5262	0.4998
7.6	0.9995	0.9089	0.8333	0.7692	0.7143	0.6666	0.6250	0.5882	0.5555	0.5262	0.4998
7.7	0.9995	0.9089	0.8333	0.7692	0.7143	0.6666	0.6250	0.5882	0.5555	0.5262	0.4998
7.8	0.9996	0.9090	0.8333	0.7692	0.7143	0.6666	0.6250	0.5882	0.5555	0.5262	0.4998
7.9	0.9996	0.9090	0.8333	0.7692	0.7143	0.6666	0.6250	0.5882	0.5555	0.5262	0.4998
8.0	0.9997	0.9090	0.8333	0.7692	0.7143	0.6666	0.6250	0.5882	0.5555	0.5262	0.4998
8.1	0.9997	0.9090	0.8333	0.7692	0.7143	0.6666	0.6250	0.5882	0.5555	0.5262	0.4998
8.2	0.9997	0.9090	0.8333	0.7692	0.7143	0.6666	0.6250	0.5882	0.5555	0.5262	0.4997
8.3	0.9998	0.9090	0.8333	0.7692	0.7143	0.6666	0.6250	0.5882	0.5555	0.5261	0.4997
8.4	0.9998	0.9090	0.8333	0.7692	0.7143	0.6666	0.6250	0.5882	0.5555	0.5261	0.4997
8.5	0.9998	0.9090	0.8333	0.7692	0.7143	0.6666	0.6250	0.5882	0.5554	0.5261	0.4997
8.6	0.9998	0.9090	0.8333	0.7692	0.7143	0.6666	0.6250	0.5882	0.5554	0.5261	0.4997
8.7	0.9998	0.9090	0.8333	0.7692	0.7143	0.6666	0.6250	0.5882	0.5554	0.5261	0.4997
8.8	0.9998	0.9091	0.8333	0.7692	0.7143	0.6666	0.6250	0.5882	0.5554	0.5261	0.4997
8.9	0.9999	0.9091	0.8333	0.7692	0.7143	0.6666	0.6250	0.5882	0.5554	0.5261	0.4997
9.0	0.9999	0.9091	0.8333	0.7692	0.7143	0.6666	0.6250	0.5882	0.5554	0.5261	0.4997
9.1	0.9999	0.9091	0.8333	0.7692	0.7143	0.6666	0.6250	0.5882	0.5554	0.5261	0.4997
9.2	0.9999	0.9091	0.8333	0.7692	0.7143	0.6666	0.6250	0.5882	0.5554	0.5261	0.4997
9.3	0.9999	0.9091	0.8333	0.7692	0.7143	0.6666	0.6250	0.5882	0.5554	0.5261	0.4997
9.4	0.9999	0.9091	0.8333	0.7692	0.7143	0.6666	0.6250	0.5882	0.5554	0.5261	0.4997
9.5	0.9999	0.9091	0.8333	0.7692	0.7143	0.6666	0.6250	0.5882	0.5554	0.5261	0.4997
9.6	0.9999	0.9091	0.8333	0.7692	0.7143	0.6666	0.6250	0.5882	0.5554	0.5261	0.4996
9.7	0.9999	0.9091	0.8333	0.7692	0.7143	0.6666	0.6250	0.5882	0.5554	0.5261	0.4996
9.8	0.9999	0.9091	0.8333	0.7692	0.7143	0.6666	0.6250	0.5882	0.5554	0.5261	0.4996
9.9	1.0000	0.9091	0.8333	0.7692	0.7143	0.6666	0.6250	0.5882	0.5554	0.5261	0.4996
10.0	1.0000	0.9091	0.8333	0.7692	0.7143	0.6666	0.6250	0.5882	0.5554	0.5261	0.4996

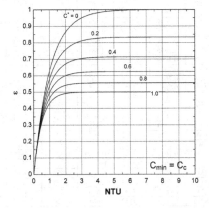

CONFIGURATION 15

1TUBE6ROW6PASS1CIR(PAR)

(b) $C_{min} = C_h$

NTU	Heat Exchanger Effectiveness										
	$C^* (C_{min} = C_h)$										
	0	0.1	0.2	0.3	0.4	0.5	0.6	0.7	0.8	0.9	1.0
0	0.0000	0.0000	0.0000	0.0000	0.0000	0.0000	0.0000	0.0000	0.0000	0.0000	0.0000
0.1	0.0952	0.0947	0.0942	0.0938	0.0933	0.0929	0.0924	0.0920	0.0915	0.0911	0.0906
0.2	0.1813	0.1795	0.1778	0.1761	0.1745	0.1728	0.1712	0.1696	0.1680	0.1664	0.1649
0.3	0.2592	0.2555	0.2520	0.2484	0.2450	0.2416	0.2383	0.2351	0.2319	0.2287	0.2257
0.4	0.3297	0.3236	0.3177	0.3120	0.3063	0.3009	0.2955	0.2903	0.2853	0.2803	0.2755
0.5	0.3935	0.3846	0.3761	0.3677	0.3597	0.3519	0.3443	0.3370	0.3299	0.3230	0.3163
0.6	0.4512	0.4393	0.4278	0.4167	0.4061	0.3958	0.3859	0.3764	0.3672	0.3583	0.3497
0.7	0.5034	0.4882	0.4737	0.4598	0.4464	0.4337	0.4214	0.4096	0.3983	0.3875	0.3771
0.8	0.5507	0.5321	0.5144	0.4976	0.4815	0.4662	0.4516	0.4377	0.4244	0.4117	0.3995
0.9	0.5934	0.5714	0.5506	0.5308	0.5121	0.4943	0.4774	0.4614	0.4461	0.4317	0.4179
1.0	0.6321	0.6066	0.5826	0.5600	0.5386	0.5184	0.4994	0.4814	0.4643	0.4482	0.4330
1.1	0.6671	0.6382	0.6110	0.5856	0.5617	0.5392	0.5181	0.4982	0.4795	0.4619	0.4453
1.2	0.6988	0.6665	0.6363	0.6081	0.5818	0.5571	0.5341	0.5125	0.4922	0.4732	0.4554
1.3	0.7275	0.6918	0.6586	0.6278	0.5992	0.5725	0.5477	0.5245	0.5028	0.4826	0.4636
1.4	0.7534	0.7145	0.6785	0.6452	0.6144	0.5858	0.5593	0.5346	0.5117	0.4903	0.4704
1.5	0.7769	0.7348	0.6961	0.6604	0.6276	0.5972	0.5691	0.5431	0.5190	0.4967	0.4759
1.6	0.7981	0.7530	0.7117	0.6738	0.6390	0.6070	0.5775	0.5503	0.5252	0.5019	0.4804
1.7	0.8173	0.7693	0.7255	0.6856	0.6490	0.6155	0.5847	0.5564	0.5303	0.5063	0.4840
1.8	0.8347	0.7839	0.7378	0.6959	0.6576	0.6227	0.5908	0.5615	0.5346	0.5099	0.4870
1.9	0.8504	0.7970	0.7487	0.7049	0.6652	0.6290	0.5960	0.5658	0.5382	0.5128	0.4895
2.0	0.8647	0.8087	0.7584	0.7129	0.6717	0.6343	0.6004	0.5694	0.5411	0.5152	0.4915
2.1	0.8775	0.8193	0.7669	0.7198	0.6774	0.6390	0.6041	0.5725	0.5436	0.5172	0.4931
2.2	0.8892	0.8287	0.7745	0.7260	0.6823	0.6429	0.6073	0.5750	0.5457	0.5189	0.4944
2.3	0.8997	0.8371	0.7812	0.7313	0.6866	0.6463	0.6100	0.5772	0.5474	0.5202	0.4955
2.4	0.9093	0.8446	0.7872	0.7360	0.6903	0.6492	0.6123	0.5790	0.5488	0.5213	0.4963
2.5	0.9179	0.8514	0.7925	0.7401	0.6935	0.6517	0.6142	0.5805	0.5499	0.5222	0.4970
2.6	0.9257	0.8574	0.7972	0.7438	0.6963	0.6539	0.6159	0.5817	0.5509	0.5230	0.4976
2.7	0.9328	0.8628	0.8013	0.7469	0.6987	0.6557	0.6173	0.5828	0.5517	0.5236	0.4981
2.8	0.9392	0.8677	0.8050	0.7497	0.7008	0.6573	0.6185	0.5837	0.5524	0.5241	0.4984
2.9	0.9450	0.8720	0.8082	0.7521	0.7026	0.6586	0.6195	0.5844	0.5529	0.5245	0.4987
3.0	0.9502	0.8759	0.8111	0.7542	0.7041	0.6598	0.6203	0.5850	0.5534	0.5248	0.4990
3.1	0.9550	0.8794	0.8136	0.7561	0.7055	0.6608	0.6210	0.5855	0.5537	0.5251	0.4992
3.2	0.9592	0.8825	0.8159	0.7577	0.7067	0.6616	0.6216	0.5860	0.5540	0.5253	0.4993
3.3	0.9631	0.8853	0.8179	0.7591	0.7077	0.6623	0.6221	0.5863	0.5543	0.5255	0.4994
3.4	0.9666	0.8878	0.8196	0.7604	0.7085	0.6629	0.6225	0.5866	0.5545	0.5256	0.4995
3.5	0.9698	0.8900	0.8212	0.7614	0.7093	0.6634	0.6229	0.5869	0.5547	0.5257	0.4996
3.6	0.9727	0.8920	0.8225	0.7624	0.7099	0.6639	0.6232	0.5871	0.5548	0.5258	0.4997
3.7	0.9753	0.8937	0.8237	0.7632	0.7105	0.6642	0.6234	0.5872	0.5549	0.5259	0.4997
3.8	0.9776	0.8953	0.8248	0.7639	0.7109	0.6645	0.6236	0.5874	0.5550	0.5259	0.4997
3.9	0.9798	0.8967	0.8257	0.7645	0.7113	0.6648	0.6238	0.5875	0.5551	0.5260	0.4998
4.0	0.9817	0.8980	0.8266	0.7650	0.7117	0.6650	0.6239	0.5876	0.5551	0.5260	0.4998
4.1	0.9834	0.8991	0.8273	0.7655	0.7120	0.6652	0.6241	0.5876	0.5552	0.5260	0.4998
4.2	0.9850	0.9001	0.8279	0.7659	0.7122	0.6654	0.6242	0.5877	0.5552	0.5261	0.4998
4.3	0.9864	0.9010	0.8285	0.7662	0.7124	0.6655	0.6242	0.5877	0.5552	0.5261	0.4998
4.4	0.9877	0.9018	0.8289	0.7665	0.7126	0.6656	0.6243	0.5878	0.5552	0.5261	0.4998
4.5	0.9889	0.9025	0.8294	0.7668	0.7127	0.6657	0.6243	0.5878	0.5552	0.5261	0.4998
4.6	0.9899	0.9031	0.8297	0.7670	0.7129	0.6657	0.6244	0.5878	0.5553	0.5261	0.4998
4.7	0.9909	0.9037	0.8300	0.7672	0.7130	0.6658	0.6244	0.5878	0.5553	0.5261	0.4999
4.8	0.9918	0.9042	0.8303	0.7673	0.7130	0.6658	0.6244	0.5878	0.5553	0.5261	0.4999
4.9	0.9926	0.9046	0.8306	0.7674	0.7131	0.6659	0.6244	0.5878	0.5553	0.5261	0.4999
5.0	0.9933	0.9050	0.8307	0.7675	0.7132	0.6659	0.6245	0.5878	0.5553	0.5261	0.4999
5.1	0.9939	0.9053	0.8309	0.7676	0.7132	0.6659	0.6245	0.5879	0.5553	0.5261	0.4999
5.2	0.9945	0.9056	0.8311	0.7677	0.7132	0.6659	0.6245	0.5878	0.5553	0.5261	0.4999
5.3	0.9950	0.9059	0.8312	0.7677	0.7132	0.6659	0.6245	0.5878	0.5553	0.5261	0.4999
5.4	0.9955	0.9061	0.8313	0.7678	0.7132	0.6659	0.6245	0.5878	0.5553	0.5261	0.4998
5.5	0.9959	0.9063	0.8313	0.7678	0.7133	0.6659	0.6245	0.5878	0.5553	0.5261	0.4998
5.6	0.9963	0.9064	0.8314	0.7678	0.7132	0.6659	0.6244	0.5878	0.5553	0.5261	0.4998
5.7	0.9967	0.9066	0.8314	0.7678	0.7132	0.6659	0.6244	0.5878	0.5553	0.5261	0.4998
5.8	0.9970	0.9067	0.8315	0.7678	0.7132	0.6659	0.6244	0.5878	0.5553	0.5261	0.4998
5.9	0.9973	0.9068	0.8315	0.7678	0.7132	0.6659	0.6244	0.5878	0.5552	0.5261	0.4998
6.0	0.9975	0.9069	0.8315	0.7678	0.7132	0.6659	0.6244	0.5878	0.5552	0.5261	0.4998

NTU	Heat Exchanger Effectiveness										
	C^* ($C_{min} = C_h$)										
	0	0.1	0.2	0.3	0.4	0.5	0.6	0.7	0.8	0.9	1.0
6.1	0.9978	0.9069	0.8315	0.7678	0.7132	0.6658	0.6244	0.5878	0.5552	0.5261	0.4998
6.2	0.9980	0.9070	0.8315	0.7677	0.7131	0.6658	0.6244	0.5878	0.5552	0.5261	0.4998
6.3	0.9982	0.9070	0.8315	0.7677	0.7131	0.6658	0.6244	0.5878	0.5552	0.5261	0.4998
6.4	0.9983	0.9070	0.8314	0.7677	0.7131	0.6658	0.6243	0.5878	0.5552	0.5261	0.4998
6.5	0.9985	0.9070	0.8314	0.7676	0.7130	0.6657	0.6243	0.5878	0.5552	0.5261	0.4998
6.6	0.9986	0.9070	0.8313	0.7676	0.7130	0.6657	0.6243	0.5877	0.5552	0.5261	0.4998
6.7	0.9988	0.9070	0.8313	0.7675	0.7130	0.6657	0.6243	0.5877	0.5552	0.5261	0.4998
6.8	0.9989	0.9070	0.8312	0.7675	0.7129	0.6657	0.6243	0.5877	0.5552	0.5261	0.4998
6.9	0.9990	0.9070	0.8312	0.7674	0.7129	0.6657	0.6243	0.5877	0.5552	0.5261	0.4998
7.0	0.9991	0.9069	0.8311	0.7674	0.7129	0.6656	0.6243	0.5877	0.5552	0.5260	0.4998
7.1	0.9992	0.9069	0.8310	0.7673	0.7128	0.6656	0.6242	0.5877	0.5552	0.5260	0.4998
7.2	0.9993	0.9068	0.8310	0.7673	0.7128	0.6656	0.6242	0.5877	0.5552	0.5260	0.4998
7.3	0.9993	0.9068	0.8309	0.7672	0.7127	0.6655	0.6242	0.5877	0.5552	0.5260	0.4998
7.4	0.9994	0.9067	0.8308	0.7671	0.7127	0.6655	0.6242	0.5877	0.5551	0.5260	0.4998
7.5	0.9994	0.9066	0.8307	0.7671	0.7127	0.6655	0.6242	0.5876	0.5551	0.5260	0.4998
7.6	0.9995	0.9066	0.8307	0.7670	0.7126	0.6655	0.6241	0.5876	0.5551	0.5260	0.4998
7.7	0.9995	0.9065	0.8306	0.7669	0.7126	0.6654	0.6241	0.5876	0.5551	0.5260	0.4998
7.8	0.9996	0.9064	0.8305	0.7669	0.7125	0.6654	0.6241	0.5876	0.5551	0.5260	0.4998
7.9	0.9996	0.9063	0.8304	0.7668	0.7125	0.6654	0.6241	0.5876	0.5551	0.5260	0.4998
8.0	0.9997	0.9062	0.8303	0.7667	0.7124	0.6653	0.6241	0.5876	0.5551	0.5260	0.4998
8.1	0.9997	0.9061	0.8302	0.7667	0.7124	0.6653	0.6240	0.5876	0.5551	0.5260	0.4998
8.2	0.9997	0.9060	0.8301	0.7666	0.7123	0.6653	0.6240	0.5876	0.5551	0.5260	0.4997
8.3	0.9998	0.9059	0.8300	0.7665	0.7123	0.6652	0.6240	0.5875	0.5551	0.5260	0.4997
8.4	0.9998	0.9058	0.8299	0.7664	0.7122	0.6652	0.6240	0.5875	0.5551	0.5260	0.4997
8.5	0.9998	0.9057	0.8298	0.7664	0.7122	0.6652	0.6240	0.5875	0.5550	0.5259	0.4997
8.6	0.9998	0.9056	0.8297	0.7663	0.7121	0.6651	0.6239	0.5875	0.5550	0.5259	0.4997
8.7	0.9998	0.9055	0.8296	0.7662	0.7121	0.6651	0.6239	0.5875	0.5550	0.5259	0.4997
8.8	0.9998	0.9054	0.8295	0.7661	0.7120	0.6651	0.6239	0.5875	0.5550	0.5259	0.4997
8.9	0.9999	0.9053	0.8294	0.7661	0.7120	0.6650	0.6239	0.5874	0.5550	0.5259	0.4997
9.0	0.9999	0.9051	0.8292	0.7660	0.7119	0.6650	0.6238	0.5874	0.5550	0.5259	0.4997
9.1	0.9999	0.9050	0.8291	0.7659	0.7119	0.6650	0.6238	0.5874	0.5550	0.5259	0.4997
9.2	0.9999	0.9049	0.8290	0.7658	0.7118	0.6649	0.6238	0.5874	0.5550	0.5259	0.4997
9.3	0.9999	0.9048	0.8289	0.7657	0.7118	0.6649	0.6238	0.5874	0.5549	0.5259	0.4997
9.4	0.9999	0.9046	0.8288	0.7656	0.7117	0.6649	0.6237	0.5874	0.5549	0.5259	0.4997
9.5	0.9999	0.9045	0.8286	0.7655	0.7116	0.6648	0.6237	0.5873	0.5549	0.5258	0.4997
9.6	0.9999	0.9044	0.8285	0.7655	0.7116	0.6648	0.6237	0.5873	0.5549	0.5258	0.4996
9.7	0.9999	0.9042	0.8284	0.7654	0.7115	0.6647	0.6237	0.5873	0.5549	0.5258	0.4996
9.8	0.9999	0.9041	0.8283	0.7653	0.7115	0.6647	0.6236	0.5873	0.5549	0.5258	0.4996
9.9	1.0000	0.9039	0.8281	0.7652	0.7114	0.6647	0.6236	0.5873	0.5549	0.5258	0.4996
10.0	1.0000	0.9038	0.8280	0.7651	0.7113	0.6646	0.6236	0.5872	0.5548	0.5258	0.4996

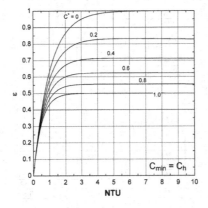

CONFIGURATION 16

1TUBE10ROW10PASS1CIR(PAR)

(a) $C_{min} = C_c$

NTU	Heat Exchanger Effectiveness										
	$C^* (C_{min} = C_c)$										
	0	0.1	0.2	0.3	0.4	0.5	0.6	0.7	0.8	0.9	1.0
0	0.0000	0.0000	0.0000	0.0000	0.0000	0.0000	0.0000	0.0000	0.0000	0.0000	0.0000
0.1	0.0952	0.0947	0.0942	0.0938	0.0933	0.0929	0.0924	0.0920	0.0915	0.0911	0.0906
0.2	0.1813	0.1795	0.1778	0.1761	0.1744	0.1728	0.1712	0.1696	0.1680	0.1664	0.1648
0.3	0.2592	0.2555	0.2519	0.2484	0.2450	0.2416	0.2383	0.2350	0.2318	0.2287	0.2256
0.4	0.3297	0.3236	0.3177	0.3119	0.3063	0.3008	0.2955	0.2903	0.2852	0.2802	0.2754
0.5	0.3935	0.3846	0.3760	0.3677	0.3596	0.3518	0.3442	0.3369	0.3298	0.3228	0.3161
0.6	0.4512	0.4392	0.4277	0.4167	0.4060	0.3957	0.3858	0.3762	0.3670	0.3581	0.3495
0.7	0.5034	0.4882	0.4736	0.4597	0.4463	0.4335	0.4212	0.4094	0.3981	0.3873	0.3768
0.8	0.5507	0.5321	0.5143	0.4974	0.4813	0.4660	0.4514	0.4374	0.4241	0.4114	0.3992
0.9	0.5934	0.5713	0.5504	0.5306	0.5118	0.4940	0.4771	0.4610	0.4458	0.4313	0.4176
1.0	0.6321	0.6065	0.5824	0.5597	0.5383	0.5181	0.4990	0.4810	0.4639	0.4478	0.4326
1.1	0.6671	0.6381	0.6108	0.5853	0.5613	0.5388	0.5177	0.4978	0.4791	0.4615	0.4448
1.2	0.6988	0.6663	0.6360	0.6078	0.5814	0.5567	0.5336	0.5120	0.4918	0.4727	0.4549
1.3	0.7275	0.6916	0.6584	0.6275	0.5988	0.5721	0.5472	0.5240	0.5023	0.4821	0.4631
1.4	0.7534	0.7143	0.6782	0.6448	0.6139	0.5853	0.5588	0.5341	0.5111	0.4898	0.4699
1.5	0.7769	0.7346	0.6958	0.6600	0.6271	0.5967	0.5686	0.5426	0.5185	0.4962	0.4754
1.6	0.7981	0.7528	0.7114	0.6734	0.6385	0.6065	0.5770	0.5498	0.5247	0.5014	0.4799
1.7	0.8173	0.7691	0.7252	0.6851	0.6485	0.6149	0.5842	0.5559	0.5298	0.5058	0.4836
1.8	0.8347	0.7837	0.7375	0.6954	0.6571	0.6222	0.5902	0.5610	0.5341	0.5094	0.4866
1.9	0.8504	0.7968	0.7483	0.7045	0.6646	0.6284	0.5954	0.5653	0.5377	0.5124	0.4891
2.0	0.8647	0.8085	0.7580	0.7124	0.6712	0.6338	0.5998	0.5689	0.5407	0.5148	0.4911
2.1	0.8775	0.8190	0.7665	0.7194	0.6769	0.6384	0.6036	0.5720	0.5432	0.5168	0.4927
2.2	0.8892	0.8284	0.7741	0.7255	0.6818	0.6424	0.6068	0.5746	0.5452	0.5185	0.4941
2.3	0.8997	0.8368	0.7808	0.7309	0.6861	0.6458	0.6095	0.5767	0.5470	0.5199	0.4952
2.4	0.9093	0.8444	0.7868	0.7356	0.6898	0.6488	0.6119	0.5786	0.5484	0.5210	0.4961
2.5	0.9179	0.8511	0.7921	0.7397	0.6930	0.6513	0.6138	0.5801	0.5496	0.5220	0.4968
2.6	0.9257	0.8572	0.7968	0.7433	0.6958	0.6535	0.6155	0.5814	0.5506	0.5227	0.4974
2.7	0.9328	0.8626	0.8010	0.7465	0.6983	0.6553	0.6169	0.5825	0.5514	0.5234	0.4979
2.8	0.9392	0.8675	0.8046	0.7493	0.7004	0.6569	0.6182	0.5834	0.5521	0.5239	0.4983
2.9	0.9450	0.8718	0.8079	0.7518	0.7022	0.6583	0.6192	0.5842	0.5527	0.5243	0.4986
3.0	0.9502	0.8757	0.8108	0.7539	0.7038	0.6595	0.6201	0.5848	0.5532	0.5247	0.4989
3.1	0.9550	0.8792	0.8134	0.7558	0.7052	0.6605	0.6208	0.5854	0.5536	0.5250	0.4991
3.2	0.9592	0.8823	0.8157	0.7575	0.7064	0.6614	0.6215	0.5858	0.5539	0.5252	0.4992
3.3	0.9631	0.8851	0.8177	0.7589	0.7075	0.6622	0.6220	0.5862	0.5542	0.5254	0.4994
3.4	0.9666	0.8877	0.8195	0.7602	0.7084	0.6628	0.6225	0.5866	0.5544	0.5256	0.4995
3.5	0.9698	0.8899	0.8211	0.7613	0.7092	0.6634	0.6228	0.5868	0.5546	0.5257	0.4996
3.6	0.9727	0.8919	0.8225	0.7623	0.7099	0.6638	0.6232	0.5871	0.5548	0.5258	0.4997
3.7	0.9753	0.8937	0.8237	0.7632	0.7105	0.6642	0.6235	0.5872	0.5549	0.5259	0.4997
3.8	0.9776	0.8953	0.8248	0.7639	0.7110	0.6646	0.6237	0.5874	0.5550	0.5260	0.4998
3.9	0.9798	0.8968	0.8258	0.7646	0.7114	0.6649	0.6239	0.5875	0.5551	0.5260	0.4998
4.0	0.9817	0.8981	0.8267	0.7652	0.7118	0.6651	0.6241	0.5877	0.5552	0.5261	0.4999
4.1	0.9834	0.8992	0.8274	0.7657	0.7121	0.6654	0.6242	0.5878	0.5553	0.5261	0.4999
4.2	0.9850	0.9003	0.8281	0.7661	0.7124	0.6656	0.6243	0.5878	0.5553	0.5262	0.4999
4.3	0.9864	0.9012	0.8287	0.7665	0.7127	0.6657	0.6244	0.5879	0.5553	0.5262	0.4999
4.4	0.9877	0.9020	0.8292	0.7668	0.7129	0.6659	0.6245	0.5880	0.5554	0.5262	0.4999
4.5	0.9889	0.9028	0.8297	0.7671	0.7131	0.6660	0.6246	0.5880	0.5554	0.5262	0.4999
4.6	0.9899	0.9034	0.8301	0.7674	0.7132	0.6661	0.6247	0.5880	0.5554	0.5262	0.5000
4.7	0.9909	0.9040	0.8305	0.7676	0.7134	0.6662	0.6247	0.5881	0.5555	0.5263	0.5000
4.8	0.9918	0.9046	0.8308	0.7678	0.7135	0.6662	0.6248	0.5881	0.5555	0.5263	0.5000
4.9	0.9926	0.9050	0.8311	0.7680	0.7136	0.6663	0.6248	0.5881	0.5555	0.5263	0.5000
5.0	0.9933	0.9055	0.8314	0.7682	0.7137	0.6664	0.6248	0.5881	0.5555	0.5263	0.5000
5.1	0.9939	0.9058	0.8316	0.7683	0.7138	0.6664	0.6249	0.5882	0.5555	0.5263	0.5000
5.2	0.9945	0.9062	0.8318	0.7684	0.7139	0.6664	0.6249	0.5882	0.5555	0.5263	0.5000
5.3	0.9950	0.9065	0.8320	0.7685	0.7139	0.6665	0.6249	0.5882	0.5555	0.5263	0.5000
5.4	0.9955	0.9068	0.8321	0.7686	0.7140	0.6665	0.6249	0.5882	0.5555	0.5263	0.5000
5.5	0.9959	0.9070	0.8323	0.7687	0.7140	0.6665	0.6249	0.5882	0.5555	0.5263	0.5000
5.6	0.9963	0.9072	0.8324	0.7688	0.7140	0.6665	0.6249	0.5882	0.5555	0.5263	0.5000
5.7	0.9967	0.9074	0.8325	0.7688	0.7141	0.6666	0.6249	0.5882	0.5555	0.5263	0.5000
5.8	0.9970	0.9076	0.8326	0.7689	0.7141	0.6666	0.6250	0.5882	0.5555	0.5263	0.5000
5.9	0.9973	0.9078	0.8327	0.7689	0.7141	0.6666	0.6250	0.5882	0.5555	0.5263	0.5000
6.0	0.9975	0.9079	0.8328	0.7690	0.7142	0.6666	0.6250	0.5882	0.5555	0.5263	0.5000

NTU	Heat Exchanger Effectiveness										
	$C^* (C_{min} = C_c)$										
	0	0.1	0.2	0.3	0.4	0.5	0.6	0.7	0.8	0.9	1.0
6.1	0.9978	0.9080	0.8328	0.7690	0.7142	0.6666	0.6250	0.5882	0.5555	0.5263	0.5000
6.2	0.9980	0.9081	0.8329	0.7690	0.7142	0.6666	0.6250	0.5882	0.5555	0.5263	0.5000
6.3	0.9982	0.9082	0.8329	0.7691	0.7142	0.6666	0.6250	0.5882	0.5555	0.5263	0.5000
6.4	0.9983	0.9083	0.8330	0.7691	0.7142	0.6666	0.6250	0.5882	0.5555	0.5263	0.5000
6.5	0.9985	0.9084	0.8330	0.7691	0.7142	0.6666	0.6250	0.5882	0.5555	0.5263	0.5000
6.6	0.9986	0.9085	0.8331	0.7691	0.7142	0.6666	0.6250	0.5882	0.5555	0.5263	0.5000
6.7	0.9988	0.9086	0.8331	0.7691	0.7142	0.6666	0.6250	0.5882	0.5555	0.5263	0.5000
6.8	0.9989	0.9086	0.8331	0.7691	0.7142	0.6666	0.6250	0.5882	0.5555	0.5263	0.5000
6.9	0.9990	0.9087	0.8331	0.7692	0.7143	0.6667	0.6250	0.5882	0.5556	0.5263	0.5000
7.0	0.9991	0.9087	0.8332	0.7692	0.7143	0.6667	0.6250	0.5882	0.5556	0.5263	0.5000
7.1	0.9992	0.9087	0.8332	0.7692	0.7143	0.6667	0.6250	0.5882	0.5556	0.5263	0.5000
7.2	0.9993	0.9088	0.8332	0.7692	0.7143	0.6667	0.6250	0.5882	0.5556	0.5263	0.5000
7.3	0.9993	0.9088	0.8332	0.7692	0.7143	0.6667	0.6250	0.5882	0.5556	0.5263	0.5000
7.4	0.9994	0.9088	0.8332	0.7692	0.7143	0.6667	0.6250	0.5882	0.5556	0.5263	0.5000
7.5	0.9994	0.9089	0.8332	0.7692	0.7143	0.6667	0.6250	0.5882	0.5556	0.5263	0.5000
7.6	0.9995	0.9089	0.8333	0.7692	0.7143	0.6667	0.6250	0.5882	0.5556	0.5263	0.5000
7.7	0.9995	0.9089	0.8333	0.7692	0.7143	0.6667	0.6250	0.5882	0.5556	0.5263	0.5000
7.8	0.9996	0.9089	0.8333	0.7692	0.7143	0.6667	0.6250	0.5882	0.5556	0.5263	0.5000
7.9	0.9996	0.9090	0.8333	0.7692	0.7143	0.6667	0.6250	0.5882	0.5556	0.5263	0.5000
8.0	0.9997	0.9090	0.8333	0.7692	0.7143	0.6667	0.6250	0.5882	0.5556	0.5263	0.5000
8.1	0.9997	0.9090	0.8333	0.7692	0.7143	0.6667	0.6250	0.5882	0.5556	0.5263	0.5000
8.2	0.9997	0.9090	0.8333	0.7692	0.7143	0.6667	0.6250	0.5882	0.5556	0.5263	0.5000
8.3	0.9998	0.9090	0.8333	0.7692	0.7143	0.6667	0.6250	0.5882	0.5556	0.5263	0.5000
8.4	0.9998	0.9090	0.8333	0.7692	0.7143	0.6667	0.6250	0.5882	0.5556	0.5263	0.5000
8.5	0.9998	0.9090	0.8333	0.7692	0.7143	0.6667	0.6250	0.5882	0.5556	0.5263	0.5000
8.6	0.9998	0.9090	0.8333	0.7692	0.7143	0.6667	0.6250	0.5882	0.5556	0.5263	0.5000
8.7	0.9998	0.9090	0.8333	0.7692	0.7143	0.6667	0.6250	0.5882	0.5556	0.5263	0.5000
8.8	0.9998	0.9090	0.8333	0.7692	0.7143	0.6667	0.6250	0.5882	0.5556	0.5263	0.5000
8.9	0.9999	0.9090	0.8333	0.7692	0.7143	0.6667	0.6250	0.5882	0.5556	0.5263	0.5000
9.0	0.9999	0.9091	0.8333	0.7692	0.7143	0.6667	0.6250	0.5882	0.5556	0.5263	0.5000
9.1	0.9999	0.9091	0.8333	0.7692	0.7143	0.6667	0.6250	0.5882	0.5556	0.5263	0.5000
9.2	0.9999	0.9091	0.8333	0.7692	0.7143	0.6667	0.6250	0.5882	0.5556	0.5263	0.5000
9.3	0.9999	0.9091	0.8333	0.7692	0.7143	0.6667	0.6250	0.5882	0.5556	0.5263	0.5000
9.4	0.9999	0.9091	0.8333	0.7692	0.7143	0.6667	0.6250	0.5882	0.5556	0.5263	0.5000
9.5	0.9999	0.9091	0.8333	0.7692	0.7143	0.6667	0.6250	0.5882	0.5556	0.5263	0.5000
9.6	0.9999	0.9091	0.8333	0.7692	0.7143	0.6667	0.6250	0.5882	0.5556	0.5263	0.5000
9.7	0.9999	0.9091	0.8333	0.7692	0.7143	0.6667	0.6250	0.5882	0.5556	0.5263	0.5000
9.8	0.9999	0.9091	0.8333	0.7692	0.7143	0.6667	0.6250	0.5882	0.5556	0.5263	0.5000
9.9	1.0000	0.9091	0.8333	0.7692	0.7143	0.6667	0.6250	0.5882	0.5556	0.5263	0.5000
10.0	1.0000	0.9091	0.8333	0.7692	0.7143	0.6667	0.6250	0.5882	0.5556	0.5263	0.5000

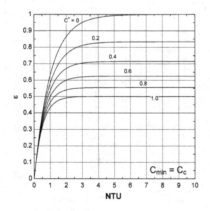

CONFIGURATION 16

1TUBE10ROW10PASS1CIR(PAR)

(b) $C_{min} = C_h$

NTU	Heat Exchanger Effectiveness										
	$C^* (C_{min} = C_h)$										
	0	0.1	0.2	0.3	0.4	0.5	0.6	0.7	0.8	0.9	1.0
0	0.0000	0.0000	0.0000	0.0000	0.0000	0.0000	0.0000	0.0000	0.0000	0.0000	0.0000
0.1	0.0952	0.0947	0.0942	0.0938	0.0933	0.0929	0.0924	0.0920	0.0915	0.0911	0.0906
0.2	0.1813	0.1795	0.1778	0.1761	0.1744	0.1728	0.1712	0.1696	0.1680	0.1664	0.1648
0.3	0.2592	0.2555	0.2519	0.2484	0.2450	0.2416	0.2383	0.2350	0.2318	0.2287	0.2256
0.4	0.3297	0.3236	0.3177	0.3119	0.3063	0.3008	0.2955	0.2903	0.2852	0.2802	0.2754
0.5	0.3935	0.3846	0.3760	0.3677	0.3596	0.3518	0.3442	0.3369	0.3298	0.3228	0.3161
0.6	0.4512	0.4392	0.4277	0.4167	0.4060	0.3957	0.3858	0.3762	0.3670	0.3581	0.3495
0.7	0.5034	0.4882	0.4736	0.4597	0.4463	0.4335	0.4212	0.4094	0.3981	0.3873	0.3768
0.8	0.5507	0.5321	0.5143	0.4974	0.4813	0.4660	0.4514	0.4374	0.4241	0.4114	0.3992
0.9	0.5934	0.5713	0.5504	0.5306	0.5118	0.4940	0.4771	0.4610	0.4458	0.4313	0.4176
1.0	0.6321	0.6065	0.5824	0.5597	0.5383	0.5181	0.4990	0.4810	0.4639	0.4478	0.4326
1.1	0.6671	0.6381	0.6108	0.5853	0.5613	0.5388	0.5177	0.4978	0.4791	0.4615	0.4448
1.2	0.6988	0.6663	0.6360	0.6078	0.5814	0.5567	0.5336	0.5120	0.4918	0.4727	0.4549
1.3	0.7275	0.6916	0.6584	0.6275	0.5988	0.5721	0.5472	0.5240	0.5023	0.4821	0.4631
1.4	0.7534	0.7143	0.6782	0.6448	0.6139	0.5853	0.5588	0.5341	0.5111	0.4898	0.4699
1.5	0.7769	0.7346	0.6958	0.6600	0.6271	0.5967	0.5686	0.5426	0.5185	0.4962	0.4754
1.6	0.7981	0.7528	0.7114	0.6734	0.6385	0.6065	0.5770	0.5498	0.5247	0.5014	0.4799
1.7	0.8173	0.7691	0.7252	0.6851	0.6485	0.6149	0.5841	0.5559	0.5298	0.5058	0.4836
1.8	0.8347	0.7837	0.7375	0.6954	0.6571	0.6222	0.5902	0.5610	0.5341	0.5094	0.4866
1.9	0.8504	0.7968	0.7483	0.7045	0.6646	0.6284	0.5954	0.5653	0.5377	0.5124	0.4891
2.0	0.8647	0.8085	0.7580	0.7124	0.6712	0.6338	0.5998	0.5689	0.5407	0.5148	0.4911
2.1	0.8775	0.8190	0.7665	0.7194	0.6768	0.6384	0.6036	0.5720	0.5432	0.5168	0.4927
2.2	0.8892	0.8284	0.7741	0.7255	0.6818	0.6424	0.6068	0.5745	0.5452	0.5185	0.4941
2.3	0.8997	0.8368	0.7808	0.7308	0.6861	0.6458	0.6095	0.5767	0.5470	0.5199	0.4952
2.4	0.9093	0.8444	0.7868	0.7356	0.6898	0.6488	0.6119	0.5786	0.5484	0.5210	0.4961
2.5	0.9179	0.8511	0.7921	0.7397	0.6930	0.6513	0.6138	0.5801	0.5496	0.5220	0.4968
2.6	0.9257	0.8572	0.7968	0.7433	0.6958	0.6535	0.6155	0.5814	0.5506	0.5227	0.4974
2.7	0.9328	0.8626	0.8009	0.7465	0.6983	0.6553	0.6169	0.5825	0.5514	0.5234	0.4979
2.8	0.9392	0.8675	0.8046	0.7493	0.7004	0.6569	0.6181	0.5834	0.5521	0.5239	0.4983
2.9	0.9450	0.8718	0.8079	0.7518	0.7022	0.6583	0.6192	0.5842	0.5527	0.5243	0.4986
3.0	0.9502	0.8757	0.8108	0.7539	0.7038	0.6595	0.6201	0.5848	0.5532	0.5247	0.4989
3.1	0.9550	0.8792	0.8134	0.7558	0.7052	0.6605	0.6208	0.5854	0.5536	0.5250	0.4991
3.2	0.9592	0.8823	0.8156	0.7575	0.7064	0.6614	0.6214	0.5858	0.5539	0.5252	0.4992
3.3	0.9631	0.8851	0.8176	0.7589	0.7075	0.6621	0.6220	0.5862	0.5542	0.5254	0.4994
3.4	0.9666	0.8876	0.8194	0.7602	0.7084	0.6628	0.6224	0.5865	0.5544	0.5256	0.4995
3.5	0.9698	0.8899	0.8210	0.7613	0.7091	0.6633	0.6228	0.5868	0.5546	0.5257	0.4996
3.6	0.9727	0.8919	0.8224	0.7623	0.7098	0.6638	0.6231	0.5870	0.5548	0.5258	0.4997
3.7	0.9753	0.8937	0.8237	0.7631	0.7104	0.6642	0.6234	0.5872	0.5549	0.5259	0.4997
3.8	0.9776	0.8953	0.8248	0.7639	0.7109	0.6646	0.6237	0.5874	0.5550	0.5260	0.4998
3.9	0.9798	0.8967	0.8257	0.7645	0.7114	0.6648	0.6239	0.5875	0.5551	0.5260	0.4998
4.0	0.9817	0.8980	0.8266	0.7651	0.7118	0.6651	0.6240	0.5876	0.5552	0.5261	0.4999
4.1	0.9834	0.8992	0.8274	0.7656	0.7121	0.6653	0.6242	0.5877	0.5552	0.5261	0.4999
4.2	0.9850	0.9002	0.8280	0.7661	0.7124	0.6655	0.6243	0.5878	0.5553	0.5262	0.4999
4.3	0.9864	0.9011	0.8286	0.7664	0.7126	0.6657	0.6244	0.5879	0.5553	0.5262	0.4999
4.4	0.9877	0.9020	0.8292	0.7668	0.7128	0.6658	0.6245	0.5879	0.5554	0.5262	0.4999
4.5	0.9889	0.9027	0.8296	0.7671	0.7130	0.6659	0.6246	0.5880	0.5554	0.5262	0.4999
4.6	0.9899	0.9034	0.8300	0.7673	0.7132	0.6660	0.6246	0.5880	0.5554	0.5262	0.5000
4.7	0.9909	0.9040	0.8304	0.7676	0.7133	0.6661	0.6247	0.5880	0.5554	0.5262	0.5000
4.8	0.9918	0.9045	0.8307	0.7678	0.7134	0.6662	0.6247	0.5881	0.5555	0.5263	0.5000
4.9	0.9926	0.9050	0.8310	0.7679	0.7135	0.6662	0.6248	0.5881	0.5555	0.5263	0.5000
5.0	0.9933	0.9054	0.8313	0.7681	0.7136	0.6663	0.6248	0.5881	0.5555	0.5263	0.5000
5.1	0.9939	0.9058	0.8315	0.7682	0.7137	0.6663	0.6248	0.5881	0.5555	0.5263	0.5000
5.2	0.9945	0.9061	0.8317	0.7683	0.7138	0.6664	0.6248	0.5881	0.5555	0.5263	0.5000
5.3	0.9950	0.9064	0.8319	0.7684	0.7138	0.6664	0.6248	0.5881	0.5555	0.5263	0.5000
5.4	0.9955	0.9067	0.8320	0.7685	0.7139	0.6664	0.6249	0.5882	0.5555	0.5263	0.5000
5.5	0.9959	0.9069	0.8321	0.7686	0.7139	0.6665	0.6249	0.5882	0.5555	0.5263	0.5000
5.6	0.9963	0.9071	0.8323	0.7686	0.7139	0.6665	0.6249	0.5882	0.5555	0.5263	0.5000
5.7	0.9967	0.9073	0.8324	0.7687	0.7140	0.6665	0.6249	0.5882	0.5555	0.5263	0.5000
5.8	0.9970	0.9075	0.8324	0.7687	0.7140	0.6665	0.6249	0.5882	0.5555	0.5263	0.5000
5.9	0.9973	0.9076	0.8325	0.7688	0.7140	0.6665	0.6249	0.5882	0.5555	0.5263	0.5000
6.0	0.9975	0.9077	0.8326	0.7688	0.7140	0.6665	0.6249	0.5882	0.5555	0.5263	0.5000

NTU	Heat Exchanger Effectiveness										
	$C^* (C_{min} = C_h)$										
	0	0.1	0.2	0.3	0.4	0.5	0.6	0.7	0.8	0.9	1.0
6.1	0.9978	0.9079	0.8326	0.7688	0.7141	0.6665	0.6249	0.5882	0.5555	0.5263	0.5000
6.2	0.9980	0.9080	0.8327	0.7689	0.7141	0.6665	0.6249	0.5882	0.5555	0.5263	0.5000
6.3	0.9982	0.9081	0.8327	0.7689	0.7141	0.6665	0.6249	0.5882	0.5555	0.5263	0.5000
6.4	0.9983	0.9081	0.8328	0.7689	0.7141	0.6665	0.6249	0.5882	0.5555	0.5263	0.5000
6.5	0.9985	0.9082	0.8328	0.7689	0.7141	0.6665	0.6249	0.5882	0.5555	0.5263	0.5000
6.6	0.9986	0.9083	0.8328	0.7689	0.7141	0.6666	0.6249	0.5882	0.5555	0.5263	0.5000
6.7	0.9988	0.9083	0.8329	0.7689	0.7141	0.6666	0.6249	0.5882	0.5555	0.5263	0.5000
6.8	0.9989	0.9084	0.8329	0.7689	0.7141	0.6666	0.6249	0.5882	0.5555	0.5263	0.5000
6.9	0.9990	0.9084	0.8329	0.7689	0.7141	0.6666	0.6249	0.5882	0.5555	0.5263	0.5000
7.0	0.9991	0.9084	0.8329	0.7690	0.7141	0.6666	0.6249	0.5882	0.5555	0.5263	0.5000
7.1	0.9992	0.9085	0.8329	0.7690	0.7141	0.6666	0.6249	0.5882	0.5555	0.5263	0.5000
7.2	0.9993	0.9085	0.8329	0.7690	0.7141	0.6666	0.6249	0.5882	0.5555	0.5263	0.5000
7.3	0.9993	0.9085	0.8329	0.7690	0.7141	0.6666	0.6249	0.5882	0.5555	0.5263	0.5000
7.4	0.9994	0.9085	0.8329	0.7690	0.7141	0.6666	0.6249	0.5882	0.5555	0.5263	0.5000
7.5	0.9994	0.9086	0.8329	0.7690	0.7141	0.6666	0.6249	0.5882	0.5555	0.5263	0.5000
7.6	0.9995	0.9086	0.8329	0.7690	0.7141	0.6666	0.6249	0.5882	0.5555	0.5263	0.5000
7.7	0.9995	0.9086	0.8329	0.7689	0.7141	0.6666	0.6249	0.5882	0.5555	0.5263	0.5000
7.8	0.9996	0.9086	0.8329	0.7689	0.7141	0.6666	0.6249	0.5882	0.5555	0.5263	0.5000
7.9	0.9996	0.9086	0.8329	0.7689	0.7141	0.6666	0.6249	0.5882	0.5555	0.5263	0.5000
8.0	0.9997	0.9086	0.8329	0.7689	0.7141	0.6666	0.6249	0.5882	0.5555	0.5263	0.5000
8.1	0.9997	0.9086	0.8329	0.7689	0.7141	0.6666	0.6249	0.5882	0.5555	0.5263	0.5000
8.2	0.9997	0.9086	0.8329	0.7689	0.7141	0.6666	0.6249	0.5882	0.5555	0.5263	0.5000
8.3	0.9998	0.9086	0.8329	0.7689	0.7141	0.6666	0.6249	0.5882	0.5555	0.5263	0.5000
8.4	0.9998	0.9086	0.8329	0.7689	0.7141	0.6666	0.6249	0.5882	0.5555	0.5263	0.5000
8.5	0.9998	0.9085	0.8329	0.7689	0.7141	0.6665	0.6249	0.5882	0.5555	0.5263	0.5000
8.6	0.9998	0.9085	0.8329	0.7689	0.7141	0.6665	0.6249	0.5882	0.5555	0.5263	0.5000
8.7	0.9998	0.9085	0.8328	0.7689	0.7141	0.6665	0.6249	0.5882	0.5555	0.5263	0.5000
8.8	0.9998	0.9085	0.8328	0.7689	0.7141	0.6665	0.6249	0.5882	0.5555	0.5263	0.5000
8.9	0.9999	0.9085	0.8328	0.7689	0.7141	0.6665	0.6249	0.5882	0.5555	0.5263	0.5000
9.0	0.9999	0.9085	0.8328	0.7689	0.7141	0.6665	0.6249	0.5882	0.5555	0.5263	0.5000
9.1	0.9999	0.9085	0.8328	0.7689	0.7141	0.6665	0.6249	0.5882	0.5555	0.5263	0.5000
9.2	0.9999	0.9085	0.8328	0.7689	0.7141	0.6665	0.6249	0.5882	0.5555	0.5263	0.5000
9.3	0.9999	0.9084	0.8328	0.7689	0.7141	0.6665	0.6249	0.5882	0.5555	0.5263	0.5000
9.4	0.9999	0.9084	0.8328	0.7689	0.7141	0.6665	0.6249	0.5882	0.5555	0.5263	0.5000
9.5	0.9999	0.9084	0.8327	0.7688	0.7141	0.6665	0.6249	0.5882	0.5555	0.5263	0.5000
9.6	0.9999	0.9084	0.8327	0.7688	0.7141	0.6665	0.6249	0.5882	0.5555	0.5263	0.5000
9.7	0.9999	0.9084	0.8327	0.7688	0.7141	0.6665	0.6249	0.5882	0.5555	0.5263	0.5000
9.8	0.9999	0.9083	0.8327	0.7688	0.7141	0.6665	0.6249	0.5882	0.5555	0.5263	0.5000
9.9	1.0000	0.9083	0.8327	0.7688	0.7141	0.6665	0.6249	0.5882	0.5555	0.5263	0.5000
10.0	1.0000	0.9083	0.8327	0.7688	0.7140	0.6665	0.6249	0.5882	0.5555	0.5263	0.5000

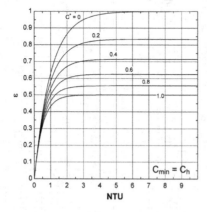

CONFIGURATION 17

1TUBE20ROW20PASS1CIR(PAR)

(a) $C_{min} = C_c$

NTU	Heat Exchanger Effectiveness										
	$C^* (C_{min} = C_c)$										
	0	0.1	0.2	0.3	0.4	0.5	0.6	0.7	0.8	0.9	1.0
0	0.0000	0.0000	0.0000	0.0000	0.0000	0.0000	0.0000	0.0000	0.0000	0.0000	0.0000
0.1	0.0952	0.0947	0.0942	0.0938	0.0933	0.0929	0.0924	0.0920	0.0915	0.0911	0.0906
0.2	0.1813	0.1795	0.1778	0.1761	0.1744	0.1728	0.1712	0.1695	0.1680	0.1664	0.1648
0.3	0.2592	0.2555	0.2519	0.2484	0.2450	0.2416	0.2383	0.2350	0.2318	0.2287	0.2256
0.4	0.3297	0.3236	0.3177	0.3119	0.3063	0.3008	0.2955	0.2902	0.2851	0.2802	0.2753
0.5	0.3935	0.3846	0.3760	0.3677	0.3596	0.3518	0.3442	0.3368	0.3297	0.3228	0.3161
0.6	0.4512	0.4392	0.4277	0.4166	0.4059	0.3956	0.3857	0.3761	0.3669	0.3580	0.3494
0.7	0.5034	0.4882	0.4736	0.4596	0.4462	0.4334	0.4211	0.4093	0.3980	0.3872	0.3767
0.8	0.5507	0.5320	0.5143	0.4974	0.4813	0.4659	0.4513	0.4373	0.4240	0.4112	0.3991
0.9	0.5934	0.5713	0.5504	0.5305	0.5117	0.4939	0.4770	0.4609	0.4457	0.4312	0.4174
1.0	0.6321	0.6065	0.5824	0.5596	0.5382	0.5180	0.4989	0.4808	0.4638	0.4477	0.4324
1.1	0.6671	0.6380	0.6108	0.5852	0.5612	0.5387	0.5175	0.4976	0.4789	0.4613	0.4447
1.2	0.6988	0.6663	0.6359	0.6076	0.5812	0.5565	0.5334	0.5118	0.4916	0.4725	0.4547
1.3	0.7275	0.6916	0.6583	0.6273	0.5986	0.5719	0.5470	0.5238	0.5021	0.4819	0.4629
1.4	0.7534	0.7142	0.6781	0.6447	0.6137	0.5851	0.5585	0.5339	0.5109	0.4896	0.4697
1.5	0.7769	0.7345	0.6956	0.6598	0.6269	0.5965	0.5684	0.5424	0.5183	0.4959	0.4752
1.6	0.7981	0.7527	0.7112	0.6732	0.6383	0.6063	0.5768	0.5496	0.5244	0.5012	0.4797
1.7	0.8173	0.7690	0.7250	0.6849	0.6483	0.6147	0.5839	0.5556	0.5296	0.5056	0.4834
1.8	0.8347	0.7836	0.7373	0.6952	0.6569	0.6219	0.5900	0.5607	0.5339	0.5092	0.4864
1.9	0.8504	0.7967	0.7482	0.7042	0.6644	0.6282	0.5952	0.5650	0.5375	0.5121	0.4889
2.0	0.8647	0.8084	0.7578	0.7122	0.6709	0.6336	0.5996	0.5687	0.5404	0.5146	0.4909
2.1	0.8775	0.8189	0.7663	0.7191	0.6766	0.6382	0.6034	0.5717	0.5429	0.5166	0.4926
2.2	0.8892	0.8283	0.7739	0.7253	0.6815	0.6422	0.6066	0.5743	0.5450	0.5183	0.4939
2.3	0.8997	0.8367	0.7807	0.7306	0.6858	0.6456	0.6093	0.5765	0.5468	0.5197	0.4950
2.4	0.9093	0.8443	0.7866	0.7353	0.6896	0.6485	0.6116	0.5784	0.5482	0.5209	0.4959
2.5	0.9179	0.8510	0.7919	0.7395	0.6928	0.6511	0.6136	0.5799	0.5494	0.5218	0.4967
2.6	0.9257	0.8571	0.7966	0.7431	0.6956	0.6532	0.6153	0.5812	0.5505	0.5226	0.4973
2.7	0.9328	0.8625	0.8008	0.7463	0.6981	0.6551	0.6168	0.5823	0.5513	0.5232	0.4978
2.8	0.9392	0.8674	0.8045	0.7491	0.7002	0.6567	0.6180	0.5833	0.5520	0.5238	0.4982
2.9	0.9450	0.8717	0.8077	0.7516	0.7020	0.6581	0.6190	0.5840	0.5526	0.5242	0.4985
3.0	0.9502	0.8756	0.8106	0.7537	0.7036	0.6593	0.6199	0.5847	0.5531	0.5246	0.4988
3.1	0.9550	0.8791	0.8132	0.7556	0.7050	0.6604	0.6207	0.5853	0.5535	0.5249	0.4990
3.2	0.9592	0.8822	0.8155	0.7573	0.7063	0.6612	0.6213	0.5857	0.5538	0.5251	0.4992
3.3	0.9631	0.8850	0.8175	0.7588	0.7073	0.6620	0.6219	0.5861	0.5541	0.5253	0.4993
3.4	0.9666	0.8875	0.8193	0.7600	0.7082	0.6627	0.6223	0.5865	0.5544	0.5255	0.4995
3.5	0.9698	0.8898	0.8209	0.7612	0.7090	0.6632	0.6227	0.5867	0.5546	0.5257	0.4996
3.6	0.9727	0.8918	0.8223	0.7621	0.7097	0.6637	0.6231	0.5870	0.5547	0.5258	0.4996
3.7	0.9753	0.8936	0.8236	0.7630	0.7103	0.6641	0.6234	0.5872	0.5549	0.5259	0.4997
3.8	0.9776	0.8952	0.8247	0.7638	0.7108	0.6645	0.6236	0.5873	0.5550	0.5259	0.4998
3.9	0.9798	0.8967	0.8256	0.7644	0.7113	0.6648	0.6238	0.5875	0.5551	0.5260	0.4998
4.0	0.9817	0.8980	0.8265	0.7650	0.7117	0.6650	0.6240	0.5876	0.5552	0.5261	0.4998
4.1	0.9834	0.8991	0.8273	0.7655	0.7120	0.6653	0.6241	0.5877	0.5552	0.5261	0.4999
4.2	0.9850	0.9002	0.8280	0.7660	0.7123	0.6655	0.6243	0.5878	0.5553	0.5261	0.4999
4.3	0.9864	0.9011	0.8286	0.7664	0.7126	0.6656	0.6244	0.5879	0.5553	0.5262	0.4999
4.4	0.9877	0.9019	0.8291	0.7667	0.7128	0.6658	0.6245	0.5879	0.5554	0.5262	0.4999
4.5	0.9889	0.9027	0.8296	0.7670	0.7130	0.6659	0.6246	0.5880	0.5554	0.5262	0.4999
4.6	0.9899	0.9033	0.8300	0.7673	0.7132	0.6660	0.6246	0.5880	0.5554	0.5262	0.5000
4.7	0.9909	0.9039	0.8304	0.7676	0.7133	0.6661	0.6247	0.5880	0.5554	0.5263	0.5000
4.8	0.9918	0.9045	0.8307	0.7678	0.7134	0.6662	0.6247	0.5881	0.5555	0.5263	0.5000
4.9	0.9926	0.9050	0.8310	0.7679	0.7136	0.6663	0.6248	0.5881	0.5555	0.5263	0.5000
5.0	0.9933	0.9054	0.8313	0.7681	0.7137	0.6663	0.6248	0.5881	0.5555	0.5263	0.5000
5.1	0.9939	0.9058	0.8315	0.7682	0.7137	0.6664	0.6248	0.5881	0.5555	0.5263	0.5000
5.2	0.9945	0.9061	0.8317	0.7684	0.7138	0.6664	0.6249	0.5882	0.5555	0.5263	0.5000
5.3	0.9950	0.9064	0.8319	0.7685	0.7139	0.6664	0.6249	0.5882	0.5555	0.5263	0.5000
5.4	0.9955	0.9067	0.8321	0.7686	0.7139	0.6665	0.6249	0.5882	0.5555	0.5263	0.5000
5.5	0.9959	0.9070	0.8322	0.7686	0.7140	0.6665	0.6249	0.5882	0.5555	0.5263	0.5000
5.6	0.9963	0.9072	0.8323	0.7687	0.7140	0.6665	0.6249	0.5882	0.5555	0.5263	0.5000
5.7	0.9967	0.9074	0.8325	0.7688	0.7141	0.6665	0.6249	0.5882	0.5555	0.5263	0.5000
5.8	0.9970	0.9076	0.8326	0.7688	0.7141	0.6666	0.6249	0.5882	0.5555	0.5263	0.5000
5.9	0.9973	0.9077	0.8326	0.7689	0.7141	0.6666	0.6250	0.5882	0.5555	0.5263	0.5000
6.0	0.9975	0.9079	0.8327	0.7689	0.7141	0.6666	0.6250	0.5882	0.5555	0.5263	0.5000

NTU	Heat Exchanger Effectiveness										
	$C^* (C_{min} = C_c)$										
	0	0.1	0.2	0.3	0.4	0.5	0.6	0.7	0.8	0.9	1.0
6.1	0.9978	0.9080	0.8328	0.7690	0.7142	0.6666	0.6250	0.5882	0.5555	0.5263	0.5000
6.2	0.9980	0.9081	0.8329	0.7690	0.7142	0.6666	0.6250	0.5882	0.5555	0.5263	0.5000
6.3	0.9982	0.9082	0.8329	0.7690	0.7142	0.6666	0.6250	0.5882	0.5556	0.5263	0.5000
6.4	0.9983	0.9083	0.8330	0.7691	0.7142	0.6666	0.6250	0.5882	0.5556	0.5263	0.5000
6.5	0.9985	0.9084	0.8330	0.7691	0.7142	0.6666	0.6250	0.5882	0.5556	0.5263	0.5000
6.6	0.9986	0.9085	0.8330	0.7691	0.7142	0.6666	0.6250	0.5882	0.5556	0.5263	0.5000
6.7	0.9988	0.9085	0.8331	0.7691	0.7142	0.6666	0.6250	0.5882	0.5556	0.5263	0.5000
6.8	0.9989	0.9086	0.8331	0.7691	0.7142	0.6666	0.6250	0.5882	0.5556	0.5263	0.5000
6.9	0.9990	0.9086	0.8331	0.7691	0.7142	0.6666	0.6250	0.5882	0.5556	0.5263	0.5000
7.0	0.9991	0.9087	0.8332	0.7692	0.7142	0.6666	0.6250	0.5882	0.5556	0.5263	0.5000
7.1	0.9992	0.9087	0.8332	0.7692	0.7143	0.6667	0.6250	0.5882	0.5556	0.5263	0.5000
7.2	0.9993	0.9088	0.8332	0.7692	0.7143	0.6667	0.6250	0.5882	0.5556	0.5263	0.5000
7.3	0.9993	0.9088	0.8332	0.7692	0.7143	0.6667	0.6250	0.5882	0.5556	0.5263	0.5000
7.4	0.9994	0.9088	0.8332	0.7692	0.7143	0.6667	0.6250	0.5882	0.5556	0.5263	0.5000
7.5	0.9994	0.9089	0.8332	0.7692	0.7143	0.6667	0.6250	0.5882	0.5556	0.5263	0.5000
7.6	0.9995	0.9089	0.8332	0.7692	0.7143	0.6667	0.6250	0.5882	0.5556	0.5263	0.5000
7.7	0.9995	0.9089	0.8333	0.7692	0.7143	0.6667	0.6250	0.5882	0.5556	0.5263	0.5000
7.8	0.9996	0.9089	0.8333	0.7692	0.7143	0.6667	0.6250	0.5882	0.5556	0.5263	0.5000
7.9	0.9996	0.9089	0.8333	0.7692	0.7143	0.6667	0.6250	0.5882	0.5556	0.5263	0.5000
8.0	0.9997	0.9090	0.8333	0.7692	0.7143	0.6667	0.6250	0.5882	0.5556	0.5263	0.5000
8.1	0.9997	0.9090	0.8333	0.7692	0.7143	0.6667	0.6250	0.5882	0.5556	0.5263	0.5000
8.2	0.9997	0.9090	0.8333	0.7692	0.7143	0.6667	0.6250	0.5882	0.5556	0.5263	0.5000
8.3	0.9998	0.9090	0.8333	0.7692	0.7143	0.6667	0.6250	0.5882	0.5556	0.5263	0.5000
8.4	0.9998	0.9090	0.8333	0.7692	0.7143	0.6667	0.6250	0.5882	0.5556	0.5263	0.5000
8.5	0.9998	0.9090	0.8333	0.7692	0.7143	0.6667	0.6250	0.5882	0.5556	0.5263	0.5000
8.6	0.9998	0.9090	0.8333	0.7692	0.7143	0.6667	0.6250	0.5882	0.5556	0.5263	0.5000
8.7	0.9998	0.9090	0.8333	0.7692	0.7143	0.6667	0.6250	0.5882	0.5556	0.5263	0.5000
8.8	0.9998	0.9090	0.8333	0.7692	0.7143	0.6667	0.6250	0.5882	0.5556	0.5263	0.5000
8.9	0.9999	0.9090	0.8333	0.7692	0.7143	0.6667	0.6250	0.5882	0.5556	0.5263	0.5000
9.0	0.9999	0.9090	0.8333	0.7692	0.7143	0.6667	0.6250	0.5882	0.5556	0.5263	0.5000
9.1	0.9999	0.9091	0.8333	0.7692	0.7143	0.6667	0.6250	0.5882	0.5556	0.5263	0.5000
9.2	0.9999	0.9091	0.8333	0.7692	0.7143	0.6667	0.6250	0.5882	0.5556	0.5263	0.5000
9.3	0.9999	0.9091	0.8333	0.7692	0.7143	0.6667	0.6250	0.5882	0.5556	0.5263	0.5000
9.4	0.9999	0.9091	0.8333	0.7692	0.7143	0.6667	0.6250	0.5882	0.5556	0.5263	0.5000
9.5	0.9999	0.9091	0.8333	0.7692	0.7143	0.6667	0.6250	0.5882	0.5556	0.5263	0.5000
9.6	0.9999	0.9091	0.8333	0.7692	0.7143	0.6667	0.6250	0.5882	0.5556	0.5263	0.5000
9.7	0.9999	0.9091	0.8333	0.7692	0.7143	0.6667	0.6250	0.5882	0.5556	0.5263	0.5000
9.8	0.9999	0.9091	0.8333	0.7692	0.7143	0.6667	0.6250	0.5882	0.5556	0.5263	0.5000
9.9	1.0000	0.9091	0.8333	0.7692	0.7143	0.6667	0.6250	0.5882	0.5556	0.5263	0.5000
10.0	1.0000	0.9091	0.8333	0.7692	0.7143	0.6667	0.6250	0.5882	0.5556	0.5263	0.5000

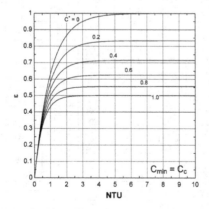

CONFIGURATION 17

1TUBE20ROW20PASS1CIR(PAR)

(b) $C_{min} = C_h$

NTU	Heat Exchanger Effectiveness										
	C^* ($C_{min} = C_h$)										
	0	0.1	0.2	0.3	0.4	0.5	0.6	0.7	0.8	0.9	1.0
0	0.0000	0.0000	0.0000	0.0000	0.0000	0.0000	0.0000	0.0000	0.0000	0.0000	0.0000
0.1	0.0952	0.0947	0.0942	0.0938	0.0933	0.0929	0.0924	0.0920	0.0915	0.0911	0.0906
0.2	0.1813	0.1795	0.1778	0.1761	0.1744	0.1728	0.1712	0.1695	0.1680	0.1664	0.1648
0.3	0.2592	0.2555	0.2519	0.2484	0.2450	0.2416	0.2383	0.2350	0.2318	0.2287	0.2256
0.4	0.3297	0.3236	0.3177	0.3119	0.3063	0.3008	0.2955	0.2902	0.2851	0.2802	0.2753
0.5	0.3935	0.3846	0.3760	0.3677	0.3596	0.3518	0.3442	0.3368	0.3297	0.3228	0.3161
0.6	0.4512	0.4392	0.4277	0.4166	0.4059	0.3956	0.3857	0.3761	0.3669	0.3580	0.3494
0.7	0.5034	0.4882	0.4736	0.4596	0.4462	0.4334	0.4211	0.4093	0.3980	0.3872	0.3767
0.8	0.5507	0.5320	0.5143	0.4974	0.4813	0.4659	0.4513	0.4373	0.4240	0.4112	0.3991
0.9	0.5934	0.5713	0.5504	0.5305	0.5117	0.4939	0.4770	0.4609	0.4457	0.4312	0.4174
1.0	0.6321	0.6065	0.5824	0.5596	0.5382	0.5180	0.4989	0.4808	0.4638	0.4477	0.4324
1.1	0.6671	0.6380	0.6108	0.5852	0.5612	0.5387	0.5175	0.4976	0.4789	0.4613	0.4447
1.2	0.6988	0.6663	0.6359	0.6076	0.5812	0.5565	0.5334	0.5118	0.4916	0.4725	0.4547
1.3	0.7275	0.6916	0.6583	0.6273	0.5986	0.5719	0.5470	0.5238	0.5021	0.4819	0.4629
1.4	0.7534	0.7142	0.6781	0.6447	0.6137	0.5851	0.5585	0.5339	0.5109	0.4896	0.4697
1.5	0.7769	0.7345	0.6956	0.6598	0.6269	0.5965	0.5684	0.5424	0.5183	0.4959	0.4752
1.6	0.7981	0.7527	0.7112	0.6732	0.6383	0.6063	0.5768	0.5496	0.5244	0.5012	0.4797
1.7	0.8173	0.7690	0.7250	0.6849	0.6483	0.6147	0.5839	0.5556	0.5296	0.5056	0.4834
1.8	0.8347	0.7836	0.7373	0.6952	0.6569	0.6219	0.5900	0.5607	0.5339	0.5092	0.4864
1.9	0.8504	0.7967	0.7482	0.7042	0.6644	0.6282	0.5952	0.5650	0.5375	0.5121	0.4889
2.0	0.8647	0.8084	0.7578	0.7122	0.6709	0.6336	0.5996	0.5687	0.5404	0.5146	0.4909
2.1	0.8775	0.8189	0.7663	0.7191	0.6766	0.6382	0.6034	0.5717	0.5429	0.5166	0.4926
2.2	0.8892	0.8283	0.7739	0.7253	0.6815	0.6422	0.6066	0.5743	0.5450	0.5183	0.4939
2.3	0.8997	0.8367	0.7807	0.7306	0.6858	0.6456	0.6093	0.5765	0.5468	0.5197	0.4950
2.4	0.9093	0.8443	0.7866	0.7353	0.6896	0.6485	0.6116	0.5784	0.5482	0.5209	0.4959
2.5	0.9179	0.8510	0.7919	0.7395	0.6928	0.6511	0.6136	0.5799	0.5494	0.5218	0.4967
2.6	0.9257	0.8571	0.7966	0.7431	0.6956	0.6532	0.6153	0.5812	0.5505	0.5226	0.4973
2.7	0.9328	0.8625	0.8008	0.7463	0.6981	0.6551	0.6168	0.5823	0.5513	0.5232	0.4978
2.8	0.9392	0.8674	0.8044	0.7491	0.7002	0.6567	0.6180	0.5833	0.5520	0.5238	0.4982
2.9	0.9450	0.8717	0.8077	0.7516	0.7020	0.6581	0.6190	0.5840	0.5526	0.5242	0.4985
3.0	0.9502	0.8756	0.8106	0.7537	0.7036	0.6593	0.6199	0.5847	0.5531	0.5246	0.4988
3.1	0.9550	0.8791	0.8132	0.7556	0.7050	0.6604	0.6207	0.5853	0.5535	0.5249	0.4990
3.2	0.9592	0.8822	0.8155	0.7573	0.7063	0.6612	0.6213	0.5857	0.5538	0.5251	0.4992
3.3	0.9631	0.8850	0.8175	0.7587	0.7073	0.6620	0.6219	0.5861	0.5541	0.5253	0.4993
3.4	0.9666	0.8875	0.8193	0.7600	0.7082	0.6627	0.6223	0.5865	0.5544	0.5255	0.4995
3.5	0.9698	0.8898	0.8209	0.7612	0.7090	0.6632	0.6227	0.5867	0.5546	0.5257	0.4996
3.6	0.9727	0.8918	0.8223	0.7621	0.7097	0.6637	0.6231	0.5870	0.5547	0.5258	0.4996
3.7	0.9753	0.8936	0.8236	0.7630	0.7103	0.6641	0.6234	0.5872	0.5549	0.5259	0.4997
3.8	0.9776	0.8952	0.8247	0.7638	0.7108	0.6645	0.6236	0.5873	0.5550	0.5259	0.4998
3.9	0.9798	0.8967	0.8256	0.7644	0.7113	0.6648	0.6238	0.5875	0.5551	0.5260	0.4998
4.0	0.9817	0.8980	0.8265	0.7650	0.7117	0.6650	0.6240	0.5876	0.5552	0.5261	0.4998
4.1	0.9834	0.8991	0.8273	0.7655	0.7120	0.6653	0.6241	0.5877	0.5552	0.5261	0.4999
4.2	0.9850	0.9002	0.8280	0.7660	0.7123	0.6655	0.6243	0.5878	0.5553	0.5261	0.4999
4.3	0.9864	0.9011	0.8286	0.7664	0.7126	0.6656	0.6244	0.5879	0.5553	0.5262	0.4999
4.4	0.9877	0.9019	0.8291	0.7667	0.7128	0.6658	0.6245	0.5879	0.5554	0.5262	0.4999
4.5	0.9889	0.9027	0.8296	0.7670	0.7130	0.6659	0.6245	0.5880	0.5554	0.5262	0.4999
4.6	0.9899	0.9033	0.8300	0.7673	0.7132	0.6660	0.6246	0.5880	0.5554	0.5262	0.5000
4.7	0.9909	0.9039	0.8304	0.7675	0.7133	0.6661	0.6247	0.5880	0.5554	0.5263	0.5000
4.8	0.9918	0.9045	0.8307	0.7678	0.7134	0.6662	0.6247	0.5881	0.5555	0.5263	0.5000
4.9	0.9926	0.9050	0.8310	0.7679	0.7136	0.6663	0.6248	0.5881	0.5555	0.5263	0.5000
5.0	0.9933	0.9054	0.8313	0.7681	0.7136	0.6663	0.6248	0.5881	0.5555	0.5263	0.5000
5.1	0.9939	0.9058	0.8315	0.7682	0.7137	0.6664	0.6248	0.5881	0.5555	0.5263	0.5000
5.2	0.9945	0.9061	0.8317	0.7684	0.7138	0.6664	0.6249	0.5882	0.5555	0.5263	0.5000
5.3	0.9950	0.9064	0.8319	0.7685	0.7139	0.6664	0.6249	0.5882	0.5555	0.5263	0.5000
5.4	0.9955	0.9067	0.8321	0.7686	0.7139	0.6665	0.6249	0.5882	0.5555	0.5263	0.5000
5.5	0.9959	0.9070	0.8322	0.7686	0.7140	0.6665	0.6249	0.5882	0.5555	0.5263	0.5000
5.6	0.9963	0.9072	0.8323	0.7687	0.7140	0.6665	0.6249	0.5882	0.5555	0.5263	0.5000
5.7	0.9967	0.9074	0.8324	0.7688	0.7140	0.6665	0.6249	0.5882	0.5555	0.5263	0.5000
5.8	0.9970	0.9076	0.8325	0.7688	0.7141	0.6666	0.6249	0.5882	0.5555	0.5263	0.5000
5.9	0.9973	0.9077	0.8326	0.7689	0.7141	0.6666	0.6250	0.5882	0.5555	0.5263	0.5000
6.0	0.9975	0.9079	0.8327	0.7689	0.7141	0.6666	0.6250	0.5882	0.5555	0.5263	0.5000

NTU	Heat Exchanger Effectiveness										
	C^* ($C_{min} = C_h$)										
	0	0.1	0.2	0.3	0.4	0.5	0.6	0.7	0.8	0.9	1.0
6.1	0.9978	0.9080	0.8328	0.7690	0.7141	0.6666	0.6250	0.5882	0.5555	0.5263	0.5000
6.2	0.9980	0.9081	0.8328	0.7690	0.7142	0.6666	0.6250	0.5882	0.5555	0.5263	0.5000
6.3	0.9982	0.9082	0.8329	0.7690	0.7142	0.6666	0.6250	0.5882	0.5555	0.5263	0.5000
6.4	0.9983	0.9083	0.8329	0.7690	0.7142	0.6666	0.6250	0.5882	0.5555	0.5263	0.5000
6.5	0.9985	0.9084	0.8330	0.7691	0.7142	0.6666	0.6250	0.5882	0.5556	0.5263	0.5000
6.6	0.9986	0.9084	0.8330	0.7691	0.7142	0.6666	0.6250	0.5882	0.5556	0.5263	0.5000
6.7	0.9988	0.9085	0.8331	0.7691	0.7142	0.6666	0.6250	0.5882	0.5556	0.5263	0.5000
6.8	0.9989	0.9086	0.8331	0.7691	0.7142	0.6666	0.6250	0.5882	0.5556	0.5263	0.5000
6.9	0.9990	0.9086	0.8331	0.7691	0.7142	0.6666	0.6250	0.5882	0.5556	0.5263	0.5000
7.0	0.9991	0.9087	0.8331	0.7691	0.7142	0.6666	0.6250	0.5882	0.5556	0.5263	0.5000
7.1	0.9992	0.9087	0.8332	0.7691	0.7142	0.6666	0.6250	0.5882	0.5556	0.5263	0.5000
7.2	0.9993	0.9087	0.8332	0.7692	0.7143	0.6666	0.6250	0.5882	0.5556	0.5263	0.5000
7.3	0.9993	0.9088	0.8332	0.7692	0.7143	0.6667	0.6250	0.5882	0.5556	0.5263	0.5000
7.4	0.9994	0.9088	0.8332	0.7692	0.7143	0.6667	0.6250	0.5882	0.5556	0.5263	0.5000
7.5	0.9994	0.9088	0.8332	0.7692	0.7143	0.6667	0.6250	0.5882	0.5556	0.5263	0.5000
7.6	0.9995	0.9089	0.8332	0.7692	0.7143	0.6667	0.6250	0.5882	0.5556	0.5263	0.5000
7.7	0.9995	0.9089	0.8332	0.7692	0.7143	0.6667	0.6250	0.5882	0.5556	0.5263	0.5000
7.8	0.9996	0.9089	0.8332	0.7692	0.7143	0.6667	0.6250	0.5882	0.5556	0.5263	0.5000
7.9	0.9996	0.9089	0.8333	0.7692	0.7143	0.6667	0.6250	0.5882	0.5556	0.5263	0.5000
8.0	0.9997	0.9089	0.8333	0.7692	0.7143	0.6667	0.6250	0.5882	0.5556	0.5263	0.5000
8.1	0.9997	0.9089	0.8333	0.7692	0.7143	0.6667	0.6250	0.5882	0.5556	0.5263	0.5000
8.2	0.9997	0.9090	0.8333	0.7692	0.7143	0.6667	0.6250	0.5882	0.5556	0.5263	0.5000
8.3	0.9998	0.9090	0.8333	0.7692	0.7143	0.6667	0.6250	0.5882	0.5556	0.5263	0.5000
8.4	0.9998	0.9090	0.8333	0.7692	0.7143	0.6667	0.6250	0.5882	0.5556	0.5263	0.5000
8.5	0.9998	0.9090	0.8333	0.7692	0.7143	0.6667	0.6250	0.5882	0.5556	0.5263	0.5000
8.6	0.9998	0.9090	0.8333	0.7692	0.7143	0.6667	0.6250	0.5882	0.5556	0.5263	0.5000
8.7	0.9998	0.9090	0.8333	0.7692	0.7143	0.6667	0.6250	0.5882	0.5556	0.5263	0.5000
8.8	0.9998	0.9090	0.8333	0.7692	0.7143	0.6667	0.6250	0.5882	0.5556	0.5263	0.5000
8.9	0.9999	0.9090	0.8333	0.7692	0.7143	0.6667	0.6250	0.5882	0.5556	0.5263	0.5000
9.0	0.9999	0.9090	0.8333	0.7692	0.7143	0.6667	0.6250	0.5882	0.5556	0.5263	0.5000
9.1	0.9999	0.9090	0.8333	0.7692	0.7143	0.6667	0.6250	0.5882	0.5556	0.5263	0.5000
9.2	0.9999	0.9090	0.8333	0.7692	0.7143	0.6667	0.6250	0.5882	0.5556	0.5263	0.5000
9.3	0.9999	0.9090	0.8333	0.7692	0.7143	0.6667	0.6250	0.5882	0.5556	0.5263	0.5000
9.4	0.9999	0.9090	0.8333	0.7692	0.7143	0.6667	0.6250	0.5882	0.5556	0.5263	0.5000
9.5	0.9999	0.9090	0.8333	0.7692	0.7143	0.6667	0.6250	0.5882	0.5556	0.5263	0.5000
9.6	0.9999	0.9090	0.8333	0.7692	0.7143	0.6667	0.6250	0.5882	0.5556	0.5263	0.5000
9.7	0.9999	0.9090	0.8333	0.7692	0.7143	0.6667	0.6250	0.5882	0.5556	0.5263	0.5000
9.8	0.9999	0.9090	0.8333	0.7692	0.7143	0.6667	0.6250	0.5882	0.5556	0.5263	0.5000
9.9	1.0000	0.9090	0.8333	0.7692	0.7143	0.6667	0.6250	0.5882	0.5556	0.5263	0.5000
10.0	1.0000	0.9090	0.8333	0.7692	0.7143	0.6667	0.6250	0.5882	0.5556	0.5263	0.5000

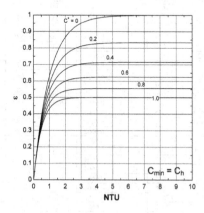

CONFIGURATION 18

1TUBE50ROW50PASS1CIR(PAR)

(a) $C_{min} = C_c$

NTU	Heat Exchanger Effectiveness										
	C^* $(C_{min} = C_c)$										
	0	0.1	0.2	0.3	0.4	0.5	0.6	0.7	0.8	0.9	1.0
0	0.0000	0.0000	0.0000	0.0000	0.0000	0.0000	0.0000	0.0000	0.0000	0.0000	0.0000
0.1	0.0952	0.0947	0.0942	0.0938	0.0933	0.0929	0.0924	0.0920	0.0915	0.0911	0.0906
0.2	0.1813	0.1795	0.1778	0.1761	0.1744	0.1728	0.1712	0.1695	0.1680	0.1664	0.1648
0.3	0.2592	0.2555	0.2519	0.2484	0.2450	0.2416	0.2383	0.2350	0.2318	0.2287	0.2256
0.4	0.3297	0.3236	0.3177	0.3119	0.3063	0.3008	0.2954	0.2902	0.2851	0.2802	0.2753
0.5	0.3935	0.3846	0.3760	0.3677	0.3596	0.3518	0.3442	0.3368	0.3297	0.3228	0.3161
0.6	0.4512	0.4392	0.4277	0.4166	0.4059	0.3956	0.3857	0.3761	0.3669	0.3580	0.3494
0.7	0.5034	0.4882	0.4736	0.4596	0.4462	0.4334	0.4211	0.4093	0.3980	0.3871	0.3767
0.8	0.5507	0.5320	0.5143	0.4973	0.4812	0.4659	0.4512	0.4373	0.4239	0.4112	0.3991
0.9	0.5934	0.5713	0.5503	0.5305	0.5117	0.4938	0.4769	0.4609	0.4456	0.4311	0.4174
1.0	0.6321	0.6065	0.5823	0.5596	0.5382	0.5179	0.4988	0.4808	0.4637	0.4476	0.4323
1.1	0.6671	0.6380	0.6107	0.5852	0.5612	0.5386	0.5175	0.4976	0.4789	0.4612	0.4446
1.2	0.6988	0.6662	0.6359	0.6076	0.5812	0.5565	0.5334	0.5118	0.4915	0.4725	0.4547
1.3	0.7275	0.6915	0.6582	0.6273	0.5986	0.5718	0.5469	0.5237	0.5021	0.4818	0.4629
1.4	0.7534	0.7142	0.6780	0.6446	0.6137	0.5850	0.5585	0.5338	0.5109	0.4895	0.4696
1.5	0.7769	0.7345	0.6956	0.6598	0.6268	0.5964	0.5683	0.5423	0.5182	0.4959	0.4751
1.6	0.7981	0.7527	0.7112	0.6731	0.6383	0.6062	0.5767	0.5495	0.5244	0.5012	0.4796
1.7	0.8173	0.7690	0.7250	0.6849	0.6482	0.6146	0.5838	0.5556	0.5295	0.5055	0.4833
1.8	0.8347	0.7836	0.7372	0.6951	0.6568	0.6219	0.5899	0.5607	0.5338	0.5091	0.4863
1.9	0.8504	0.7967	0.7481	0.7042	0.6643	0.6281	0.5951	0.5650	0.5374	0.5121	0.4888
2.0	0.8647	0.8084	0.7577	0.7121	0.6709	0.6335	0.5995	0.5686	0.5404	0.5146	0.4909
2.1	0.8775	0.8189	0.7663	0.7191	0.6765	0.6381	0.6033	0.5717	0.5429	0.5166	0.4925
2.2	0.8892	0.8283	0.7739	0.7252	0.6815	0.6421	0.6065	0.5743	0.5450	0.5183	0.4939
2.3	0.8997	0.8367	0.7806	0.7306	0.6858	0.6455	0.6092	0.5765	0.5467	0.5197	0.4950
2.4	0.9093	0.8442	0.7866	0.7353	0.6895	0.6485	0.6116	0.5783	0.5482	0.5208	0.4959
2.5	0.9179	0.8510	0.7919	0.7394	0.6927	0.6510	0.6136	0.5799	0.5494	0.5218	0.4966
2.6	0.9257	0.8570	0.7965	0.7431	0.6955	0.6532	0.6153	0.5812	0.5504	0.5226	0.4972
2.7	0.9328	0.8625	0.8007	0.7462	0.6980	0.6551	0.6167	0.5823	0.5513	0.5232	0.4977
2.8	0.9392	0.8673	0.8044	0.7490	0.7001	0.6567	0.6179	0.5832	0.5520	0.5237	0.4982
2.9	0.9450	0.8717	0.8077	0.7515	0.7020	0.6581	0.6190	0.5840	0.5526	0.5242	0.4985
3.0	0.9502	0.8756	0.8106	0.7537	0.7036	0.6593	0.6199	0.5847	0.5531	0.5246	0.4988
3.1	0.9550	0.8791	0.8131	0.7556	0.7050	0.6603	0.6206	0.5852	0.5535	0.5249	0.4990
3.2	0.9592	0.8822	0.8154	0.7572	0.7062	0.6612	0.6213	0.5857	0.5538	0.5251	0.4992
3.3	0.9631	0.8850	0.8175	0.7587	0.7073	0.6620	0.6218	0.5861	0.5541	0.5253	0.4993
3.4	0.9666	0.8875	0.8193	0.7600	0.7082	0.6626	0.6223	0.5864	0.5543	0.5255	0.4994
3.5	0.9698	0.8898	0.8208	0.7611	0.7090	0.6632	0.6227	0.5867	0.5545	0.5256	0.4995
3.6	0.9727	0.8918	0.8223	0.7621	0.7097	0.6637	0.6230	0.5869	0.5547	0.5258	0.4996
3.7	0.9753	0.8936	0.8235	0.7630	0.7103	0.6641	0.6233	0.5871	0.5548	0.5259	0.4997
3.8	0.9776	0.8952	0.8246	0.7637	0.7108	0.6644	0.6236	0.5873	0.5550	0.5259	0.4998
3.9	0.9798	0.8966	0.8256	0.7644	0.7113	0.6648	0.6238	0.5875	0.5551	0.5260	0.4998
4.0	0.9817	0.8979	0.8265	0.7650	0.7117	0.6650	0.6240	0.5876	0.5551	0.5261	0.4998
4.1	0.9834	0.8991	0.8273	0.7655	0.7120	0.6652	0.6241	0.5877	0.5552	0.5261	0.4999
4.2	0.9850	0.9001	0.8279	0.7660	0.7123	0.6654	0.6242	0.5878	0.5553	0.5261	0.4999
4.3	0.9864	0.9011	0.8286	0.7664	0.7126	0.6656	0.6244	0.5878	0.5553	0.5262	0.4999
4.4	0.9877	0.9019	0.8291	0.7667	0.7128	0.6658	0.6245	0.5879	0.5554	0.5262	0.4999
4.5	0.9889	0.9027	0.8296	0.7670	0.7130	0.6659	0.6245	0.5880	0.5554	0.5262	0.4999
4.6	0.9899	0.9033	0.8300	0.7673	0.7131	0.6660	0.6246	0.5880	0.5554	0.5262	0.5000
4.7	0.9909	0.9039	0.8304	0.7675	0.7133	0.6661	0.6247	0.5880	0.5554	0.5262	0.5000
4.8	0.9918	0.9045	0.8307	0.7677	0.7134	0.6662	0.6247	0.5881	0.5555	0.5263	0.5000
4.9	0.9926	0.9049	0.8310	0.7679	0.7135	0.6662	0.6248	0.5881	0.5555	0.5263	0.5000
5.0	0.9933	0.9054	0.8313	0.7681	0.7136	0.6663	0.6248	0.5881	0.5555	0.5263	0.5000
5.1	0.9939	0.9058	0.8315	0.7682	0.7137	0.6664	0.6248	0.5881	0.5555	0.5263	0.5000
5.2	0.9945	0.9061	0.8317	0.7683	0.7138	0.6664	0.6248	0.5882	0.5555	0.5263	0.5000
5.3	0.9950	0.9064	0.8319	0.7685	0.7139	0.6664	0.6249	0.5882	0.5555	0.5263	0.5000
5.4	0.9955	0.9067	0.8321	0.7685	0.7139	0.6665	0.6249	0.5882	0.5555	0.5263	0.5000
5.5	0.9959	0.9070	0.8322	0.7686	0.7140	0.6665	0.6249	0.5882	0.5555	0.5263	0.5000
5.6	0.9963	0.9072	0.8323	0.7687	0.7140	0.6665	0.6249	0.5882	0.5555	0.5263	0.5000
5.7	0.9967	0.9074	0.8324	0.7688	0.7140	0.6665	0.6249	0.5882	0.5555	0.5263	0.5000
5.8	0.9970	0.9076	0.8325	0.7688	0.7141	0.6666	0.6249	0.5882	0.5555	0.5263	0.5000
5.9	0.9973	0.9077	0.8326	0.7689	0.7141	0.6666	0.6250	0.5882	0.5555	0.5263	0.5000
6.0	0.9975	0.9079	0.8327	0.7689	0.7141	0.6666	0.6250	0.5882	0.5555	0.5263	0.5000

NTU	Heat Exchanger Effectiveness										
	C^* ($C_{min} = C_c$)										
	0	0.1	0.2	0.3	0.4	0.5	0.6	0.7	0.8	0.9	1.0
6.1	0.9978	0.9080	0.8328	0.7690	0.7141	0.6666	0.6250	0.5882	0.5555	0.5263	0.5000
6.2	0.9980	0.9081	0.8328	0.7690	0.7142	0.6666	0.6250	0.5882	0.5555	0.5263	0.5000
6.3	0.9982	0.9082	0.8329	0.7690	0.7142	0.6666	0.6250	0.5882	0.5555	0.5263	0.5000
6.4	0.9983	0.9083	0.8330	0.7690	0.7142	0.6666	0.6250	0.5882	0.5556	0.5263	0.5000
6.5	0.9985	0.9084	0.8330	0.7691	0.7142	0.6666	0.6250	0.5882	0.5556	0.5263	0.5000
6.6	0.9986	0.9085	0.8330	0.7691	0.7142	0.6666	0.6250	0.5882	0.5556	0.5263	0.5000
6.7	0.9988	0.9085	0.8331	0.7691	0.7142	0.6666	0.6250	0.5882	0.5556	0.5263	0.5000
6.8	0.9989	0.9086	0.8331	0.7691	0.7142	0.6666	0.6250	0.5882	0.5556	0.5263	0.5000
6.9	0.9990	0.9086	0.8331	0.7691	0.7142	0.6666	0.6250	0.5882	0.5556	0.5263	0.5000
7.0	0.9991	0.9087	0.8331	0.7691	0.7142	0.6666	0.6250	0.5882	0.5556	0.5263	0.5000
7.1	0.9992	0.9087	0.8332	0.7692	0.7143	0.6667	0.6250	0.5882	0.5556	0.5263	0.5000
7.2	0.9993	0.9088	0.8332	0.7692	0.7143	0.6667	0.6250	0.5882	0.5556	0.5263	0.5000
7.3	0.9993	0.9088	0.8332	0.7692	0.7143	0.6667	0.6250	0.5882	0.5556	0.5263	0.5000
7.4	0.9994	0.9088	0.8332	0.7692	0.7143	0.6667	0.6250	0.5882	0.5556	0.5263	0.5000
7.5	0.9994	0.9089	0.8332	0.7692	0.7143	0.6667	0.6250	0.5882	0.5556	0.5263	0.5000
7.6	0.9995	0.9089	0.8332	0.7692	0.7143	0.6667	0.6250	0.5882	0.5556	0.5263	0.5000
7.7	0.9995	0.9089	0.8333	0.7692	0.7143	0.6667	0.6250	0.5882	0.5556	0.5263	0.5000
7.8	0.9996	0.9089	0.8333	0.7692	0.7143	0.6667	0.6250	0.5882	0.5556	0.5263	0.5000
7.9	0.9996	0.9089	0.8333	0.7692	0.7143	0.6667	0.6250	0.5882	0.5556	0.5263	0.5000
8.0	0.9997	0.9090	0.8333	0.7692	0.7143	0.6667	0.6250	0.5882	0.5556	0.5263	0.5000
8.1	0.9997	0.9090	0.8333	0.7692	0.7143	0.6667	0.6250	0.5882	0.5556	0.5263	0.5000
8.2	0.9997	0.9090	0.8333	0.7692	0.7143	0.6667	0.6250	0.5882	0.5556	0.5263	0.5000
8.3	0.9998	0.9090	0.8333	0.7692	0.7143	0.6667	0.6250	0.5882	0.5556	0.5263	0.5000
8.4	0.9998	0.9090	0.8333	0.7692	0.7143	0.6667	0.6250	0.5882	0.5556	0.5263	0.5000
8.5	0.9998	0.9090	0.8333	0.7692	0.7143	0.6667	0.6250	0.5882	0.5556	0.5263	0.5000
8.6	0.9998	0.9090	0.8333	0.7692	0.7143	0.6667	0.6250	0.5882	0.5556	0.5263	0.5000
8.7	0.9998	0.9090	0.8333	0.7692	0.7143	0.6667	0.6250	0.5882	0.5556	0.5263	0.5000
8.8	0.9998	0.9090	0.8333	0.7692	0.7143	0.6667	0.6250	0.5882	0.5556	0.5263	0.5000
8.9	0.9999	0.9090	0.8333	0.7692	0.7143	0.6667	0.6250	0.5882	0.5556	0.5263	0.5000
9.0	0.9999	0.9090	0.8333	0.7692	0.7143	0.6667	0.6250	0.5882	0.5556	0.5263	0.5000
9.1	0.9999	0.9091	0.8333	0.7692	0.7143	0.6667	0.6250	0.5882	0.5556	0.5263	0.5000
9.2	0.9999	0.9091	0.8333	0.7692	0.7143	0.6667	0.6250	0.5882	0.5556	0.5263	0.5000
9.3	0.9999	0.9091	0.8333	0.7692	0.7143	0.6667	0.6250	0.5882	0.5556	0.5263	0.5000
9.4	0.9999	0.9091	0.8333	0.7692	0.7143	0.6667	0.6250	0.5882	0.5556	0.5263	0.5000
9.5	0.9999	0.9091	0.8333	0.7692	0.7143	0.6667	0.6250	0.5882	0.5556	0.5263	0.5000
9.6	0.9999	0.9091	0.8333	0.7692	0.7143	0.6667	0.6250	0.5882	0.5556	0.5263	0.5000
9.7	0.9999	0.9091	0.8333	0.7692	0.7143	0.6667	0.6250	0.5882	0.5556	0.5263	0.5000
9.8	0.9999	0.9091	0.8333	0.7692	0.7143	0.6667	0.6250	0.5882	0.5556	0.5263	0.5000
9.9	1.0000	0.9091	0.8333	0.7692	0.7143	0.6667	0.6250	0.5882	0.5556	0.5263	0.5000
10.0	1.0000	0.9091	0.8333	0.7692	0.7143	0.6667	0.6250	0.5882	0.5556	0.5263	0.5000

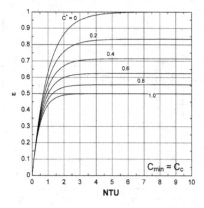

CONFIGURATION 18

1TUBE50ROW50PASS1CIR(PAR)

(b) $C_{min} = C_h$

NTU	Heat Exchanger Effectiveness										
	$C^* (C_{min} = C_h)$										
	0	0.1	0.2	0.3	0.4	0.5	0.6	0.7	0.8	0.9	1.0
0	0.0000	0.0000	0.0000	0.0000	0.0000	0.0000	0.0000	0.0000	0.0000	0.0000	0.0000
0.1	0.0952	0.0947	0.0942	0.0938	0.0933	0.0929	0.0924	0.0920	0.0915	0.0911	0.0906
0.2	0.1813	0.1795	0.1778	0.1761	0.1744	0.1728	0.1712	0.1695	0.1680	0.1664	0.1648
0.3	0.2592	0.2555	0.2519	0.2484	0.2450	0.2416	0.2383	0.2350	0.2318	0.2287	0.2256
0.4	0.3297	0.3236	0.3177	0.3119	0.3063	0.3008	0.2954	0.2902	0.2851	0.2802	0.2753
0.5	0.3935	0.3846	0.3760	0.3677	0.3596	0.3518	0.3442	0.3368	0.3297	0.3228	0.3161
0.6	0.4512	0.4392	0.4277	0.4166	0.4059	0.3956	0.3857	0.3761	0.3669	0.3580	0.3494
0.7	0.5034	0.4882	0.4736	0.4596	0.4462	0.4334	0.4211	0.4093	0.3980	0.3871	0.3767
0.8	0.5507	0.5320	0.5143	0.4973	0.4812	0.4659	0.4512	0.4373	0.4239	0.4112	0.3991
0.9	0.5934	0.5713	0.5503	0.5305	0.5117	0.4938	0.4769	0.4609	0.4456	0.4311	0.4174
1.0	0.6321	0.6065	0.5823	0.5596	0.5382	0.5179	0.4988	0.4808	0.4637	0.4476	0.4323
1.1	0.6671	0.6380	0.6107	0.5852	0.5612	0.5386	0.5175	0.4976	0.4789	0.4612	0.4446
1.2	0.6988	0.6662	0.6359	0.6076	0.5812	0.5565	0.5334	0.5118	0.4915	0.4725	0.4547
1.3	0.7275	0.6915	0.6582	0.6273	0.5986	0.5718	0.5469	0.5237	0.5021	0.4818	0.4629
1.4	0.7534	0.7142	0.6780	0.6446	0.6137	0.5850	0.5585	0.5338	0.5109	0.4895	0.4696
1.5	0.7769	0.7345	0.6956	0.6598	0.6268	0.5964	0.5683	0.5423	0.5182	0.4959	0.4751
1.6	0.7981	0.7527	0.7112	0.6731	0.6383	0.6062	0.5767	0.5495	0.5244	0.5012	0.4796
1.7	0.8173	0.7690	0.7250	0.6849	0.6482	0.6146	0.5838	0.5556	0.5295	0.5055	0.4833
1.8	0.8347	0.7836	0.7372	0.6951	0.6568	0.6219	0.5899	0.5607	0.5338	0.5091	0.4863
1.9	0.8504	0.7967	0.7481	0.7042	0.6643	0.6281	0.5951	0.5650	0.5374	0.5121	0.4888
2.0	0.8647	0.8084	0.7577	0.7121	0.6709	0.6335	0.5995	0.5686	0.5404	0.5146	0.4909
2.1	0.8775	0.8189	0.7663	0.7191	0.6765	0.6381	0.6033	0.5717	0.5429	0.5166	0.4925
2.2	0.8892	0.8283	0.7739	0.7252	0.6815	0.6421	0.6065	0.5743	0.5450	0.5183	0.4939
2.3	0.8997	0.8367	0.7806	0.7306	0.6858	0.6455	0.6092	0.5765	0.5467	0.5197	0.4950
2.4	0.9093	0.8442	0.7866	0.7353	0.6895	0.6485	0.6116	0.5783	0.5482	0.5208	0.4959
2.5	0.9179	0.8510	0.7919	0.7394	0.6927	0.6510	0.6136	0.5799	0.5494	0.5218	0.4966
2.6	0.9257	0.8570	0.7965	0.7431	0.6955	0.6532	0.6153	0.5812	0.5504	0.5226	0.4972
2.7	0.9328	0.8625	0.8007	0.7462	0.6980	0.6551	0.6167	0.5823	0.5513	0.5232	0.4977
2.8	0.9392	0.8673	0.8044	0.7490	0.7001	0.6567	0.6179	0.5832	0.5520	0.5237	0.4982
2.9	0.9450	0.8717	0.8077	0.7515	0.7020	0.6581	0.6190	0.5840	0.5526	0.5242	0.4985
3.0	0.9502	0.8756	0.8106	0.7537	0.7036	0.6593	0.6199	0.5847	0.5531	0.5246	0.4988
3.1	0.9550	0.8791	0.8131	0.7556	0.7050	0.6603	0.6206	0.5852	0.5535	0.5249	0.4990
3.2	0.9592	0.8822	0.8154	0.7572	0.7062	0.6612	0.6213	0.5857	0.5538	0.5251	0.4992
3.3	0.9631	0.8850	0.8175	0.7587	0.7073	0.6620	0.6218	0.5861	0.5541	0.5253	0.4993
3.4	0.9666	0.8875	0.8193	0.7600	0.7082	0.6626	0.6223	0.5864	0.5543	0.5255	0.4994
3.5	0.9698	0.8898	0.8208	0.7611	0.7090	0.6632	0.6227	0.5867	0.5545	0.5256	0.4995
3.6	0.9727	0.8918	0.8223	0.7621	0.7097	0.6637	0.6230	0.5869	0.5547	0.5258	0.4996
3.7	0.9753	0.8936	0.8235	0.7630	0.7103	0.6641	0.6233	0.5871	0.5548	0.5259	0.4997
3.8	0.9776	0.8952	0.8246	0.7637	0.7108	0.6644	0.6236	0.5873	0.5550	0.5259	0.4998
3.9	0.9798	0.8966	0.8256	0.7644	0.7113	0.6648	0.6238	0.5875	0.5551	0.5260	0.4998
4.0	0.9817	0.8979	0.8265	0.7650	0.7117	0.6650	0.6240	0.5876	0.5551	0.5261	0.4998
4.1	0.9834	0.8991	0.8273	0.7655	0.7120	0.6652	0.6241	0.5877	0.5552	0.5261	0.4999
4.2	0.9850	0.9001	0.8279	0.7660	0.7123	0.6654	0.6242	0.5878	0.5553	0.5261	0.4999
4.3	0.9864	0.9011	0.8286	0.7664	0.7126	0.6656	0.6244	0.5878	0.5553	0.5262	0.4999
4.4	0.9877	0.9019	0.8291	0.7667	0.7128	0.6658	0.6245	0.5879	0.5554	0.5262	0.4999
4.5	0.9889	0.9027	0.8296	0.7670	0.7130	0.6659	0.6245	0.5880	0.5554	0.5262	0.4999
4.6	0.9899	0.9033	0.8300	0.7673	0.7131	0.6660	0.6246	0.5880	0.5554	0.5262	0.5000
4.7	0.9909	0.9039	0.8304	0.7675	0.7133	0.6661	0.6247	0.5880	0.5554	0.5262	0.5000
4.8	0.9918	0.9045	0.8307	0.7677	0.7134	0.6662	0.6247	0.5881	0.5555	0.5263	0.5000
4.9	0.9926	0.9049	0.8310	0.7679	0.7135	0.6662	0.6248	0.5881	0.5555	0.5263	0.5000
5.0	0.9933	0.9054	0.8313	0.7681	0.7136	0.6663	0.6248	0.5881	0.5555	0.5263	0.5000
5.1	0.9939	0.9058	0.8315	0.7682	0.7137	0.6664	0.6248	0.5881	0.5555	0.5263	0.5000
5.2	0.9945	0.9061	0.8317	0.7683	0.7138	0.6664	0.6248	0.5882	0.5555	0.5263	0.5000
5.3	0.9950	0.9064	0.8319	0.7685	0.7139	0.6664	0.6249	0.5882	0.5555	0.5263	0.5000
5.4	0.9955	0.9067	0.8321	0.7685	0.7139	0.6665	0.6249	0.5882	0.5555	0.5263	0.5000
5.5	0.9959	0.9070	0.8322	0.7686	0.7140	0.6665	0.6249	0.5882	0.5555	0.5263	0.5000
5.6	0.9963	0.9072	0.8323	0.7687	0.7140	0.6665	0.6249	0.5882	0.5555	0.5263	0.5000
5.7	0.9967	0.9074	0.8324	0.7688	0.7140	0.6665	0.6249	0.5882	0.5555	0.5263	0.5000
5.8	0.9970	0.9076	0.8325	0.7688	0.7141	0.6666	0.6249	0.5882	0.5555	0.5263	0.5000
5.9	0.9973	0.9077	0.8326	0.7689	0.7141	0.6666	0.6250	0.5882	0.5555	0.5263	0.5000
6.0	0.9975	0.9079	0.8327	0.7689	0.7141	0.6666	0.6250	0.5882	0.5555	0.5263	0.5000

NTU	Heat Exchanger Effectiveness										
	C^* ($C_{min} = C_h$)										
	0	0.1	0.2	0.3	0.4	0.5	0.6	0.7	0.8	0.9	1.0
6.1	0.9978	0.9080	0.8328	0.7690	0.7141	0.6666	0.6250	0.5882	0.5555	0.5263	0.5000
6.2	0.9980	0.9081	0.8328	0.7690	0.7142	0.6666	0.6250	0.5882	0.5555	0.5263	0.5000
6.3	0.9982	0.9082	0.8329	0.7690	0.7142	0.6666	0.6250	0.5882	0.5555	0.5263	0.5000
6.4	0.9983	0.9083	0.8330	0.7690	0.7142	0.6666	0.6250	0.5882	0.5556	0.5263	0.5000
6.5	0.9985	0.9084	0.8330	0.7691	0.7142	0.6666	0.6250	0.5882	0.5556	0.5263	0.5000
6.6	0.9986	0.9085	0.8330	0.7691	0.7142	0.6666	0.6250	0.5882	0.5556	0.5263	0.5000
6.7	0.9988	0.9085	0.8331	0.7691	0.7142	0.6666	0.6250	0.5882	0.5556	0.5263	0.5000
6.8	0.9989	0.9086	0.8331	0.7691	0.7142	0.6666	0.6250	0.5882	0.5556	0.5263	0.5000
6.9	0.9990	0.9086	0.8331	0.7691	0.7142	0.6666	0.6250	0.5882	0.5556	0.5263	0.5000
7.0	0.9991	0.9087	0.8331	0.7691	0.7142	0.6666	0.6250	0.5882	0.5556	0.5263	0.5000
7.1	0.9992	0.9087	0.8332	0.7692	0.7143	0.6667	0.6250	0.5882	0.5556	0.5263	0.5000
7.2	0.9993	0.9088	0.8332	0.7692	0.7143	0.6667	0.6250	0.5882	0.5556	0.5263	0.5000
7.3	0.9993	0.9088	0.8332	0.7692	0.7143	0.6667	0.6250	0.5882	0.5556	0.5263	0.5000
7.4	0.9994	0.9088	0.8332	0.7692	0.7143	0.6667	0.6250	0.5882	0.5556	0.5263	0.5000
7.5	0.9994	0.9089	0.8332	0.7692	0.7143	0.6667	0.6250	0.5882	0.5556	0.5263	0.5000
7.6	0.9995	0.9089	0.8332	0.7692	0.7143	0.6667	0.6250	0.5882	0.5556	0.5263	0.5000
7.7	0.9995	0.9089	0.8333	0.7692	0.7143	0.6667	0.6250	0.5882	0.5556	0.5263	0.5000
7.8	0.9996	0.9089	0.8333	0.7692	0.7143	0.6667	0.6250	0.5882	0.5556	0.5263	0.5000
7.9	0.9996	0.9089	0.8333	0.7692	0.7143	0.6667	0.6250	0.5882	0.5556	0.5263	0.5000
8.0	0.9997	0.9090	0.8333	0.7692	0.7143	0.6667	0.6250	0.5882	0.5556	0.5263	0.5000
8.1	0.9997	0.9090	0.8333	0.7692	0.7143	0.6667	0.6250	0.5882	0.5556	0.5263	0.5000
8.2	0.9997	0.9090	0.8333	0.7692	0.7143	0.6667	0.6250	0.5882	0.5556	0.5263	0.5000
8.3	0.9998	0.9090	0.8333	0.7692	0.7143	0.6667	0.6250	0.5882	0.5556	0.5263	0.5000
8.4	0.9998	0.9090	0.8333	0.7692	0.7143	0.6667	0.6250	0.5882	0.5556	0.5263	0.5000
8.5	0.9998	0.9090	0.8333	0.7692	0.7143	0.6667	0.6250	0.5882	0.5556	0.5263	0.5000
8.6	0.9998	0.9090	0.8333	0.7692	0.7143	0.6667	0.6250	0.5882	0.5556	0.5263	0.5000
8.7	0.9998	0.9090	0.8333	0.7692	0.7143	0.6667	0.6250	0.5882	0.5556	0.5263	0.5000
8.8	0.9998	0.9090	0.8333	0.7692	0.7143	0.6667	0.6250	0.5882	0.5556	0.5263	0.5000
8.9	0.9999	0.9090	0.8333	0.7692	0.7143	0.6667	0.6250	0.5882	0.5556	0.5263	0.5000
9.0	0.9999	0.9090	0.8333	0.7692	0.7143	0.6667	0.6250	0.5882	0.5556	0.5263	0.5000
9.1	0.9999	0.9090	0.8333	0.7692	0.7143	0.6667	0.6250	0.5882	0.5556	0.5263	0.5000
9.2	0.9999	0.9091	0.8333	0.7692	0.7143	0.6667	0.6250	0.5882	0.5556	0.5263	0.5000
9.3	0.9999	0.9091	0.8333	0.7692	0.7143	0.6667	0.6250	0.5882	0.5556	0.5263	0.5000
9.4	0.9999	0.9091	0.8333	0.7692	0.7143	0.6667	0.6250	0.5882	0.5556	0.5263	0.5000
9.5	0.9999	0.9091	0.8333	0.7692	0.7143	0.6667	0.6250	0.5882	0.5556	0.5263	0.5000
9.6	0.9999	0.9091	0.8333	0.7692	0.7143	0.6667	0.6250	0.5882	0.5556	0.5263	0.5000
9.7	0.9999	0.9091	0.8333	0.7692	0.7143	0.6667	0.6250	0.5882	0.5556	0.5263	0.5000
9.8	0.9999	0.9091	0.8333	0.7692	0.7143	0.6667	0.6250	0.5882	0.5556	0.5263	0.5000
9.9	1.0000	0.9091	0.8333	0.7692	0.7143	0.6667	0.6250	0.5882	0.5556	0.5263	0.5000
10.0	1.0000	0.9091	0.8333	0.7692	0.7143	0.6667	0.6250	0.5882	0.5556	0.5263	0.5000

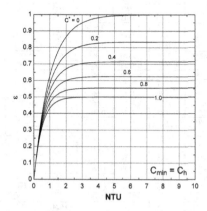

CONFIGURATION 19

1TUBE100ROW100PASS1CIR(PAR)

(a) $C_{min} = C_c$

NTU	Heat Exchanger Effectiveness										
	C^* ($C_{min} = C_c$)										
	0	0.1	0.2	0.3	0.4	0.5	0.6	0.7	0.8	0.9	1.0
0	0.0000	0.0000	0.0000	0.0000	0.0000	0.0000	0.0000	0.0000	0.0000	0.0000	0.0000
0.1	0.0952	0.0947	0.0942	0.0938	0.0933	0.0929	0.0924	0.0920	0.0915	0.0911	0.0906
0.2	0.1813	0.1795	0.1778	0.1761	0.1744	0.1728	0.1712	0.1695	0.1680	0.1664	0.1648
0.3	0.2592	0.2555	0.2519	0.2484	0.2450	0.2416	0.2383	0.2350	0.2318	0.2287	0.2256
0.4	0.3297	0.3236	0.3177	0.3119	0.3063	0.3008	0.2954	0.2902	0.2851	0.2802	0.2753
0.5	0.3935	0.3846	0.3760	0.3677	0.3596	0.3518	0.3442	0.3368	0.3297	0.3228	0.3161
0.6	0.4512	0.4392	0.4277	0.4166	0.4059	0.3956	0.3857	0.3761	0.3669	0.3580	0.3494
0.7	0.5034	0.4882	0.4736	0.4596	0.4462	0.4334	0.4211	0.4093	0.3980	0.3871	0.3767
0.8	0.5507	0.5320	0.5143	0.4973	0.4812	0.4659	0.4512	0.4373	0.4239	0.4112	0.3991
0.9	0.5934	0.5713	0.5503	0.5305	0.5117	0.4938	0.4769	0.4609	0.4456	0.4311	0.4174
1.0	0.6321	0.6065	0.5823	0.5596	0.5381	0.5179	0.4988	0.4808	0.4637	0.4476	0.4323
1.1	0.6671	0.6380	0.6107	0.5851	0.5612	0.5386	0.5175	0.4976	0.4789	0.4612	0.4446
1.2	0.6988	0.6662	0.6359	0.6076	0.5812	0.5565	0.5334	0.5118	0.4915	0.4725	0.4546
1.3	0.7275	0.6915	0.6582	0.6273	0.5986	0.5718	0.5469	0.5237	0.5020	0.4818	0.4629
1.4	0.7534	0.7142	0.6780	0.6446	0.6137	0.5850	0.5585	0.5338	0.5109	0.4895	0.4696
1.5	0.7769	0.7345	0.6956	0.6598	0.6268	0.5964	0.5683	0.5423	0.5182	0.4959	0.4751
1.6	0.7981	0.7527	0.7112	0.6731	0.6382	0.6062	0.5767	0.5495	0.5244	0.5011	0.4796
1.7	0.8173	0.7690	0.7250	0.6848	0.6482	0.6146	0.5838	0.5555	0.5295	0.5055	0.4833
1.8	0.8347	0.7836	0.7372	0.6951	0.6568	0.6219	0.5899	0.5607	0.5338	0.5091	0.4863
1.9	0.8504	0.7966	0.7481	0.7042	0.6643	0.6281	0.5951	0.5650	0.5374	0.5121	0.4888
2.0	0.8647	0.8084	0.7577	0.7121	0.6709	0.6335	0.5995	0.5686	0.5404	0.5145	0.4908
2.1	0.8775	0.8189	0.7663	0.7191	0.6765	0.6381	0.6033	0.5717	0.5429	0.5166	0.4925
2.2	0.8892	0.8283	0.7739	0.7252	0.6815	0.6421	0.6065	0.5743	0.5450	0.5183	0.4939
2.3	0.8997	0.8367	0.7806	0.7306	0.6858	0.6455	0.6092	0.5765	0.5467	0.5197	0.4950
2.4	0.9093	0.8442	0.7866	0.7353	0.6895	0.6485	0.6116	0.5783	0.5482	0.5208	0.4959
2.5	0.9179	0.8510	0.7918	0.7394	0.6927	0.6510	0.6136	0.5798	0.5494	0.5218	0.4966
2.6	0.9257	0.8570	0.7965	0.7430	0.6955	0.6532	0.6152	0.5812	0.5504	0.5226	0.4972
2.7	0.9328	0.8625	0.8007	0.7462	0.6980	0.6551	0.6167	0.5823	0.5513	0.5232	0.4977
2.8	0.9392	0.8673	0.8044	0.7490	0.7001	0.6567	0.6179	0.5832	0.5520	0.5237	0.4982
2.9	0.9450	0.8717	0.8077	0.7515	0.7020	0.6581	0.6190	0.5840	0.5526	0.5242	0.4985
3.0	0.9502	0.8756	0.8106	0.7537	0.7036	0.6593	0.6199	0.5847	0.5530	0.5246	0.4988
3.1	0.9550	0.8791	0.8131	0.7556	0.7050	0.6603	0.6206	0.5852	0.5535	0.5249	0.4990
3.2	0.9592	0.8822	0.8154	0.7572	0.7062	0.6612	0.6213	0.5857	0.5538	0.5251	0.4992
3.3	0.9631	0.8850	0.8175	0.7587	0.7073	0.6619	0.6218	0.5861	0.5541	0.5253	0.4993
3.4	0.9666	0.8875	0.8192	0.7600	0.7082	0.6626	0.6223	0.5864	0.5543	0.5255	0.4994
3.5	0.9698	0.8897	0.8208	0.7611	0.7090	0.6632	0.6227	0.5867	0.5545	0.5256	0.4995
3.6	0.9727	0.8918	0.8223	0.7621	0.7097	0.6637	0.6230	0.5869	0.5547	0.5258	0.4996
3.7	0.9753	0.8936	0.8235	0.7630	0.7103	0.6641	0.6233	0.5871	0.5548	0.5259	0.4997
3.8	0.9776	0.8952	0.8246	0.7637	0.7108	0.6644	0.6236	0.5873	0.5550	0.5259	0.4998
3.9	0.9798	0.8966	0.8256	0.7644	0.7112	0.6647	0.6238	0.5875	0.5551	0.5260	0.4998
4.0	0.9817	0.8979	0.8265	0.7650	0.7116	0.6650	0.6240	0.5876	0.5551	0.5261	0.4998
4.1	0.9834	0.8991	0.8273	0.7655	0.7120	0.6652	0.6241	0.5877	0.5552	0.5261	0.4999
4.2	0.9850	0.9001	0.8279	0.7660	0.7123	0.6654	0.6242	0.5878	0.5553	0.5261	0.4999
4.3	0.9864	0.9011	0.8286	0.7664	0.7126	0.6656	0.6244	0.5878	0.5553	0.5262	0.4999
4.4	0.9877	0.9019	0.8291	0.7667	0.7128	0.6658	0.6245	0.5879	0.5554	0.5262	0.4999
4.5	0.9889	0.9027	0.8296	0.7670	0.7130	0.6659	0.6245	0.5880	0.5554	0.5262	0.4999
4.6	0.9899	0.9033	0.8300	0.7673	0.7131	0.6660	0.6246	0.5880	0.5554	0.5262	0.5000
4.7	0.9909	0.9039	0.8304	0.7675	0.7133	0.6661	0.6247	0.5880	0.5554	0.5262	0.5000
4.8	0.9918	0.9045	0.8307	0.7677	0.7134	0.6662	0.6247	0.5881	0.5555	0.5263	0.5000
4.9	0.9926	0.9049	0.8310	0.7679	0.7135	0.6662	0.6248	0.5881	0.5555	0.5263	0.5000
5.0	0.9933	0.9054	0.8313	0.7681	0.7136	0.6663	0.6248	0.5881	0.5555	0.5263	0.5000
5.1	0.9939	0.9058	0.8315	0.7682	0.7137	0.6664	0.6248	0.5881	0.5555	0.5263	0.5000
5.2	0.9945	0.9061	0.8317	0.7683	0.7138	0.6664	0.6248	0.5881	0.5555	0.5263	0.5000
5.3	0.9950	0.9064	0.8319	0.7684	0.7139	0.6664	0.6249	0.5882	0.5555	0.5263	0.5000
5.4	0.9955	0.9067	0.8321	0.7685	0.7139	0.6665	0.6249	0.5882	0.5555	0.5263	0.5000
5.5	0.9959	0.9069	0.8322	0.7686	0.7140	0.6665	0.6249	0.5882	0.5555	0.5263	0.5000
5.6	0.9963	0.9072	0.8323	0.7687	0.7140	0.6665	0.6249	0.5882	0.5555	0.5263	0.5000
5.7	0.9967	0.9074	0.8324	0.7688	0.7140	0.6665	0.6249	0.5882	0.5555	0.5263	0.5000
5.8	0.9970	0.9076	0.8325	0.7688	0.7141	0.6666	0.6249	0.5882	0.5555	0.5263	0.5000
5.9	0.9973	0.9077	0.8326	0.7689	0.7141	0.6666	0.6250	0.5882	0.5555	0.5263	0.5000
6.0	0.9975	0.9079	0.8327	0.7689	0.7141	0.6666	0.6250	0.5882	0.5555	0.5263	0.5000

NTU	Heat Exchanger Effectiveness										
	C^* ($C_{min} = C_c$)										
	0	0.1	0.2	0.3	0.4	0.5	0.6	0.7	0.8	0.9	1.0
6.1	0.9978	0.9080	0.8328	0.7690	0.7141	0.6666	0.6250	0.5882	0.5555	0.5263	0.5000
6.2	0.9980	0.9081	0.8328	0.7690	0.7142	0.6666	0.6250	0.5882	0.5555	0.5263	0.5000
6.3	0.9982	0.9082	0.8329	0.7690	0.7142	0.6666	0.6250	0.5882	0.5555	0.5263	0.5000
6.4	0.9983	0.9083	0.8329	0.7690	0.7142	0.6666	0.6250	0.5882	0.5556	0.5263	0.5000
6.5	0.9985	0.9084	0.8330	0.7691	0.7142	0.6666	0.6250	0.5882	0.5556	0.5263	0.5000
6.6	0.9986	0.9085	0.8330	0.7691	0.7142	0.6666	0.6250	0.5882	0.5556	0.5263	0.5000
6.7	0.9988	0.9085	0.8331	0.7691	0.7142	0.6666	0.6250	0.5882	0.5556	0.5263	0.5000
6.8	0.9989	0.9086	0.8331	0.7691	0.7142	0.6666	0.6250	0.5882	0.5556	0.5263	0.5000
6.9	0.9990	0.9086	0.8331	0.7691	0.7142	0.6666	0.6250	0.5882	0.5556	0.5263	0.5000
7.0	0.9991	0.9087	0.8331	0.7691	0.7142	0.6666	0.6250	0.5882	0.5556	0.5263	0.5000
7.1	0.9992	0.9087	0.8332	0.7692	0.7143	0.6667	0.6250	0.5882	0.5556	0.5263	0.5000
7.2	0.9993	0.9088	0.8332	0.7692	0.7143	0.6667	0.6250	0.5882	0.5556	0.5263	0.5000
7.3	0.9993	0.9088	0.8332	0.7692	0.7143	0.6667	0.6250	0.5882	0.5556	0.5263	0.5000
7.4	0.9994	0.9088	0.8332	0.7692	0.7143	0.6667	0.6250	0.5882	0.5556	0.5263	0.5000
7.5	0.9994	0.9089	0.8332	0.7692	0.7143	0.6667	0.6250	0.5882	0.5556	0.5263	0.5000
7.6	0.9995	0.9089	0.8332	0.7692	0.7143	0.6667	0.6250	0.5882	0.5556	0.5263	0.5000
7.7	0.9995	0.9089	0.8333	0.7692	0.7143	0.6667	0.6250	0.5882	0.5556	0.5263	0.5000
7.8	0.9996	0.9089	0.8333	0.7692	0.7143	0.6667	0.6250	0.5882	0.5556	0.5263	0.5000
7.9	0.9996	0.9089	0.8333	0.7692	0.7143	0.6667	0.6250	0.5882	0.5556	0.5263	0.5000
8.0	0.9997	0.9090	0.8333	0.7692	0.7143	0.6667	0.6250	0.5882	0.5556	0.5263	0.5000
8.1	0.9997	0.9090	0.8333	0.7692	0.7143	0.6667	0.6250	0.5882	0.5556	0.5263	0.5000
8.2	0.9997	0.9090	0.8333	0.7692	0.7143	0.6667	0.6250	0.5882	0.5556	0.5263	0.5000
8.3	0.9998	0.9090	0.8333	0.7692	0.7143	0.6667	0.6250	0.5882	0.5556	0.5263	0.5000
8.4	0.9998	0.9090	0.8333	0.7692	0.7143	0.6667	0.6250	0.5882	0.5556	0.5263	0.5000
8.5	0.9998	0.9090	0.8333	0.7692	0.7143	0.6667	0.6250	0.5882	0.5556	0.5263	0.5000
8.6	0.9998	0.9090	0.8333	0.7692	0.7143	0.6667	0.6250	0.5882	0.5556	0.5263	0.5000
8.7	0.9998	0.9090	0.8333	0.7692	0.7143	0.6667	0.6250	0.5882	0.5556	0.5263	0.5000
8.8	0.9998	0.9090	0.8333	0.7692	0.7143	0.6667	0.6250	0.5882	0.5556	0.5263	0.5000
8.9	0.9999	0.9090	0.8333	0.7692	0.7143	0.6667	0.6250	0.5882	0.5556	0.5263	0.5000
9.0	0.9999	0.9090	0.8333	0.7692	0.7143	0.6667	0.6250	0.5882	0.5556	0.5263	0.5000
9.1	0.9999	0.9091	0.8333	0.7692	0.7143	0.6667	0.6250	0.5882	0.5556	0.5263	0.5000
9.2	0.9999	0.9091	0.8333	0.7692	0.7143	0.6667	0.6250	0.5882	0.5556	0.5263	0.5000
9.3	0.9999	0.9091	0.8333	0.7692	0.7143	0.6667	0.6250	0.5882	0.5556	0.5263	0.5000
9.4	0.9999	0.9091	0.8333	0.7692	0.7143	0.6667	0.6250	0.5882	0.5556	0.5263	0.5000
9.5	0.9999	0.9091	0.8333	0.7692	0.7143	0.6667	0.6250	0.5882	0.5556	0.5263	0.5000
9.6	0.9999	0.9091	0.8333	0.7692	0.7143	0.6667	0.6250	0.5882	0.5556	0.5263	0.5000
9.7	0.9999	0.9091	0.8333	0.7692	0.7143	0.6667	0.6250	0.5882	0.5556	0.5263	0.5000
9.8	0.9999	0.9091	0.8333	0.7692	0.7143	0.6667	0.6250	0.5882	0.5556	0.5263	0.5000
9.9	1.0000	0.9091	0.8333	0.7692	0.7143	0.6667	0.6250	0.5882	0.5556	0.5263	0.5000
10.0	1.0000	0.9091	0.8333	0.7692	0.7143	0.6667	0.6250	0.5882	0.5556	0.5263	0.5000

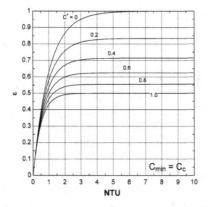

CONFIGURATION 19

1TUBE100ROW100PASS1CIR(PAR)

(b) $C_{min} = C_h$

NTU	Heat Exchanger Effectiveness										
	C^* ($C_{min} = C_h$)										
	0	0.1	0.2	0.3	0.4	0.5	0.6	0.7	0.8	0.9	1.0
0	0.0000	0.0000	0.0000	0.0000	0.0000	0.0000	0.0000	0.0000	0.0000	0.0000	0.0000
0.1	0.0952	0.0947	0.0942	0.0938	0.0933	0.0929	0.0924	0.0920	0.0915	0.0911	0.0906
0.2	0.1813	0.1795	0.1778	0.1761	0.1744	0.1728	0.1712	0.1695	0.1680	0.1664	0.1648
0.3	0.2592	0.2555	0.2519	0.2484	0.2450	0.2416	0.2383	0.2350	0.2318	0.2287	0.2256
0.4	0.3297	0.3236	0.3177	0.3119	0.3063	0.3008	0.2954	0.2902	0.2851	0.2802	0.2753
0.5	0.3935	0.3846	0.3760	0.3677	0.3596	0.3518	0.3442	0.3368	0.3297	0.3228	0.3161
0.6	0.4512	0.4392	0.4277	0.4166	0.4059	0.3956	0.3857	0.3761	0.3669	0.3580	0.3494
0.7	0.5034	0.4882	0.4736	0.4596	0.4462	0.4334	0.4211	0.4093	0.3980	0.3871	0.3767
0.8	0.5507	0.5320	0.5143	0.4973	0.4812	0.4659	0.4512	0.4373	0.4239	0.4112	0.3991
0.9	0.5934	0.5713	0.5503	0.5305	0.5117	0.4938	0.4769	0.4609	0.4456	0.4311	0.4174
1.0	0.6321	0.6065	0.5823	0.5596	0.5381	0.5179	0.4988	0.4808	0.4637	0.4476	0.4323
1.1	0.6671	0.6380	0.6107	0.5851	0.5612	0.5386	0.5175	0.4976	0.4789	0.4612	0.4446
1.2	0.6988	0.6662	0.6359	0.6076	0.5812	0.5565	0.5334	0.5118	0.4915	0.4725	0.4546
1.3	0.7275	0.6915	0.6582	0.6273	0.5986	0.5718	0.5469	0.5237	0.5020	0.4818	0.4629
1.4	0.7534	0.7142	0.6780	0.6446	0.6137	0.5850	0.5585	0.5338	0.5109	0.4895	0.4696
1.5	0.7769	0.7345	0.6956	0.6598	0.6268	0.5964	0.5683	0.5423	0.5182	0.4959	0.4751
1.6	0.7981	0.7527	0.7112	0.6731	0.6382	0.6062	0.5767	0.5495	0.5244	0.5011	0.4796
1.7	0.8173	0.7690	0.7250	0.6848	0.6482	0.6146	0.5838	0.5555	0.5295	0.5055	0.4833
1.8	0.8347	0.7836	0.7372	0.6951	0.6568	0.6219	0.5899	0.5607	0.5338	0.5091	0.4863
1.9	0.8504	0.7966	0.7481	0.7042	0.6643	0.6281	0.5951	0.5650	0.5374	0.5121	0.4888
2.0	0.8647	0.8084	0.7577	0.7121	0.6709	0.6335	0.5995	0.5686	0.5404	0.5145	0.4908
2.1	0.8775	0.8189	0.7663	0.7191	0.6765	0.6381	0.6033	0.5717	0.5429	0.5166	0.4925
2.2	0.8892	0.8283	0.7739	0.7252	0.6815	0.6421	0.6065	0.5743	0.5450	0.5183	0.4939
2.3	0.8997	0.8367	0.7806	0.7306	0.6858	0.6455	0.6092	0.5765	0.5467	0.5197	0.4950
2.4	0.9093	0.8442	0.7866	0.7353	0.6895	0.6485	0.6116	0.5783	0.5482	0.5208	0.4959
2.5	0.9179	0.8510	0.7918	0.7394	0.6927	0.6510	0.6136	0.5798	0.5494	0.5218	0.4966
2.6	0.9257	0.8570	0.7965	0.7430	0.6955	0.6532	0.6152	0.5812	0.5504	0.5226	0.4972
2.7	0.9328	0.8625	0.8007	0.7462	0.6980	0.6551	0.6167	0.5823	0.5513	0.5232	0.4977
2.8	0.9392	0.8673	0.8044	0.7490	0.7001	0.6567	0.6179	0.5832	0.5520	0.5237	0.4982
2.9	0.9450	0.8717	0.8077	0.7515	0.7020	0.6581	0.6190	0.5840	0.5526	0.5242	0.4985
3.0	0.9502	0.8756	0.8106	0.7537	0.7036	0.6593	0.6199	0.5847	0.5530	0.5246	0.4988
3.1	0.9550	0.8791	0.8131	0.7556	0.7050	0.6603	0.6206	0.5852	0.5535	0.5249	0.4990
3.2	0.9592	0.8822	0.8154	0.7572	0.7062	0.6612	0.6213	0.5857	0.5538	0.5251	0.4992
3.3	0.9631	0.8850	0.8175	0.7587	0.7073	0.6619	0.6218	0.5861	0.5541	0.5253	0.4993
3.4	0.9666	0.8875	0.8192	0.7600	0.7082	0.6626	0.6223	0.5864	0.5543	0.5255	0.4994
3.5	0.9698	0.8897	0.8208	0.7611	0.7090	0.6632	0.6227	0.5867	0.5545	0.5256	0.4995
3.6	0.9727	0.8918	0.8223	0.7621	0.7097	0.6637	0.6230	0.5869	0.5547	0.5258	0.4996
3.7	0.9753	0.8936	0.8235	0.7630	0.7103	0.6641	0.6233	0.5871	0.5548	0.5259	0.4997
3.8	0.9776	0.8952	0.8246	0.7637	0.7108	0.6644	0.6236	0.5873	0.5550	0.5259	0.4998
3.9	0.9798	0.8966	0.8256	0.7644	0.7112	0.6647	0.6238	0.5875	0.5551	0.5260	0.4998
4.0	0.9817	0.8979	0.8265	0.7650	0.7116	0.6650	0.6240	0.5876	0.5551	0.5261	0.4998
4.1	0.9834	0.8991	0.8273	0.7655	0.7120	0.6652	0.6241	0.5877	0.5552	0.5261	0.4999
4.2	0.9850	0.9001	0.8279	0.7660	0.7123	0.6654	0.6242	0.5878	0.5553	0.5261	0.4999
4.3	0.9864	0.9011	0.8286	0.7664	0.7126	0.6656	0.6244	0.5878	0.5553	0.5262	0.4999
4.4	0.9877	0.9019	0.8291	0.7667	0.7128	0.6658	0.6245	0.5879	0.5554	0.5262	0.4999
4.5	0.9889	0.9027	0.8296	0.7670	0.7130	0.6659	0.6245	0.5880	0.5554	0.5262	0.4999
4.6	0.9899	0.9033	0.8300	0.7673	0.7131	0.6660	0.6246	0.5880	0.5554	0.5262	0.5000
4.7	0.9909	0.9039	0.8304	0.7675	0.7133	0.6661	0.6247	0.5880	0.5554	0.5262	0.5000
4.8	0.9918	0.9045	0.8307	0.7677	0.7134	0.6662	0.6247	0.5881	0.5555	0.5263	0.5000
4.9	0.9926	0.9049	0.8310	0.7679	0.7135	0.6662	0.6248	0.5881	0.5555	0.5263	0.5000
5.0	0.9933	0.9054	0.8313	0.7681	0.7136	0.6663	0.6248	0.5881	0.5555	0.5263	0.5000
5.1	0.9939	0.9058	0.8315	0.7682	0.7137	0.6664	0.6248	0.5881	0.5555	0.5263	0.5000
5.2	0.9945	0.9061	0.8317	0.7683	0.7138	0.6664	0.6248	0.5881	0.5555	0.5263	0.5000
5.3	0.9950	0.9064	0.8319	0.7684	0.7139	0.6664	0.6249	0.5882	0.5555	0.5263	0.5000
5.4	0.9955	0.9067	0.8321	0.7685	0.7139	0.6665	0.6249	0.5882	0.5555	0.5263	0.5000
5.5	0.9959	0.9069	0.8322	0.7686	0.7140	0.6665	0.6249	0.5882	0.5555	0.5263	0.5000
5.6	0.9963	0.9072	0.8323	0.7687	0.7140	0.6665	0.6249	0.5882	0.5555	0.5263	0.5000
5.7	0.9967	0.9074	0.8324	0.7688	0.7140	0.6665	0.6249	0.5882	0.5555	0.5263	0.5000
5.8	0.9970	0.9076	0.8325	0.7688	0.7141	0.6666	0.6249	0.5882	0.5555	0.5263	0.5000
5.9	0.9973	0.9077	0.8326	0.7689	0.7141	0.6666	0.6250	0.5882	0.5555	0.5263	0.5000
6.0	0.9975	0.9079	0.8327	0.7689	0.7141	0.6666	0.6250	0.5882	0.5555	0.5263	0.5000

NTU	Heat Exchanger Effectiveness										
	$C^* (C_{min} = C_h)$										
	0	0.1	0.2	0.3	0.4	0.5	0.6	0.7	0.8	0.9	1.0
6.1	0.9978	0.9080	0.8328	0.7690	0.7141	0.6666	0.6250	0.5882	0.5555	0.5263	0.5000
6.2	0.9980	0.9081	0.8328	0.7690	0.7142	0.6666	0.6250	0.5882	0.5555	0.5263	0.5000
6.3	0.9982	0.9082	0.8329	0.7690	0.7142	0.6666	0.6250	0.5882	0.5555	0.5263	0.5000
6.4	0.9983	0.9083	0.8329	0.7690	0.7142	0.6666	0.6250	0.5882	0.5556	0.5263	0.5000
6.5	0.9985	0.9084	0.8330	0.7691	0.7142	0.6666	0.6250	0.5882	0.5556	0.5263	0.5000
6.6	0.9986	0.9085	0.8330	0.7691	0.7142	0.6666	0.6250	0.5882	0.5556	0.5263	0.5000
6.7	0.9988	0.9085	0.8331	0.7691	0.7142	0.6666	0.6250	0.5882	0.5556	0.5263	0.5000
6.8	0.9989	0.9086	0.8331	0.7691	0.7142	0.6666	0.6250	0.5882	0.5556	0.5263	0.5000
6.9	0.9990	0.9086	0.8331	0.7691	0.7142	0.6666	0.6250	0.5882	0.5556	0.5263	0.5000
7.0	0.9991	0.9087	0.8331	0.7691	0.7142	0.6666	0.6250	0.5882	0.5556	0.5263	0.5000
7.1	0.9992	0.9087	0.8332	0.7692	0.7143	0.6667	0.6250	0.5882	0.5556	0.5263	0.5000
7.2	0.9993	0.9088	0.8332	0.7692	0.7143	0.6667	0.6250	0.5882	0.5556	0.5263	0.5000
7.3	0.9993	0.9088	0.8332	0.7692	0.7143	0.6667	0.6250	0.5882	0.5556	0.5263	0.5000
7.4	0.9994	0.9088	0.8332	0.7692	0.7143	0.6667	0.6250	0.5882	0.5556	0.5263	0.5000
7.5	0.9994	0.9089	0.8332	0.7692	0.7143	0.6667	0.6250	0.5882	0.5556	0.5263	0.5000
7.6	0.9995	0.9089	0.8332	0.7692	0.7143	0.6667	0.6250	0.5882	0.5556	0.5263	0.5000
7.7	0.9995	0.9089	0.8333	0.7692	0.7143	0.6667	0.6250	0.5882	0.5556	0.5263	0.5000
7.8	0.9996	0.9089	0.8333	0.7692	0.7143	0.6667	0.6250	0.5882	0.5556	0.5263	0.5000
7.9	0.9996	0.9089	0.8333	0.7692	0.7143	0.6667	0.6250	0.5882	0.5556	0.5263	0.5000
8.0	0.9997	0.9090	0.8333	0.7692	0.7143	0.6667	0.6250	0.5882	0.5556	0.5263	0.5000
8.1	0.9997	0.9090	0.8333	0.7692	0.7143	0.6667	0.6250	0.5882	0.5556	0.5263	0.5000
8.2	0.9997	0.9090	0.8333	0.7692	0.7143	0.6667	0.6250	0.5882	0.5556	0.5263	0.5000
8.3	0.9998	0.9090	0.8333	0.7692	0.7143	0.6667	0.6250	0.5882	0.5556	0.5263	0.5000
8.4	0.9998	0.9090	0.8333	0.7692	0.7143	0.6667	0.6250	0.5882	0.5556	0.5263	0.5000
8.5	0.9998	0.9090	0.8333	0.7692	0.7143	0.6667	0.6250	0.5882	0.5556	0.5263	0.5000
8.6	0.9998	0.9090	0.8333	0.7692	0.7143	0.6667	0.6250	0.5882	0.5556	0.5263	0.5000
8.7	0.9998	0.9090	0.8333	0.7692	0.7143	0.6667	0.6250	0.5882	0.5556	0.5263	0.5000
8.8	0.9998	0.9090	0.8333	0.7692	0.7143	0.6667	0.6250	0.5882	0.5556	0.5263	0.5000
8.9	0.9999	0.9090	0.8333	0.7692	0.7143	0.6667	0.6250	0.5882	0.5556	0.5263	0.5000
9.0	0.9999	0.9090	0.8333	0.7692	0.7143	0.6667	0.6250	0.5882	0.5556	0.5263	0.5000
9.1	0.9999	0.9091	0.8333	0.7692	0.7143	0.6667	0.6250	0.5882	0.5556	0.5263	0.5000
9.2	0.9999	0.9091	0.8333	0.7692	0.7143	0.6667	0.6250	0.5882	0.5556	0.5263	0.5000
9.3	0.9999	0.9091	0.8333	0.7692	0.7143	0.6667	0.6250	0.5882	0.5556	0.5263	0.5000
9.4	0.9999	0.9091	0.8333	0.7692	0.7143	0.6667	0.6250	0.5882	0.5556	0.5263	0.5000
9.5	0.9999	0.9091	0.8333	0.7692	0.7143	0.6667	0.6250	0.5882	0.5556	0.5263	0.5000
9.6	0.9999	0.9091	0.8333	0.7692	0.7143	0.6667	0.6250	0.5882	0.5556	0.5263	0.5000
9.7	0.9999	0.9091	0.8333	0.7692	0.7143	0.6667	0.6250	0.5882	0.5556	0.5263	0.5000
9.8	0.9999	0.9091	0.8333	0.7692	0.7143	0.6667	0.6250	0.5882	0.5556	0.5263	0.5000
9.9	1.0000	0.9091	0.8333	0.7692	0.7143	0.6667	0.6250	0.5882	0.5556	0.5263	0.5000
10.0	1.0000	0.9091	0.8333	0.7692	0.7143	0.6667	0.6250	0.5882	0.5556	0.5263	0.5000

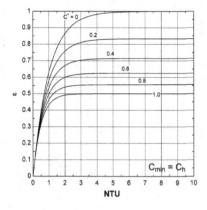

CONFIGURATION 20

1TUBE2ROW2PASS1CIR(CC)

(a) $C_{min} = C_c$

NTU	Heat Exchanger Effectiveness										
	C^* ($C_{min} = C_c$)										
	0	0.1	0.2	0.3	0.4	0.5	0.6	0.7	0.8	0.9	1.0
0	0.0000	0.0000	0.0000	0.0000	0.0000	0.0000	0.0000	0.0000	0.0000	0.0000	0.0000
0.1	0.0952	0.0947	0.0943	0.0939	0.0934	0.0930	0.0926	0.0921	0.0917	0.0913	0.0909
0.2	0.1813	0.1797	0.1782	0.1767	0.1752	0.1737	0.1722	0.1707	0.1693	0.1679	0.1664
0.3	0.2592	0.2561	0.2530	0.2500	0.2471	0.2442	0.2413	0.2384	0.2356	0.2328	0.2301
0.4	0.3297	0.3248	0.3200	0.3153	0.3107	0.3061	0.3016	0.2972	0.2929	0.2886	0.2844
0.5	0.3935	0.3867	0.3801	0.3736	0.3672	0.3609	0.3547	0.3486	0.3427	0.3368	0.3311
0.6	0.4512	0.4426	0.4341	0.4258	0.4176	0.4096	0.4017	0.3939	0.3863	0.3789	0.3716
0.7	0.5034	0.4930	0.4828	0.4727	0.4627	0.4530	0.4434	0.4340	0.4248	0.4158	0.4070
0.8	0.5507	0.5386	0.5266	0.5149	0.5033	0.4919	0.4807	0.4697	0.4589	0.4484	0.4381
0.9	0.5934	0.5798	0.5663	0.5529	0.5398	0.5268	0.5141	0.5016	0.4893	0.4774	0.4656
1.0	0.6321	0.6171	0.6021	0.5873	0.5727	0.5583	0.5441	0.5302	0.5166	0.5032	0.4901
1.1	0.6671	0.6509	0.6346	0.6185	0.6026	0.5868	0.5713	0.5560	0.5410	0.5263	0.5120
1.2	0.6988	0.6815	0.6641	0.6468	0.6296	0.6126	0.5958	0.5793	0.5630	0.5471	0.5316
1.3	0.7275	0.7092	0.6909	0.6725	0.6542	0.6361	0.6181	0.6004	0.5830	0.5659	0.5492
1.4	0.7534	0.7344	0.7152	0.6959	0.6766	0.6574	0.6384	0.6196	0.6010	0.5829	0.5652
1.5	0.7769	0.7573	0.7374	0.7173	0.6971	0.6769	0.6568	0.6370	0.6175	0.5983	0.5796
1.6	0.7981	0.7781	0.7576	0.7367	0.7157	0.6947	0.6737	0.6530	0.6325	0.6124	0.5927
1.7	0.8173	0.7969	0.7759	0.7545	0.7328	0.7110	0.6892	0.6676	0.6462	0.6252	0.6047
1.8	0.8347	0.8141	0.7927	0.7708	0.7485	0.7259	0.7034	0.6810	0.6588	0.6370	0.6156
1.9	0.8504	0.8297	0.8081	0.7857	0.7628	0.7397	0.7164	0.6933	0.6703	0.6477	0.6255
2.0	0.8647	0.8439	0.8221	0.7993	0.7760	0.7523	0.7284	0.7046	0.6809	0.6576	0.6347
2.1	0.8775	0.8568	0.8349	0.8119	0.7881	0.7639	0.7395	0.7150	0.6907	0.6667	0.6431
2.2	0.8892	0.8686	0.8466	0.8234	0.7993	0.7747	0.7497	0.7246	0.6997	0.6750	0.6509
2.3	0.8997	0.8793	0.8573	0.8340	0.8096	0.7846	0.7591	0.7335	0.7080	0.6828	0.6580
2.4	0.9093	0.8891	0.8671	0.8437	0.8191	0.7937	0.7678	0.7418	0.7157	0.6899	0.6646
2.5	0.9179	0.8980	0.8762	0.8526	0.8279	0.8022	0.7759	0.7494	0.7229	0.6966	0.6707
2.6	0.9257	0.9062	0.8844	0.8609	0.8360	0.8100	0.7834	0.7565	0.7295	0.7027	0.6763
2.7	0.9328	0.9136	0.8920	0.8685	0.8435	0.8173	0.7904	0.7630	0.7356	0.7084	0.6816
2.8	0.9392	0.9204	0.8990	0.8755	0.8504	0.8240	0.7968	0.7691	0.7413	0.7137	0.6864
2.9	0.9450	0.9265	0.9054	0.8820	0.8568	0.8303	0.8028	0.7748	0.7467	0.7186	0.6910
3.0	0.9502	0.9322	0.9113	0.8880	0.8628	0.8361	0.8084	0.7801	0.7516	0.7232	0.6952
3.1	0.9550	0.9373	0.9167	0.8936	0.8683	0.8415	0.8136	0.7851	0.7562	0.7275	0.6991
3.2	0.9592	0.9421	0.9217	0.8987	0.8735	0.8466	0.8185	0.7897	0.7606	0.7315	0.7028
3.3	0.9631	0.9464	0.9263	0.9034	0.8782	0.8513	0.8230	0.7940	0.7646	0.7352	0.7062
3.4	0.9666	0.9503	0.9306	0.9078	0.8827	0.8556	0.8272	0.7980	0.7684	0.7387	0.7094
3.5	0.9698	0.9539	0.9345	0.9119	0.8868	0.8597	0.8312	0.8018	0.7719	0.7420	0.7124
3.6	0.9727	0.9572	0.9381	0.9157	0.8907	0.8635	0.8349	0.8053	0.7752	0.7450	0.7152
3.7	0.9753	0.9603	0.9414	0.9192	0.8943	0.8671	0.8384	0.8086	0.7783	0.7479	0.7178
3.8	0.9776	0.9631	0.9445	0.9225	0.8976	0.8704	0.8416	0.8117	0.7812	0.7506	0.7202
3.9	0.9798	0.9656	0.9474	0.9255	0.9007	0.8735	0.8446	0.8146	0.7839	0.7531	0.7225
4.0	0.9817	0.9679	0.9500	0.9284	0.9036	0.8765	0.8475	0.8173	0.7865	0.7554	0.7247
4.1	0.9834	0.9701	0.9524	0.9310	0.9064	0.8792	0.8501	0.8198	0.7889	0.7577	0.7267
4.2	0.9850	0.9721	0.9547	0.9334	0.9089	0.8818	0.8527	0.8222	0.7911	0.7598	0.7286
4.3	0.9864	0.9739	0.9568	0.9357	0.9113	0.8842	0.8550	0.8245	0.7932	0.7617	0.7304
4.4	0.9877	0.9755	0.9587	0.9379	0.9135	0.8864	0.8572	0.8266	0.7952	0.7636	0.7321
4.5	0.9889	0.9771	0.9605	0.9398	0.9156	0.8885	0.8593	0.8286	0.7971	0.7653	0.7337
4.6	0.9899	0.9785	0.9622	0.9417	0.9176	0.8905	0.8613	0.8305	0.7989	0.7669	0.7352
4.7	0.9909	0.9798	0.9638	0.9434	0.9194	0.8924	0.8631	0.8323	0.8005	0.7685	0.7366
4.8	0.9918	0.9810	0.9652	0.9451	0.9211	0.8941	0.8648	0.8339	0.8021	0.7699	0.7379
4.9	0.9926	0.9821	0.9666	0.9466	0.9227	0.8958	0.8665	0.8355	0.8036	0.7713	0.7392
5.0	0.9933	0.9831	0.9678	0.9480	0.9242	0.8973	0.8680	0.8370	0.8050	0.7726	0.7404
5.1	0.9939	0.9840	0.9690	0.9493	0.9257	0.8988	0.8694	0.8384	0.8063	0.7738	0.7415
5.2	0.9945	0.9849	0.9701	0.9506	0.9270	0.9002	0.8708	0.8397	0.8076	0.7750	0.7425
5.3	0.9950	0.9857	0.9711	0.9517	0.9283	0.9014	0.8721	0.8409	0.8087	0.7761	0.7435
5.4	0.9955	0.9864	0.9720	0.9528	0.9294	0.9027	0.8733	0.8421	0.8098	0.7771	0.7445
5.5	0.9959	0.9871	0.9729	0.9538	0.9306	0.9038	0.8744	0.8432	0.8109	0.7781	0.7454
5.6	0.9963	0.9878	0.9738	0.9548	0.9316	0.9049	0.8755	0.8443	0.8119	0.7790	0.7462
5.7	0.9967	0.9883	0.9745	0.9557	0.9326	0.9059	0.8766	0.8453	0.8128	0.7799	0.7470
5.8	0.9970	0.9889	0.9752	0.9566	0.9335	0.9069	0.8775	0.8462	0.8137	0.7807	0.7477
5.9	0.9973	0.9894	0.9759	0.9574	0.9344	0.9078	0.8784	0.8471	0.8146	0.7815	0.7484
6.0	0.9975	0.9899	0.9766	0.9581	0.9352	0.9087	0.8793	0.8479	0.8153	0.7822	0.7491

NTU	Heat Exchanger Effectiveness										
	$C^* (C_{min} = C_c)$										
	0	0.1	0.2	0.3	0.4	0.5	0.6	0.7	0.8	0.9	1.0
6.1	0.9978	0.9903	0.9771	0.9588	0.9360	0.9095	0.8801	0.8487	0.8161	0.7829	0.7498
6.2	0.9980	0.9907	0.9777	0.9595	0.9367	0.9102	0.8809	0.8495	0.8168	0.7836	0.7504
6.3	0.9982	0.9911	0.9782	0.9601	0.9374	0.9110	0.8816	0.8502	0.8175	0.7842	0.7509
6.4	0.9983	0.9914	0.9787	0.9607	0.9381	0.9117	0.8823	0.8509	0.8181	0.7848	0.7515
6.5	0.9985	0.9917	0.9792	0.9613	0.9387	0.9123	0.8830	0.8515	0.8187	0.7854	0.7520
6.6	0.9986	0.9921	0.9796	0.9618	0.9393	0.9129	0.8836	0.8521	0.8193	0.7859	0.7525
6.7	0.9988	0.9923	0.9800	0.9623	0.9398	0.9135	0.8842	0.8527	0.8198	0.7864	0.7529
6.8	0.9989	0.9926	0.9804	0.9627	0.9404	0.9141	0.8847	0.8532	0.8204	0.7869	0.7534
6.9	0.9990	0.9928	0.9807	0.9632	0.9408	0.9146	0.8852	0.8537	0.8208	0.7873	0.7538
7.0	0.9991	0.9931	0.9811	0.9636	0.9413	0.9151	0.8857	0.8542	0.8213	0.7878	0.7542
7.1	0.9992	0.9933	0.9814	0.9640	0.9418	0.9155	0.8862	0.8547	0.8218	0.7882	0.7545
7.2	0.9993	0.9935	0.9817	0.9643	0.9422	0.9160	0.8867	0.8551	0.8222	0.7886	0.7549
7.3	0.9993	0.9937	0.9820	0.9647	0.9426	0.9164	0.8871	0.8555	0.8226	0.7889	0.7552
7.4	0.9994	0.9938	0.9822	0.9650	0.9429	0.9168	0.8875	0.8559	0.8229	0.7893	0.7556
7.5	0.9994	0.9940	0.9825	0.9653	0.9433	0.9172	0.8879	0.8563	0.8233	0.7896	0.7559
7.6	0.9995	0.9942	0.9827	0.9656	0.9436	0.9175	0.8882	0.8567	0.8236	0.7899	0.7561
7.7	0.9995	0.9943	0.9829	0.9659	0.9439	0.9179	0.8886	0.8570	0.8239	0.7902	0.7564
7.8	0.9996	0.9944	0.9831	0.9662	0.9442	0.9182	0.8889	0.8573	0.8243	0.7905	0.7567
7.9	0.9996	0.9946	0.9833	0.9664	0.9445	0.9185	0.8892	0.8576	0.8245	0.7908	0.7569
8.0	0.9997	0.9947	0.9835	0.9667	0.9448	0.9188	0.8895	0.8579	0.8248	0.7910	0.7571
8.1	0.9997	0.9948	0.9837	0.9669	0.9451	0.9191	0.8898	0.8582	0.8251	0.7913	0.7574
8.2	0.9997	0.9949	0.9839	0.9671	0.9453	0.9193	0.8900	0.8584	0.8253	0.7915	0.7576
8.3	0.9998	0.9950	0.9840	0.9673	0.9455	0.9196	0.8903	0.8587	0.8256	0.7917	0.7578
8.4	0.9998	0.9951	0.9842	0.9675	0.9457	0.9198	0.8905	0.8589	0.8258	0.7919	0.7580
8.5	0.9998	0.9952	0.9843	0.9677	0.9460	0.9200	0.8908	0.8591	0.8260	0.7921	0.7581
8.6	0.9998	0.9953	0.9844	0.9678	0.9462	0.9202	0.8910	0.8593	0.8262	0.7923	0.7583
8.7	0.9998	0.9953	0.9846	0.9680	0.9463	0.9204	0.8912	0.8595	0.8264	0.7925	0.7585
8.8	0.9998	0.9954	0.9847	0.9682	0.9465	0.9206	0.8914	0.8597	0.8266	0.7926	0.7586
8.9	0.9999	0.9955	0.9848	0.9683	0.9467	0.9208	0.8916	0.8599	0.8267	0.7928	0.7588
9.0	0.9999	0.9955	0.9849	0.9684	0.9468	0.9210	0.8917	0.8601	0.8269	0.7930	0.7589
9.1	0.9999	0.9956	0.9850	0.9686	0.9470	0.9211	0.8919	0.8602	0.8271	0.7931	0.7590
9.2	0.9999	0.9957	0.9851	0.9687	0.9471	0.9213	0.8921	0.8604	0.8272	0.7932	0.7592
9.3	0.9999	0.9957	0.9852	0.9688	0.9473	0.9214	0.8922	0.8605	0.8273	0.7934	0.7593
9.4	0.9999	0.9958	0.9853	0.9689	0.9474	0.9216	0.8923	0.8607	0.8275	0.7935	0.7594
9.5	0.9999	0.9958	0.9854	0.9690	0.9475	0.9217	0.8925	0.8608	0.8276	0.7936	0.7595
9.6	0.9999	0.9959	0.9854	0.9691	0.9476	0.9218	0.8926	0.8609	0.8277	0.7937	0.7596
9.7	0.9999	0.9959	0.9855	0.9692	0.9478	0.9219	0.8927	0.8611	0.8278	0.7938	0.7597
9.8	0.9999	0.9959	0.9856	0.9693	0.9479	0.9221	0.8928	0.8612	0.8279	0.7939	0.7598
9.9	1.0000	0.9960	0.9856	0.9694	0.9480	0.9222	0.8930	0.8613	0.8280	0.7940	0.7599
10.0	1.0000	0.9960	0.9857	0.9695	0.9480	0.9223	0.8931	0.8614	0.8281	0.7941	0.7600

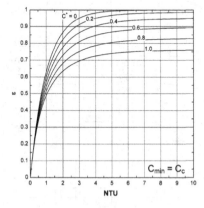

CONFIGURATION 20

1TUBE2ROW2PASS1CIR(CC)

(b) $C_{min} = C_h$

NTU	Heat Exchanger Effectiveness										
	C^* $(C_{min} = C_h)$										
	0	0.1	0.2	0.3	0.4	0.5	0.6	0.7	0.8	0.9	1.0
0	0.0000	0.0000	0.0000	0.0000	0.0000	0.0000	0.0000	0.0000	0.0000	0.0000	0.0000
0.1	0.0952	0.0947	0.0943	0.0939	0.0934	0.0930	0.0926	0.0921	0.0917	0.0913	0.0909
0.2	0.1813	0.1797	0.1782	0.1767	0.1752	0.1737	0.1722	0.1707	0.1693	0.1679	0.1664
0.3	0.2592	0.2561	0.2531	0.2500	0.2471	0.2442	0.2413	0.2384	0.2356	0.2328	0.2301
0.4	0.3297	0.3248	0.3200	0.3153	0.3107	0.3061	0.3016	0.2972	0.2929	0.2886	0.2844
0.5	0.3935	0.3867	0.3801	0.3736	0.3672	0.3609	0.3547	0.3487	0.3427	0.3368	0.3311
0.6	0.4512	0.4426	0.4341	0.4258	0.4176	0.4096	0.4017	0.3940	0.3863	0.3789	0.3716
0.7	0.5034	0.4930	0.4828	0.4727	0.4628	0.4530	0.4435	0.4341	0.4249	0.4158	0.4070
0.8	0.5507	0.5386	0.5267	0.5149	0.5034	0.4920	0.4808	0.4698	0.4590	0.4484	0.4381
0.9	0.5934	0.5798	0.5664	0.5530	0.5399	0.5270	0.5142	0.5017	0.4894	0.4774	0.4656
1.0	0.6321	0.6171	0.6023	0.5875	0.5729	0.5585	0.5443	0.5304	0.5167	0.5033	0.4901
1.1	0.6671	0.6510	0.6348	0.6188	0.6029	0.5871	0.5715	0.5562	0.5412	0.5264	0.5120
1.2	0.6988	0.6816	0.6644	0.6472	0.6300	0.6130	0.5962	0.5796	0.5633	0.5473	0.5316
1.3	0.7275	0.7094	0.6912	0.6730	0.6548	0.6366	0.6186	0.6008	0.5833	0.5661	0.5492
1.4	0.7534	0.7347	0.7157	0.6965	0.6773	0.6581	0.6390	0.6201	0.6015	0.5831	0.5652
1.5	0.7769	0.7576	0.7379	0.7180	0.6979	0.6778	0.6577	0.6377	0.6180	0.5986	0.5796
1.6	0.7981	0.7784	0.7582	0.7376	0.7168	0.6958	0.6748	0.6539	0.6332	0.6127	0.5927
1.7	0.8173	0.7974	0.7768	0.7556	0.7341	0.7123	0.6905	0.6686	0.6470	0.6256	0.6047
1.8	0.8347	0.8146	0.7937	0.7721	0.7500	0.7275	0.7049	0.6822	0.6597	0.6374	0.6156
1.9	0.8504	0.8303	0.8092	0.7872	0.7646	0.7415	0.7182	0.6947	0.6714	0.6483	0.6255
2.0	0.8647	0.8446	0.8234	0.8011	0.7781	0.7545	0.7305	0.7063	0.6822	0.6582	0.6347
2.1	0.8775	0.8577	0.8364	0.8139	0.7905	0.7664	0.7418	0.7170	0.6921	0.6674	0.6431
2.2	0.8892	0.8695	0.8483	0.8257	0.8020	0.7775	0.7523	0.7269	0.7013	0.6759	0.6509
2.3	0.8997	0.8804	0.8593	0.8366	0.8127	0.7877	0.7621	0.7360	0.7098	0.6837	0.6580
2.4	0.9093	0.8903	0.8693	0.8467	0.8225	0.7973	0.7712	0.7446	0.7177	0.6910	0.6646
2.5	0.9179	0.8993	0.8786	0.8559	0.8317	0.8061	0.7796	0.7525	0.7251	0.6977	0.6707
2.6	0.9257	0.9075	0.8871	0.8645	0.8402	0.8144	0.7875	0.7599	0.7319	0.7040	0.6763
2.7	0.9328	0.9151	0.8949	0.8725	0.8481	0.8221	0.7948	0.7668	0.7383	0.7098	0.6816
2.8	0.9392	0.9219	0.9021	0.8799	0.8555	0.8293	0.8017	0.7732	0.7443	0.7152	0.6864
2.9	0.9450	0.9282	0.9088	0.8867	0.8623	0.8360	0.8081	0.7793	0.7498	0.7203	0.6910
3.0	0.9502	0.9340	0.9149	0.8931	0.8687	0.8423	0.8142	0.7849	0.7550	0.7250	0.6952
3.1	0.9550	0.9393	0.9206	0.8990	0.8747	0.8481	0.8198	0.7902	0.7599	0.7294	0.6991
3.2	0.9592	0.9441	0.9258	0.9045	0.8803	0.8537	0.8251	0.7952	0.7645	0.7335	0.7028
3.3	0.9631	0.9485	0.9307	0.9096	0.8855	0.8588	0.8301	0.7999	0.7688	0.7374	0.7062
3.4	0.9666	0.9526	0.9352	0.9144	0.8904	0.8637	0.8348	0.8043	0.7728	0.7410	0.7094
3.5	0.9698	0.9563	0.9393	0.9188	0.8950	0.8683	0.8392	0.8084	0.7766	0.7444	0.7124
3.6	0.9727	0.9597	0.9432	0.9230	0.8993	0.8726	0.8433	0.8123	0.7801	0.7476	0.7152
3.7	0.9753	0.9629	0.9468	0.9269	0.9034	0.8766	0.8473	0.8160	0.7835	0.7506	0.7178
3.8	0.9776	0.9657	0.9501	0.9305	0.9072	0.8805	0.8510	0.8194	0.7867	0.7534	0.7202
3.9	0.9798	0.9684	0.9532	0.9339	0.9108	0.8841	0.8545	0.8227	0.7896	0.7560	0.7225
4.0	0.9817	0.9708	0.9560	0.9371	0.9142	0.8875	0.8578	0.8258	0.7925	0.7585	0.7247
4.1	0.9834	0.9730	0.9587	0.9401	0.9173	0.8907	0.8609	0.8288	0.7951	0.7609	0.7267
4.2	0.9850	0.9751	0.9612	0.9429	0.9203	0.8938	0.8639	0.8315	0.7976	0.7631	0.7286
4.3	0.9864	0.9769	0.9635	0.9456	0.9232	0.8967	0.8667	0.8342	0.8000	0.7652	0.7304
4.4	0.9877	0.9787	0.9656	0.9480	0.9259	0.8994	0.8694	0.8367	0.8023	0.7671	0.7321
4.5	0.9889	0.9803	0.9676	0.9504	0.9284	0.9020	0.8719	0.8390	0.8044	0.7690	0.7337
4.6	0.9899	0.9817	0.9695	0.9526	0.9308	0.9045	0.8743	0.8413	0.8064	0.7708	0.7352
4.7	0.9909	0.9831	0.9712	0.9546	0.9331	0.9068	0.8766	0.8434	0.8083	0.7724	0.7366
4.8	0.9918	0.9843	0.9728	0.9566	0.9352	0.9091	0.8788	0.8455	0.8102	0.7740	0.7379
4.9	0.9926	0.9855	0.9744	0.9584	0.9373	0.9112	0.8809	0.8474	0.8119	0.7755	0.7392
5.0	0.9933	0.9865	0.9758	0.9601	0.9392	0.9132	0.8829	0.8492	0.8135	0.7769	0.7404
5.1	0.9939	0.9875	0.9771	0.9618	0.9410	0.9151	0.8847	0.8510	0.8151	0.7782	0.7415
5.2	0.9945	0.9884	0.9783	0.9633	0.9428	0.9170	0.8865	0.8527	0.8166	0.7795	0.7425
5.3	0.9950	0.9893	0.9795	0.9648	0.9444	0.9187	0.8883	0.8542	0.8180	0.7807	0.7435
5.4	0.9955	0.9900	0.9806	0.9661	0.9460	0.9204	0.8899	0.8558	0.8193	0.7819	0.7445
5.5	0.9959	0.9907	0.9816	0.9674	0.9475	0.9220	0.8914	0.8572	0.8206	0.7829	0.7454
5.6	0.9963	0.9914	0.9826	0.9687	0.9490	0.9235	0.8929	0.8586	0.8218	0.7840	0.7462
5.7	0.9967	0.9920	0.9835	0.9698	0.9503	0.9249	0.8944	0.8599	0.8230	0.7849	0.7470
5.8	0.9970	0.9926	0.9843	0.9710	0.9516	0.9263	0.8957	0.8611	0.8241	0.7859	0.7477
5.9	0.9973	0.9931	0.9851	0.9720	0.9529	0.9276	0.8970	0.8623	0.8251	0.7867	0.7484
6.0	0.9975	0.9936	0.9859	0.9730	0.9541	0.9289	0.8983	0.8635	0.8261	0.7876	0.7491

NTU	Heat Exchanger Effectiveness										
	$C^* (C_{min} = C_h)$										
	0	0.1	0.2	0.3	0.4	0.5	0.6	0.7	0.8	0.9	1.0
6.1	0.9978	0.9940	0.9866	0.9740	0.9552	0.9301	0.8995	0.8646	0.8270	0.7884	0.7498
6.2	0.9980	0.9944	0.9872	0.9749	0.9563	0.9313	0.9006	0.8656	0.8280	0.7891	0.7504
6.3	0.9982	0.9948	0.9879	0.9757	0.9573	0.9324	0.9017	0.8666	0.8288	0.7898	0.7509
6.4	0.9983	0.9952	0.9884	0.9765	0.9583	0.9335	0.9028	0.8676	0.8297	0.7905	0.7515
6.5	0.9985	0.9955	0.9890	0.9773	0.9592	0.9345	0.9038	0.8685	0.8304	0.7912	0.7520
6.6	0.9986	0.9958	0.9895	0.9781	0.9602	0.9355	0.9047	0.8694	0.8312	0.7918	0.7525
6.7	0.9988	0.9961	0.9900	0.9788	0.9610	0.9364	0.9057	0.8702	0.8319	0.7924	0.7529
6.8	0.9989	0.9963	0.9905	0.9794	0.9619	0.9373	0.9066	0.8711	0.8326	0.7929	0.7534
6.9	0.9990	0.9966	0.9909	0.9801	0.9627	0.9382	0.9074	0.8718	0.8333	0.7934	0.7538
7.0	0.9991	0.9968	0.9913	0.9807	0.9634	0.9390	0.9082	0.8726	0.8339	0.7939	0.7542
7.1	0.9992	0.9970	0.9917	0.9813	0.9642	0.9398	0.9090	0.8733	0.8345	0.7944	0.7545
7.2	0.9993	0.9972	0.9921	0.9818	0.9649	0.9406	0.9098	0.8740	0.8351	0.7949	0.7549
7.3	0.9993	0.9974	0.9924	0.9824	0.9655	0.9414	0.9105	0.8746	0.8356	0.7953	0.7552
7.4	0.9994	0.9976	0.9928	0.9829	0.9662	0.9421	0.9112	0.8752	0.8361	0.7957	0.7556
7.5	0.9994	0.9977	0.9931	0.9834	0.9668	0.9428	0.9119	0.8759	0.8366	0.7961	0.7559
7.6	0.9995	0.9979	0.9934	0.9838	0.9674	0.9434	0.9125	0.8764	0.8371	0.7965	0.7561
7.7	0.9995	0.9980	0.9936	0.9843	0.9680	0.9441	0.9132	0.8770	0.8376	0.7969	0.7564
7.8	0.9996	0.9981	0.9939	0.9847	0.9686	0.9447	0.9138	0.8775	0.8380	0.7972	0.7567
7.9	0.9996	0.9983	0.9942	0.9851	0.9691	0.9453	0.9144	0.8780	0.8384	0.7975	0.7569
8.0	0.9997	0.9984	0.9944	0.9855	0.9696	0.9459	0.9149	0.8785	0.8388	0.7979	0.7571
8.1	0.9997	0.9985	0.9946	0.9859	0.9701	0.9464	0.9155	0.8790	0.8392	0.7982	0.7574
8.2	0.9997	0.9986	0.9949	0.9862	0.9706	0.9469	0.9160	0.8794	0.8396	0.7984	0.7576
8.3	0.9998	0.9987	0.9951	0.9866	0.9710	0.9475	0.9165	0.8799	0.8399	0.7987	0.7578
8.4	0.9998	0.9987	0.9953	0.9869	0.9715	0.9480	0.9169	0.8803	0.8403	0.7990	0.7580
8.5	0.9998	0.9988	0.9954	0.9872	0.9719	0.9484	0.9174	0.8807	0.8406	0.7992	0.7581
8.6	0.9998	0.9989	0.9956	0.9876	0.9723	0.9489	0.9179	0.8811	0.8409	0.7995	0.7583
8.7	0.9998	0.9990	0.9958	0.9878	0.9727	0.9493	0.9183	0.8815	0.8412	0.7997	0.7585
8.8	0.9998	0.9990	0.9959	0.9881	0.9731	0.9498	0.9187	0.8818	0.8415	0.7999	0.7586
8.9	0.9999	0.9991	0.9961	0.9884	0.9735	0.9502	0.9191	0.8822	0.8418	0.8001	0.7588
9.0	0.9999	0.9991	0.9962	0.9887	0.9739	0.9506	0.9195	0.8825	0.8420	0.8003	0.7589
9.1	0.9999	0.9992	0.9964	0.9889	0.9742	0.9510	0.9199	0.8828	0.8423	0.8005	0.7590
9.2	0.9999	0.9992	0.9965	0.9892	0.9745	0.9514	0.9202	0.8831	0.8425	0.8007	0.7592
9.3	0.9999	0.9993	0.9966	0.9894	0.9749	0.9517	0.9206	0.8834	0.8428	0.8008	0.7593
9.4	0.9999	0.9993	0.9968	0.9896	0.9752	0.9521	0.9209	0.8837	0.8430	0.8010	0.7594
9.5	0.9999	0.9994	0.9969	0.9898	0.9755	0.9524	0.9212	0.8840	0.8432	0.8011	0.7595
9.6	0.9999	0.9994	0.9970	0.9901	0.9758	0.9528	0.9216	0.8843	0.8434	0.8013	0.7596
9.7	0.9999	0.9994	0.9971	0.9903	0.9761	0.9531	0.9219	0.8845	0.8436	0.8014	0.7597
9.8	0.9999	0.9995	0.9972	0.9905	0.9764	0.9534	0.9222	0.8848	0.8438	0.8016	0.7598
9.9	1.0000	0.9995	0.9973	0.9906	0.9766	0.9537	0.9224	0.8850	0.8440	0.8017	0.7599
10.0	1.0000	0.9995	0.9974	0.9908	0.9769	0.9540	0.9227	0.8852	0.8441	0.8018	0.7600

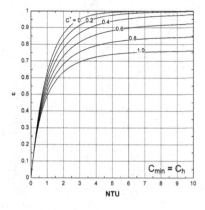

CONFIGURATION 21

1TUBE3ROW3PASS1CIR(PAR)

(a) $C_{min} = C_c$

NTU	Heat Exchanger Effectiveness										
	C^* ($C_{min} = C_c$)										
	0	0.1	0.2	0.3	0.4	0.5	0.6	0.7	0.8	0.9	1.0
0	0.0000	0.0000	0.0000	0.0000	0.0000	0.0000	0.0000	0.0000	0.0000	0.0000	0.0000
0.1	0.0952	0.0947	0.0943	0.0939	0.0934	0.0930	0.0926	0.0921	0.0917	0.0913	0.0909
0.2	0.1813	0.1797	0.1782	0.1767	0.1752	0.1737	0.1723	0.1708	0.1694	0.1680	0.1666
0.3	0.2592	0.2561	0.2531	0.2502	0.2472	0.2444	0.2415	0.2387	0.2359	0.2332	0.2305
0.4	0.3297	0.3249	0.3202	0.3156	0.3110	0.3066	0.3021	0.2978	0.2935	0.2893	0.2851
0.5	0.3935	0.3869	0.3805	0.3741	0.3678	0.3617	0.3556	0.3496	0.3438	0.3380	0.3323
0.6	0.4512	0.4429	0.4346	0.4266	0.4186	0.4107	0.4030	0.3954	0.3880	0.3807	0.3735
0.7	0.5034	0.4934	0.4835	0.4737	0.4641	0.4547	0.4453	0.4362	0.4271	0.4183	0.4096
0.8	0.5507	0.5391	0.5276	0.5163	0.5051	0.4941	0.4832	0.4725	0.4620	0.4517	0.4416
0.9	0.5934	0.5805	0.5676	0.5548	0.5422	0.5297	0.5174	0.5052	0.4933	0.4816	0.4701
1.0	0.6321	0.6179	0.6038	0.5897	0.5757	0.5619	0.5482	0.5347	0.5214	0.5084	0.4955
1.1	0.6671	0.6519	0.6366	0.6214	0.6062	0.5911	0.5762	0.5614	0.5468	0.5325	0.5184
1.2	0.6988	0.6827	0.6664	0.6502	0.6339	0.6177	0.6016	0.5857	0.5699	0.5544	0.5391
1.3	0.7275	0.7106	0.6936	0.6764	0.6592	0.6420	0.6248	0.6078	0.5909	0.5743	0.5579
1.4	0.7534	0.7360	0.7183	0.7004	0.6824	0.6642	0.6461	0.6280	0.6101	0.5924	0.5749
1.5	0.7769	0.7591	0.7409	0.7223	0.7035	0.6846	0.6656	0.6466	0.6277	0.6090	0.5905
1.6	0.7981	0.7801	0.7614	0.7424	0.7230	0.7033	0.6835	0.6636	0.6439	0.6242	0.6048
1.7	0.8173	0.7991	0.7802	0.7608	0.7408	0.7205	0.7000	0.6794	0.6587	0.6382	0.6179
1.8	0.8347	0.8165	0.7974	0.7776	0.7572	0.7364	0.7152	0.6939	0.6725	0.6511	0.6300
1.9	0.8504	0.8323	0.8131	0.7931	0.7724	0.7510	0.7293	0.7073	0.6852	0.6631	0.6411
2.0	0.8647	0.8467	0.8275	0.8073	0.7863	0.7646	0.7424	0.7197	0.6969	0.6741	0.6514
2.1	0.8775	0.8598	0.8407	0.8205	0.7992	0.7772	0.7545	0.7313	0.7079	0.6844	0.6610
2.2	0.8892	0.8717	0.8528	0.8325	0.8111	0.7888	0.7657	0.7420	0.7180	0.6939	0.6698
2.3	0.8997	0.8826	0.8638	0.8437	0.8222	0.7996	0.7762	0.7520	0.7275	0.7028	0.6781
2.4	0.9093	0.8925	0.8740	0.8539	0.8324	0.8097	0.7859	0.7614	0.7364	0.7111	0.6858
2.5	0.9179	0.9015	0.8833	0.8634	0.8419	0.8190	0.7950	0.7701	0.7446	0.7188	0.6929
2.6	0.9257	0.9098	0.8919	0.8722	0.8507	0.8277	0.8035	0.7783	0.7523	0.7261	0.6996
2.7	0.9328	0.9173	0.8998	0.8803	0.8589	0.8358	0.8114	0.7859	0.7596	0.7328	0.7059
2.8	0.9392	0.9242	0.9070	0.8878	0.8665	0.8434	0.8188	0.7930	0.7664	0.7392	0.7118
2.9	0.9450	0.9305	0.9137	0.8947	0.8736	0.8505	0.8258	0.7998	0.7728	0.7452	0.7173
3.0	0.9502	0.9362	0.9199	0.9011	0.8802	0.8571	0.8323	0.8061	0.7788	0.7508	0.7225
3.1	0.9550	0.9415	0.9255	0.9071	0.8863	0.8633	0.8384	0.8120	0.7844	0.7561	0.7274
3.2	0.9592	0.9463	0.9307	0.9126	0.8920	0.8691	0.8442	0.8176	0.7897	0.7610	0.7319
3.3	0.9631	0.9506	0.9355	0.9178	0.8974	0.8745	0.8496	0.8228	0.7947	0.7657	0.7363
3.4	0.9666	0.9546	0.9400	0.9225	0.9024	0.8796	0.8546	0.8278	0.7994	0.7701	0.7403
3.5	0.9698	0.9583	0.9441	0.9270	0.9070	0.8844	0.8594	0.8324	0.8039	0.7743	0.7442
3.6	0.9727	0.9617	0.9478	0.9311	0.9114	0.8889	0.8639	0.8368	0.8081	0.7783	0.7478
3.7	0.9753	0.9647	0.9513	0.9349	0.9155	0.8931	0.8682	0.8410	0.8121	0.7820	0.7512
3.8	0.9776	0.9675	0.9545	0.9385	0.9193	0.8971	0.8722	0.8449	0.8159	0.7855	0.7545
3.9	0.9798	0.9701	0.9575	0.9418	0.9229	0.9008	0.8760	0.8487	0.8194	0.7889	0.7576
4.0	0.9817	0.9724	0.9603	0.9449	0.9262	0.9044	0.8795	0.8522	0.8228	0.7921	0.7605
4.1	0.9834	0.9746	0.9628	0.9478	0.9294	0.9077	0.8829	0.8555	0.8260	0.7951	0.7633
4.2	0.9850	0.9766	0.9652	0.9505	0.9323	0.9108	0.8861	0.8587	0.8291	0.7979	0.7659
4.3	0.9864	0.9784	0.9674	0.9530	0.9351	0.9137	0.8891	0.8617	0.8320	0.8007	0.7684
4.4	0.9877	0.9801	0.9694	0.9554	0.9377	0.9165	0.8920	0.8645	0.8347	0.8032	0.7707
4.5	0.9889	0.9816	0.9713	0.9576	0.9402	0.9192	0.8947	0.8672	0.8373	0.8057	0.7730
4.6	0.9899	0.9830	0.9730	0.9596	0.9425	0.9216	0.8973	0.8698	0.8398	0.8080	0.7751
4.7	0.9909	0.9843	0.9746	0.9615	0.9447	0.9240	0.8997	0.8722	0.8422	0.8102	0.7772
4.8	0.9918	0.9854	0.9761	0.9633	0.9467	0.9262	0.9020	0.8745	0.8444	0.8124	0.7791
4.9	0.9926	0.9865	0.9775	0.9650	0.9486	0.9283	0.9042	0.8767	0.8466	0.8144	0.7810
5.0	0.9933	0.9875	0.9788	0.9666	0.9504	0.9303	0.9063	0.8788	0.8486	0.8163	0.7827
5.1	0.9939	0.9884	0.9800	0.9680	0.9521	0.9322	0.9082	0.8808	0.8505	0.8181	0.7844
5.2	0.9945	0.9893	0.9811	0.9694	0.9537	0.9339	0.9101	0.8827	0.8524	0.8199	0.7860
5.3	0.9950	0.9900	0.9822	0.9707	0.9553	0.9356	0.9119	0.8845	0.8542	0.8216	0.7875
5.4	0.9955	0.9908	0.9831	0.9719	0.9567	0.9372	0.9136	0.8863	0.8559	0.8231	0.7890
5.5	0.9959	0.9914	0.9840	0.9731	0.9581	0.9387	0.9152	0.8879	0.8575	0.8247	0.7904
5.6	0.9963	0.9920	0.9849	0.9742	0.9593	0.9402	0.9168	0.8895	0.8590	0.8261	0.7917
5.7	0.9967	0.9926	0.9857	0.9752	0.9606	0.9416	0.9182	0.8910	0.8605	0.8275	0.7930
5.8	0.9970	0.9931	0.9864	0.9761	0.9617	0.9429	0.9196	0.8924	0.8619	0.8288	0.7942
5.9	0.9973	0.9936	0.9871	0.9770	0.9628	0.9441	0.9210	0.8938	0.8632	0.8301	0.7954
6.0	0.9975	0.9940	0.9877	0.9779	0.9638	0.9453	0.9222	0.8951	0.8645	0.8313	0.7965

NTU	Heat Exchanger Effectiveness										
	$C^* (C_{min} = C_c)$										
	0	0.1	0.2	0.3	0.4	0.5	0.6	0.7	0.8	0.9	1.0
6.1	0.9978	0.9944	0.9883	0.9787	0.9648	0.9464	0.9234	0.8963	0.8657	0.8325	0.7976
6.2	0.9980	0.9948	0.9889	0.9794	0.9657	0.9474	0.9246	0.8975	0.8669	0.8336	0.7986
6.3	0.9982	0.9951	0.9894	0.9801	0.9666	0.9485	0.9257	0.8987	0.8680	0.8347	0.7996
6.4	0.9983	0.9954	0.9899	0.9808	0.9674	0.9494	0.9267	0.8998	0.8691	0.8357	0.8005
6.5	0.9985	0.9957	0.9903	0.9814	0.9682	0.9503	0.9278	0.9008	0.8701	0.8367	0.8014
6.6	0.9986	0.9960	0.9908	0.9820	0.9690	0.9512	0.9287	0.9018	0.8711	0.8376	0.8023
6.7	0.9988	0.9962	0.9912	0.9826	0.9697	0.9521	0.9296	0.9027	0.8721	0.8385	0.8031
6.8	0.9989	0.9965	0.9915	0.9831	0.9703	0.9528	0.9305	0.9037	0.8730	0.8394	0.8039
6.9	0.9990	0.9967	0.9919	0.9836	0.9710	0.9536	0.9313	0.9045	0.8739	0.8402	0.8047
7.0	0.9991	0.9969	0.9922	0.9841	0.9716	0.9543	0.9322	0.9054	0.8747	0.8410	0.8054
7.1	0.9992	0.9971	0.9925	0.9845	0.9722	0.9550	0.9329	0.9062	0.8755	0.8418	0.8061
7.2	0.9993	0.9973	0.9928	0.9849	0.9727	0.9557	0.9337	0.9070	0.8763	0.8425	0.8068
7.3	0.9993	0.9974	0.9931	0.9853	0.9733	0.9563	0.9344	0.9077	0.8770	0.8432	0.8074
7.4	0.9994	0.9976	0.9934	0.9857	0.9738	0.9569	0.9350	0.9084	0.8777	0.8439	0.8080
7.5	0.9994	0.9977	0.9936	0.9861	0.9742	0.9575	0.9357	0.9091	0.8784	0.8445	0.8086
7.6	0.9995	0.9978	0.9938	0.9864	0.9747	0.9580	0.9363	0.9097	0.8790	0.8452	0.8092
7.7	0.9995	0.9979	0.9940	0.9867	0.9751	0.9586	0.9369	0.9104	0.8797	0.8458	0.8098
7.8	0.9996	0.9981	0.9942	0.9871	0.9755	0.9591	0.9375	0.9110	0.8803	0.8464	0.8103
7.9	0.9996	0.9982	0.9944	0.9874	0.9759	0.9596	0.9380	0.9115	0.8809	0.8469	0.8108
8.0	0.9997	0.9983	0.9946	0.9876	0.9763	0.9600	0.9385	0.9121	0.8814	0.8474	0.8113
8.1	0.9997	0.9983	0.9948	0.9879	0.9767	0.9605	0.9390	0.9126	0.8820	0.8480	0.8118
8.2	0.9997	0.9984	0.9950	0.9881	0.9770	0.9609	0.9395	0.9131	0.8825	0.8484	0.8122
8.3	0.9998	0.9985	0.9951	0.9884	0.9773	0.9613	0.9400	0.9136	0.8830	0.8489	0.8127
8.4	0.9998	0.9986	0.9953	0.9886	0.9777	0.9617	0.9404	0.9141	0.8834	0.8494	0.8131
8.5	0.9998	0.9986	0.9954	0.9888	0.9780	0.9621	0.9408	0.9146	0.8839	0.8498	0.8135
8.6	0.9998	0.9987	0.9955	0.9890	0.9782	0.9624	0.9413	0.9150	0.8843	0.8502	0.8139
8.7	0.9998	0.9988	0.9956	0.9892	0.9785	0.9628	0.9416	0.9154	0.8848	0.8507	0.8143
8.8	0.9998	0.9988	0.9958	0.9894	0.9788	0.9631	0.9420	0.9158	0.8852	0.8510	0.8146
8.9	0.9999	0.9989	0.9959	0.9896	0.9790	0.9634	0.9424	0.9162	0.8856	0.8514	0.8150
9.0	0.9999	0.9989	0.9960	0.9898	0.9793	0.9637	0.9427	0.9166	0.8859	0.8518	0.8153
9.1	0.9999	0.9990	0.9961	0.9899	0.9795	0.9640	0.9431	0.9170	0.8863	0.8521	0.8156
9.2	0.9999	0.9990	0.9962	0.9901	0.9797	0.9643	0.9434	0.9173	0.8867	0.8525	0.8159
9.3	0.9999	0.9991	0.9963	0.9902	0.9799	0.9645	0.9437	0.9176	0.8870	0.8528	0.8162
9.4	0.9999	0.9991	0.9963	0.9904	0.9801	0.9648	0.9440	0.9180	0.8873	0.8531	0.8165
9.5	0.9999	0.9991	0.9964	0.9905	0.9803	0.9650	0.9443	0.9183	0.8876	0.8534	0.8168
9.6	0.9999	0.9992	0.9965	0.9907	0.9805	0.9653	0.9446	0.9186	0.8879	0.8537	0.8171
9.7	0.9999	0.9992	0.9966	0.9908	0.9807	0.9655	0.9448	0.9189	0.8882	0.8540	0.8173
9.8	0.9999	0.9992	0.9966	0.9909	0.9809	0.9657	0.9451	0.9191	0.8885	0.8543	0.8176
9.9	1.0000	0.9993	0.9967	0.9910	0.9810	0.9659	0.9453	0.9194	0.8888	0.8545	0.8178
10.0	1.0000	0.9993	0.9968	0.9911	0.9812	0.9661	0.9456	0.9197	0.8890	0.8548	0.8180

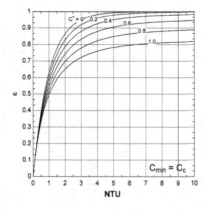

CONFIGURATION 21

1TUBE3ROW3PASS1CIR(CC)

(b) $C_{min} = C_h$

NTU	Heat Exchanger Effectiveness										
	C^* ($C_{min} = C_h$)										
	0	0.1	0.2	0.3	0.4	0.5	0.6	0.7	0.8	0.9	1.0
0	0.0000	0.0000	0.0000	0.0000	0.0000	0.0000	0.0000	0.0000	0.0000	0.0000	0.0000
0.1	0.0952	0.0947	0.0943	0.0939	0.0934	0.0930	0.0926	0.0921	0.0917	0.0913	0.0909
0.2	0.1813	0.1797	0.1782	0.1767	0.1752	0.1737	0.1723	0.1708	0.1694	0.1680	0.1666
0.3	0.2592	0.2561	0.2531	0.2502	0.2472	0.2444	0.2415	0.2387	0.2359	0.2332	0.2305
0.4	0.3297	0.3249	0.3202	0.3156	0.3111	0.3066	0.3021	0.2978	0.2935	0.2893	0.2851
0.5	0.3935	0.3869	0.3805	0.3741	0.3678	0.3617	0.3556	0.3496	0.3438	0.3380	0.3323
0.6	0.4512	0.4429	0.4347	0.4266	0.4186	0.4107	0.4030	0.3954	0.3880	0.3807	0.3735
0.7	0.5034	0.4934	0.4835	0.4738	0.4641	0.4547	0.4453	0.4362	0.4272	0.4183	0.4096
0.8	0.5507	0.5391	0.5277	0.5163	0.5052	0.4941	0.4833	0.4726	0.4621	0.4517	0.4416
0.9	0.5934	0.5805	0.5676	0.5548	0.5422	0.5297	0.5174	0.5053	0.4933	0.4816	0.4701
1.0	0.6321	0.6179	0.6038	0.5897	0.5758	0.5619	0.5483	0.5348	0.5215	0.5084	0.4955
1.1	0.6671	0.6519	0.6367	0.6214	0.6063	0.5912	0.5762	0.5615	0.5469	0.5325	0.5184
1.2	0.6988	0.6827	0.6665	0.6503	0.6340	0.6178	0.6017	0.5858	0.5700	0.5544	0.5391
1.3	0.7275	0.7107	0.6937	0.6765	0.6593	0.6421	0.6250	0.6079	0.5910	0.5743	0.5579
1.4	0.7534	0.7361	0.7184	0.7005	0.6825	0.6644	0.6462	0.6282	0.6102	0.5925	0.5749
1.5	0.7769	0.7592	0.7410	0.7225	0.7037	0.6848	0.6658	0.6467	0.6278	0.6091	0.5905
1.6	0.7981	0.7801	0.7616	0.7426	0.7232	0.7035	0.6837	0.6638	0.6440	0.6243	0.6048
1.7	0.8173	0.7992	0.7804	0.7610	0.7411	0.7208	0.7003	0.6796	0.6589	0.6383	0.6179
1.8	0.8347	0.8166	0.7976	0.7779	0.7576	0.7367	0.7155	0.6941	0.6727	0.6512	0.6300
1.9	0.8504	0.8324	0.8134	0.7934	0.7728	0.7514	0.7297	0.7076	0.6854	0.6632	0.6411
2.0	0.8647	0.8468	0.8278	0.8077	0.7868	0.7651	0.7428	0.7201	0.6972	0.6743	0.6514
2.1	0.8775	0.8599	0.8410	0.8209	0.7997	0.7777	0.7550	0.7317	0.7082	0.6846	0.6610
2.2	0.8892	0.8719	0.8531	0.8330	0.8117	0.7894	0.7663	0.7426	0.7184	0.6941	0.6698
2.3	0.8997	0.8828	0.8642	0.8442	0.8229	0.8003	0.7768	0.7526	0.7280	0.7030	0.6781
2.4	0.9093	0.8927	0.8745	0.8546	0.8332	0.8105	0.7867	0.7621	0.7369	0.7113	0.6858
2.5	0.9179	0.9018	0.8838	0.8641	0.8427	0.8199	0.7959	0.7709	0.7452	0.7191	0.6929
2.6	0.9257	0.9101	0.8925	0.8729	0.8516	0.8287	0.8045	0.7791	0.7530	0.7264	0.6996
2.7	0.9328	0.9176	0.9004	0.8811	0.8599	0.8370	0.8125	0.7868	0.7603	0.7332	0.7059
2.8	0.9392	0.9245	0.9077	0.8887	0.8676	0.8447	0.8200	0.7941	0.7672	0.7396	0.7118
2.9	0.9450	0.9309	0.9144	0.8957	0.8748	0.8519	0.8271	0.8009	0.7736	0.7456	0.7173
3.0	0.9502	0.9366	0.9206	0.9022	0.8815	0.8586	0.8338	0.8073	0.7797	0.7513	0.7225
3.1	0.9550	0.9419	0.9263	0.9083	0.8878	0.8649	0.8400	0.8134	0.7854	0.7566	0.7274
3.2	0.9592	0.9467	0.9316	0.9139	0.8936	0.8709	0.8459	0.8191	0.7908	0.7616	0.7319
3.3	0.9631	0.9511	0.9365	0.9191	0.8991	0.8764	0.8514	0.8245	0.7959	0.7664	0.7363
3.4	0.9666	0.9551	0.9410	0.9240	0.9042	0.8817	0.8567	0.8295	0.8008	0.7709	0.7403
3.5	0.9698	0.9588	0.9451	0.9285	0.9090	0.8866	0.8616	0.8344	0.8053	0.7751	0.7442
3.6	0.9727	0.9622	0.9489	0.9327	0.9135	0.8913	0.8663	0.8389	0.8097	0.7791	0.7478
3.7	0.9753	0.9653	0.9525	0.9367	0.9177	0.8956	0.8707	0.8432	0.8137	0.7829	0.7512
3.8	0.9776	0.9681	0.9558	0.9403	0.9217	0.8998	0.8749	0.8473	0.8176	0.7865	0.7545
3.9	0.9798	0.9707	0.9588	0.9438	0.9254	0.9037	0.8788	0.8512	0.8213	0.7899	0.7576
4.0	0.9817	0.9730	0.9616	0.9470	0.9289	0.9074	0.8825	0.8548	0.8248	0.7931	0.7605
4.1	0.9834	0.9752	0.9642	0.9499	0.9322	0.9108	0.8861	0.8583	0.8281	0.7962	0.7633
4.2	0.9850	0.9772	0.9666	0.9527	0.9353	0.9141	0.8895	0.8617	0.8313	0.7991	0.7659
4.3	0.9864	0.9790	0.9688	0.9553	0.9382	0.9173	0.8927	0.8648	0.8343	0.8019	0.7684
4.4	0.9877	0.9807	0.9709	0.9578	0.9409	0.9202	0.8957	0.8678	0.8372	0.8046	0.7707
4.5	0.9889	0.9822	0.9728	0.9600	0.9435	0.9230	0.8986	0.8707	0.8399	0.8071	0.7730
4.6	0.9899	0.9837	0.9746	0.9622	0.9460	0.9257	0.9014	0.8734	0.8425	0.8095	0.7751
4.7	0.9909	0.9849	0.9763	0.9642	0.9483	0.9282	0.9040	0.8760	0.8450	0.8118	0.7772
4.8	0.9918	0.9861	0.9778	0.9661	0.9504	0.9306	0.9065	0.8785	0.8474	0.8139	0.7791
4.9	0.9926	0.9872	0.9792	0.9678	0.9525	0.9328	0.9089	0.8809	0.8497	0.8160	0.7810
5.0	0.9933	0.9882	0.9805	0.9695	0.9544	0.9350	0.9111	0.8832	0.8518	0.8180	0.7827
5.1	0.9939	0.9891	0.9818	0.9710	0.9563	0.9371	0.9133	0.8854	0.8539	0.8199	0.7844
5.2	0.9945	0.9900	0.9829	0.9725	0.9580	0.9390	0.9154	0.8874	0.8559	0.8217	0.7860
5.3	0.9950	0.9908	0.9840	0.9739	0.9597	0.9409	0.9173	0.8894	0.8578	0.8235	0.7875
5.4	0.9955	0.9915	0.9850	0.9752	0.9612	0.9426	0.9192	0.8913	0.8596	0.8251	0.7890
5.5	0.9959	0.9921	0.9859	0.9764	0.9627	0.9443	0.9210	0.8931	0.8614	0.8267	0.7904
5.6	0.9963	0.9927	0.9868	0.9775	0.9641	0.9460	0.9228	0.8949	0.8630	0.8283	0.7917
5.7	0.9967	0.9933	0.9876	0.9786	0.9655	0.9475	0.9244	0.8966	0.8646	0.8297	0.7930
5.8	0.9970	0.9938	0.9884	0.9796	0.9667	0.9490	0.9260	0.8982	0.8662	0.8311	0.7942
5.9	0.9973	0.9943	0.9891	0.9806	0.9679	0.9504	0.9275	0.8997	0.8676	0.8324	0.7954
6.0	0.9975	0.9947	0.9897	0.9815	0.9691	0.9517	0.9290	0.9012	0.8691	0.8337	0.7965

NTU	Heat Exchanger Effectiveness										
	C^* ($C_{min} = C_h$)										
	0	0.1	0.2	0.3	0.4	0.5	0.6	0.7	0.8	0.9	1.0
6.1	0.9978	0.9951	0.9903	0.9824	0.9702	0.9530	0.9304	0.9026	0.8704	0.8350	0.7976
6.2	0.9980	0.9955	0.9909	0.9832	0.9712	0.9542	0.9317	0.9040	0.8717	0.8361	0.7986
6.3	0.9982	0.9958	0.9914	0.9839	0.9722	0.9554	0.9330	0.9053	0.8730	0.8373	0.7996
6.4	0.9983	0.9961	0.9919	0.9846	0.9732	0.9565	0.9343	0.9066	0.8742	0.8384	0.8005
6.5	0.9985	0.9964	0.9924	0.9853	0.9741	0.9576	0.9355	0.9078	0.8753	0.8394	0.8014
6.6	0.9986	0.9967	0.9928	0.9860	0.9749	0.9586	0.9366	0.9089	0.8765	0.8404	0.8023
6.7	0.9988	0.9969	0.9932	0.9866	0.9757	0.9596	0.9377	0.9101	0.8775	0.8414	0.8031
6.8	0.9989	0.9972	0.9936	0.9872	0.9765	0.9606	0.9388	0.9112	0.8786	0.8423	0.8039
6.9	0.9990	0.9974	0.9940	0.9877	0.9773	0.9615	0.9398	0.9122	0.8796	0.8432	0.8047
7.0	0.9991	0.9976	0.9943	0.9882	0.9780	0.9624	0.9408	0.9132	0.8805	0.8441	0.8054
7.1	0.9992	0.9977	0.9946	0.9887	0.9786	0.9632	0.9417	0.9142	0.8815	0.8449	0.8061
7.2	0.9993	0.9979	0.9949	0.9892	0.9793	0.9640	0.9426	0.9151	0.8823	0.8457	0.8068
7.3	0.9993	0.9981	0.9952	0.9896	0.9799	0.9648	0.9435	0.9160	0.8832	0.8465	0.8074
7.4	0.9994	0.9982	0.9955	0.9901	0.9805	0.9655	0.9443	0.9169	0.8840	0.8472	0.8080
7.5	0.9994	0.9983	0.9957	0.9905	0.9811	0.9663	0.9451	0.9177	0.8848	0.8479	0.8086
7.6	0.9995	0.9985	0.9960	0.9908	0.9816	0.9669	0.9459	0.9185	0.8856	0.8486	0.8092
7.7	0.9995	0.9986	0.9962	0.9912	0.9821	0.9676	0.9467	0.9193	0.8863	0.8492	0.8098
7.8	0.9996	0.9987	0.9964	0.9915	0.9826	0.9683	0.9474	0.9201	0.8871	0.8499	0.8103
7.9	0.9996	0.9988	0.9966	0.9919	0.9831	0.9689	0.9481	0.9208	0.8878	0.8505	0.8108
8.0	0.9997	0.9989	0.9968	0.9922	0.9836	0.9695	0.9488	0.9215	0.8884	0.8511	0.8113
8.1	0.9997	0.9989	0.9969	0.9925	0.9840	0.9700	0.9495	0.9222	0.8891	0.8517	0.8118
8.2	0.9997	0.9990	0.9971	0.9928	0.9844	0.9706	0.9501	0.9229	0.8897	0.8522	0.8122
8.3	0.9998	0.9991	0.9972	0.9930	0.9849	0.9711	0.9507	0.9235	0.8903	0.8527	0.8127
8.4	0.9998	0.9991	0.9974	0.9933	0.9852	0.9716	0.9513	0.9241	0.8909	0.8532	0.8131
8.5	0.9998	0.9992	0.9975	0.9935	0.9856	0.9721	0.9519	0.9247	0.8914	0.8537	0.8135
8.6	0.9998	0.9993	0.9976	0.9938	0.9860	0.9726	0.9525	0.9253	0.8920	0.8542	0.8139
8.7	0.9998	0.9993	0.9978	0.9940	0.9863	0.9731	0.9530	0.9258	0.8925	0.8547	0.8143
8.8	0.9998	0.9994	0.9979	0.9942	0.9867	0.9735	0.9535	0.9264	0.8930	0.8551	0.8146
8.9	0.9999	0.9994	0.9980	0.9944	0.9870	0.9740	0.9540	0.9269	0.8935	0.8555	0.8150
9.0	0.9999	0.9994	0.9981	0.9946	0.9873	0.9744	0.9545	0.9274	0.8940	0.8559	0.8153
9.1	0.9999	0.9995	0.9982	0.9948	0.9876	0.9748	0.9550	0.9279	0.8944	0.8563	0.8156
9.2	0.9999	0.9995	0.9983	0.9949	0.9879	0.9752	0.9554	0.9284	0.8949	0.8567	0.8159
9.3	0.9999	0.9996	0.9983	0.9951	0.9882	0.9756	0.9559	0.9288	0.8953	0.8571	0.8162
9.4	0.9999	0.9996	0.9984	0.9953	0.9884	0.9759	0.9563	0.9293	0.8957	0.8575	0.8165
9.5	0.9999	0.9996	0.9985	0.9954	0.9887	0.9763	0.9567	0.9297	0.8961	0.8578	0.8168
9.6	0.9999	0.9996	0.9986	0.9956	0.9889	0.9766	0.9571	0.9301	0.8965	0.8581	0.8171
9.7	0.9999	0.9997	0.9986	0.9957	0.9892	0.9769	0.9575	0.9306	0.8969	0.8585	0.8173
9.8	0.9999	0.9997	0.9987	0.9959	0.9894	0.9773	0.9579	0.9310	0.8973	0.8588	0.8176
9.9	1.0000	0.9997	0.9988	0.9960	0.9896	0.9776	0.9583	0.9313	0.8976	0.8591	0.8178
10.0	1.0000	0.9997	0.9988	0.9961	0.9898	0.9779	0.9587	0.9317	0.8980	0.8594	0.8180

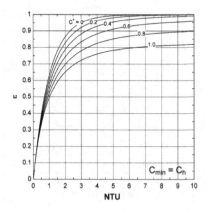

CONFIGURATION 22

1TUBE4ROW4PASS1CIR(CC)

(a) $C_{min} = C_c$

NTU	Heat Exchanger Effectiveness										
	C^* ($C_{min} = C_c$)										
	0	0.1	0.2	0.3	0.4	0.5	0.6	0.7	0.8	0.9	1.0
0	0.0000	0.0000	0.0000	0.0000	0.0000	0.0000	0.0000	0.0000	0.0000	0.0000	0.0000
0.1	0.0952	0.0947	0.0943	0.0939	0.0934	0.0930	0.0926	0.0922	0.0917	0.0913	0.0909
0.2	0.1813	0.1797	0.1782	0.1767	0.1752	0.1738	0.1723	0.1709	0.1694	0.1680	0.1666
0.3	0.2592	0.2562	0.2532	0.2502	0.2473	0.2444	0.2416	0.2388	0.2360	0.2333	0.2306
0.4	0.3297	0.3250	0.3203	0.3157	0.3112	0.3067	0.3023	0.2980	0.2937	0.2895	0.2854
0.5	0.3935	0.3870	0.3806	0.3743	0.3680	0.3619	0.3559	0.3500	0.3441	0.3384	0.3328
0.6	0.4512	0.4430	0.4348	0.4268	0.4189	0.4111	0.4035	0.3960	0.3886	0.3813	0.3741
0.7	0.5034	0.4935	0.4838	0.4741	0.4646	0.4552	0.4460	0.4369	0.4280	0.4192	0.4105
0.8	0.5507	0.5393	0.5280	0.5168	0.5058	0.4949	0.4841	0.4735	0.4631	0.4529	0.4428
0.9	0.5934	0.5807	0.5680	0.5554	0.5430	0.5307	0.5185	0.5065	0.4947	0.4830	0.4716
1.0	0.6321	0.6182	0.6043	0.5905	0.5767	0.5631	0.5496	0.5363	0.5231	0.5102	0.4974
1.1	0.6671	0.6522	0.6373	0.6223	0.6074	0.5926	0.5779	0.5633	0.5489	0.5347	0.5207
1.2	0.6988	0.6831	0.6672	0.6513	0.6354	0.6195	0.6036	0.5879	0.5723	0.5570	0.5418
1.3	0.7275	0.7111	0.6945	0.6778	0.6609	0.6441	0.6272	0.6104	0.5937	0.5772	0.5610
1.4	0.7534	0.7366	0.7194	0.7019	0.6843	0.6666	0.6488	0.6310	0.6133	0.5958	0.5785
1.5	0.7769	0.7597	0.7421	0.7241	0.7058	0.6872	0.6686	0.6500	0.6313	0.6128	0.5945
1.6	0.7981	0.7807	0.7628	0.7443	0.7254	0.7063	0.6869	0.6674	0.6479	0.6285	0.6093
1.7	0.8173	0.7999	0.7817	0.7629	0.7436	0.7238	0.7038	0.6835	0.6632	0.6430	0.6228
1.8	0.8347	0.8173	0.7990	0.7799	0.7602	0.7400	0.7194	0.6985	0.6774	0.6563	0.6354
1.9	0.8504	0.8331	0.8148	0.7956	0.7756	0.7550	0.7338	0.7123	0.6906	0.6688	0.6470
2.0	0.8647	0.8476	0.8293	0.8101	0.7899	0.7689	0.7473	0.7252	0.7028	0.6803	0.6578
2.1	0.8775	0.8607	0.8426	0.8234	0.8030	0.7818	0.7598	0.7372	0.7142	0.6910	0.6678
2.2	0.8892	0.8727	0.8548	0.8356	0.8152	0.7937	0.7714	0.7484	0.7249	0.7011	0.6772
2.3	0.8997	0.8836	0.8660	0.8469	0.8265	0.8049	0.7823	0.7588	0.7348	0.7104	0.6859
2.4	0.9093	0.8936	0.8763	0.8574	0.8370	0.8152	0.7924	0.7686	0.7441	0.7192	0.6941
2.5	0.9179	0.9027	0.8857	0.8670	0.8467	0.8249	0.8019	0.7778	0.7529	0.7274	0.7017
2.6	0.9257	0.9110	0.8944	0.8759	0.8557	0.8339	0.8107	0.7863	0.7611	0.7351	0.7089
2.7	0.9328	0.9185	0.9023	0.8842	0.8641	0.8424	0.8190	0.7944	0.7688	0.7424	0.7156
2.8	0.9392	0.9254	0.9097	0.8918	0.8720	0.8502	0.8268	0.8020	0.7760	0.7493	0.7220
2.9	0.9450	0.9318	0.9164	0.8989	0.8793	0.8576	0.8342	0.8091	0.7829	0.7557	0.7280
3.0	0.9502	0.9375	0.9226	0.9055	0.8861	0.8645	0.8411	0.8159	0.7893	0.7618	0.7336
3.1	0.9550	0.9428	0.9283	0.9115	0.8924	0.8710	0.8475	0.8222	0.7954	0.7676	0.7390
3.2	0.9592	0.9476	0.9336	0.9172	0.8983	0.8771	0.8536	0.8282	0.8012	0.7730	0.7440
3.3	0.9631	0.9520	0.9385	0.9224	0.9039	0.8828	0.8594	0.8339	0.8067	0.7782	0.7488
3.4	0.9666	0.9560	0.9429	0.9273	0.9091	0.8882	0.8648	0.8392	0.8118	0.7830	0.7533
3.5	0.9698	0.9596	0.9471	0.9318	0.9139	0.8932	0.8699	0.8443	0.8167	0.7877	0.7576
3.6	0.9727	0.9630	0.9509	0.9361	0.9184	0.8980	0.8748	0.8491	0.8214	0.7921	0.7617
3.7	0.9753	0.9661	0.9544	0.9400	0.9227	0.9024	0.8793	0.8537	0.8258	0.7963	0.7655
3.8	0.9776	0.9689	0.9576	0.9436	0.9266	0.9066	0.8837	0.8580	0.8300	0.8002	0.7692
3.9	0.9798	0.9714	0.9606	0.9470	0.9304	0.9106	0.8878	0.8621	0.8340	0.8040	0.7727
4.0	0.9817	0.9738	0.9634	0.9502	0.9339	0.9143	0.8916	0.8660	0.8378	0.8076	0.7760
4.1	0.9834	0.9759	0.9659	0.9531	0.9371	0.9179	0.8953	0.8697	0.8414	0.8111	0.7792
4.2	0.9850	0.9779	0.9683	0.9559	0.9402	0.9212	0.8988	0.8732	0.8449	0.8143	0.7822
4.3	0.9864	0.9797	0.9705	0.9584	0.9431	0.9244	0.9021	0.8766	0.8482	0.8175	0.7851
4.4	0.9877	0.9813	0.9725	0.9608	0.9458	0.9273	0.9053	0.8798	0.8513	0.8204	0.7879
4.5	0.9889	0.9828	0.9744	0.9631	0.9484	0.9301	0.9082	0.8828	0.8543	0.8233	0.7905
4.6	0.9899	0.9842	0.9761	0.9651	0.9508	0.9328	0.9111	0.8857	0.8572	0.8260	0.7930
4.7	0.9909	0.9855	0.9777	0.9671	0.9531	0.9353	0.9138	0.8885	0.8599	0.8286	0.7954
4.8	0.9918	0.9867	0.9792	0.9689	0.9552	0.9377	0.9163	0.8911	0.8625	0.8311	0.7977
4.9	0.9926	0.9877	0.9806	0.9706	0.9572	0.9400	0.9188	0.8936	0.8650	0.8335	0.7999
5.0	0.9933	0.9887	0.9819	0.9722	0.9591	0.9421	0.9211	0.8961	0.8674	0.8358	0.8020
5.1	0.9939	0.9896	0.9831	0.9737	0.9609	0.9442	0.9233	0.8984	0.8697	0.8380	0.8041
5.2	0.9945	0.9904	0.9842	0.9751	0.9626	0.9461	0.9254	0.9006	0.8719	0.8401	0.8060
5.3	0.9950	0.9912	0.9852	0.9764	0.9641	0.9479	0.9274	0.9027	0.8740	0.8421	0.8079
5.4	0.9955	0.9919	0.9861	0.9776	0.9656	0.9497	0.9293	0.9047	0.8760	0.8440	0.8096
5.5	0.9959	0.9925	0.9870	0.9788	0.9671	0.9513	0.9312	0.9066	0.8779	0.8459	0.8114
5.6	0.9963	0.9931	0.9878	0.9798	0.9684	0.9529	0.9329	0.9084	0.8797	0.8477	0.8130
5.7	0.9967	0.9936	0.9886	0.9809	0.9697	0.9544	0.9346	0.9102	0.8816	0.8494	0.8146
5.8	0.9970	0.9941	0.9893	0.9818	0.9709	0.9558	0.9362	0.9119	0.8833	0.8510	0.8161
5.9	0.9973	0.9945	0.9899	0.9827	0.9720	0.9571	0.9377	0.9135	0.8849	0.8526	0.8176
6.0	0.9975	0.9950	0.9906	0.9835	0.9731	0.9584	0.9391	0.9151	0.8865	0.8541	0.8190

NTU	Heat Exchanger Effectiveness										
	C^* ($C_{min} = C_c$)										
	0	0.1	0.2	0.3	0.4	0.5	0.6	0.7	0.8	0.9	1.0
6.1	0.9978	0.9953	0.9911	0.9843	0.9741	0.9597	0.9405	0.9166	0.8880	0.8556	0.8203
6.2	0.9980	0.9957	0.9917	0.9851	0.9750	0.9608	0.9419	0.9180	0.8895	0.8570	0.8216
6.3	0.9982	0.9960	0.9921	0.9857	0.9759	0.9619	0.9431	0.9194	0.8909	0.8584	0.8229
6.4	0.9983	0.9963	0.9926	0.9864	0.9768	0.9630	0.9444	0.9207	0.8922	0.8597	0.8241
6.5	0.9985	0.9966	0.9930	0.9870	0.9776	0.9640	0.9455	0.9220	0.8935	0.8609	0.8252
6.6	0.9986	0.9968	0.9934	0.9876	0.9784	0.9650	0.9467	0.9232	0.8948	0.8621	0.8263
6.7	0.9988	0.9971	0.9938	0.9881	0.9791	0.9659	0.9477	0.9243	0.8960	0.8633	0.8274
6.8	0.9989	0.9973	0.9942	0.9887	0.9798	0.9668	0.9488	0.9255	0.8971	0.8644	0.8285
6.9	0.9990	0.9975	0.9945	0.9891	0.9805	0.9676	0.9497	0.9266	0.8982	0.8655	0.8295
7.0	0.9991	0.9977	0.9948	0.9896	0.9811	0.9684	0.9507	0.9276	0.8993	0.8666	0.8304
7.1	0.9992	0.9978	0.9951	0.9900	0.9817	0.9692	0.9516	0.9286	0.9004	0.8676	0.8314
7.2	0.9993	0.9980	0.9953	0.9904	0.9823	0.9699	0.9525	0.9296	0.9014	0.8685	0.8323
7.3	0.9993	0.9981	0.9956	0.9908	0.9828	0.9706	0.9533	0.9305	0.9023	0.8695	0.8331
7.4	0.9994	0.9983	0.9958	0.9912	0.9834	0.9713	0.9541	0.9314	0.9032	0.8704	0.8340
7.5	0.9994	0.9984	0.9960	0.9915	0.9839	0.9719	0.9549	0.9322	0.9041	0.8713	0.8348
7.6	0.9995	0.9985	0.9962	0.9919	0.9843	0.9725	0.9556	0.9331	0.9050	0.8721	0.8356
7.7	0.9995	0.9986	0.9964	0.9922	0.9848	0.9731	0.9563	0.9339	0.9058	0.8729	0.8363
7.8	0.9996	0.9987	0.9966	0.9925	0.9852	0.9737	0.9570	0.9346	0.9066	0.8737	0.8370
7.9	0.9996	0.9988	0.9968	0.9927	0.9856	0.9742	0.9577	0.9354	0.9074	0.8745	0.8377
8.0	0.9997	0.9989	0.9969	0.9930	0.9860	0.9747	0.9583	0.9361	0.9082	0.8752	0.8384
8.1	0.9997	0.9989	0.9971	0.9932	0.9864	0.9752	0.9589	0.9368	0.9089	0.8759	0.8391
8.2	0.9997	0.9990	0.9972	0.9935	0.9867	0.9757	0.9595	0.9375	0.9096	0.8766	0.8397
8.3	0.9998	0.9991	0.9974	0.9937	0.9870	0.9762	0.9601	0.9381	0.9103	0.8773	0.8403
8.4	0.9998	0.9991	0.9975	0.9939	0.9874	0.9766	0.9606	0.9387	0.9109	0.8779	0.8409
8.5	0.9998	0.9992	0.9976	0.9941	0.9877	0.9770	0.9611	0.9393	0.9116	0.8786	0.8415
8.6	0.9998	0.9992	0.9977	0.9943	0.9880	0.9774	0.9616	0.9399	0.9122	0.8792	0.8421
8.7	0.9998	0.9993	0.9978	0.9945	0.9882	0.9778	0.9621	0.9405	0.9128	0.8798	0.8426
8.8	0.9998	0.9993	0.9979	0.9947	0.9885	0.9782	0.9626	0.9410	0.9133	0.8803	0.8432
8.9	0.9999	0.9994	0.9980	0.9948	0.9888	0.9786	0.9630	0.9415	0.9139	0.8809	0.8437
9.0	0.9999	0.9994	0.9981	0.9950	0.9890	0.9789	0.9635	0.9420	0.9144	0.8814	0.8442
9.1	0.9999	0.9995	0.9982	0.9951	0.9892	0.9792	0.9639	0.9425	0.9150	0.8819	0.8446
9.2	0.9999	0.9995	0.9982	0.9953	0.9895	0.9795	0.9643	0.9430	0.9155	0.8824	0.8451
9.3	0.9999	0.9995	0.9983	0.9954	0.9897	0.9799	0.9647	0.9434	0.9159	0.8829	0.8455
9.4	0.9999	0.9995	0.9984	0.9955	0.9899	0.9801	0.9651	0.9439	0.9164	0.8834	0.8460
9.5	0.9999	0.9996	0.9984	0.9957	0.9901	0.9804	0.9654	0.9443	0.9169	0.8838	0.8464
9.6	0.9999	0.9996	0.9985	0.9958	0.9903	0.9807	0.9658	0.9447	0.9173	0.8843	0.8468
9.7	0.9999	0.9996	0.9986	0.9959	0.9905	0.9810	0.9661	0.9451	0.9177	0.8847	0.8472
9.8	0.9999	0.9996	0.9986	0.9960	0.9906	0.9812	0.9665	0.9455	0.9182	0.8851	0.8476
9.9	1.0000	0.9997	0.9987	0.9961	0.9908	0.9815	0.9668	0.9459	0.9186	0.8855	0.8480
10.0	1.0000	0.9997	0.9987	0.9962	0.9910	0.9817	0.9671	0.9462	0.9190	0.8859	0.8483

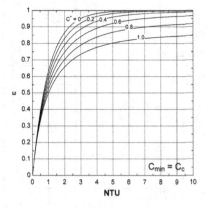

CONFIGURATION 22

1TUBE4ROW4PASS1CIR(CC)

(b) $C_{min} = C_h$

NTU	Heat Exchanger Effectiveness										
	C^* $(C_{min} = C_h)$										
	0	0.1	0.2	0.3	0.4	0.5	0.6	0.7	0.8	0.9	1.0
0	0.0000	0.0000	0.0000	0.0000	0.0000	0.0000	0.0000	0.0000	0.0000	0.0000	0.0000
0.1	0.0952	0.0947	0.0943	0.0939	0.0934	0.0930	0.0926	0.0922	0.0917	0.0913	0.0909
0.2	0.1813	0.1797	0.1782	0.1767	0.1752	0.1738	0.1723	0.1709	0.1694	0.1680	0.1666
0.3	0.2592	0.2562	0.2532	0.2502	0.2473	0.2444	0.2416	0.2388	0.2360	0.2333	0.2306
0.4	0.3297	0.3250	0.3203	0.3157	0.3112	0.3067	0.3023	0.2980	0.2937	0.2895	0.2854
0.5	0.3935	0.3870	0.3806	0.3743	0.3680	0.3619	0.3559	0.3500	0.3441	0.3384	0.3328
0.6	0.4512	0.4430	0.4348	0.4268	0.4189	0.4111	0.4035	0.3960	0.3886	0.3813	0.3741
0.7	0.5034	0.4935	0.4838	0.4741	0.4646	0.4552	0.4460	0.4369	0.4280	0.4192	0.4105
0.8	0.5507	0.5393	0.5280	0.5168	0.5058	0.4949	0.4841	0.4735	0.4631	0.4529	0.4428
0.9	0.5934	0.5807	0.5680	0.5555	0.5430	0.5307	0.5185	0.5065	0.4947	0.4830	0.4716
1.0	0.6321	0.6182	0.6043	0.5905	0.5768	0.5631	0.5496	0.5363	0.5231	0.5102	0.4974
1.1	0.6671	0.6522	0.6373	0.6224	0.6075	0.5926	0.5779	0.5633	0.5489	0.5347	0.5207
1.2	0.6988	0.6831	0.6673	0.6514	0.6354	0.6195	0.6037	0.5879	0.5724	0.5570	0.5418
1.3	0.7275	0.7111	0.6945	0.6778	0.6610	0.6441	0.6272	0.6104	0.5938	0.5773	0.5610
1.4	0.7534	0.7366	0.7194	0.7020	0.6844	0.6666	0.6488	0.6311	0.6134	0.5958	0.5785
1.5	0.7769	0.7597	0.7421	0.7241	0.7058	0.6873	0.6687	0.6500	0.6314	0.6129	0.5945
1.6	0.7981	0.7808	0.7628	0.7444	0.7255	0.7064	0.6870	0.6675	0.6480	0.6285	0.6093
1.7	0.8173	0.7999	0.7817	0.7630	0.7436	0.7239	0.7039	0.6836	0.6633	0.6430	0.6228
1.8	0.8347	0.8173	0.7991	0.7800	0.7604	0.7401	0.7195	0.6986	0.6775	0.6564	0.6354
1.9	0.8504	0.8332	0.8149	0.7957	0.7758	0.7551	0.7340	0.7124	0.6907	0.6688	0.6470
2.0	0.8647	0.8476	0.8294	0.8102	0.7900	0.7691	0.7474	0.7253	0.7029	0.6803	0.6578
2.1	0.8775	0.8608	0.8427	0.8235	0.8032	0.7820	0.7600	0.7374	0.7143	0.6911	0.6678
2.2	0.8892	0.8728	0.8549	0.8358	0.8154	0.7940	0.7716	0.7486	0.7250	0.7011	0.6772
2.3	0.8997	0.8837	0.8661	0.8471	0.8267	0.8051	0.7825	0.7590	0.7350	0.7105	0.6859
2.4	0.9093	0.8937	0.8764	0.8576	0.8372	0.8155	0.7927	0.7688	0.7443	0.7193	0.6941
2.5	0.9179	0.9028	0.8859	0.8672	0.8470	0.8252	0.8022	0.7780	0.7531	0.7275	0.7017
2.6	0.9257	0.9111	0.8946	0.8762	0.8561	0.8343	0.8111	0.7867	0.7613	0.7353	0.7089
2.7	0.9328	0.9186	0.9025	0.8845	0.8645	0.8428	0.8194	0.7948	0.7690	0.7426	0.7156
2.8	0.9392	0.9256	0.9099	0.8922	0.8724	0.8507	0.8273	0.8024	0.7763	0.7494	0.7220
2.9	0.9450	0.9319	0.9167	0.8993	0.8797	0.8581	0.8347	0.8096	0.7832	0.7559	0.7280
3.0	0.9502	0.9376	0.9229	0.9059	0.8866	0.8651	0.8416	0.8164	0.7897	0.7620	0.7336
3.1	0.9550	0.9429	0.9286	0.9120	0.8930	0.8716	0.8481	0.8228	0.7958	0.7678	0.7390
3.2	0.9592	0.9477	0.9339	0.9177	0.8989	0.8777	0.8543	0.8288	0.8017	0.7732	0.7440
3.3	0.9631	0.9521	0.9388	0.9229	0.9045	0.8835	0.8601	0.8345	0.8072	0.7784	0.7488
3.4	0.9666	0.9561	0.9433	0.9278	0.9097	0.8889	0.8656	0.8399	0.8124	0.7833	0.7533
3.5	0.9698	0.9598	0.9474	0.9324	0.9146	0.8940	0.8708	0.8451	0.8173	0.7880	0.7576
3.6	0.9727	0.9632	0.9512	0.9366	0.9192	0.8988	0.8757	0.8499	0.8220	0.7924	0.7617
3.7	0.9753	0.9662	0.9548	0.9406	0.9235	0.9034	0.8803	0.8545	0.8265	0.7966	0.7655
3.8	0.9776	0.9691	0.9580	0.9443	0.9275	0.9076	0.8847	0.8589	0.8307	0.8006	0.7692
3.9	0.9798	0.9716	0.9611	0.9477	0.9313	0.9117	0.8889	0.8631	0.8348	0.8044	0.7727
4.0	0.9817	0.9740	0.9638	0.9509	0.9348	0.9155	0.8928	0.8671	0.8386	0.8081	0.7760
4.1	0.9834	0.9761	0.9664	0.9539	0.9382	0.9191	0.8966	0.8708	0.8423	0.8115	0.7792
4.2	0.9850	0.9781	0.9688	0.9567	0.9413	0.9225	0.9001	0.8744	0.8458	0.8148	0.7822
4.3	0.9864	0.9799	0.9710	0.9593	0.9442	0.9257	0.9035	0.8779	0.8492	0.8180	0.7851
4.4	0.9877	0.9816	0.9730	0.9617	0.9470	0.9287	0.9067	0.8811	0.8524	0.8210	0.7879
4.5	0.9889	0.9831	0.9749	0.9639	0.9496	0.9316	0.9098	0.8842	0.8554	0.8239	0.7905
4.6	0.9899	0.9845	0.9767	0.9661	0.9521	0.9343	0.9127	0.8872	0.8583	0.8267	0.7930
4.7	0.9909	0.9857	0.9783	0.9680	0.9544	0.9369	0.9155	0.8901	0.8611	0.8293	0.7954
4.8	0.9918	0.9869	0.9798	0.9699	0.9566	0.9394	0.9181	0.8928	0.8638	0.8318	0.7977
4.9	0.9926	0.9880	0.9812	0.9716	0.9586	0.9417	0.9206	0.8954	0.8664	0.8343	0.7999
5.0	0.9933	0.9889	0.9825	0.9732	0.9606	0.9439	0.9230	0.8979	0.8688	0.8366	0.8020
5.1	0.9939	0.9898	0.9837	0.9748	0.9624	0.9461	0.9253	0.9003	0.8712	0.8388	0.8041
5.2	0.9945	0.9907	0.9848	0.9762	0.9641	0.9480	0.9275	0.9025	0.8734	0.8409	0.8060
5.3	0.9950	0.9914	0.9858	0.9775	0.9658	0.9500	0.9296	0.9047	0.8756	0.8430	0.8079
5.4	0.9955	0.9921	0.9868	0.9788	0.9673	0.9518	0.9316	0.9068	0.8777	0.8450	0.8096
5.5	0.9959	0.9927	0.9876	0.9799	0.9688	0.9535	0.9335	0.9088	0.8797	0.8469	0.8114
5.6	0.9963	0.9933	0.9885	0.9810	0.9702	0.9551	0.9354	0.9108	0.8816	0.8487	0.8130
5.7	0.9967	0.9938	0.9892	0.9821	0.9715	0.9567	0.9371	0.9126	0.8835	0.8504	0.8146
5.8	0.9970	0.9943	0.9899	0.9830	0.9727	0.9582	0.9388	0.9144	0.8853	0.8521	0.8161
5.9	0.9973	0.9948	0.9906	0.9839	0.9739	0.9596	0.9404	0.9161	0.8870	0.8537	0.8176
6.0	0.9975	0.9952	0.9912	0.9848	0.9750	0.9610	0.9420	0.9178	0.8886	0.8553	0.8190

NTU	Heat Exchanger Effectiveness										
	$C^* (C_{min} = C_h)$										
	0	0.1	0.2	0.3	0.4	0.5	0.6	0.7	0.8	0.9	1.0
6.1	0.9978	0.9956	0.9918	0.9856	0.9761	0.9622	0.9435	0.9194	0.8902	0.8568	0.8203
6.2	0.9980	0.9959	0.9923	0.9864	0.9771	0.9635	0.9449	0.9209	0.8917	0.8583	0.8216
6.3	0.9982	0.9963	0.9928	0.9871	0.9780	0.9647	0.9462	0.9223	0.8932	0.8596	0.8229
6.4	0.9983	0.9965	0.9933	0.9877	0.9789	0.9658	0.9475	0.9238	0.8946	0.8610	0.8241
6.5	0.9985	0.9968	0.9937	0.9884	0.9798	0.9669	0.9488	0.9251	0.8960	0.8623	0.8252
6.6	0.9986	0.9971	0.9941	0.9890	0.9806	0.9679	0.9500	0.9264	0.8973	0.8635	0.8263
6.7	0.9988	0.9973	0.9945	0.9895	0.9814	0.9689	0.9512	0.9277	0.8986	0.8648	0.8274
6.8	0.9989	0.9975	0.9948	0.9900	0.9821	0.9698	0.9523	0.9289	0.8999	0.8659	0.8285
6.9	0.9990	0.9977	0.9952	0.9905	0.9828	0.9707	0.9534	0.9301	0.9010	0.8670	0.8295
7.0	0.9991	0.9979	0.9955	0.9910	0.9835	0.9716	0.9544	0.9312	0.9022	0.8681	0.8304
7.1	0.9992	0.9980	0.9957	0.9915	0.9841	0.9724	0.9554	0.9323	0.9033	0.8692	0.8314
7.2	0.9993	0.9982	0.9960	0.9919	0.9847	0.9732	0.9563	0.9334	0.9044	0.8702	0.8323
7.3	0.9993	0.9983	0.9963	0.9923	0.9853	0.9740	0.9573	0.9344	0.9054	0.8712	0.8331
7.4	0.9994	0.9985	0.9965	0.9926	0.9858	0.9747	0.9581	0.9354	0.9064	0.8721	0.8340
7.5	0.9994	0.9986	0.9967	0.9930	0.9863	0.9754	0.9590	0.9363	0.9074	0.8730	0.8348
7.6	0.9995	0.9987	0.9969	0.9933	0.9868	0.9761	0.9598	0.9372	0.9083	0.8739	0.8356
7.7	0.9995	0.9988	0.9971	0.9936	0.9873	0.9767	0.9606	0.9381	0.9092	0.8748	0.8363
7.8	0.9996	0.9989	0.9973	0.9939	0.9878	0.9774	0.9614	0.9390	0.9101	0.8756	0.8370
7.9	0.9996	0.9990	0.9974	0.9942	0.9882	0.9779	0.9621	0.9398	0.9110	0.8764	0.8377
8.0	0.9997	0.9990	0.9976	0.9945	0.9886	0.9785	0.9628	0.9406	0.9118	0.8772	0.8384
8.1	0.9997	0.9991	0.9977	0.9947	0.9890	0.9791	0.9635	0.9414	0.9126	0.8780	0.8391
8.2	0.9997	0.9992	0.9979	0.9950	0.9894	0.9796	0.9642	0.9422	0.9134	0.8787	0.8397
8.3	0.9998	0.9992	0.9980	0.9952	0.9897	0.9801	0.9649	0.9429	0.9141	0.8794	0.8403
8.4	0.9998	0.9993	0.9981	0.9954	0.9901	0.9806	0.9655	0.9436	0.9149	0.8801	0.8409
8.5	0.9998	0.9994	0.9982	0.9956	0.9904	0.9811	0.9661	0.9443	0.9156	0.8808	0.8415
8.6	0.9998	0.9994	0.9983	0.9958	0.9907	0.9815	0.9667	0.9450	0.9163	0.8814	0.8421
8.7	0.9998	0.9994	0.9984	0.9960	0.9910	0.9820	0.9672	0.9456	0.9169	0.8820	0.8426
8.8	0.9998	0.9995	0.9985	0.9962	0.9913	0.9824	0.9678	0.9462	0.9176	0.8826	0.8432
8.9	0.9999	0.9995	0.9986	0.9963	0.9916	0.9828	0.9683	0.9468	0.9182	0.8832	0.8437
9.0	0.9999	0.9996	0.9987	0.9965	0.9919	0.9832	0.9688	0.9474	0.9188	0.8838	0.8442
9.1	0.9999	0.9996	0.9987	0.9966	0.9921	0.9836	0.9693	0.9480	0.9194	0.8843	0.8446
9.2	0.9999	0.9996	0.9988	0.9968	0.9924	0.9839	0.9698	0.9486	0.9200	0.8849	0.8451
9.3	0.9999	0.9996	0.9989	0.9969	0.9926	0.9843	0.9702	0.9491	0.9205	0.8854	0.8455
9.4	0.9999	0.9997	0.9989	0.9970	0.9928	0.9846	0.9707	0.9496	0.9211	0.8859	0.8460
9.5	0.9999	0.9997	0.9990	0.9972	0.9930	0.9850	0.9711	0.9501	0.9216	0.8864	0.8464
9.6	0.9999	0.9997	0.9991	0.9973	0.9932	0.9853	0.9716	0.9506	0.9221	0.8869	0.8468
9.7	0.9999	0.9997	0.9991	0.9974	0.9934	0.9856	0.9720	0.9511	0.9226	0.8873	0.8472
9.8	0.9999	0.9998	0.9992	0.9975	0.9936	0.9859	0.9724	0.9516	0.9231	0.8878	0.8476
9.9	1.0000	0.9998	0.9992	0.9976	0.9938	0.9862	0.9727	0.9520	0.9235	0.8882	0.8480
10.0	1.0000	0.9998	0.9992	0.9977	0.9940	0.9865	0.9731	0.9525	0.9240	0.8886	0.8483

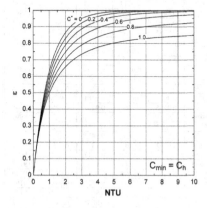

CONFIGURATION 23

1TUBE5ROW5PASS1CIR(CC)

(a) $C_{min} = C_c$

NTU	Heat Exchanger Effectiveness $C^* (C_{min} = C_c)$										
	0	0.1	0.2	0.3	0.4	0.5	0.6	0.7	0.8	0.9	1.0
0	0.0000	0.0000	0.0000	0.0000	0.0000	0.0000	0.0000	0.0000	0.0000	0.0000	0.0000
0.1	0.0952	0.0947	0.0943	0.0939	0.0934	0.0930	0.0926	0.0922	0.0917	0.0913	0.0909
0.2	0.1813	0.1797	0.1782	0.1767	0.1752	0.1738	0.1723	0.1709	0.1694	0.1680	0.1666
0.3	0.2592	0.2562	0.2532	0.2502	0.2473	0.2445	0.2416	0.2388	0.2361	0.2334	0.2307
0.4	0.3297	0.3250	0.3203	0.3157	0.3112	0.3068	0.3024	0.2981	0.2938	0.2896	0.2855
0.5	0.3935	0.3870	0.3806	0.3743	0.3681	0.3620	0.3560	0.3501	0.3443	0.3386	0.3330
0.6	0.4512	0.4430	0.4349	0.4269	0.4191	0.4113	0.4037	0.3962	0.3888	0.3816	0.3744
0.7	0.5034	0.4936	0.4839	0.4743	0.4648	0.4555	0.4463	0.4372	0.4283	0.4196	0.4110
0.8	0.5507	0.5394	0.5281	0.5170	0.5061	0.4952	0.4845	0.4740	0.4636	0.4534	0.4434
0.9	0.5934	0.5808	0.5682	0.5558	0.5434	0.5311	0.5190	0.5071	0.4953	0.4837	0.4724
1.0	0.6321	0.6183	0.6046	0.5909	0.5772	0.5637	0.5503	0.5370	0.5239	0.5110	0.4984
1.1	0.6671	0.6524	0.6376	0.6228	0.6080	0.5933	0.5787	0.5642	0.5499	0.5357	0.5218
1.2	0.6988	0.6833	0.6676	0.6519	0.6361	0.6203	0.6046	0.5890	0.5735	0.5582	0.5431
1.3	0.7275	0.7113	0.6950	0.6784	0.6618	0.6450	0.6283	0.6117	0.5951	0.5787	0.5625
1.4	0.7534	0.7368	0.7199	0.7027	0.6853	0.6677	0.6501	0.6325	0.6149	0.5975	0.5802
1.5	0.7769	0.7600	0.7426	0.7249	0.7068	0.6885	0.6701	0.6516	0.6331	0.6147	0.5965
1.6	0.7981	0.7810	0.7634	0.7452	0.7266	0.7077	0.6885	0.6692	0.6499	0.6306	0.6114
1.7	0.8173	0.8002	0.7824	0.7639	0.7449	0.7254	0.7056	0.6856	0.6654	0.6453	0.6252
1.8	0.8347	0.8176	0.7997	0.7810	0.7617	0.7418	0.7214	0.7007	0.6798	0.6589	0.6380
1.9	0.8504	0.8335	0.8156	0.7968	0.7772	0.7569	0.7360	0.7148	0.6932	0.6716	0.6499
2.0	0.8647	0.8480	0.8302	0.8113	0.7916	0.7709	0.7497	0.7279	0.7057	0.6833	0.6609
2.1	0.8775	0.8612	0.8435	0.8247	0.8048	0.7840	0.7623	0.7401	0.7173	0.6943	0.6712
2.2	0.8892	0.8732	0.8558	0.8371	0.8171	0.7961	0.7742	0.7515	0.7282	0.7046	0.6808
2.3	0.8997	0.8841	0.8670	0.8485	0.8285	0.8074	0.7852	0.7621	0.7384	0.7142	0.6898
2.4	0.9093	0.8941	0.8773	0.8590	0.8391	0.8179	0.7955	0.7721	0.7479	0.7232	0.6982
2.5	0.9179	0.9032	0.8868	0.8687	0.8490	0.8278	0.8052	0.7815	0.7569	0.7317	0.7061
2.6	0.9257	0.9115	0.8955	0.8777	0.8581	0.8369	0.8143	0.7903	0.7654	0.7397	0.7135
2.7	0.9328	0.9191	0.9035	0.8860	0.8666	0.8455	0.8227	0.7986	0.7733	0.7472	0.7205
2.8	0.9392	0.9260	0.9109	0.8937	0.8746	0.8535	0.8307	0.8064	0.7808	0.7543	0.7271
2.9	0.9450	0.9323	0.9177	0.9009	0.8820	0.8611	0.8382	0.8138	0.7879	0.7610	0.7334
3.0	0.9502	0.9381	0.9239	0.9075	0.8889	0.8681	0.8453	0.8207	0.7946	0.7673	0.7393
3.1	0.9550	0.9434	0.9296	0.9136	0.8953	0.8747	0.8520	0.8273	0.8009	0.7733	0.7449
3.2	0.9592	0.9482	0.9349	0.9193	0.9013	0.8809	0.8582	0.8335	0.8069	0.7790	0.7502
3.3	0.9631	0.9526	0.9398	0.9246	0.9070	0.8868	0.8642	0.8393	0.8126	0.7844	0.7552
3.4	0.9666	0.9566	0.9443	0.9295	0.9122	0.8923	0.8698	0.8449	0.8180	0.7896	0.7599
3.5	0.9698	0.9602	0.9484	0.9341	0.9171	0.8974	0.8750	0.8502	0.8232	0.7944	0.7645
3.6	0.9727	0.9636	0.9522	0.9384	0.9217	0.9023	0.8800	0.8552	0.8281	0.7991	0.7688
3.7	0.9753	0.9667	0.9558	0.9423	0.9260	0.9069	0.8848	0.8599	0.8327	0.8035	0.7729
3.8	0.9776	0.9695	0.9590	0.9460	0.9301	0.9112	0.8892	0.8645	0.8371	0.8077	0.7768
3.9	0.9798	0.9720	0.9620	0.9494	0.9339	0.9152	0.8935	0.8688	0.8414	0.8118	0.7805
4.0	0.9817	0.9744	0.9648	0.9526	0.9374	0.9191	0.8975	0.8729	0.8454	0.8156	0.7841
4.1	0.9834	0.9765	0.9673	0.9555	0.9408	0.9227	0.9013	0.8767	0.8492	0.8193	0.7875
4.2	0.9850	0.9785	0.9697	0.9583	0.9439	0.9261	0.9050	0.8805	0.8529	0.8228	0.7908
4.3	0.9864	0.9803	0.9719	0.9609	0.9468	0.9294	0.9084	0.8840	0.8564	0.8261	0.7939
4.4	0.9877	0.9819	0.9739	0.9633	0.9496	0.9324	0.9117	0.8874	0.8597	0.8294	0.7969
4.5	0.9889	0.9834	0.9758	0.9655	0.9522	0.9353	0.9148	0.8906	0.8630	0.8324	0.7997
4.6	0.9899	0.9848	0.9775	0.9676	0.9546	0.9381	0.9178	0.8937	0.8660	0.8354	0.8024
4.7	0.9909	0.9860	0.9791	0.9696	0.9569	0.9407	0.9206	0.8966	0.8690	0.8382	0.8051
4.8	0.9918	0.9872	0.9806	0.9714	0.9591	0.9431	0.9233	0.8994	0.8718	0.8409	0.8076
4.9	0.9926	0.9882	0.9819	0.9731	0.9611	0.9455	0.9258	0.9021	0.8745	0.8435	0.8100
5.0	0.9933	0.9892	0.9832	0.9747	0.9630	0.9477	0.9283	0.9047	0.8771	0.8460	0.8123
5.1	0.9939	0.9901	0.9844	0.9762	0.9648	0.9498	0.9306	0.9071	0.8796	0.8484	0.8146
5.2	0.9945	0.9909	0.9854	0.9775	0.9665	0.9517	0.9328	0.9095	0.8819	0.8507	0.8167
5.3	0.9950	0.9916	0.9865	0.9788	0.9681	0.9536	0.9349	0.9117	0.8842	0.8530	0.8188
5.4	0.9955	0.9923	0.9874	0.9801	0.9696	0.9554	0.9369	0.9139	0.8864	0.8551	0.8207
5.5	0.9959	0.9929	0.9882	0.9812	0.9711	0.9571	0.9389	0.9160	0.8886	0.8572	0.8227
5.6	0.9963	0.9935	0.9890	0.9823	0.9724	0.9587	0.9407	0.9179	0.8906	0.8591	0.8245
5.7	0.9967	0.9940	0.9898	0.9833	0.9737	0.9603	0.9425	0.9199	0.8926	0.8611	0.8263
5.8	0.9970	0.9945	0.9905	0.9842	0.9749	0.9617	0.9441	0.9217	0.8944	0.8629	0.8280
5.9	0.9973	0.9949	0.9911	0.9851	0.9760	0.9631	0.9457	0.9234	0.8963	0.8647	0.8296
6.0	0.9975	0.9953	0.9917	0.9859	0.9771	0.9645	0.9473	0.9251	0.8980	0.8664	0.8312

NTU	Heat Exchanger Effectiveness										
	C^* ($C_{min} = C_c$)										
	0	0.1	0.2	0.3	0.4	0.5	0.6	0.7	0.8	0.9	1.0
6.1	0.9978	0.9957	0.9922	0.9866	0.9781	0.9657	0.9488	0.9268	0.8997	0.8680	0.8327
6.2	0.9980	0.9961	0.9928	0.9874	0.9790	0.9669	0.9502	0.9283	0.9013	0.8696	0.8342
6.3	0.9982	0.9964	0.9932	0.9880	0.9800	0.9681	0.9515	0.9298	0.9029	0.8712	0.8356
6.4	0.9983	0.9967	0.9937	0.9887	0.9808	0.9691	0.9528	0.9313	0.9044	0.8726	0.8370
6.5	0.9985	0.9969	0.9941	0.9893	0.9816	0.9702	0.9541	0.9326	0.9058	0.8741	0.8383
6.6	0.9986	0.9972	0.9945	0.9898	0.9824	0.9712	0.9552	0.9340	0.9072	0.8755	0.8396
6.7	0.9988	0.9974	0.9948	0.9904	0.9831	0.9721	0.9564	0.9353	0.9086	0.8768	0.8409
6.8	0.9989	0.9976	0.9951	0.9909	0.9838	0.9730	0.9575	0.9365	0.9099	0.8781	0.8421
6.9	0.9990	0.9978	0.9955	0.9913	0.9845	0.9739	0.9585	0.9377	0.9112	0.8793	0.8432
7.0	0.9991	0.9980	0.9957	0.9918	0.9851	0.9747	0.9595	0.9389	0.9124	0.8806	0.8444
7.1	0.9992	0.9981	0.9960	0.9922	0.9857	0.9755	0.9605	0.9400	0.9136	0.8817	0.8455
7.2	0.9993	0.9983	0.9963	0.9926	0.9862	0.9762	0.9614	0.9410	0.9147	0.8829	0.8465
7.3	0.9993	0.9984	0.9965	0.9929	0.9868	0.9769	0.9623	0.9421	0.9158	0.8840	0.8475
7.4	0.9994	0.9985	0.9967	0.9933	0.9873	0.9776	0.9632	0.9431	0.9169	0.8850	0.8485
7.5	0.9994	0.9986	0.9969	0.9936	0.9878	0.9783	0.9640	0.9440	0.9179	0.8860	0.8495
7.6	0.9995	0.9987	0.9971	0.9939	0.9882	0.9789	0.9648	0.9449	0.9189	0.8870	0.8504
7.7	0.9995	0.9988	0.9973	0.9942	0.9886	0.9795	0.9656	0.9458	0.9199	0.8880	0.8513
7.8	0.9996	0.9989	0.9974	0.9944	0.9891	0.9801	0.9663	0.9467	0.9208	0.8889	0.8522
7.9	0.9996	0.9990	0.9976	0.9947	0.9895	0.9806	0.9670	0.9475	0.9217	0.8899	0.8530
8.0	0.9997	0.9991	0.9977	0.9949	0.9898	0.9812	0.9677	0.9483	0.9226	0.8907	0.8538
8.1	0.9997	0.9991	0.9978	0.9952	0.9902	0.9817	0.9683	0.9491	0.9234	0.8916	0.8546
8.2	0.9997	0.9992	0.9980	0.9954	0.9905	0.9822	0.9690	0.9499	0.9243	0.8924	0.8554
8.3	0.9998	0.9993	0.9981	0.9956	0.9908	0.9826	0.9696	0.9506	0.9250	0.8932	0.8562
8.4	0.9998	0.9993	0.9982	0.9958	0.9911	0.9831	0.9702	0.9513	0.9258	0.8940	0.8569
8.5	0.9998	0.9994	0.9983	0.9960	0.9914	0.9835	0.9707	0.9520	0.9266	0.8947	0.8576
8.6	0.9998	0.9994	0.9984	0.9961	0.9917	0.9839	0.9713	0.9526	0.9273	0.8955	0.8583
8.7	0.9998	0.9995	0.9985	0.9963	0.9920	0.9843	0.9718	0.9533	0.9280	0.8962	0.8589
8.8	0.9998	0.9995	0.9986	0.9965	0.9922	0.9847	0.9723	0.9539	0.9287	0.8969	0.8596
8.9	0.9999	0.9995	0.9986	0.9966	0.9925	0.9850	0.9728	0.9545	0.9293	0.8976	0.8602
9.0	0.9999	0.9996	0.9987	0.9967	0.9927	0.9854	0.9733	0.9550	0.9300	0.8982	0.8608
9.1	0.9999	0.9996	0.9988	0.9969	0.9929	0.9857	0.9737	0.9556	0.9306	0.8988	0.8614
9.2	0.9999	0.9996	0.9988	0.9970	0.9931	0.9860	0.9741	0.9561	0.9312	0.8995	0.8620
9.3	0.9999	0.9996	0.9989	0.9971	0.9933	0.9863	0.9746	0.9567	0.9318	0.9000	0.8626
9.4	0.9999	0.9997	0.9990	0.9972	0.9935	0.9866	0.9750	0.9572	0.9324	0.9006	0.8631
9.5	0.9999	0.9997	0.9990	0.9973	0.9937	0.9869	0.9754	0.9577	0.9329	0.9012	0.8636
9.6	0.9999	0.9997	0.9991	0.9974	0.9939	0.9872	0.9758	0.9581	0.9335	0.9017	0.8641
9.7	0.9999	0.9997	0.9991	0.9975	0.9941	0.9875	0.9761	0.9586	0.9340	0.9023	0.8646
9.8	0.9999	0.9998	0.9992	0.9976	0.9942	0.9877	0.9765	0.9591	0.9345	0.9028	0.8651
9.9	1.0000	0.9998	0.9992	0.9977	0.9944	0.9880	0.9768	0.9595	0.9350	0.9033	0.8656
10.0	1.0000	0.9998	0.9992	0.9978	0.9945	0.9882	0.9772	0.9599	0.9354	0.9038	0.8661

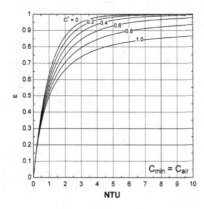

CONFIGURATION 23

1TUBE5ROW5PASS1CIR(CC)

(b) $C_{min} = C_h$

NTU	Heat Exchanger Effectiveness										
	$C^* (C_{min} = C_h)$										
	0	0.1	0.2	0.3	0.4	0.5	0.6	0.7	0.8	0.9	1.0
0	0.0000	0.0000	0.0000	0.0000	0.0000	0.0000	0.0000	0.0000	0.0000	0.0000	0.0000
0.1	0.0952	0.0947	0.0943	0.0939	0.0934	0.0930	0.0926	0.0922	0.0917	0.0913	0.0909
0.2	0.1813	0.1797	0.1782	0.1767	0.1752	0.1738	0.1723	0.1709	0.1694	0.1680	0.1666
0.3	0.2592	0.2562	0.2532	0.2502	0.2473	0.2445	0.2416	0.2388	0.2361	0.2334	0.2307
0.4	0.3297	0.3250	0.3203	0.3157	0.3112	0.3068	0.3024	0.2981	0.2938	0.2896	0.2855
0.5	0.3935	0.3870	0.3806	0.3743	0.3681	0.3620	0.3560	0.3501	0.3443	0.3386	0.3330
0.6	0.4512	0.4430	0.4349	0.4269	0.4191	0.4113	0.4037	0.3962	0.3888	0.3816	0.3744
0.7	0.5034	0.4936	0.4839	0.4743	0.4648	0.4555	0.4463	0.4372	0.4283	0.4196	0.4110
0.8	0.5507	0.5394	0.5282	0.5171	0.5061	0.4952	0.4845	0.4740	0.4636	0.4534	0.4434
0.9	0.5934	0.5808	0.5682	0.5558	0.5434	0.5311	0.5190	0.5071	0.4953	0.4837	0.4724
1.0	0.6321	0.6183	0.6046	0.5909	0.5772	0.5637	0.5503	0.5370	0.5239	0.5110	0.4984
1.1	0.6671	0.6524	0.6376	0.6228	0.6080	0.5933	0.5787	0.5642	0.5499	0.5358	0.5218
1.2	0.6988	0.6833	0.6676	0.6519	0.6361	0.6203	0.6046	0.5890	0.5735	0.5582	0.5431
1.3	0.7275	0.7113	0.6950	0.6784	0.6618	0.6451	0.6283	0.6117	0.5951	0.5787	0.5625
1.4	0.7534	0.7368	0.7199	0.7027	0.6853	0.6677	0.6501	0.6325	0.6149	0.5975	0.5802
1.5	0.7769	0.7600	0.7426	0.7249	0.7068	0.6886	0.6701	0.6516	0.6331	0.6147	0.5965
1.6	0.7981	0.7811	0.7634	0.7452	0.7267	0.7077	0.6886	0.6693	0.6499	0.6306	0.6114
1.7	0.8173	0.8002	0.7824	0.7639	0.7449	0.7254	0.7056	0.6856	0.6655	0.6453	0.6252
1.8	0.8347	0.8177	0.7998	0.7811	0.7617	0.7418	0.7214	0.7008	0.6799	0.6589	0.6380
1.9	0.8504	0.8335	0.8157	0.7969	0.7773	0.7570	0.7361	0.7148	0.6933	0.6716	0.6499
2.0	0.8647	0.8480	0.8302	0.8114	0.7916	0.7710	0.7497	0.7279	0.7057	0.6833	0.6609
2.1	0.8775	0.8612	0.8436	0.8248	0.8049	0.7841	0.7624	0.7401	0.7174	0.6943	0.6712
2.2	0.8892	0.8732	0.8558	0.8371	0.8172	0.7962	0.7743	0.7515	0.7283	0.7046	0.6808
2.3	0.8997	0.8842	0.8671	0.8485	0.8286	0.8075	0.7853	0.7622	0.7385	0.7142	0.6898
2.4	0.9093	0.8941	0.8774	0.8591	0.8392	0.8180	0.7956	0.7722	0.7480	0.7233	0.6982
2.5	0.9179	0.9032	0.8869	0.8688	0.8491	0.8279	0.8053	0.7816	0.7570	0.7317	0.7061
2.6	0.9257	0.9116	0.8956	0.8778	0.8583	0.8371	0.8144	0.7904	0.7655	0.7397	0.7135
2.7	0.9328	0.9191	0.9036	0.8861	0.8668	0.8457	0.8229	0.7987	0.7734	0.7473	0.7205
2.8	0.9392	0.9261	0.9110	0.8939	0.8748	0.8537	0.8309	0.8066	0.7809	0.7544	0.7271
2.9	0.9450	0.9324	0.9178	0.9010	0.8822	0.8613	0.8384	0.8139	0.7880	0.7611	0.7334
3.0	0.9502	0.9382	0.9240	0.9077	0.8891	0.8683	0.8455	0.8209	0.7948	0.7674	0.7393
3.1	0.9550	0.9434	0.9297	0.9138	0.8955	0.8750	0.8522	0.8275	0.8011	0.7734	0.7449
3.2	0.9592	0.9482	0.9350	0.9195	0.9016	0.8812	0.8585	0.8337	0.8071	0.7791	0.7502
3.3	0.9631	0.9526	0.9399	0.9248	0.9072	0.8871	0.8645	0.8396	0.8128	0.7845	0.7552
3.4	0.9666	0.9566	0.9444	0.9298	0.9125	0.8926	0.8701	0.8452	0.8183	0.7897	0.7599
3.5	0.9698	0.9603	0.9486	0.9343	0.9174	0.8978	0.8754	0.8505	0.8234	0.7946	0.7645
3.6	0.9727	0.9637	0.9524	0.9386	0.9220	0.9026	0.8804	0.8555	0.8283	0.7992	0.7688
3.7	0.9753	0.9667	0.9559	0.9426	0.9264	0.9073	0.8852	0.8603	0.8330	0.8037	0.7729
3.8	0.9776	0.9695	0.9592	0.9462	0.9304	0.9116	0.8897	0.8649	0.8375	0.8079	0.7768
3.9	0.9798	0.9721	0.9622	0.9497	0.9342	0.9157	0.8940	0.8692	0.8417	0.8119	0.7805
4.0	0.9817	0.9744	0.9650	0.9529	0.9378	0.9196	0.8980	0.8733	0.8458	0.8158	0.7841
4.1	0.9834	0.9766	0.9675	0.9559	0.9412	0.9232	0.9019	0.8772	0.8496	0.8195	0.7875
4.2	0.9850	0.9785	0.9699	0.9586	0.9443	0.9267	0.9055	0.8810	0.8533	0.8230	0.7908
4.3	0.9864	0.9803	0.9721	0.9612	0.9473	0.9299	0.9090	0.8846	0.8568	0.8264	0.7939
4.4	0.9877	0.9820	0.9741	0.9636	0.9501	0.9330	0.9123	0.8880	0.8602	0.8296	0.7969
4.5	0.9889	0.9835	0.9760	0.9659	0.9527	0.9360	0.9155	0.8912	0.8635	0.8327	0.7997
4.6	0.9899	0.9849	0.9777	0.9680	0.9551	0.9387	0.9185	0.8943	0.8666	0.8357	0.8024
4.7	0.9909	0.9861	0.9793	0.9700	0.9575	0.9414	0.9213	0.8973	0.8695	0.8385	0.8051
4.8	0.9918	0.9873	0.9808	0.9718	0.9596	0.9439	0.9241	0.9002	0.8724	0.8413	0.8076
4.9	0.9926	0.9883	0.9822	0.9735	0.9617	0.9462	0.9267	0.9029	0.8751	0.8439	0.8100
5.0	0.9933	0.9893	0.9834	0.9751	0.9636	0.9485	0.9291	0.9055	0.8777	0.8464	0.8123
5.1	0.9939	0.9902	0.9846	0.9766	0.9655	0.9506	0.9315	0.9080	0.8803	0.8488	0.8146
5.2	0.9945	0.9910	0.9857	0.9780	0.9672	0.9526	0.9338	0.9104	0.8827	0.8512	0.8167
5.3	0.9950	0.9917	0.9867	0.9793	0.9688	0.9545	0.9359	0.9127	0.8850	0.8534	0.8188
5.4	0.9955	0.9924	0.9876	0.9805	0.9703	0.9564	0.9380	0.9149	0.8873	0.8556	0.8207
5.5	0.9959	0.9930	0.9885	0.9817	0.9718	0.9581	0.9399	0.9170	0.8894	0.8576	0.8227
5.6	0.9963	0.9936	0.9893	0.9827	0.9732	0.9597	0.9418	0.9191	0.8915	0.8597	0.8245
5.7	0.9967	0.9941	0.9900	0.9838	0.9744	0.9613	0.9436	0.9210	0.8935	0.8616	0.8263
5.8	0.9970	0.9946	0.9907	0.9847	0.9757	0.9628	0.9454	0.9229	0.8954	0.8634	0.8280
5.9	0.9973	0.9950	0.9914	0.9856	0.9768	0.9642	0.9470	0.9247	0.8973	0.8652	0.8296
6.0	0.9975	0.9954	0.9920	0.9864	0.9779	0.9656	0.9486	0.9264	0.8991	0.8670	0.8312

NTU	Heat Exchanger Effectiveness										
	C^* ($C_{min} = C_h$)										
	0	0.1	0.2	0.3	0.4	0.5	0.6	0.7	0.8	0.9	1.0
6.1	0.9978	0.9958	0.9925	0.9872	0.9789	0.9669	0.9501	0.9281	0.9008	0.8686	0.8327
6.2	0.9980	0.9962	0.9930	0.9879	0.9799	0.9681	0.9516	0.9297	0.9025	0.8703	0.8342
6.3	0.9982	0.9965	0.9935	0.9886	0.9808	0.9693	0.9530	0.9313	0.9041	0.8718	0.8356
6.4	0.9983	0.9968	0.9939	0.9892	0.9817	0.9704	0.9543	0.9328	0.9056	0.8733	0.8370
6.5	0.9985	0.9970	0.9944	0.9898	0.9825	0.9715	0.9556	0.9342	0.9071	0.8748	0.8383
6.6	0.9986	0.9973	0.9947	0.9904	0.9833	0.9725	0.9568	0.9356	0.9086	0.8762	0.8396
6.7	0.9988	0.9975	0.9951	0.9909	0.9841	0.9735	0.9580	0.9369	0.9100	0.8776	0.8409
6.8	0.9989	0.9977	0.9954	0.9914	0.9848	0.9744	0.9591	0.9382	0.9113	0.8789	0.8421
6.9	0.9990	0.9979	0.9957	0.9919	0.9855	0.9753	0.9602	0.9395	0.9126	0.8802	0.8432
7.0	0.9991	0.9980	0.9960	0.9923	0.9861	0.9761	0.9613	0.9407	0.9139	0.8814	0.8444
7.1	0.9992	0.9982	0.9963	0.9927	0.9867	0.9769	0.9623	0.9418	0.9151	0.8826	0.8455
7.2	0.9993	0.9983	0.9965	0.9931	0.9873	0.9777	0.9633	0.9429	0.9163	0.8838	0.8465
7.3	0.9993	0.9985	0.9967	0.9935	0.9878	0.9785	0.9642	0.9440	0.9174	0.8849	0.8475
7.4	0.9994	0.9986	0.9970	0.9939	0.9883	0.9792	0.9651	0.9451	0.9186	0.8860	0.8485
7.5	0.9994	0.9987	0.9972	0.9942	0.9888	0.9799	0.9660	0.9461	0.9196	0.8870	0.8495
7.6	0.9995	0.9988	0.9973	0.9945	0.9893	0.9805	0.9668	0.9471	0.9207	0.8881	0.8504
7.7	0.9995	0.9989	0.9975	0.9948	0.9897	0.9811	0.9676	0.9480	0.9217	0.8890	0.8513
7.8	0.9996	0.9990	0.9977	0.9950	0.9901	0.9817	0.9684	0.9489	0.9227	0.8900	0.8522
7.9	0.9996	0.9991	0.9978	0.9953	0.9906	0.9823	0.9692	0.9498	0.9236	0.8909	0.8530
8.0	0.9997	0.9991	0.9980	0.9955	0.9909	0.9829	0.9699	0.9507	0.9245	0.8918	0.8538
8.1	0.9997	0.9992	0.9981	0.9958	0.9913	0.9834	0.9706	0.9515	0.9254	0.8927	0.8546
8.2	0.9997	0.9993	0.9982	0.9960	0.9916	0.9839	0.9712	0.9523	0.9263	0.8936	0.8554
8.3	0.9998	0.9993	0.9983	0.9962	0.9920	0.9844	0.9719	0.9531	0.9272	0.8944	0.8562
8.4	0.9998	0.9994	0.9984	0.9964	0.9923	0.9849	0.9725	0.9538	0.9280	0.8952	0.8569
8.5	0.9998	0.9994	0.9985	0.9966	0.9926	0.9853	0.9731	0.9546	0.9288	0.8960	0.8576
8.6	0.9998	0.9995	0.9986	0.9967	0.9929	0.9857	0.9737	0.9553	0.9295	0.8968	0.8583
8.7	0.9998	0.9995	0.9987	0.9969	0.9931	0.9862	0.9743	0.9559	0.9303	0.8975	0.8589
8.8	0.9998	0.9996	0.9988	0.9970	0.9934	0.9866	0.9748	0.9566	0.9310	0.8982	0.8596
8.9	0.9999	0.9996	0.9989	0.9972	0.9937	0.9869	0.9753	0.9573	0.9317	0.8989	0.8602
9.0	0.9999	0.9996	0.9989	0.9973	0.9939	0.9873	0.9759	0.9579	0.9324	0.8996	0.8608
9.1	0.9999	0.9996	0.9990	0.9975	0.9941	0.9877	0.9763	0.9585	0.9331	0.9003	0.8614
9.2	0.9999	0.9997	0.9991	0.9976	0.9943	0.9880	0.9768	0.9591	0.9337	0.9009	0.8620
9.3	0.9999	0.9997	0.9991	0.9977	0.9945	0.9883	0.9773	0.9597	0.9344	0.9015	0.8626
9.4	0.9999	0.9997	0.9992	0.9978	0.9947	0.9886	0.9777	0.9602	0.9350	0.9021	0.8631
9.5	0.9999	0.9997	0.9992	0.9979	0.9949	0.9889	0.9782	0.9608	0.9356	0.9027	0.8636
9.6	0.9999	0.9998	0.9993	0.9980	0.9951	0.9892	0.9786	0.9613	0.9362	0.9033	0.8641
9.7	0.9999	0.9998	0.9993	0.9981	0.9953	0.9895	0.9790	0.9618	0.9367	0.9039	0.8646
9.8	0.9999	0.9998	0.9994	0.9982	0.9955	0.9898	0.9794	0.9623	0.9373	0.9044	0.8651
9.9	1.0000	0.9998	0.9994	0.9983	0.9956	0.9901	0.9798	0.9628	0.9378	0.9049	0.8656
10.0	1.0000	0.9998	0.9994	0.9983	0.9958	0.9903	0.9801	0.9633	0.9384	0.9055	0.8661

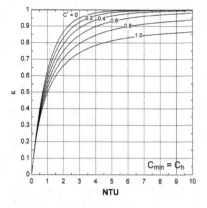

CONFIGURATION 24

1TUBE6ROW6PASS1CIR(CC)

(a) $C_{min} = C_c$

NTU	Heat Exchanger Effectiveness										
	$C^* (C_{min} = C_c)$										
	0	0.1	0.2	0.3	0.4	0.5	0.6	0.7	0.8	0.9	1.0
0	0.0000	0.0000	0.0000	0.0000	0.0000	0.0000	0.0000	0.0000	0.0000	0.0000	0.0000
0.1	0.0952	0.0947	0.0943	0.0939	0.0934	0.0930	0.0926	0.0922	0.0917	0.0913	0.0909
0.2	0.1813	0.1797	0.1782	0.1767	0.1752	0.1738	0.1723	0.1709	0.1695	0.1680	0.1666
0.3	0.2592	0.2562	0.2532	0.2503	0.2473	0.2445	0.2417	0.2389	0.2361	0.2334	0.2307
0.4	0.3297	0.3250	0.3203	0.3158	0.3113	0.3068	0.3024	0.2981	0.2939	0.2897	0.2856
0.5	0.3935	0.3870	0.3807	0.3744	0.3682	0.3621	0.3561	0.3502	0.3444	0.3387	0.3331
0.6	0.4512	0.4430	0.4350	0.4270	0.4192	0.4114	0.4038	0.3963	0.3890	0.3817	0.3746
0.7	0.5034	0.4936	0.4839	0.4744	0.4649	0.4556	0.4465	0.4374	0.4285	0.4198	0.4112
0.8	0.5507	0.5394	0.5282	0.5172	0.5062	0.4954	0.4848	0.4743	0.4639	0.4537	0.4437
0.9	0.5934	0.5809	0.5683	0.5559	0.5436	0.5314	0.5193	0.5074	0.4957	0.4841	0.4728
1.0	0.6321	0.6184	0.6047	0.5911	0.5775	0.5640	0.5507	0.5374	0.5244	0.5115	0.4989
1.1	0.6671	0.6525	0.6378	0.6230	0.6083	0.5937	0.5791	0.5647	0.5504	0.5363	0.5224
1.2	0.6988	0.6834	0.6678	0.6522	0.6365	0.6208	0.6051	0.5896	0.5741	0.5589	0.5438
1.3	0.7275	0.7115	0.6952	0.6788	0.6622	0.6456	0.6289	0.6123	0.5958	0.5795	0.5633
1.4	0.7534	0.7369	0.7201	0.7031	0.6858	0.6683	0.6508	0.6332	0.6157	0.5984	0.5812
1.5	0.7769	0.7601	0.7429	0.7253	0.7074	0.6892	0.6709	0.6525	0.6341	0.6157	0.5975
1.6	0.7981	0.7812	0.7637	0.7457	0.7273	0.7085	0.6894	0.6702	0.6510	0.6317	0.6126
1.7	0.8173	0.8004	0.7827	0.7644	0.7456	0.7262	0.7066	0.6867	0.6666	0.6466	0.6266
1.8	0.8347	0.8178	0.8001	0.7816	0.7624	0.7427	0.7225	0.7019	0.6812	0.6603	0.6395
1.9	0.8504	0.8337	0.8161	0.7975	0.7780	0.7579	0.7372	0.7161	0.6947	0.6731	0.6515
2.0	0.8647	0.8482	0.8306	0.8120	0.7925	0.7721	0.7510	0.7293	0.7073	0.6850	0.6626
2.1	0.8775	0.8614	0.8440	0.8255	0.8058	0.7852	0.7637	0.7416	0.7190	0.6961	0.6730
2.2	0.8892	0.8734	0.8563	0.8378	0.8182	0.7974	0.7757	0.7531	0.7300	0.7065	0.6828
2.3	0.8997	0.8844	0.8676	0.8493	0.8296	0.8088	0.7868	0.7639	0.7403	0.7163	0.6919
2.4	0.9093	0.8944	0.8779	0.8598	0.8403	0.8194	0.7972	0.7740	0.7500	0.7254	0.7005
2.5	0.9179	0.9035	0.8874	0.8696	0.8502	0.8293	0.8070	0.7835	0.7591	0.7340	0.7085
2.6	0.9257	0.9118	0.8961	0.8786	0.8594	0.8385	0.8162	0.7925	0.7677	0.7422	0.7161
2.7	0.9328	0.9194	0.9041	0.8870	0.8680	0.8472	0.8248	0.8009	0.7758	0.7498	0.7233
2.8	0.9392	0.9263	0.9115	0.8947	0.8760	0.8553	0.8328	0.8088	0.7834	0.7571	0.7300
2.9	0.9450	0.9326	0.9183	0.9019	0.8834	0.8629	0.8404	0.8163	0.7907	0.7639	0.7364
3.0	0.9502	0.9384	0.9246	0.9086	0.8904	0.8700	0.8476	0.8233	0.7975	0.7704	0.7424
3.1	0.9550	0.9437	0.9303	0.9147	0.8969	0.8767	0.8544	0.8300	0.8040	0.7766	0.7482
3.2	0.9592	0.9485	0.9356	0.9205	0.9029	0.8830	0.8607	0.8363	0.8101	0.7824	0.7536
3.3	0.9631	0.9529	0.9405	0.9258	0.9086	0.8889	0.8667	0.8423	0.8159	0.7879	0.7588
3.4	0.9666	0.9569	0.9450	0.9307	0.9139	0.8944	0.8724	0.8480	0.8215	0.7932	0.7637
3.5	0.9698	0.9606	0.9491	0.9353	0.9188	0.8997	0.8778	0.8534	0.8268	0.7982	0.7684
3.6	0.9727	0.9639	0.9530	0.9396	0.9235	0.9046	0.8829	0.8585	0.8318	0.8030	0.7728
3.7	0.9753	0.9670	0.9565	0.9435	0.9278	0.9092	0.8877	0.8634	0.8366	0.8076	0.7771
3.8	0.9776	0.9698	0.9597	0.9472	0.9319	0.9136	0.8923	0.8680	0.8411	0.8120	0.7811
3.9	0.9798	0.9723	0.9627	0.9506	0.9357	0.9177	0.8966	0.8724	0.8455	0.8161	0.7850
4.0	0.9817	0.9747	0.9655	0.9538	0.9393	0.9216	0.9007	0.8766	0.8496	0.8201	0.7887
4.1	0.9834	0.9768	0.9681	0.9568	0.9427	0.9253	0.9046	0.8806	0.8536	0.8239	0.7923
4.2	0.9850	0.9788	0.9704	0.9596	0.9458	0.9288	0.9083	0.8844	0.8574	0.8276	0.7957
4.3	0.9864	0.9805	0.9726	0.9622	0.9488	0.9321	0.9118	0.8881	0.8610	0.8311	0.7989
4.4	0.9877	0.9822	0.9746	0.9646	0.9515	0.9352	0.9152	0.8916	0.8645	0.8344	0.8020
4.5	0.9889	0.9837	0.9765	0.9668	0.9542	0.9381	0.9184	0.8949	0.8678	0.8377	0.8050
4.6	0.9899	0.9850	0.9782	0.9689	0.9566	0.9409	0.9214	0.8981	0.8710	0.8407	0.8079
4.7	0.9909	0.9863	0.9798	0.9708	0.9589	0.9435	0.9243	0.9011	0.8741	0.8437	0.8107
4.8	0.9918	0.9874	0.9812	0.9727	0.9611	0.9460	0.9270	0.9040	0.8770	0.8466	0.8133
4.9	0.9926	0.9885	0.9826	0.9743	0.9631	0.9484	0.9297	0.9068	0.8798	0.8493	0.8159
5.0	0.9933	0.9894	0.9839	0.9759	0.9650	0.9506	0.9322	0.9094	0.8826	0.8519	0.8183
5.1	0.9939	0.9903	0.9850	0.9774	0.9669	0.9527	0.9345	0.9120	0.8852	0.8545	0.8207
5.2	0.9945	0.9911	0.9861	0.9788	0.9686	0.9548	0.9368	0.9144	0.8877	0.8569	0.8230
5.3	0.9950	0.9919	0.9871	0.9801	0.9702	0.9567	0.9390	0.9168	0.8901	0.8593	0.8252
5.4	0.9955	0.9925	0.9880	0.9813	0.9717	0.9585	0.9411	0.9190	0.8924	0.8615	0.8273
5.5	0.9959	0.9932	0.9889	0.9824	0.9731	0.9602	0.9430	0.9212	0.8946	0.8637	0.8293
5.6	0.9963	0.9937	0.9896	0.9835	0.9745	0.9618	0.9449	0.9233	0.8968	0.8658	0.8313
5.7	0.9967	0.9942	0.9904	0.9845	0.9757	0.9634	0.9467	0.9252	0.8988	0.8679	0.8332
5.8	0.9970	0.9947	0.9911	0.9854	0.9769	0.9649	0.9485	0.9272	0.9008	0.8698	0.8350
5.9	0.9973	0.9951	0.9917	0.9862	0.9781	0.9663	0.9501	0.9290	0.9027	0.8717	0.8368
6.0	0.9975	0.9955	0.9923	0.9870	0.9791	0.9676	0.9517	0.9308	0.9046	0.8736	0.8385

NTU	Heat Exchanger Effectiveness										
	C^* ($C_{min} = C_c$)										
	0	0.1	0.2	0.3	0.4	0.5	0.6	0.7	0.8	0.9	1.0
6.1	0.9978	0.9959	0.9928	0.9878	0.9801	0.9689	0.9532	0.9325	0.9064	0.8753	0.8401
6.2	0.9980	0.9962	0.9933	0.9885	0.9811	0.9701	0.9547	0.9341	0.9081	0.8770	0.8417
6.3	0.9982	0.9965	0.9938	0.9892	0.9820	0.9713	0.9561	0.9357	0.9098	0.8787	0.8433
6.4	0.9983	0.9968	0.9942	0.9898	0.9828	0.9724	0.9574	0.9372	0.9114	0.8803	0.8448
6.5	0.9985	0.9971	0.9946	0.9904	0.9836	0.9734	0.9587	0.9386	0.9129	0.8818	0.8462
6.6	0.9986	0.9973	0.9950	0.9909	0.9844	0.9744	0.9599	0.9400	0.9144	0.8833	0.8476
6.7	0.9988	0.9975	0.9953	0.9914	0.9851	0.9753	0.9611	0.9414	0.9159	0.8848	0.8490
6.8	0.9989	0.9977	0.9956	0.9919	0.9858	0.9762	0.9622	0.9427	0.9173	0.8862	0.8503
6.9	0.9990	0.9979	0.9959	0.9924	0.9864	0.9771	0.9633	0.9440	0.9187	0.8876	0.8516
7.0	0.9991	0.9981	0.9962	0.9928	0.9871	0.9779	0.9643	0.9452	0.9200	0.8889	0.8528
7.1	0.9992	0.9982	0.9964	0.9932	0.9876	0.9787	0.9653	0.9463	0.9212	0.8902	0.8540
7.2	0.9993	0.9984	0.9967	0.9936	0.9882	0.9795	0.9663	0.9475	0.9225	0.8914	0.8552
7.3	0.9993	0.9985	0.9969	0.9939	0.9887	0.9802	0.9672	0.9485	0.9237	0.8926	0.8563
7.4	0.9994	0.9986	0.9971	0.9942	0.9892	0.9809	0.9681	0.9496	0.9248	0.8938	0.8574
7.5	0.9994	0.9987	0.9973	0.9945	0.9897	0.9815	0.9689	0.9506	0.9259	0.8949	0.8584
7.6	0.9995	0.9988	0.9975	0.9948	0.9901	0.9822	0.9697	0.9516	0.9270	0.8960	0.8595
7.7	0.9995	0.9989	0.9976	0.9951	0.9905	0.9828	0.9705	0.9525	0.9281	0.8971	0.8605
7.8	0.9996	0.9990	0.9978	0.9954	0.9909	0.9833	0.9713	0.9535	0.9291	0.8981	0.8614
7.9	0.9996	0.9991	0.9979	0.9956	0.9913	0.9839	0.9720	0.9543	0.9301	0.8991	0.8624
8.0	0.9997	0.9992	0.9981	0.9958	0.9917	0.9844	0.9727	0.9552	0.9310	0.9001	0.8633
8.1	0.9997	0.9992	0.9982	0.9960	0.9920	0.9849	0.9734	0.9560	0.9320	0.9010	0.8642
8.2	0.9997	0.9993	0.9983	0.9962	0.9923	0.9854	0.9740	0.9568	0.9328	0.9020	0.8651
8.3	0.9998	0.9993	0.9984	0.9964	0.9926	0.9858	0.9746	0.9576	0.9337	0.9029	0.8659
8.4	0.9998	0.9994	0.9985	0.9966	0.9929	0.9863	0.9752	0.9583	0.9346	0.9037	0.8667
8.5	0.9998	0.9994	0.9986	0.9968	0.9932	0.9867	0.9758	0.9591	0.9354	0.9046	0.8675
8.6	0.9998	0.9995	0.9987	0.9969	0.9935	0.9871	0.9764	0.9598	0.9362	0.9054	0.8683
8.7	0.9998	0.9995	0.9988	0.9971	0.9937	0.9875	0.9769	0.9604	0.9370	0.9062	0.8690
8.8	0.9998	0.9996	0.9988	0.9972	0.9940	0.9878	0.9774	0.9611	0.9377	0.9070	0.8698
8.9	0.9999	0.9996	0.9989	0.9974	0.9942	0.9882	0.9779	0.9617	0.9384	0.9077	0.8705
9.0	0.9999	0.9996	0.9990	0.9975	0.9944	0.9885	0.9784	0.9624	0.9392	0.9085	0.8712
9.1	0.9999	0.9997	0.9990	0.9976	0.9946	0.9889	0.9789	0.9630	0.9398	0.9092	0.8719
9.2	0.9999	0.9997	0.9991	0.9977	0.9948	0.9892	0.9793	0.9635	0.9405	0.9099	0.8725
9.3	0.9999	0.9997	0.9991	0.9978	0.9950	0.9895	0.9798	0.9641	0.9412	0.9106	0.8732
9.4	0.9999	0.9997	0.9992	0.9979	0.9952	0.9898	0.9802	0.9646	0.9418	0.9112	0.8738
9.5	0.9999	0.9997	0.9992	0.9980	0.9953	0.9901	0.9806	0.9652	0.9424	0.9119	0.8744
9.6	0.9999	0.9998	0.9993	0.9981	0.9955	0.9903	0.9810	0.9657	0.9430	0.9125	0.8750
9.7	0.9999	0.9998	0.9993	0.9982	0.9957	0.9906	0.9813	0.9662	0.9436	0.9131	0.8756
9.8	0.9999	0.9998	0.9994	0.9983	0.9958	0.9908	0.9817	0.9667	0.9442	0.9137	0.8761
9.9	1.0000	0.9998	0.9994	0.9984	0.9960	0.9911	0.9821	0.9671	0.9447	0.9143	0.8767
10.0	1.0000	0.9998	0.9994	0.9984	0.9961	0.9913	0.9824	0.9676	0.9453	0.9149	0.8772

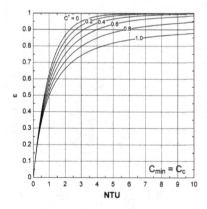

CONFIGURATION 24

1TUBE6ROW6PASS1CIR(CC)

(b) $C_{min} = C_h$

NTU	Heat Exchanger Effectiveness										
	$C^* (C_{min} = C_h)$										
	0	0.1	0.2	0.3	0.4	0.5	0.6	0.7	0.8	0.9	1.0
0	0.0000	0.0000	0.0000	0.0000	0.0000	0.0000	0.0000	0.0000	0.0000	0.0000	0.0000
0.1	0.0952	0.0947	0.0943	0.0939	0.0934	0.0930	0.0926	0.0922	0.0917	0.0913	0.0909
0.2	0.1813	0.1797	0.1782	0.1767	0.1752	0.1738	0.1723	0.1709	0.1695	0.1680	0.1666
0.3	0.2592	0.2562	0.2532	0.2503	0.2473	0.2445	0.2417	0.2389	0.2361	0.2334	0.2307
0.4	0.3297	0.3250	0.3203	0.3158	0.3113	0.3068	0.3024	0.2981	0.2939	0.2897	0.2856
0.5	0.3935	0.3870	0.3807	0.3744	0.3682	0.3621	0.3561	0.3502	0.3444	0.3387	0.3331
0.6	0.4512	0.4430	0.4350	0.4270	0.4192	0.4114	0.4038	0.3963	0.3890	0.3817	0.3746
0.7	0.5034	0.4936	0.4839	0.4744	0.4649	0.4556	0.4465	0.4374	0.4285	0.4198	0.4112
0.8	0.5507	0.5394	0.5282	0.5172	0.5062	0.4954	0.4848	0.4743	0.4639	0.4537	0.4437
0.9	0.5934	0.5809	0.5683	0.5559	0.5436	0.5314	0.5193	0.5074	0.4957	0.4841	0.4728
1.0	0.6321	0.6184	0.6047	0.5911	0.5775	0.5640	0.5507	0.5374	0.5244	0.5115	0.4989
1.1	0.6671	0.6525	0.6378	0.6230	0.6083	0.5937	0.5791	0.5647	0.5504	0.5363	0.5224
1.2	0.6988	0.6834	0.6678	0.6522	0.6365	0.6208	0.6051	0.5896	0.5741	0.5589	0.5438
1.3	0.7275	0.7115	0.6952	0.6788	0.6622	0.6456	0.6289	0.6123	0.5958	0.5795	0.5633
1.4	0.7534	0.7370	0.7202	0.7031	0.6858	0.6683	0.6508	0.6332	0.6157	0.5984	0.5812
1.5	0.7769	0.7601	0.7429	0.7253	0.7074	0.6892	0.6709	0.6525	0.6341	0.6157	0.5975
1.6	0.7981	0.7812	0.7637	0.7457	0.7273	0.7085	0.6894	0.6702	0.6510	0.6317	0.6126
1.7	0.8173	0.8004	0.7827	0.7644	0.7456	0.7263	0.7066	0.6867	0.6666	0.6466	0.6266
1.8	0.8347	0.8179	0.8001	0.7816	0.7625	0.7427	0.7225	0.7019	0.6812	0.6603	0.6395
1.9	0.8504	0.8337	0.8161	0.7975	0.7781	0.7580	0.7373	0.7161	0.6947	0.6731	0.6515
2.0	0.8647	0.8482	0.8307	0.8121	0.7925	0.7721	0.7510	0.7293	0.7073	0.6850	0.6626
2.1	0.8775	0.8614	0.8440	0.8255	0.8058	0.7852	0.7638	0.7417	0.7190	0.6961	0.6730
2.2	0.8892	0.8734	0.8563	0.8379	0.8182	0.7974	0.7757	0.7532	0.7301	0.7065	0.6828
2.3	0.8997	0.8844	0.8676	0.8493	0.8297	0.8088	0.7869	0.7640	0.7404	0.7163	0.6919
2.4	0.9093	0.8944	0.8779	0.8599	0.8403	0.8194	0.7973	0.7741	0.7501	0.7255	0.7005
2.5	0.9179	0.9035	0.8874	0.8697	0.8503	0.8293	0.8071	0.7836	0.7592	0.7341	0.7085
2.6	0.9257	0.9118	0.8962	0.8787	0.8595	0.8386	0.8162	0.7925	0.7678	0.7422	0.7161
2.7	0.9328	0.9194	0.9042	0.8871	0.8681	0.8473	0.8248	0.8009	0.7759	0.7499	0.7233
2.8	0.9392	0.9263	0.9116	0.8948	0.8761	0.8554	0.8329	0.8089	0.7835	0.7571	0.7300
2.9	0.9450	0.9327	0.9184	0.9020	0.8835	0.8630	0.8405	0.8164	0.7907	0.7640	0.7364
3.0	0.9502	0.9384	0.9246	0.9086	0.8905	0.8701	0.8477	0.8234	0.7976	0.7704	0.7424
3.1	0.9550	0.9437	0.9304	0.9148	0.8970	0.8768	0.8545	0.8301	0.8041	0.7766	0.7482
3.2	0.9592	0.9485	0.9357	0.9206	0.9030	0.8831	0.8609	0.8365	0.8102	0.7824	0.7536
3.3	0.9631	0.9529	0.9406	0.9259	0.9087	0.8890	0.8669	0.8425	0.8161	0.7880	0.7588
3.4	0.9666	0.9569	0.9451	0.9308	0.9140	0.8946	0.8726	0.8482	0.8216	0.7933	0.7637
3.5	0.9698	0.9606	0.9492	0.9354	0.9190	0.8998	0.8780	0.8536	0.8269	0.7983	0.7684
3.6	0.9727	0.9639	0.9530	0.9397	0.9236	0.9048	0.8831	0.8587	0.8319	0.8031	0.7728
3.7	0.9753	0.9670	0.9566	0.9437	0.9280	0.9094	0.8879	0.8636	0.8367	0.8077	0.7771
3.8	0.9776	0.9698	0.9598	0.9474	0.9321	0.9138	0.8925	0.8682	0.8413	0.8121	0.7811
3.9	0.9798	0.9724	0.9628	0.9508	0.9359	0.9180	0.8969	0.8727	0.8456	0.8162	0.7850
4.0	0.9817	0.9747	0.9656	0.9540	0.9395	0.9219	0.9010	0.8769	0.8498	0.8202	0.7887
4.1	0.9834	0.9768	0.9682	0.9570	0.9429	0.9256	0.9049	0.8809	0.8538	0.8240	0.7923
4.2	0.9850	0.9788	0.9705	0.9597	0.9460	0.9291	0.9086	0.8847	0.8576	0.8277	0.7957
4.3	0.9864	0.9806	0.9727	0.9623	0.9490	0.9324	0.9122	0.8884	0.8613	0.8312	0.7989
4.4	0.9877	0.9822	0.9747	0.9647	0.9518	0.9355	0.9155	0.8919	0.8647	0.8346	0.8020
4.5	0.9889	0.9837	0.9766	0.9670	0.9544	0.9384	0.9187	0.8952	0.8681	0.8378	0.8050
4.6	0.9899	0.9851	0.9783	0.9691	0.9569	0.9412	0.9218	0.8984	0.8713	0.8409	0.8079
4.7	0.9909	0.9863	0.9799	0.9710	0.9592	0.9439	0.9247	0.9015	0.8744	0.8439	0.8107
4.8	0.9918	0.9875	0.9814	0.9729	0.9614	0.9464	0.9275	0.9044	0.8773	0.8467	0.8133
4.9	0.9926	0.9885	0.9827	0.9746	0.9634	0.9488	0.9301	0.9072	0.8802	0.8495	0.8159
5.0	0.9933	0.9895	0.9840	0.9762	0.9654	0.9510	0.9326	0.9099	0.8829	0.8521	0.8183
5.1	0.9939	0.9904	0.9851	0.9776	0.9672	0.9532	0.9350	0.9125	0.8855	0.8547	0.8207
5.2	0.9945	0.9912	0.9862	0.9790	0.9689	0.9552	0.9373	0.9149	0.8881	0.8571	0.8230
5.3	0.9950	0.9919	0.9872	0.9803	0.9705	0.9571	0.9395	0.9173	0.8905	0.8595	0.8252
5.4	0.9955	0.9926	0.9881	0.9815	0.9720	0.9590	0.9416	0.9196	0.8928	0.8618	0.8273
5.5	0.9959	0.9932	0.9890	0.9827	0.9735	0.9607	0.9436	0.9218	0.8951	0.8640	0.8293
5.6	0.9963	0.9938	0.9898	0.9837	0.9748	0.9623	0.9455	0.9238	0.8972	0.8661	0.8313
5.7	0.9967	0.9943	0.9905	0.9847	0.9761	0.9639	0.9473	0.9259	0.8993	0.8682	0.8332
5.8	0.9970	0.9948	0.9912	0.9856	0.9773	0.9654	0.9491	0.9278	0.9014	0.8701	0.8350
5.9	0.9973	0.9952	0.9918	0.9865	0.9785	0.9668	0.9508	0.9297	0.9033	0.8720	0.8368
6.0	0.9975	0.9956	0.9924	0.9873	0.9795	0.9682	0.9524	0.9315	0.9052	0.8739	0.8385

NTU	Heat Exchanger Effectiveness										
	C^* ($C_{min} = C_h$)										
	0	0.1	0.2	0.3	0.4	0.5	0.6	0.7	0.8	0.9	1.0
6.1	0.9978	0.9960	0.9929	0.9881	0.9806	0.9695	0.9539	0.9332	0.9070	0.8757	0.8401
6.2	0.9980	0.9963	0.9934	0.9888	0.9815	0.9707	0.9554	0.9348	0.9087	0.8774	0.8417
6.3	0.9982	0.9966	0.9939	0.9894	0.9824	0.9719	0.9568	0.9364	0.9104	0.8791	0.8433
6.4	0.9983	0.9969	0.9943	0.9901	0.9833	0.9730	0.9582	0.9380	0.9121	0.8807	0.8448
6.5	0.9985	0.9971	0.9947	0.9906	0.9841	0.9741	0.9595	0.9395	0.9136	0.8822	0.8462
6.6	0.9986	0.9974	0.9951	0.9912	0.9849	0.9751	0.9607	0.9409	0.9152	0.8838	0.8476
6.7	0.9988	0.9976	0.9954	0.9917	0.9856	0.9760	0.9619	0.9423	0.9167	0.8852	0.8490
6.8	0.9989	0.9978	0.9957	0.9922	0.9863	0.9770	0.9631	0.9436	0.9181	0.8867	0.8503
6.9	0.9990	0.9980	0.9960	0.9926	0.9869	0.9778	0.9642	0.9449	0.9195	0.8880	0.8516
7.0	0.9991	0.9981	0.9963	0.9931	0.9875	0.9787	0.9652	0.9461	0.9208	0.8894	0.8528
7.1	0.9992	0.9983	0.9966	0.9935	0.9881	0.9795	0.9662	0.9473	0.9221	0.8907	0.8540
7.2	0.9993	0.9984	0.9968	0.9938	0.9887	0.9802	0.9672	0.9485	0.9234	0.8919	0.8552
7.3	0.9993	0.9985	0.9970	0.9942	0.9892	0.9810	0.9682	0.9496	0.9246	0.8931	0.8563
7.4	0.9994	0.9987	0.9972	0.9945	0.9897	0.9817	0.9691	0.9507	0.9258	0.8943	0.8574
7.5	0.9994	0.9988	0.9974	0.9948	0.9902	0.9823	0.9699	0.9517	0.9269	0.8955	0.8584
7.6	0.9995	0.9989	0.9976	0.9951	0.9906	0.9830	0.9708	0.9528	0.9280	0.8966	0.8595
7.7	0.9995	0.9990	0.9978	0.9954	0.9910	0.9836	0.9716	0.9537	0.9291	0.8977	0.8605
7.8	0.9996	0.9990	0.9979	0.9956	0.9915	0.9842	0.9724	0.9547	0.9301	0.8987	0.8614
7.9	0.9996	0.9991	0.9980	0.9959	0.9918	0.9847	0.9731	0.9556	0.9312	0.8998	0.8624
8.0	0.9997	0.9992	0.9982	0.9961	0.9922	0.9853	0.9738	0.9565	0.9321	0.9008	0.8633
8.1	0.9997	0.9993	0.9983	0.9963	0.9925	0.9858	0.9745	0.9573	0.9331	0.9017	0.8642
8.2	0.9997	0.9993	0.9984	0.9965	0.9929	0.9863	0.9752	0.9582	0.9340	0.9027	0.8651
8.3	0.9998	0.9994	0.9985	0.9967	0.9932	0.9867	0.9759	0.9590	0.9349	0.9036	0.8659
8.4	0.9998	0.9994	0.9986	0.9969	0.9935	0.9872	0.9765	0.9597	0.9358	0.9045	0.8667
8.5	0.9998	0.9995	0.9987	0.9971	0.9938	0.9876	0.9771	0.9605	0.9367	0.9053	0.8675
8.6	0.9998	0.9995	0.9988	0.9972	0.9940	0.9880	0.9777	0.9612	0.9375	0.9062	0.8683
8.7	0.9998	0.9996	0.9989	0.9974	0.9943	0.9884	0.9782	0.9619	0.9383	0.9070	0.8690
8.8	0.9998	0.9996	0.9989	0.9975	0.9945	0.9888	0.9788	0.9626	0.9391	0.9078	0.8698
8.9	0.9999	0.9996	0.9990	0.9976	0.9948	0.9892	0.9793	0.9633	0.9398	0.9086	0.8705
9.0	0.9999	0.9996	0.9991	0.9978	0.9950	0.9895	0.9798	0.9639	0.9406	0.9093	0.8712
9.1	0.9999	0.9997	0.9991	0.9979	0.9952	0.9899	0.9803	0.9646	0.9413	0.9101	0.8719
9.2	0.9999	0.9997	0.9992	0.9980	0.9954	0.9902	0.9807	0.9652	0.9420	0.9108	0.8725
9.3	0.9999	0.9997	0.9992	0.9981	0.9956	0.9905	0.9812	0.9658	0.9427	0.9115	0.8732
9.4	0.9999	0.9997	0.9993	0.9982	0.9957	0.9908	0.9816	0.9663	0.9434	0.9122	0.8738
9.5	0.9999	0.9998	0.9993	0.9983	0.9959	0.9911	0.9821	0.9669	0.9440	0.9128	0.8744
9.6	0.9999	0.9998	0.9994	0.9984	0.9961	0.9913	0.9825	0.9674	0.9446	0.9135	0.8750
9.7	0.9999	0.9998	0.9994	0.9985	0.9962	0.9916	0.9829	0.9680	0.9452	0.9141	0.8756
9.8	0.9999	0.9998	0.9995	0.9985	0.9964	0.9919	0.9832	0.9685	0.9458	0.9147	0.8761
9.9	1.0000	0.9998	0.9995	0.9986	0.9965	0.9921	0.9836	0.9690	0.9464	0.9153	0.8767
10.0	1.0000	0.9998	0.9995	0.9987	0.9967	0.9923	0.9840	0.9695	0.9470	0.9159	0.8772

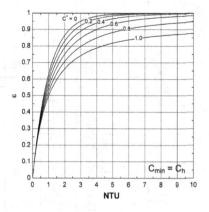

CONFIGURATION 25

1TUBE10ROW10PASS1CIR(CC)

(a) $C_{min} = C_c$

NTU	Heat Exchanger Effectiveness										
	C^* $(C_{min} = C_c)$										
	0	0.1	0.2	0.3	0.4	0.5	0.6	0.7	0.8	0.9	1.0
0	0.0000	0.0000	0.0000	0.0000	0.0000	0.0000	0.0000	0.0000	0.0000	0.0000	0.0000
0.1	0.0952	0.0947	0.0943	0.0939	0.0934	0.0930	0.0926	0.0922	0.0917	0.0913	0.0909
0.2	0.1813	0.1797	0.1782	0.1767	0.1753	0.1738	0.1723	0.1709	0.1695	0.1681	0.1667
0.3	0.2592	0.2562	0.2532	0.2503	0.2474	0.2445	0.2417	0.2389	0.2361	0.2334	0.2307
0.4	0.3297	0.3250	0.3204	0.3158	0.3113	0.3069	0.3025	0.2982	0.2940	0.2898	0.2857
0.5	0.3935	0.3870	0.3807	0.3744	0.3683	0.3622	0.3562	0.3503	0.3446	0.3388	0.3332
0.6	0.4512	0.4431	0.4350	0.4271	0.4193	0.4116	0.4040	0.3965	0.3892	0.3820	0.3749
0.7	0.5034	0.4937	0.4840	0.4745	0.4651	0.4558	0.4467	0.4377	0.4288	0.4201	0.4116
0.8	0.5507	0.5395	0.5284	0.5174	0.5065	0.4957	0.4851	0.4746	0.4643	0.4542	0.4442
0.9	0.5934	0.5809	0.5685	0.5562	0.5439	0.5318	0.5198	0.5079	0.4962	0.4847	0.4733
1.0	0.6321	0.6185	0.6049	0.5914	0.5779	0.5645	0.5512	0.5380	0.5250	0.5122	0.4996
1.1	0.6671	0.6526	0.6380	0.6234	0.6088	0.5943	0.5798	0.5654	0.5512	0.5372	0.5233
1.2	0.6988	0.6835	0.6681	0.6526	0.6370	0.6214	0.6059	0.5904	0.5751	0.5599	0.5449
1.3	0.7275	0.7116	0.6955	0.6793	0.6628	0.6463	0.6298	0.6133	0.5969	0.5806	0.5645
1.4	0.7534	0.7371	0.7205	0.7036	0.6865	0.6692	0.6518	0.6344	0.6170	0.5997	0.5825
1.5	0.7769	0.7604	0.7434	0.7260	0.7082	0.6902	0.6720	0.6538	0.6355	0.6172	0.5991
1.6	0.7981	0.7815	0.7642	0.7464	0.7282	0.7096	0.6907	0.6717	0.6526	0.6334	0.6144
1.7	0.8173	0.8007	0.7833	0.7652	0.7466	0.7275	0.7080	0.6883	0.6684	0.6484	0.6285
1.8	0.8347	0.8181	0.8007	0.7825	0.7636	0.7441	0.7241	0.7037	0.6831	0.6624	0.6416
1.9	0.8504	0.8340	0.8167	0.7984	0.7793	0.7594	0.7390	0.7181	0.6968	0.6753	0.6538
2.0	0.8647	0.8485	0.8313	0.8130	0.7938	0.7737	0.7529	0.7314	0.7096	0.6875	0.6652
2.1	0.8775	0.8618	0.8447	0.8265	0.8072	0.7869	0.7658	0.7439	0.7215	0.6988	0.6758
2.2	0.8892	0.8738	0.8570	0.8390	0.8197	0.7993	0.7779	0.7556	0.7327	0.7094	0.6858
2.3	0.8997	0.8848	0.8683	0.8505	0.8313	0.8108	0.7892	0.7666	0.7433	0.7194	0.6951
2.4	0.9093	0.8948	0.8787	0.8611	0.8420	0.8215	0.7997	0.7769	0.7531	0.7287	0.7039
2.5	0.9179	0.9039	0.8882	0.8709	0.8520	0.8315	0.8097	0.7866	0.7625	0.7376	0.7122
2.6	0.9257	0.9122	0.8970	0.8800	0.8613	0.8409	0.8190	0.7957	0.7712	0.7459	0.7200
2.7	0.9328	0.9198	0.9051	0.8884	0.8700	0.8497	0.8277	0.8043	0.7795	0.7538	0.7274
2.8	0.9392	0.9268	0.9125	0.8962	0.8780	0.8579	0.8360	0.8124	0.7874	0.7613	0.7343
2.9	0.9450	0.9331	0.9193	0.9034	0.8855	0.8656	0.8437	0.8200	0.7948	0.7683	0.7409
3.0	0.9502	0.9389	0.9255	0.9101	0.8926	0.8728	0.8510	0.8273	0.8018	0.7750	0.7472
3.1	0.9550	0.9441	0.9313	0.9163	0.8991	0.8796	0.8579	0.8341	0.8085	0.7814	0.7532
3.2	0.9592	0.9489	0.9366	0.9221	0.9053	0.8860	0.8644	0.8406	0.8149	0.7875	0.7588
3.3	0.9631	0.9533	0.9415	0.9275	0.9110	0.8920	0.8706	0.8468	0.8209	0.7932	0.7642
3.4	0.9666	0.9574	0.9460	0.9324	0.9163	0.8977	0.8764	0.8526	0.8266	0.7987	0.7694
3.5	0.9698	0.9610	0.9502	0.9370	0.9213	0.9030	0.8819	0.8582	0.8321	0.8040	0.7743
3.6	0.9727	0.9644	0.9540	0.9413	0.9260	0.9080	0.8871	0.8635	0.8373	0.8090	0.7790
3.7	0.9753	0.9674	0.9575	0.9453	0.9304	0.9127	0.8921	0.8685	0.8423	0.8138	0.7835
3.8	0.9776	0.9702	0.9608	0.9490	0.9345	0.9172	0.8968	0.8733	0.8471	0.8184	0.7878
3.9	0.9798	0.9728	0.9638	0.9524	0.9384	0.9214	0.9012	0.8779	0.8516	0.8228	0.7919
4.0	0.9817	0.9751	0.9666	0.9556	0.9420	0.9253	0.9054	0.8823	0.8560	0.8270	0.7958
4.1	0.9834	0.9772	0.9691	0.9586	0.9454	0.9291	0.9095	0.8864	0.8602	0.8310	0.7996
4.2	0.9850	0.9792	0.9715	0.9614	0.9486	0.9326	0.9133	0.8904	0.8642	0.8349	0.8032
4.3	0.9864	0.9810	0.9736	0.9640	0.9516	0.9360	0.9169	0.8942	0.8680	0.8386	0.8067
4.4	0.9877	0.9826	0.9756	0.9664	0.9544	0.9391	0.9203	0.8978	0.8717	0.8422	0.8100
4.5	0.9889	0.9841	0.9775	0.9686	0.9570	0.9421	0.9236	0.9013	0.8752	0.8457	0.8133
4.6	0.9899	0.9854	0.9792	0.9707	0.9595	0.9450	0.9268	0.9046	0.8786	0.8490	0.8164
4.7	0.9909	0.9867	0.9808	0.9726	0.9618	0.9476	0.9297	0.9078	0.8818	0.8521	0.8194
4.8	0.9918	0.9878	0.9822	0.9745	0.9640	0.9502	0.9326	0.9109	0.8850	0.8552	0.8222
4.9	0.9926	0.9889	0.9836	0.9761	0.9660	0.9526	0.9353	0.9138	0.8880	0.8582	0.8250
5.0	0.9933	0.9898	0.9848	0.9777	0.9679	0.9549	0.9379	0.9166	0.8909	0.8610	0.8277
5.1	0.9939	0.9907	0.9859	0.9792	0.9698	0.9570	0.9403	0.9193	0.8937	0.8638	0.8303
5.2	0.9945	0.9915	0.9870	0.9805	0.9715	0.9591	0.9427	0.9218	0.8964	0.8664	0.8328
5.3	0.9950	0.9922	0.9880	0.9818	0.9731	0.9610	0.9449	0.9243	0.8989	0.8690	0.8352
5.4	0.9955	0.9929	0.9889	0.9830	0.9746	0.9628	0.9471	0.9267	0.9014	0.8715	0.8375
5.5	0.9959	0.9935	0.9897	0.9841	0.9760	0.9646	0.9491	0.9290	0.9039	0.8739	0.8398
5.6	0.9963	0.9940	0.9905	0.9851	0.9773	0.9662	0.9511	0.9312	0.9062	0.8762	0.8420
5.7	0.9967	0.9945	0.9912	0.9861	0.9786	0.9678	0.9530	0.9333	0.9084	0.8785	0.8441
5.8	0.9970	0.9950	0.9918	0.9870	0.9798	0.9693	0.9547	0.9353	0.9106	0.8806	0.8461
5.9	0.9973	0.9954	0.9925	0.9878	0.9809	0.9707	0.9565	0.9373	0.9127	0.8827	0.8481
6.0	0.9975	0.9958	0.9930	0.9886	0.9819	0.9721	0.9581	0.9391	0.9147	0.8848	0.8500

NTU	Heat Exchanger Effectiveness										
	$C^* (C_{min} = C_c)$										
	0	0.1	0.2	0.3	0.4	0.5	0.6	0.7	0.8	0.9	1.0
6.1	0.9978	0.9962	0.9935	0.9894	0.9829	0.9734	0.9597	0.9410	0.9167	0.8868	0.8519
6.2	0.9980	0.9965	0.9940	0.9900	0.9839	0.9746	0.9612	0.9427	0.9186	0.8887	0.8537
6.3	0.9982	0.9968	0.9945	0.9907	0.9847	0.9757	0.9626	0.9444	0.9204	0.8905	0.8555
6.4	0.9983	0.9970	0.9949	0.9913	0.9856	0.9768	0.9640	0.9460	0.9222	0.8923	0.8572
6.5	0.9985	0.9973	0.9952	0.9918	0.9864	0.9779	0.9653	0.9475	0.9239	0.8941	0.8588
6.6	0.9986	0.9975	0.9956	0.9924	0.9871	0.9789	0.9665	0.9490	0.9255	0.8958	0.8604
6.7	0.9988	0.9977	0.9959	0.9928	0.9878	0.9798	0.9678	0.9505	0.9271	0.8974	0.8620
6.8	0.9989	0.9979	0.9962	0.9933	0.9884	0.9807	0.9689	0.9519	0.9287	0.8990	0.8635
6.9	0.9990	0.9981	0.9965	0.9937	0.9891	0.9816	0.9700	0.9532	0.9302	0.9006	0.8650
7.0	0.9991	0.9983	0.9968	0.9941	0.9896	0.9824	0.9711	0.9545	0.9317	0.9021	0.8664
7.1	0.9992	0.9984	0.9970	0.9945	0.9902	0.9831	0.9721	0.9558	0.9331	0.9036	0.8678
7.2	0.9993	0.9985	0.9972	0.9948	0.9907	0.9839	0.9731	0.9570	0.9345	0.9050	0.8692
7.3	0.9993	0.9987	0.9974	0.9952	0.9912	0.9846	0.9740	0.9582	0.9358	0.9064	0.8705
7.4	0.9994	0.9988	0.9976	0.9955	0.9917	0.9853	0.9749	0.9593	0.9371	0.9078	0.8718
7.5	0.9994	0.9989	0.9978	0.9957	0.9921	0.9859	0.9758	0.9604	0.9383	0.9091	0.8731
7.6	0.9995	0.9990	0.9979	0.9960	0.9925	0.9865	0.9766	0.9614	0.9396	0.9104	0.8743
7.7	0.9995	0.9991	0.9981	0.9963	0.9929	0.9871	0.9774	0.9625	0.9408	0.9116	0.8755
7.8	0.9996	0.9991	0.9982	0.9965	0.9933	0.9876	0.9782	0.9634	0.9419	0.9128	0.8767
7.9	0.9996	0.9992	0.9984	0.9967	0.9936	0.9882	0.9789	0.9644	0.9430	0.9140	0.8778
8.0	0.9997	0.9993	0.9985	0.9969	0.9940	0.9887	0.9796	0.9653	0.9441	0.9152	0.8789
8.1	0.9997	0.9993	0.9986	0.9971	0.9943	0.9891	0.9803	0.9662	0.9452	0.9163	0.8800
8.2	0.9997	0.9994	0.9987	0.9973	0.9946	0.9896	0.9810	0.9671	0.9462	0.9174	0.8810
8.3	0.9998	0.9994	0.9988	0.9974	0.9948	0.9900	0.9816	0.9679	0.9472	0.9185	0.8820
8.4	0.9998	0.9995	0.9989	0.9976	0.9951	0.9904	0.9822	0.9687	0.9482	0.9195	0.8830
8.5	0.9998	0.9995	0.9990	0.9977	0.9954	0.9908	0.9828	0.9695	0.9491	0.9205	0.8840
8.6	0.9998	0.9996	0.9990	0.9979	0.9956	0.9912	0.9834	0.9702	0.9500	0.9215	0.8850
8.7	0.9998	0.9996	0.9991	0.9980	0.9958	0.9916	0.9839	0.9710	0.9509	0.9225	0.8859
8.8	0.9998	0.9996	0.9992	0.9981	0.9960	0.9919	0.9844	0.9717	0.9518	0.9235	0.8868
8.9	0.9999	0.9997	0.9992	0.9982	0.9962	0.9922	0.9849	0.9724	0.9526	0.9244	0.8877
9.0	0.9999	0.9997	0.9993	0.9984	0.9964	0.9926	0.9854	0.9730	0.9535	0.9253	0.8886
9.1	0.9999	0.9997	0.9993	0.9984	0.9966	0.9929	0.9859	0.9737	0.9543	0.9262	0.8894
9.2	0.9999	0.9997	0.9994	0.9985	0.9968	0.9931	0.9863	0.9743	0.9550	0.9270	0.8903
9.3	0.9999	0.9998	0.9994	0.9986	0.9969	0.9934	0.9867	0.9749	0.9558	0.9279	0.8911
9.4	0.9999	0.9998	0.9995	0.9987	0.9971	0.9937	0.9871	0.9755	0.9565	0.9287	0.8919
9.5	0.9999	0.9998	0.9995	0.9988	0.9972	0.9939	0.9875	0.9761	0.9573	0.9295	0.8927
9.6	0.9999	0.9998	0.9995	0.9989	0.9973	0.9942	0.9879	0.9766	0.9580	0.9303	0.8934
9.7	0.9999	0.9998	0.9996	0.9989	0.9975	0.9944	0.9883	0.9771	0.9587	0.9310	0.8942
9.8	0.9999	0.9998	0.9996	0.9990	0.9976	0.9946	0.9886	0.9777	0.9593	0.9318	0.8949
9.9	1.0000	0.9999	0.9996	0.9990	0.9977	0.9948	0.9890	0.9782	0.9600	0.9325	0.8956
10.0	1.0000	0.9999	0.9997	0.9991	0.9978	0.9950	0.9893	0.9787	0.9606	0.9332	0.8963

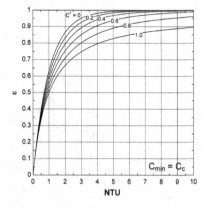

CONFIGURATION 25

1TUBE10ROW10PASS1CIR(CC)

(b) $C_{min} = C_h$

NTU	Heat Exchanger Effectiveness										
	C^* $(C_{min} = C_h)$										
	0	0.1	0.2	0.3	0.4	0.5	0.6	0.7	0.8	0.9	1.0
0	0.0000	0.0000	0.0000	0.0000	0.0000	0.0000	0.0000	0.0000	0.0000	0.0000	0.0000
0.1	0.0952	0.0947	0.0943	0.0939	0.0934	0.0930	0.0926	0.0922	0.0917	0.0913	0.0909
0.2	0.1813	0.1797	0.1782	0.1767	0.1753	0.1738	0.1723	0.1709	0.1695	0.1681	0.1667
0.3	0.2592	0.2562	0.2532	0.2503	0.2474	0.2445	0.2417	0.2389	0.2361	0.2334	0.2307
0.4	0.3297	0.3250	0.3204	0.3158	0.3113	0.3069	0.3025	0.2982	0.2940	0.2898	0.2857
0.5	0.3935	0.3870	0.3807	0.3744	0.3683	0.3622	0.3562	0.3503	0.3446	0.3388	0.3332
0.6	0.4512	0.4431	0.4350	0.4271	0.4193	0.4116	0.4040	0.3965	0.3892	0.3820	0.3749
0.7	0.5034	0.4937	0.4840	0.4745	0.4651	0.4558	0.4467	0.4377	0.4288	0.4201	0.4116
0.8	0.5507	0.5395	0.5284	0.5174	0.5065	0.4957	0.4851	0.4746	0.4643	0.4542	0.4442
0.9	0.5934	0.5809	0.5685	0.5562	0.5439	0.5318	0.5198	0.5079	0.4962	0.4847	0.4733
1.0	0.6321	0.6185	0.6049	0.5914	0.5779	0.5645	0.5512	0.5380	0.5250	0.5122	0.4996
1.1	0.6671	0.6526	0.6380	0.6234	0.6088	0.5943	0.5798	0.5654	0.5512	0.5372	0.5233
1.2	0.6988	0.6835	0.6681	0.6526	0.6370	0.6214	0.6059	0.5904	0.5751	0.5599	0.5449
1.3	0.7275	0.7116	0.6955	0.6793	0.6628	0.6463	0.6298	0.6133	0.5969	0.5806	0.5645
1.4	0.7534	0.7371	0.7205	0.7036	0.6865	0.6692	0.6518	0.6344	0.6170	0.5997	0.5825
1.5	0.7769	0.7604	0.7434	0.7260	0.7082	0.6902	0.6720	0.6538	0.6355	0.6172	0.5991
1.6	0.7981	0.7815	0.7642	0.7464	0.7282	0.7096	0.6907	0.6717	0.6526	0.6334	0.6144
1.7	0.8173	0.8007	0.7833	0.7652	0.7466	0.7275	0.7080	0.6883	0.6684	0.6484	0.6285
1.8	0.8347	0.8181	0.8007	0.7825	0.7636	0.7441	0.7241	0.7037	0.6831	0.6624	0.6416
1.9	0.8504	0.8340	0.8167	0.7984	0.7793	0.7594	0.7390	0.7181	0.6968	0.6753	0.6538
2.0	0.8647	0.8485	0.8313	0.8130	0.7938	0.7737	0.7529	0.7314	0.7096	0.6875	0.6652
2.1	0.8775	0.8618	0.8447	0.8265	0.8072	0.7869	0.7658	0.7439	0.7215	0.6988	0.6758
2.2	0.8892	0.8738	0.8570	0.8390	0.8197	0.7993	0.7779	0.7556	0.7327	0.7094	0.6858
2.3	0.8997	0.8848	0.8683	0.8505	0.8313	0.8108	0.7892	0.7666	0.7433	0.7194	0.6951
2.4	0.9093	0.8948	0.8787	0.8611	0.8420	0.8215	0.7997	0.7769	0.7532	0.7287	0.7039
2.5	0.9179	0.9039	0.8882	0.8709	0.8520	0.8315	0.8097	0.7866	0.7625	0.7376	0.7122
2.6	0.9257	0.9122	0.8970	0.8800	0.8613	0.8409	0.8190	0.7957	0.7712	0.7459	0.7200
2.7	0.9328	0.9198	0.9051	0.8884	0.8700	0.8497	0.8277	0.8043	0.7795	0.7538	0.7274
2.8	0.9392	0.9268	0.9125	0.8962	0.8780	0.8579	0.8360	0.8124	0.7874	0.7613	0.7343
2.9	0.9450	0.9331	0.9193	0.9035	0.8856	0.8656	0.8437	0.8200	0.7948	0.7683	0.7409
3.0	0.9502	0.9389	0.9255	0.9101	0.8926	0.8729	0.8510	0.8273	0.8018	0.7750	0.7472
3.1	0.9550	0.9441	0.9313	0.9164	0.8991	0.8796	0.8579	0.8341	0.8085	0.7814	0.7532
3.2	0.9592	0.9489	0.9366	0.9221	0.9053	0.8860	0.8644	0.8406	0.8149	0.7875	0.7588
3.3	0.9631	0.9533	0.9415	0.9275	0.9110	0.8920	0.8706	0.8468	0.8209	0.7933	0.7642
3.4	0.9666	0.9574	0.9460	0.9324	0.9164	0.8977	0.8764	0.8527	0.8267	0.7988	0.7694
3.5	0.9698	0.9610	0.9502	0.9370	0.9214	0.9030	0.8819	0.8582	0.8321	0.8040	0.7743
3.6	0.9727	0.9644	0.9540	0.9413	0.9261	0.9080	0.8872	0.8635	0.8374	0.8090	0.7790
3.7	0.9753	0.9674	0.9575	0.9453	0.9305	0.9127	0.8921	0.8686	0.8424	0.8138	0.7835
3.8	0.9776	0.9702	0.9608	0.9490	0.9346	0.9172	0.8968	0.8734	0.8471	0.8184	0.7878
3.9	0.9798	0.9728	0.9638	0.9525	0.9384	0.9214	0.9012	0.8779	0.8517	0.8228	0.7919
4.0	0.9817	0.9751	0.9666	0.9557	0.9420	0.9254	0.9055	0.8823	0.8560	0.8270	0.7958
4.1	0.9834	0.9772	0.9691	0.9586	0.9454	0.9291	0.9095	0.8865	0.8602	0.8311	0.7996
4.2	0.9850	0.9792	0.9715	0.9614	0.9486	0.9327	0.9133	0.8904	0.8642	0.8349	0.8032
4.3	0.9864	0.9810	0.9736	0.9640	0.9516	0.9360	0.9169	0.8942	0.8680	0.8387	0.8067
4.4	0.9877	0.9826	0.9756	0.9664	0.9544	0.9392	0.9204	0.8979	0.8717	0.8422	0.8100
4.5	0.9889	0.9841	0.9775	0.9686	0.9570	0.9422	0.9237	0.9014	0.8752	0.8457	0.8133
4.6	0.9899	0.9854	0.9792	0.9707	0.9595	0.9450	0.9268	0.9047	0.8786	0.8490	0.8164
4.7	0.9909	0.9867	0.9808	0.9727	0.9618	0.9477	0.9298	0.9079	0.8819	0.8522	0.8194
4.8	0.9918	0.9878	0.9822	0.9745	0.9640	0.9502	0.9326	0.9109	0.8850	0.8552	0.8222
4.9	0.9926	0.9889	0.9836	0.9762	0.9661	0.9526	0.9354	0.9138	0.8880	0.8582	0.8250
5.0	0.9933	0.9898	0.9848	0.9777	0.9680	0.9549	0.9379	0.9166	0.8909	0.8611	0.8277
5.1	0.9939	0.9907	0.9860	0.9792	0.9698	0.9571	0.9404	0.9193	0.8937	0.8638	0.8303
5.2	0.9945	0.9915	0.9870	0.9806	0.9715	0.9591	0.9428	0.9219	0.8964	0.8665	0.8328
5.3	0.9950	0.9922	0.9880	0.9818	0.9731	0.9611	0.9450	0.9244	0.8990	0.8691	0.8352
5.4	0.9955	0.9929	0.9889	0.9830	0.9746	0.9629	0.9472	0.9268	0.9015	0.8715	0.8375
5.5	0.9959	0.9935	0.9897	0.9841	0.9760	0.9647	0.9492	0.9291	0.9039	0.8739	0.8398
5.6	0.9963	0.9940	0.9905	0.9852	0.9774	0.9663	0.9512	0.9313	0.9063	0.8763	0.8420
5.7	0.9967	0.9945	0.9912	0.9861	0.9786	0.9679	0.9530	0.9334	0.9085	0.8785	0.8441
5.8	0.9970	0.9950	0.9919	0.9870	0.9798	0.9694	0.9548	0.9354	0.9107	0.8807	0.8461
5.9	0.9973	0.9954	0.9925	0.9879	0.9809	0.9708	0.9565	0.9374	0.9128	0.8828	0.8481
6.0	0.9975	0.9958	0.9930	0.9887	0.9820	0.9722	0.9582	0.9392	0.9148	0.8848	0.8500

| NTU | Heat Exchanger Effectiveness | | | | | | | | | | |
| | C^* ($C_{min} = C_h$) | | | | | | | | | | |
	0	0.1	0.2	0.3	0.4	0.5	0.6	0.7	0.8	0.9	1.0
6.1	0.9978	0.9962	0.9936	0.9894	0.9830	0.9734	0.9598	0.9411	0.9168	0.8868	0.8519
6.2	0.9980	0.9965	0.9940	0.9901	0.9839	0.9747	0.9613	0.9428	0.9186	0.8887	0.8537
6.3	0.9982	0.9968	0.9945	0.9907	0.9848	0.9758	0.9627	0.9445	0.9205	0.8906	0.8555
6.4	0.9983	0.9971	0.9949	0.9913	0.9856	0.9769	0.9641	0.9461	0.9223	0.8924	0.8572
6.5	0.9985	0.9973	0.9953	0.9919	0.9864	0.9780	0.9654	0.9477	0.9240	0.8942	0.8588
6.6	0.9986	0.9975	0.9956	0.9924	0.9872	0.9789	0.9667	0.9492	0.9256	0.8959	0.8604
6.7	0.9988	0.9977	0.9959	0.9929	0.9878	0.9799	0.9679	0.9506	0.9273	0.8975	0.8620
6.8	0.9989	0.9979	0.9962	0.9933	0.9885	0.9808	0.9690	0.9520	0.9288	0.8991	0.8635
6.9	0.9990	0.9981	0.9965	0.9938	0.9891	0.9817	0.9702	0.9534	0.9303	0.9007	0.8650
7.0	0.9991	0.9983	0.9968	0.9942	0.9897	0.9825	0.9712	0.9547	0.9318	0.9022	0.8664
7.1	0.9992	0.9984	0.9970	0.9945	0.9903	0.9832	0.9722	0.9559	0.9332	0.9037	0.8678
7.2	0.9993	0.9985	0.9972	0.9949	0.9908	0.9840	0.9732	0.9572	0.9346	0.9051	0.8692
7.3	0.9993	0.9987	0.9974	0.9952	0.9913	0.9847	0.9742	0.9583	0.9359	0.9065	0.8705
7.4	0.9994	0.9988	0.9976	0.9955	0.9917	0.9854	0.9751	0.9595	0.9372	0.9079	0.8718
7.5	0.9994	0.9989	0.9978	0.9958	0.9922	0.9860	0.9760	0.9606	0.9385	0.9092	0.8731
7.6	0.9995	0.9990	0.9980	0.9960	0.9926	0.9866	0.9768	0.9616	0.9397	0.9105	0.8743
7.7	0.9995	0.9991	0.9981	0.9963	0.9930	0.9872	0.9776	0.9626	0.9409	0.9117	0.8755
7.8	0.9996	0.9991	0.9982	0.9965	0.9933	0.9877	0.9784	0.9636	0.9421	0.9130	0.8766
7.9	0.9996	0.9992	0.9984	0.9967	0.9937	0.9883	0.9791	0.9646	0.9432	0.9141	0.8778
8.0	0.9997	0.9993	0.9985	0.9969	0.9940	0.9888	0.9798	0.9655	0.9443	0.9153	0.8789
8.1	0.9997	0.9993	0.9986	0.9971	0.9943	0.9892	0.9805	0.9664	0.9454	0.9164	0.8800
8.2	0.9997	0.9994	0.9987	0.9973	0.9946	0.9897	0.9812	0.9673	0.9464	0.9175	0.8810
8.3	0.9998	0.9994	0.9988	0.9975	0.9949	0.9901	0.9818	0.9681	0.9474	0.9186	0.8820
8.4	0.9998	0.9995	0.9989	0.9976	0.9952	0.9906	0.9824	0.9689	0.9484	0.9197	0.8830
8.5	0.9998	0.9995	0.9990	0.9978	0.9954	0.9910	0.9830	0.9697	0.9493	0.9207	0.8840
8.6	0.9998	0.9996	0.9990	0.9979	0.9957	0.9913	0.9835	0.9705	0.9503	0.9217	0.8850
8.7	0.9998	0.9996	0.9991	0.9980	0.9959	0.9917	0.9841	0.9712	0.9512	0.9227	0.8859
8.8	0.9998	0.9996	0.9992	0.9982	0.9961	0.9920	0.9846	0.9719	0.9520	0.9236	0.8868
8.9	0.9999	0.9997	0.9992	0.9983	0.9963	0.9924	0.9851	0.9726	0.9529	0.9245	0.8877
9.0	0.9999	0.9997	0.9993	0.9984	0.9965	0.9927	0.9856	0.9733	0.9537	0.9254	0.8886
9.1	0.9999	0.9997	0.9993	0.9985	0.9967	0.9930	0.9861	0.9739	0.9545	0.9263	0.8894
9.2	0.9999	0.9997	0.9994	0.9986	0.9968	0.9933	0.9865	0.9746	0.9553	0.9272	0.8903
9.3	0.9999	0.9998	0.9994	0.9987	0.9970	0.9935	0.9869	0.9752	0.9561	0.9280	0.8911
9.4	0.9999	0.9998	0.9995	0.9987	0.9971	0.9938	0.9874	0.9758	0.9568	0.9289	0.8919
9.5	0.9999	0.9998	0.9995	0.9988	0.9973	0.9941	0.9878	0.9763	0.9576	0.9297	0.8927
9.6	0.9999	0.9998	0.9995	0.9989	0.9974	0.9943	0.9881	0.9769	0.9583	0.9305	0.8934
9.7	0.9999	0.9998	0.9996	0.9990	0.9975	0.9945	0.9885	0.9774	0.9590	0.9312	0.8942
9.8	0.9999	0.9998	0.9996	0.9990	0.9977	0.9947	0.9889	0.9780	0.9596	0.9320	0.8949
9.9	1.0000	0.9999	0.9996	0.9991	0.9978	0.9950	0.9892	0.9785	0.9603	0.9327	0.8956
10.0	1.0000	0.9999	0.9997	0.9991	0.9979	0.9952	0.9895	0.9790	0.9609	0.9335	0.8963

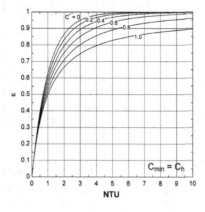

CONFIGURATION 26

1TUBE20ROW20PASS1CIR(CC)

(a) $C_{min} = C_c$

NTU	Heat Exchanger Effectiveness										
	C^* ($C_{min} = C_c$)										
	0	0.1	0.2	0.3	0.4	0.5	0.6	0.7	0.8	0.9	1.0
0	0.0000	0.0000	0.0000	0.0000	0.0000	0.0000	0.0000	0.0000	0.0000	0.0000	0.0000
0.1	0.0952	0.0947	0.0943	0.0939	0.0934	0.0930	0.0926	0.0922	0.0917	0.0913	0.0909
0.2	0.1813	0.1797	0.1782	0.1767	0.1753	0.1738	0.1723	0.1709	0.1695	0.1681	0.1667
0.3	0.2592	0.2562	0.2532	0.2503	0.2474	0.2445	0.2417	0.2389	0.2362	0.2334	0.2308
0.4	0.3297	0.3250	0.3204	0.3158	0.3113	0.3069	0.3025	0.2982	0.2940	0.2898	0.2857
0.5	0.3935	0.3870	0.3807	0.3745	0.3683	0.3623	0.3563	0.3504	0.3446	0.3389	0.3333
0.6	0.4512	0.4431	0.4350	0.4271	0.4193	0.4116	0.4041	0.3966	0.3893	0.3821	0.3750
0.7	0.5034	0.4937	0.4841	0.4746	0.4652	0.4559	0.4468	0.4378	0.4290	0.4203	0.4117
0.8	0.5507	0.5395	0.5284	0.5174	0.5066	0.4958	0.4852	0.4748	0.4645	0.4543	0.4444
0.9	0.5934	0.5810	0.5686	0.5563	0.5440	0.5319	0.5199	0.5081	0.4964	0.4849	0.4736
1.0	0.6321	0.6186	0.6050	0.5915	0.5780	0.5647	0.5514	0.5383	0.5253	0.5125	0.4999
1.1	0.6671	0.6527	0.6381	0.6236	0.6090	0.5945	0.5801	0.5657	0.5515	0.5375	0.5237
1.2	0.6988	0.6836	0.6682	0.6528	0.6373	0.6217	0.6062	0.5908	0.5755	0.5603	0.5453
1.3	0.7275	0.7117	0.6957	0.6795	0.6631	0.6467	0.6302	0.6137	0.5974	0.5811	0.5650
1.4	0.7534	0.7372	0.7207	0.7039	0.6868	0.6696	0.6522	0.6349	0.6175	0.6002	0.5831
1.5	0.7769	0.7604	0.7435	0.7262	0.7086	0.6906	0.6725	0.6543	0.6361	0.6179	0.5998
1.6	0.7981	0.7816	0.7644	0.7467	0.7286	0.7101	0.6913	0.6723	0.6532	0.6341	0.6151
1.7	0.8173	0.8008	0.7835	0.7655	0.7470	0.7280	0.7086	0.6890	0.6691	0.6492	0.6293
1.8	0.8347	0.8183	0.8009	0.7829	0.7641	0.7447	0.7248	0.7045	0.6839	0.6633	0.6425
1.9	0.8504	0.8342	0.8169	0.7988	0.7798	0.7601	0.7397	0.7189	0.6977	0.6763	0.6548
2.0	0.8647	0.8487	0.8316	0.8135	0.7943	0.7744	0.7537	0.7323	0.7106	0.6885	0.6663
2.1	0.8775	0.8619	0.8450	0.8270	0.8078	0.7877	0.7667	0.7449	0.7226	0.6999	0.6770
2.2	0.8892	0.8739	0.8574	0.8395	0.8203	0.8001	0.7788	0.7567	0.7339	0.7106	0.6871
2.3	0.8997	0.8849	0.8687	0.8510	0.8319	0.8116	0.7902	0.7677	0.7445	0.7207	0.6965
2.4	0.9093	0.8949	0.8791	0.8616	0.8427	0.8224	0.8008	0.7781	0.7545	0.7302	0.7054
2.5	0.9179	0.9041	0.8886	0.8715	0.8528	0.8325	0.8108	0.7879	0.7639	0.7391	0.7138
2.6	0.9257	0.9124	0.8974	0.8806	0.8621	0.8419	0.8202	0.7970	0.7727	0.7475	0.7217
2.7	0.9328	0.9200	0.9054	0.8890	0.8708	0.8507	0.8290	0.8057	0.7811	0.7555	0.7291
2.8	0.9392	0.9269	0.9129	0.8968	0.8789	0.8590	0.8373	0.8139	0.7891	0.7631	0.7362
2.9	0.9450	0.9333	0.9197	0.9041	0.8864	0.8667	0.8451	0.8216	0.7966	0.7702	0.7429
3.0	0.9502	0.9390	0.9259	0.9108	0.8935	0.8740	0.8525	0.8289	0.8037	0.7770	0.7493
3.1	0.9550	0.9443	0.9317	0.9170	0.9001	0.8809	0.8594	0.8359	0.8105	0.7835	0.7554
3.2	0.9592	0.9491	0.9370	0.9228	0.9062	0.8873	0.8660	0.8425	0.8169	0.7897	0.7611
3.3	0.9631	0.9535	0.9419	0.9282	0.9120	0.8933	0.8722	0.8487	0.8230	0.7955	0.7666
3.4	0.9666	0.9575	0.9465	0.9331	0.9174	0.8990	0.8781	0.8546	0.8289	0.8012	0.7719
3.5	0.9698	0.9612	0.9506	0.9378	0.9224	0.9044	0.8837	0.8603	0.8344	0.8065	0.7769
3.6	0.9727	0.9646	0.9544	0.9420	0.9271	0.9094	0.8889	0.8656	0.8398	0.8116	0.7817
3.7	0.9753	0.9676	0.9580	0.9460	0.9315	0.9142	0.8939	0.8707	0.8448	0.8165	0.7863
3.8	0.9776	0.9704	0.9612	0.9497	0.9356	0.9187	0.8987	0.8756	0.8497	0.8212	0.7907
3.9	0.9798	0.9730	0.9642	0.9532	0.9395	0.9229	0.9032	0.8803	0.8543	0.8257	0.7949
4.0	0.9817	0.9753	0.9670	0.9564	0.9431	0.9269	0.9074	0.8847	0.8588	0.8300	0.7989
4.1	0.9834	0.9774	0.9695	0.9594	0.9465	0.9307	0.9115	0.8889	0.8630	0.8342	0.8028
4.2	0.9850	0.9794	0.9719	0.9621	0.9497	0.9342	0.9154	0.8930	0.8671	0.8381	0.8066
4.3	0.9864	0.9811	0.9740	0.9647	0.9527	0.9376	0.9190	0.8968	0.8710	0.8420	0.8101
4.4	0.9877	0.9828	0.9760	0.9671	0.9555	0.9408	0.9225	0.9005	0.8748	0.8456	0.8136
4.5	0.9889	0.9842	0.9779	0.9694	0.9582	0.9438	0.9259	0.9041	0.8784	0.8492	0.8169
4.6	0.9899	0.9856	0.9796	0.9714	0.9607	0.9467	0.9290	0.9074	0.8819	0.8526	0.8201
4.7	0.9909	0.9868	0.9812	0.9734	0.9630	0.9494	0.9320	0.9107	0.8852	0.8559	0.8232
4.8	0.9918	0.9880	0.9826	0.9752	0.9652	0.9519	0.9349	0.9138	0.8884	0.8590	0.8262
4.9	0.9926	0.9890	0.9839	0.9769	0.9672	0.9543	0.9377	0.9168	0.8915	0.8621	0.8291
5.0	0.9933	0.9899	0.9852	0.9784	0.9691	0.9566	0.9403	0.9196	0.8945	0.8651	0.8319
5.1	0.9939	0.9908	0.9863	0.9799	0.9709	0.9588	0.9428	0.9224	0.8974	0.8679	0.8346
5.2	0.9945	0.9916	0.9874	0.9812	0.9726	0.9608	0.9452	0.9250	0.9002	0.8707	0.8372
5.3	0.9950	0.9923	0.9883	0.9825	0.9742	0.9628	0.9474	0.9276	0.9028	0.8734	0.8397
5.4	0.9955	0.9930	0.9892	0.9837	0.9757	0.9646	0.9496	0.9300	0.9054	0.8759	0.8422
5.5	0.9959	0.9936	0.9900	0.9848	0.9772	0.9664	0.9517	0.9323	0.9079	0.8784	0.8445
5.6	0.9963	0.9941	0.9908	0.9858	0.9785	0.9680	0.9537	0.9346	0.9103	0.8809	0.8468
5.7	0.9967	0.9946	0.9915	0.9868	0.9797	0.9696	0.9556	0.9367	0.9126	0.8832	0.8490
5.8	0.9970	0.9951	0.9922	0.9877	0.9809	0.9711	0.9574	0.9388	0.9149	0.8855	0.8512
5.9	0.9973	0.9955	0.9928	0.9885	0.9820	0.9725	0.9591	0.9408	0.9170	0.8877	0.8533
6.0	0.9975	0.9959	0.9933	0.9893	0.9831	0.9739	0.9607	0.9427	0.9191	0.8898	0.8553

NTU	Heat Exchanger Effectiveness										
	C^* ($C_{min} = C_c$)										
	0	0.1	0.2	0.3	0.4	0.5	0.6	0.7	0.8	0.9	1.0
6.1	0.9978	0.9963	0.9938	0.9900	0.9840	0.9752	0.9623	0.9446	0.9212	0.8919	0.8573
6.2	0.9980	0.9966	0.9943	0.9906	0.9850	0.9764	0.9638	0.9464	0.9231	0.8939	0.8592
6.3	0.9982	0.9969	0.9947	0.9913	0.9858	0.9775	0.9653	0.9481	0.9250	0.8959	0.8611
6.4	0.9983	0.9971	0.9951	0.9919	0.9866	0.9786	0.9667	0.9497	0.9269	0.8978	0.8629
6.5	0.9985	0.9974	0.9955	0.9924	0.9874	0.9797	0.9680	0.9513	0.9287	0.8996	0.8647
6.6	0.9986	0.9976	0.9958	0.9929	0.9881	0.9806	0.9693	0.9529	0.9304	0.9014	0.8664
6.7	0.9988	0.9978	0.9962	0.9934	0.9888	0.9816	0.9705	0.9544	0.9321	0.9032	0.8680
6.8	0.9989	0.9980	0.9965	0.9938	0.9895	0.9825	0.9717	0.9558	0.9337	0.9049	0.8697
6.9	0.9990	0.9982	0.9967	0.9942	0.9901	0.9833	0.9728	0.9572	0.9353	0.9065	0.8712
7.0	0.9991	0.9983	0.9970	0.9946	0.9906	0.9841	0.9739	0.9585	0.9368	0.9081	0.8728
7.1	0.9992	0.9985	0.9972	0.9950	0.9912	0.9849	0.9749	0.9598	0.9383	0.9097	0.8743
7.2	0.9993	0.9986	0.9974	0.9953	0.9917	0.9856	0.9759	0.9610	0.9397	0.9112	0.8758
7.3	0.9993	0.9987	0.9976	0.9956	0.9922	0.9863	0.9768	0.9622	0.9411	0.9127	0.8772
7.4	0.9994	0.9988	0.9978	0.9959	0.9926	0.9869	0.9777	0.9634	0.9425	0.9142	0.8786
7.5	0.9994	0.9989	0.9980	0.9962	0.9930	0.9876	0.9786	0.9645	0.9438	0.9156	0.8800
7.6	0.9995	0.9990	0.9981	0.9964	0.9934	0.9882	0.9794	0.9656	0.9451	0.9170	0.8813
7.7	0.9995	0.9991	0.9983	0.9967	0.9938	0.9887	0.9802	0.9666	0.9464	0.9183	0.8826
7.8	0.9996	0.9992	0.9984	0.9969	0.9941	0.9893	0.9810	0.9676	0.9476	0.9196	0.8838
7.9	0.9996	0.9993	0.9985	0.9971	0.9945	0.9898	0.9817	0.9686	0.9487	0.9209	0.8851
8.0	0.9997	0.9993	0.9986	0.9973	0.9948	0.9903	0.9824	0.9695	0.9499	0.9222	0.8863
8.1	0.9997	0.9994	0.9987	0.9975	0.9951	0.9907	0.9831	0.9705	0.9510	0.9234	0.8875
8.2	0.9997	0.9994	0.9988	0.9976	0.9954	0.9912	0.9837	0.9713	0.9521	0.9246	0.8886
8.3	0.9998	0.9995	0.9989	0.9978	0.9956	0.9916	0.9843	0.9722	0.9531	0.9257	0.8898
8.4	0.9998	0.9995	0.9990	0.9979	0.9959	0.9920	0.9849	0.9730	0.9542	0.9269	0.8909
8.5	0.9998	0.9996	0.9991	0.9981	0.9961	0.9923	0.9855	0.9738	0.9552	0.9280	0.8919
8.6	0.9998	0.9996	0.9991	0.9982	0.9963	0.9927	0.9861	0.9746	0.9561	0.9291	0.8930
8.7	0.9998	0.9996	0.9992	0.9983	0.9965	0.9931	0.9866	0.9753	0.9571	0.9301	0.8940
8.8	0.9998	0.9997	0.9993	0.9984	0.9967	0.9934	0.9871	0.9760	0.9580	0.9311	0.8951
8.9	0.9999	0.9997	0.9993	0.9985	0.9969	0.9937	0.9876	0.9767	0.9589	0.9322	0.8960
9.0	0.9999	0.9997	0.9994	0.9986	0.9971	0.9940	0.9881	0.9774	0.9598	0.9331	0.8970
9.1	0.9999	0.9997	0.9994	0.9987	0.9973	0.9943	0.9885	0.9781	0.9606	0.9341	0.8980
9.2	0.9999	0.9998	0.9995	0.9988	0.9974	0.9945	0.9889	0.9787	0.9615	0.9351	0.8989
9.3	0.9999	0.9998	0.9995	0.9989	0.9976	0.9948	0.9894	0.9793	0.9623	0.9360	0.8998
9.4	0.9999	0.9998	0.9995	0.9990	0.9977	0.9950	0.9897	0.9799	0.9631	0.9369	0.9007
9.5	0.9999	0.9998	0.9996	0.9990	0.9978	0.9953	0.9901	0.9805	0.9638	0.9378	0.9016
9.6	0.9999	0.9998	0.9996	0.9991	0.9979	0.9955	0.9905	0.9810	0.9646	0.9387	0.9024
9.7	0.9999	0.9999	0.9996	0.9992	0.9981	0.9957	0.9909	0.9816	0.9653	0.9395	0.9033
9.8	0.9999	0.9999	0.9997	0.9992	0.9982	0.9959	0.9912	0.9821	0.9660	0.9403	0.9041
9.9	1.0000	0.9999	0.9997	0.9993	0.9983	0.9961	0.9915	0.9826	0.9667	0.9412	0.9049
10.0	1.0000	0.9999	0.9997	0.9993	0.9984	0.9963	0.9918	0.9831	0.9674	0.9420	0.9057

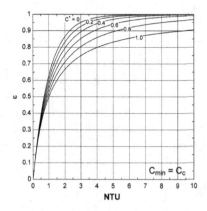

CONFIGURATION 26

1TUBE20ROW20PASS1CIR(PAR)

(b) $C_{min} = C_h$

NTU	Heat Exchanger Effectiveness										
	C^* $(C_{min} = C_h)$										
	0	0.1	0.2	0.3	0.4	0.5	0.6	0.7	0.8	0.9	1.0
0	0.0000	0.0000	0.0000	0.0000	0.0000	0.0000	0.0000	0.0000	0.0000	0.0000	0.0000
0.1	0.0952	0.0947	0.0943	0.0939	0.0934	0.0930	0.0926	0.0922	0.0917	0.0913	0.0909
0.2	0.1813	0.1797	0.1782	0.1767	0.1753	0.1738	0.1723	0.1709	0.1695	0.1681	0.1667
0.3	0.2592	0.2562	0.2532	0.2503	0.2474	0.2445	0.2417	0.2389	0.2362	0.2334	0.2308
0.4	0.3297	0.3250	0.3204	0.3158	0.3113	0.3069	0.3025	0.2982	0.2940	0.2898	0.2857
0.5	0.3935	0.3870	0.3807	0.3745	0.3683	0.3623	0.3563	0.3504	0.3446	0.3389	0.3333
0.6	0.4512	0.4431	0.4350	0.4271	0.4193	0.4116	0.4041	0.3966	0.3893	0.3821	0.3750
0.7	0.5034	0.4937	0.4841	0.4746	0.4652	0.4559	0.4468	0.4378	0.4290	0.4203	0.4117
0.8	0.5507	0.5395	0.5284	0.5174	0.5066	0.4958	0.4852	0.4748	0.4645	0.4543	0.4444
0.9	0.5934	0.5810	0.5686	0.5563	0.5440	0.5319	0.5199	0.5081	0.4964	0.4849	0.4736
1.0	0.6321	0.6186	0.6050	0.5915	0.5780	0.5647	0.5514	0.5383	0.5253	0.5125	0.4999
1.1	0.6671	0.6527	0.6381	0.6236	0.6090	0.5945	0.5801	0.5657	0.5515	0.5375	0.5237
1.2	0.6988	0.6836	0.6682	0.6528	0.6373	0.6217	0.6062	0.5908	0.5755	0.5603	0.5453
1.3	0.7275	0.7117	0.6957	0.6795	0.6631	0.6467	0.6302	0.6137	0.5974	0.5811	0.5650
1.4	0.7534	0.7372	0.7207	0.7039	0.6868	0.6696	0.6522	0.6349	0.6175	0.6002	0.5831
1.5	0.7769	0.7604	0.7435	0.7262	0.7086	0.6906	0.6725	0.6543	0.6361	0.6179	0.5998
1.6	0.7981	0.7816	0.7644	0.7467	0.7286	0.7101	0.6913	0.6723	0.6532	0.6341	0.6151
1.7	0.8173	0.8008	0.7835	0.7655	0.7470	0.7280	0.7086	0.6890	0.6691	0.6492	0.6293
1.8	0.8347	0.8183	0.8009	0.7829	0.7641	0.7447	0.7248	0.7045	0.6839	0.6633	0.6425
1.9	0.8504	0.8342	0.8169	0.7988	0.7798	0.7601	0.7397	0.7189	0.6977	0.6763	0.6548
2.0	0.8647	0.8487	0.8316	0.8135	0.7943	0.7744	0.7537	0.7323	0.7106	0.6885	0.6663
2.1	0.8775	0.8619	0.8450	0.8270	0.8078	0.7877	0.7667	0.7449	0.7226	0.6999	0.6770
2.2	0.8892	0.8739	0.8574	0.8395	0.8203	0.8001	0.7788	0.7567	0.7339	0.7106	0.6871
2.3	0.8997	0.8849	0.8687	0.8510	0.8319	0.8116	0.7902	0.7677	0.7445	0.7207	0.6965
2.4	0.9093	0.8949	0.8791	0.8616	0.8427	0.8224	0.8008	0.7781	0.7545	0.7302	0.7054
2.5	0.9179	0.9041	0.8886	0.8715	0.8528	0.8325	0.8108	0.7879	0.7639	0.7391	0.7138
2.6	0.9257	0.9124	0.8974	0.8806	0.8621	0.8419	0.8202	0.7970	0.7727	0.7475	0.7217
2.7	0.9328	0.9200	0.9054	0.8890	0.8708	0.8507	0.8290	0.8057	0.7811	0.7555	0.7291
2.8	0.9392	0.9269	0.9129	0.8968	0.8789	0.8590	0.8373	0.8139	0.7891	0.7631	0.7362
2.9	0.9450	0.9333	0.9197	0.9041	0.8864	0.8668	0.8451	0.8216	0.7966	0.7702	0.7429
3.0	0.9502	0.9390	0.9259	0.9108	0.8935	0.8740	0.8525	0.8289	0.8037	0.7770	0.7493
3.1	0.9550	0.9443	0.9317	0.9170	0.9001	0.8809	0.8594	0.8359	0.8105	0.7835	0.7554
3.2	0.9592	0.9491	0.9370	0.9228	0.9062	0.8873	0.8660	0.8425	0.8169	0.7897	0.7611
3.3	0.9631	0.9535	0.9419	0.9282	0.9120	0.8933	0.8722	0.8487	0.8230	0.7955	0.7666
3.4	0.9666	0.9575	0.9465	0.9331	0.9174	0.8990	0.8781	0.8546	0.8289	0.8012	0.7719
3.5	0.9698	0.9612	0.9506	0.9378	0.9224	0.9044	0.8837	0.8603	0.8344	0.8065	0.7769
3.6	0.9727	0.9646	0.9544	0.9420	0.9271	0.9094	0.8889	0.8656	0.8398	0.8116	0.7817
3.7	0.9753	0.9676	0.9580	0.9460	0.9315	0.9142	0.8939	0.8707	0.8448	0.8165	0.7863
3.8	0.9776	0.9704	0.9612	0.9497	0.9356	0.9187	0.8987	0.8756	0.8497	0.8212	0.7907
3.9	0.9798	0.9730	0.9642	0.9532	0.9395	0.9229	0.9032	0.8803	0.8543	0.8257	0.7949
4.0	0.9817	0.9753	0.9670	0.9564	0.9431	0.9269	0.9074	0.8847	0.8588	0.8300	0.7989
4.1	0.9834	0.9774	0.9695	0.9594	0.9465	0.9307	0.9115	0.8889	0.8630	0.8342	0.8028
4.2	0.9850	0.9794	0.9719	0.9622	0.9497	0.9342	0.9154	0.8930	0.8671	0.8381	0.8066
4.3	0.9864	0.9811	0.9740	0.9647	0.9527	0.9376	0.9190	0.8968	0.8710	0.8420	0.8101
4.4	0.9877	0.9828	0.9760	0.9671	0.9555	0.9408	0.9225	0.9005	0.8748	0.8456	0.8136
4.5	0.9889	0.9842	0.9779	0.9694	0.9582	0.9438	0.9259	0.9041	0.8784	0.8492	0.8169
4.6	0.9899	0.9856	0.9796	0.9714	0.9607	0.9467	0.9290	0.9075	0.8819	0.8526	0.8201
4.7	0.9909	0.9868	0.9812	0.9734	0.9630	0.9494	0.9320	0.9107	0.8852	0.8559	0.8232
4.8	0.9918	0.9880	0.9826	0.9752	0.9652	0.9519	0.9349	0.9138	0.8884	0.8591	0.8262
4.9	0.9926	0.9890	0.9839	0.9769	0.9672	0.9543	0.9377	0.9168	0.8915	0.8621	0.8291
5.0	0.9933	0.9899	0.9852	0.9784	0.9691	0.9566	0.9403	0.9196	0.8945	0.8651	0.8319
5.1	0.9939	0.9908	0.9863	0.9799	0.9709	0.9588	0.9428	0.9224	0.8974	0.8679	0.8346
5.2	0.9945	0.9916	0.9874	0.9812	0.9726	0.9608	0.9452	0.9250	0.9002	0.8707	0.8372
5.3	0.9950	0.9923	0.9883	0.9825	0.9742	0.9628	0.9474	0.9276	0.9028	0.8734	0.8397
5.4	0.9955	0.9930	0.9892	0.9837	0.9757	0.9646	0.9496	0.9300	0.9054	0.8759	0.8422
5.5	0.9959	0.9936	0.9900	0.9848	0.9772	0.9664	0.9517	0.9323	0.9079	0.8784	0.8445
5.6	0.9963	0.9941	0.9908	0.9858	0.9785	0.9681	0.9537	0.9346	0.9103	0.8809	0.8468
5.7	0.9967	0.9946	0.9915	0.9868	0.9797	0.9696	0.9556	0.9367	0.9126	0.8832	0.8490
5.8	0.9970	0.9951	0.9922	0.9877	0.9809	0.9711	0.9574	0.9388	0.9149	0.8855	0.8512
5.9	0.9973	0.9955	0.9928	0.9885	0.9820	0.9725	0.9591	0.9408	0.9170	0.8877	0.8533
6.0	0.9975	0.9959	0.9933	0.9893	0.9831	0.9739	0.9608	0.9427	0.9191	0.8898	0.8553

NTU	Heat Exchanger Effectiveness										
	C^* ($C_{min} = C_h$)										
	0	0.1	0.2	0.3	0.4	0.5	0.6	0.7	0.8	0.9	1.0
6.1	0.9978	0.9963	0.9938	0.9900	0.9840	0.9752	0.9623	0.9446	0.9212	0.8919	0.8573
6.2	0.9980	0.9966	0.9943	0.9906	0.9850	0.9764	0.9639	0.9464	0.9231	0.8939	0.8592
6.3	0.9982	0.9969	0.9947	0.9913	0.9858	0.9775	0.9653	0.9481	0.9250	0.8959	0.8611
6.4	0.9983	0.9971	0.9951	0.9919	0.9866	0.9786	0.9667	0.9497	0.9269	0.8978	0.8629
6.5	0.9985	0.9974	0.9955	0.9924	0.9874	0.9797	0.9680	0.9513	0.9287	0.8996	0.8647
6.6	0.9986	0.9976	0.9958	0.9929	0.9881	0.9806	0.9693	0.9529	0.9304	0.9014	0.8664
6.7	0.9988	0.9978	0.9962	0.9934	0.9888	0.9816	0.9705	0.9544	0.9321	0.9032	0.8680
6.8	0.9989	0.9980	0.9965	0.9938	0.9895	0.9825	0.9717	0.9558	0.9337	0.9049	0.8697
6.9	0.9990	0.9982	0.9967	0.9942	0.9901	0.9833	0.9728	0.9572	0.9353	0.9065	0.8712
7.0	0.9991	0.9983	0.9970	0.9946	0.9906	0.9841	0.9739	0.9585	0.9368	0.9081	0.8728
7.1	0.9992	0.9985	0.9972	0.9950	0.9912	0.9849	0.9749	0.9598	0.9383	0.9097	0.8743
7.2	0.9993	0.9986	0.9974	0.9953	0.9917	0.9856	0.9759	0.9610	0.9398	0.9112	0.8758
7.3	0.9993	0.9987	0.9976	0.9956	0.9922	0.9863	0.9768	0.9622	0.9412	0.9127	0.8772
7.4	0.9994	0.9988	0.9978	0.9959	0.9926	0.9870	0.9777	0.9634	0.9425	0.9142	0.8786
7.5	0.9994	0.9989	0.9980	0.9962	0.9930	0.9876	0.9786	0.9645	0.9438	0.9156	0.8800
7.6	0.9995	0.9990	0.9981	0.9964	0.9934	0.9882	0.9794	0.9656	0.9451	0.9170	0.8813
7.7	0.9995	0.9991	0.9983	0.9967	0.9938	0.9887	0.9802	0.9666	0.9464	0.9183	0.8826
7.8	0.9996	0.9992	0.9984	0.9969	0.9941	0.9893	0.9810	0.9676	0.9476	0.9196	0.8838
7.9	0.9996	0.9993	0.9985	0.9971	0.9945	0.9898	0.9817	0.9686	0.9488	0.9209	0.8851
8.0	0.9997	0.9993	0.9986	0.9973	0.9948	0.9903	0.9824	0.9696	0.9499	0.9222	0.8863
8.1	0.9997	0.9994	0.9987	0.9975	0.9951	0.9907	0.9831	0.9705	0.9510	0.9234	0.8875
8.2	0.9997	0.9994	0.9988	0.9976	0.9954	0.9912	0.9837	0.9713	0.9521	0.9246	0.8886
8.3	0.9998	0.9995	0.9989	0.9978	0.9956	0.9916	0.9844	0.9722	0.9532	0.9257	0.8898
8.4	0.9998	0.9995	0.9990	0.9979	0.9959	0.9920	0.9850	0.9730	0.9542	0.9269	0.8909
8.5	0.9998	0.9996	0.9991	0.9981	0.9961	0.9924	0.9855	0.9738	0.9552	0.9280	0.8919
8.6	0.9998	0.9996	0.9991	0.9982	0.9963	0.9927	0.9861	0.9746	0.9562	0.9291	0.8930
8.7	0.9998	0.9996	0.9992	0.9983	0.9965	0.9931	0.9866	0.9753	0.9571	0.9301	0.8940
8.8	0.9998	0.9997	0.9993	0.9984	0.9967	0.9934	0.9871	0.9761	0.9580	0.9312	0.8950
8.9	0.9999	0.9997	0.9993	0.9985	0.9969	0.9937	0.9876	0.9768	0.9589	0.9322	0.8960
9.0	0.9999	0.9997	0.9994	0.9986	0.9971	0.9940	0.9881	0.9774	0.9598	0.9332	0.8970
9.1	0.9999	0.9997	0.9994	0.9987	0.9972	0.9943	0.9885	0.9781	0.9606	0.9341	0.8980
9.2	0.9999	0.9998	0.9995	0.9988	0.9974	0.9945	0.9889	0.9787	0.9615	0.9351	0.8989
9.3	0.9999	0.9998	0.9995	0.9989	0.9975	0.9948	0.9894	0.9793	0.9623	0.9360	0.8998
9.4	0.9999	0.9998	0.9995	0.9990	0.9977	0.9950	0.9898	0.9799	0.9631	0.9369	0.9007
9.5	0.9999	0.9998	0.9996	0.9990	0.9978	0.9953	0.9901	0.9805	0.9638	0.9378	0.9016
9.6	0.9999	0.9998	0.9996	0.9991	0.9979	0.9955	0.9905	0.9811	0.9646	0.9387	0.9024
9.7	0.9999	0.9999	0.9996	0.9991	0.9981	0.9957	0.9909	0.9816	0.9653	0.9395	0.9033
9.8	0.9999	0.9999	0.9997	0.9992	0.9982	0.9959	0.9912	0.9821	0.9660	0.9404	0.9041
9.9	1.0000	0.9999	0.9997	0.9993	0.9983	0.9961	0.9915	0.9826	0.9667	0.9412	0.9049
10.0	1.0000	0.9999	0.9997	0.9993	0.9984	0.9963	0.9918	0.9831	0.9674	0.9420	0.9057

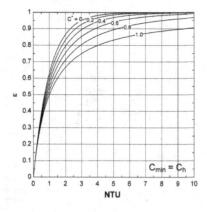

CONFIGURATION 27

1TUBE50ROW50PASS1CIR(CC)

(a) $C_{min} = C_c$

NTU	Heat Exchanger Effectiveness										
	C^* ($C_{min} = C_c$)										
	0	0.1	0.2	0.3	0.4	0.5	0.6	0.7	0.8	0.9	1.0
0	0.0000	0.0000	0.0000	0.0000	0.0000	0.0000	0.0000	0.0000	0.0000	0.0000	0.0000
0.1	0.0952	0.0947	0.0943	0.0939	0.0934	0.0930	0.0926	0.0922	0.0917	0.0913	0.0909
0.2	0.1813	0.1797	0.1782	0.1767	0.1753	0.1738	0.1723	0.1709	0.1695	0.1681	0.1667
0.3	0.2592	0.2562	0.2532	0.2503	0.2474	0.2445	0.2417	0.2389	0.2362	0.2334	0.2308
0.4	0.3297	0.3250	0.3204	0.3158	0.3113	0.3069	0.3025	0.2982	0.2940	0.2898	0.2857
0.5	0.3935	0.3871	0.3807	0.3745	0.3683	0.3623	0.3563	0.3504	0.3446	0.3389	0.3333
0.6	0.4512	0.4431	0.4351	0.4271	0.4194	0.4117	0.4041	0.3966	0.3893	0.3821	0.3750
0.7	0.5034	0.4937	0.4841	0.4746	0.4652	0.4560	0.4468	0.4379	0.4290	0.4203	0.4118
0.8	0.5507	0.5395	0.5284	0.5175	0.5066	0.4959	0.4853	0.4748	0.4645	0.4544	0.4444
0.9	0.5934	0.5810	0.5686	0.5563	0.5441	0.5320	0.5200	0.5082	0.4965	0.4850	0.4737
1.0	0.6321	0.6186	0.6050	0.5915	0.5781	0.5647	0.5515	0.5383	0.5254	0.5126	0.5000
1.1	0.6671	0.6527	0.6382	0.6236	0.6091	0.5946	0.5801	0.5658	0.5516	0.5376	0.5238
1.2	0.6988	0.6836	0.6683	0.6528	0.6373	0.6218	0.6063	0.5909	0.5756	0.5604	0.5454
1.3	0.7275	0.7117	0.6957	0.6795	0.6632	0.6468	0.6303	0.6139	0.5975	0.5813	0.5652
1.4	0.7534	0.7373	0.7207	0.7039	0.6869	0.6697	0.6524	0.6350	0.6177	0.6004	0.5833
1.5	0.7769	0.7605	0.7436	0.7263	0.7087	0.6908	0.6727	0.6545	0.6362	0.6180	0.6000
1.6	0.7981	0.7816	0.7645	0.7468	0.7287	0.7102	0.6914	0.6725	0.6534	0.6343	0.6153
1.7	0.8173	0.8008	0.7835	0.7656	0.7472	0.7282	0.7088	0.6892	0.6694	0.6495	0.6296
1.8	0.8347	0.8183	0.8010	0.7830	0.7642	0.7448	0.7249	0.7047	0.6842	0.6635	0.6428
1.9	0.8504	0.8342	0.8170	0.7989	0.7799	0.7602	0.7399	0.7191	0.6980	0.6766	0.6551
2.0	0.8647	0.8487	0.8317	0.8136	0.7945	0.7746	0.7539	0.7326	0.7109	0.6888	0.6666
2.1	0.8775	0.8619	0.8451	0.8271	0.8080	0.7879	0.7669	0.7452	0.7229	0.7002	0.6774
2.2	0.8892	0.8740	0.8574	0.8396	0.8205	0.8003	0.7791	0.7570	0.7342	0.7110	0.6874
2.3	0.8997	0.8850	0.8688	0.8511	0.8321	0.8119	0.7904	0.7680	0.7449	0.7211	0.6969
2.4	0.9093	0.8950	0.8792	0.8618	0.8429	0.8227	0.8011	0.7784	0.7549	0.7306	0.7058
2.5	0.9179	0.9041	0.8887	0.8716	0.8530	0.8327	0.8111	0.7882	0.7643	0.7395	0.7142
2.6	0.9257	0.9125	0.8975	0.8808	0.8623	0.8422	0.8205	0.7974	0.7732	0.7480	0.7221
2.7	0.9328	0.9201	0.9055	0.8892	0.8710	0.8510	0.8293	0.8061	0.7816	0.7560	0.7296
2.8	0.9392	0.9270	0.9130	0.8970	0.8791	0.8593	0.8376	0.8143	0.7895	0.7636	0.7367
2.9	0.9450	0.9333	0.9198	0.9043	0.8867	0.8671	0.8455	0.8221	0.7971	0.7708	0.7435
3.0	0.9502	0.9391	0.9261	0.9110	0.8938	0.8744	0.8529	0.8294	0.8042	0.7776	0.7499
3.1	0.9550	0.9444	0.9318	0.9172	0.9003	0.8812	0.8598	0.8364	0.8110	0.7841	0.7560
3.2	0.9592	0.9492	0.9372	0.9230	0.9065	0.8877	0.8664	0.8430	0.8175	0.7903	0.7618
3.3	0.9631	0.9536	0.9421	0.9283	0.9123	0.8937	0.8727	0.8492	0.8236	0.7962	0.7673
3.4	0.9666	0.9576	0.9466	0.9333	0.9176	0.8994	0.8786	0.8552	0.8295	0.8018	0.7726
3.5	0.9698	0.9613	0.9507	0.9380	0.9227	0.9048	0.8841	0.8609	0.8351	0.8072	0.7776
3.6	0.9727	0.9646	0.9546	0.9422	0.9274	0.9098	0.8894	0.8662	0.8404	0.8124	0.7825
3.7	0.9753	0.9677	0.9581	0.9462	0.9318	0.9146	0.8944	0.8714	0.8455	0.8173	0.7871
3.8	0.9776	0.9705	0.9613	0.9499	0.9360	0.9191	0.8992	0.8763	0.8504	0.8220	0.7915
3.9	0.9798	0.9730	0.9643	0.9534	0.9398	0.9233	0.9037	0.8809	0.8551	0.8265	0.7958
4.0	0.9817	0.9753	0.9671	0.9566	0.9435	0.9273	0.9080	0.8854	0.8596	0.8309	0.7998
4.1	0.9834	0.9775	0.9696	0.9596	0.9469	0.9311	0.9121	0.8896	0.8638	0.8350	0.8037
4.2	0.9850	0.9794	0.9720	0.9624	0.9501	0.9347	0.9160	0.8937	0.8679	0.8390	0.8075
4.3	0.9864	0.9812	0.9742	0.9649	0.9531	0.9381	0.9196	0.8976	0.8719	0.8429	0.8111
4.4	0.9877	0.9828	0.9762	0.9673	0.9559	0.9413	0.9231	0.9013	0.8757	0.8466	0.8146
4.5	0.9889	0.9843	0.9780	0.9696	0.9585	0.9443	0.9265	0.9048	0.8793	0.8502	0.8180
4.6	0.9899	0.9856	0.9797	0.9717	0.9610	0.9471	0.9297	0.9082	0.8828	0.8536	0.8212
4.7	0.9909	0.9869	0.9813	0.9736	0.9633	0.9498	0.9327	0.9115	0.8862	0.8569	0.8244
4.8	0.9918	0.9880	0.9827	0.9754	0.9655	0.9524	0.9356	0.9146	0.8894	0.8601	0.8274
4.9	0.9926	0.9890	0.9840	0.9771	0.9675	0.9548	0.9383	0.9176	0.8925	0.8632	0.8303
5.0	0.9933	0.9900	0.9853	0.9786	0.9695	0.9571	0.9410	0.9205	0.8955	0.8662	0.8331
5.1	0.9939	0.9908	0.9864	0.9801	0.9713	0.9593	0.9435	0.9233	0.8984	0.8691	0.8358
5.2	0.9945	0.9916	0.9875	0.9814	0.9730	0.9613	0.9458	0.9259	0.9012	0.8719	0.8385
5.3	0.9950	0.9924	0.9884	0.9827	0.9746	0.9633	0.9481	0.9285	0.9039	0.8746	0.8410
5.4	0.9955	0.9930	0.9893	0.9839	0.9761	0.9651	0.9503	0.9309	0.9065	0.8772	0.8435
5.5	0.9959	0.9936	0.9901	0.9850	0.9775	0.9669	0.9524	0.9333	0.9090	0.8797	0.8459
5.6	0.9963	0.9942	0.9909	0.9860	0.9788	0.9686	0.9544	0.9355	0.9115	0.8822	0.8482
5.7	0.9967	0.9947	0.9916	0.9869	0.9801	0.9701	0.9563	0.9377	0.9138	0.8846	0.8505
5.8	0.9970	0.9951	0.9922	0.9878	0.9812	0.9716	0.9581	0.9398	0.9161	0.8869	0.8527
5.9	0.9973	0.9955	0.9928	0.9887	0.9823	0.9730	0.9598	0.9418	0.9183	0.8891	0.8548
6.0	0.9975	0.9959	0.9934	0.9894	0.9834	0.9744	0.9615	0.9437	0.9204	0.8913	0.8569

NTU	Heat Exchanger Effectiveness										
	$C^* (C_{min} = C_c)$										
	0	0.1	0.2	0.3	0.4	0.5	0.6	0.7	0.8	0.9	1.0
6.1	0.9978	0.9963	0.9939	0.9901	0.9843	0.9757	0.9631	0.9456	0.9225	0.8934	0.8589
6.2	0.9980	0.9966	0.9944	0.9908	0.9853	0.9769	0.9646	0.9474	0.9244	0.8954	0.8608
6.3	0.9982	0.9969	0.9948	0.9914	0.9861	0.9780	0.9660	0.9491	0.9264	0.8974	0.8627
6.4	0.9983	0.9972	0.9952	0.9920	0.9869	0.9791	0.9674	0.9508	0.9282	0.8993	0.8645
6.5	0.9985	0.9974	0.9956	0.9926	0.9877	0.9801	0.9688	0.9524	0.9300	0.9012	0.8663
6.6	0.9986	0.9976	0.9959	0.9931	0.9884	0.9811	0.9700	0.9539	0.9318	0.9031	0.8681
6.7	0.9988	0.9978	0.9962	0.9935	0.9891	0.9821	0.9713	0.9554	0.9335	0.9048	0.8698
6.8	0.9989	0.9980	0.9965	0.9940	0.9897	0.9829	0.9724	0.9569	0.9351	0.9066	0.8715
6.9	0.9990	0.9982	0.9968	0.9944	0.9903	0.9838	0.9735	0.9583	0.9367	0.9082	0.8731
7.0	0.9991	0.9983	0.9970	0.9948	0.9909	0.9846	0.9746	0.9596	0.9383	0.9099	0.8746
7.1	0.9992	0.9985	0.9973	0.9951	0.9914	0.9853	0.9756	0.9609	0.9398	0.9115	0.8762
7.2	0.9993	0.9986	0.9975	0.9954	0.9919	0.9861	0.9766	0.9622	0.9413	0.9130	0.8777
7.3	0.9993	0.9987	0.9977	0.9958	0.9924	0.9868	0.9776	0.9634	0.9427	0.9146	0.8791
7.4	0.9994	0.9988	0.9978	0.9960	0.9929	0.9874	0.9785	0.9645	0.9440	0.9160	0.8806
7.5	0.9994	0.9989	0.9980	0.9963	0.9933	0.9880	0.9793	0.9657	0.9454	0.9175	0.8820
7.6	0.9995	0.9990	0.9982	0.9966	0.9937	0.9886	0.9802	0.9667	0.9467	0.9189	0.8833
7.7	0.9995	0.9991	0.9983	0.9968	0.9940	0.9892	0.9810	0.9678	0.9479	0.9203	0.8847
7.8	0.9996	0.9992	0.9984	0.9970	0.9944	0.9897	0.9817	0.9688	0.9492	0.9216	0.8860
7.9	0.9996	0.9993	0.9986	0.9972	0.9947	0.9902	0.9825	0.9698	0.9504	0.9229	0.8872
8.0	0.9997	0.9993	0.9987	0.9974	0.9950	0.9907	0.9832	0.9707	0.9515	0.9242	0.8885
8.1	0.9997	0.9994	0.9988	0.9976	0.9953	0.9911	0.9838	0.9716	0.9527	0.9254	0.8897
8.2	0.9997	0.9994	0.9989	0.9977	0.9956	0.9916	0.9845	0.9725	0.9538	0.9266	0.8909
8.3	0.9998	0.9995	0.9990	0.9979	0.9958	0.9920	0.9851	0.9734	0.9548	0.9278	0.8920
8.4	0.9998	0.9995	0.9990	0.9980	0.9961	0.9924	0.9857	0.9742	0.9559	0.9290	0.8932
8.5	0.9998	0.9996	0.9991	0.9982	0.9963	0.9928	0.9863	0.9750	0.9569	0.9301	0.8943
8.6	0.9998	0.9996	0.9992	0.9983	0.9965	0.9931	0.9868	0.9758	0.9579	0.9312	0.8954
8.7	0.9998	0.9996	0.9992	0.9984	0.9967	0.9934	0.9873	0.9765	0.9588	0.9323	0.8964
8.8	0.9998	0.9997	0.9993	0.9985	0.9969	0.9938	0.9878	0.9772	0.9598	0.9334	0.8975
8.9	0.9999	0.9997	0.9994	0.9986	0.9971	0.9941	0.9883	0.9779	0.9607	0.9344	0.8985
9.0	0.9999	0.9997	0.9994	0.9987	0.9973	0.9944	0.9888	0.9786	0.9616	0.9354	0.8995
9.1	0.9999	0.9998	0.9995	0.9988	0.9974	0.9946	0.9892	0.9793	0.9624	0.9364	0.9005
9.2	0.9999	0.9998	0.9995	0.9989	0.9976	0.9949	0.9896	0.9799	0.9633	0.9374	0.9015
9.3	0.9999	0.9998	0.9995	0.9990	0.9977	0.9951	0.9900	0.9805	0.9641	0.9384	0.9024
9.4	0.9999	0.9998	0.9996	0.9990	0.9979	0.9954	0.9904	0.9811	0.9649	0.9393	0.9033
9.5	0.9999	0.9998	0.9996	0.9991	0.9980	0.9956	0.9908	0.9817	0.9657	0.9402	0.9042
9.6	0.9999	0.9998	0.9996	0.9992	0.9981	0.9958	0.9912	0.9823	0.9664	0.9411	0.9051
9.7	0.9999	0.9999	0.9997	0.9992	0.9982	0.9960	0.9915	0.9828	0.9672	0.9420	0.9060
9.8	0.9999	0.9999	0.9997	0.9993	0.9983	0.9962	0.9919	0.9833	0.9679	0.9428	0.9069
9.9	1.0000	0.9999	0.9997	0.9993	0.9984	0.9964	0.9922	0.9838	0.9686	0.9437	0.9077
10.0	1.0000	0.9999	0.9997	0.9994	0.9985	0.9966	0.9925	0.9843	0.9693	0.9445	0.9085

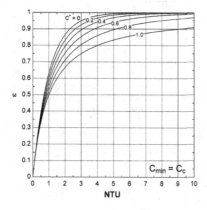

CONFIGURATION 27

1TUBE50ROW50PASS1CIR(CC)

(b) $C_{min} = C_h$

NTU	Heat Exchanger Effectiveness										
	C^* $(C_{min} = C_h)$										
	0	0.1	0.2	0.3	0.4	0.5	0.6	0.7	0.8	0.9	1.0
0	0.0000	0.0000	0.0000	0.0000	0.0000	0.0000	0.0000	0.0000	0.0000	0.0000	0.0000
0.1	0.0952	0.0947	0.0943	0.0939	0.0934	0.0930	0.0926	0.0922	0.0917	0.0913	0.0909
0.2	0.1813	0.1797	0.1782	0.1767	0.1753	0.1738	0.1723	0.1709	0.1695	0.1681	0.1667
0.3	0.2592	0.2562	0.2532	0.2503	0.2474	0.2445	0.2417	0.2389	0.2362	0.2334	0.2308
0.4	0.3297	0.3250	0.3204	0.3158	0.3113	0.3069	0.3025	0.2982	0.2940	0.2898	0.2857
0.5	0.3935	0.3871	0.3807	0.3745	0.3683	0.3623	0.3563	0.3504	0.3446	0.3389	0.3333
0.6	0.4512	0.4431	0.4351	0.4271	0.4194	0.4117	0.4041	0.3966	0.3893	0.3821	0.3750
0.7	0.5034	0.4937	0.4841	0.4746	0.4652	0.4560	0.4468	0.4379	0.4290	0.4203	0.4118
0.8	0.5507	0.5395	0.5284	0.5175	0.5066	0.4959	0.4853	0.4748	0.4645	0.4544	0.4444
0.9	0.5934	0.5810	0.5686	0.5563	0.5441	0.5320	0.5200	0.5082	0.4965	0.4850	0.4737
1.0	0.6321	0.6186	0.6050	0.5915	0.5781	0.5647	0.5515	0.5383	0.5254	0.5126	0.5000
1.1	0.6671	0.6527	0.6382	0.6236	0.6091	0.5946	0.5801	0.5658	0.5516	0.5376	0.5238
1.2	0.6988	0.6836	0.6683	0.6528	0.6373	0.6218	0.6063	0.5909	0.5756	0.5604	0.5454
1.3	0.7275	0.7117	0.6957	0.6795	0.6632	0.6468	0.6303	0.6139	0.5975	0.5813	0.5652
1.4	0.7534	0.7373	0.7207	0.7039	0.6869	0.6697	0.6524	0.6350	0.6177	0.6004	0.5833
1.5	0.7769	0.7605	0.7436	0.7263	0.7087	0.6908	0.6727	0.6545	0.6362	0.6180	0.6000
1.6	0.7981	0.7816	0.7645	0.7468	0.7287	0.7102	0.6914	0.6725	0.6534	0.6343	0.6153
1.7	0.8173	0.8008	0.7835	0.7656	0.7472	0.7282	0.7088	0.6892	0.6694	0.6495	0.6296
1.8	0.8347	0.8183	0.8010	0.7830	0.7642	0.7448	0.7249	0.7047	0.6842	0.6635	0.6428
1.9	0.8504	0.8342	0.8170	0.7989	0.7799	0.7602	0.7399	0.7191	0.6980	0.6766	0.6551
2.0	0.8647	0.8487	0.8317	0.8136	0.7945	0.7746	0.7539	0.7326	0.7109	0.6888	0.6666
2.1	0.8775	0.8619	0.8451	0.8271	0.8080	0.7879	0.7669	0.7452	0.7229	0.7002	0.6774
2.2	0.8892	0.8740	0.8574	0.8396	0.8205	0.8003	0.7791	0.7570	0.7342	0.7110	0.6874
2.3	0.8997	0.8850	0.8688	0.8511	0.8321	0.8119	0.7904	0.7680	0.7449	0.7211	0.6969
2.4	0.9093	0.8950	0.8792	0.8618	0.8429	0.8227	0.8011	0.7784	0.7549	0.7306	0.7058
2.5	0.9179	0.9041	0.8887	0.8716	0.8530	0.8327	0.8111	0.7882	0.7643	0.7395	0.7142
2.6	0.9257	0.9125	0.8975	0.8808	0.8623	0.8422	0.8205	0.7974	0.7732	0.7480	0.7221
2.7	0.9328	0.9201	0.9055	0.8892	0.8710	0.8510	0.8293	0.8061	0.7816	0.7560	0.7296
2.8	0.9392	0.9270	0.9130	0.8970	0.8791	0.8593	0.8376	0.8143	0.7895	0.7636	0.7367
2.9	0.9450	0.9333	0.9198	0.9043	0.8867	0.8671	0.8455	0.8221	0.7971	0.7708	0.7435
3.0	0.9502	0.9391	0.9261	0.9110	0.8938	0.8744	0.8529	0.8294	0.8042	0.7776	0.7499
3.1	0.9550	0.9444	0.9318	0.9172	0.9003	0.8812	0.8598	0.8364	0.8110	0.7841	0.7560
3.2	0.9592	0.9492	0.9372	0.9230	0.9065	0.8876	0.8664	0.8430	0.8175	0.7903	0.7618
3.3	0.9631	0.9536	0.9421	0.9283	0.9123	0.8937	0.8727	0.8492	0.8236	0.7962	0.7673
3.4	0.9666	0.9576	0.9466	0.9333	0.9176	0.8994	0.8786	0.8552	0.8295	0.8018	0.7726
3.5	0.9698	0.9613	0.9507	0.9380	0.9227	0.9048	0.8841	0.8609	0.8351	0.8072	0.7776
3.6	0.9727	0.9646	0.9546	0.9422	0.9274	0.9098	0.8894	0.8662	0.8404	0.8124	0.7825
3.7	0.9753	0.9677	0.9581	0.9462	0.9318	0.9146	0.8944	0.8714	0.8455	0.8173	0.7871
3.8	0.9776	0.9705	0.9613	0.9499	0.9360	0.9191	0.8992	0.8763	0.8504	0.8220	0.7915
3.9	0.9798	0.9730	0.9643	0.9534	0.9398	0.9233	0.9037	0.8809	0.8551	0.8265	0.7958
4.0	0.9817	0.9753	0.9671	0.9566	0.9435	0.9273	0.9080	0.8854	0.8596	0.8309	0.7998
4.1	0.9834	0.9775	0.9696	0.9596	0.9469	0.9311	0.9121	0.8896	0.8638	0.8350	0.8037
4.2	0.9850	0.9794	0.9720	0.9624	0.9501	0.9347	0.9160	0.8937	0.8679	0.8390	0.8075
4.3	0.9864	0.9812	0.9742	0.9649	0.9531	0.9381	0.9196	0.8976	0.8719	0.8429	0.8111
4.4	0.9877	0.9828	0.9762	0.9673	0.9559	0.9413	0.9231	0.9013	0.8757	0.8466	0.8146
4.5	0.9889	0.9843	0.9780	0.9696	0.9585	0.9443	0.9265	0.9048	0.8793	0.8502	0.8180
4.6	0.9899	0.9856	0.9797	0.9717	0.9610	0.9471	0.9297	0.9082	0.8828	0.8536	0.8212
4.7	0.9909	0.9869	0.9813	0.9736	0.9633	0.9498	0.9327	0.9115	0.8862	0.8569	0.8243
4.8	0.9918	0.9880	0.9827	0.9754	0.9655	0.9524	0.9356	0.9146	0.8894	0.8601	0.8274
4.9	0.9926	0.9890	0.9840	0.9771	0.9675	0.9548	0.9383	0.9176	0.8925	0.8632	0.8303
5.0	0.9933	0.9900	0.9853	0.9786	0.9695	0.9571	0.9410	0.9205	0.8955	0.8662	0.8331
5.1	0.9939	0.9908	0.9864	0.9801	0.9713	0.9593	0.9435	0.9233	0.8984	0.8691	0.8358
5.2	0.9945	0.9916	0.9875	0.9814	0.9730	0.9613	0.9458	0.9259	0.9012	0.8719	0.8385
5.3	0.9950	0.9924	0.9884	0.9827	0.9746	0.9633	0.9481	0.9285	0.9039	0.8746	0.8410
5.4	0.9955	0.9930	0.9893	0.9839	0.9761	0.9651	0.9503	0.9309	0.9065	0.8772	0.8435
5.5	0.9959	0.9936	0.9901	0.9850	0.9775	0.9669	0.9524	0.9333	0.9090	0.8797	0.8459
5.6	0.9963	0.9942	0.9909	0.9860	0.9788	0.9685	0.9544	0.9355	0.9115	0.8822	0.8482
5.7	0.9967	0.9947	0.9916	0.9869	0.9801	0.9701	0.9563	0.9377	0.9138	0.8846	0.8505
5.8	0.9970	0.9951	0.9922	0.9878	0.9812	0.9716	0.9581	0.9398	0.9161	0.8869	0.8527
5.9	0.9973	0.9955	0.9928	0.9887	0.9823	0.9730	0.9598	0.9418	0.9183	0.8891	0.8548
6.0	0.9975	0.9959	0.9934	0.9894	0.9834	0.9744	0.9615	0.9437	0.9204	0.8913	0.8569

NTU	Heat Exchanger Effectiveness $C^* (C_{min} = C_h)$										
	0	0.1	0.2	0.3	0.4	0.5	0.6	0.7	0.8	0.9	1.0
6.1	0.9978	0.9963	0.9939	0.9901	0.9843	0.9757	0.9631	0.9456	0.9225	0.8934	0.8589
6.2	0.9980	0.9966	0.9944	0.9908	0.9853	0.9769	0.9646	0.9474	0.9244	0.8954	0.8608
6.3	0.9982	0.9969	0.9948	0.9914	0.9861	0.9780	0.9660	0.9491	0.9264	0.8974	0.8627
6.4	0.9983	0.9972	0.9952	0.9920	0.9869	0.9791	0.9674	0.9508	0.9282	0.8993	0.8645
6.5	0.9985	0.9974	0.9956	0.9926	0.9877	0.9801	0.9688	0.9524	0.9300	0.9012	0.8663
6.6	0.9986	0.9976	0.9959	0.9931	0.9884	0.9811	0.9700	0.9539	0.9318	0.9031	0.8681
6.7	0.9988	0.9978	0.9962	0.9935	0.9891	0.9821	0.9713	0.9554	0.9335	0.9048	0.8698
6.8	0.9989	0.9980	0.9965	0.9940	0.9897	0.9829	0.9724	0.9569	0.9351	0.9066	0.8715
6.9	0.9990	0.9982	0.9968	0.9944	0.9903	0.9838	0.9735	0.9583	0.9367	0.9082	0.8731
7.0	0.9991	0.9983	0.9970	0.9948	0.9909	0.9846	0.9746	0.9596	0.9383	0.9099	0.8746
7.1	0.9992	0.9985	0.9973	0.9951	0.9914	0.9853	0.9756	0.9609	0.9398	0.9115	0.8762
7.2	0.9993	0.9986	0.9975	0.9954	0.9919	0.9861	0.9766	0.9622	0.9412	0.9130	0.8777
7.3	0.9993	0.9987	0.9977	0.9957	0.9924	0.9867	0.9776	0.9634	0.9427	0.9146	0.8791
7.4	0.9994	0.9988	0.9978	0.9960	0.9928	0.9874	0.9785	0.9645	0.9440	0.9160	0.8806
7.5	0.9994	0.9989	0.9980	0.9963	0.9933	0.9880	0.9793	0.9656	0.9454	0.9175	0.8820
7.6	0.9995	0.9990	0.9982	0.9965	0.9937	0.9886	0.9802	0.9667	0.9467	0.9189	0.8833
7.7	0.9995	0.9991	0.9983	0.9968	0.9940	0.9892	0.9810	0.9678	0.9479	0.9202	0.8847
7.8	0.9996	0.9992	0.9984	0.9970	0.9944	0.9897	0.9817	0.9688	0.9492	0.9216	0.8860
7.9	0.9996	0.9993	0.9986	0.9972	0.9947	0.9902	0.9824	0.9698	0.9504	0.9229	0.8872
8.0	0.9997	0.9993	0.9987	0.9974	0.9950	0.9907	0.9831	0.9707	0.9515	0.9242	0.8885
8.1	0.9997	0.9994	0.9988	0.9976	0.9953	0.9911	0.9838	0.9716	0.9527	0.9254	0.8897
8.2	0.9997	0.9994	0.9989	0.9977	0.9956	0.9916	0.9845	0.9725	0.9538	0.9266	0.8909
8.3	0.9998	0.9995	0.9989	0.9979	0.9958	0.9920	0.9851	0.9734	0.9548	0.9278	0.8920
8.4	0.9998	0.9995	0.9990	0.9980	0.9961	0.9924	0.9857	0.9742	0.9559	0.9290	0.8932
8.5	0.9998	0.9996	0.9991	0.9982	0.9963	0.9927	0.9862	0.9750	0.9569	0.9301	0.8943
8.6	0.9998	0.9996	0.9992	0.9983	0.9965	0.9931	0.9868	0.9758	0.9579	0.9312	0.8954
8.7	0.9998	0.9996	0.9992	0.9984	0.9967	0.9934	0.9873	0.9765	0.9588	0.9323	0.8964
8.8	0.9998	0.9997	0.9993	0.9985	0.9969	0.9938	0.9878	0.9772	0.9598	0.9334	0.8975
8.9	0.9999	0.9997	0.9993	0.9986	0.9971	0.9941	0.9883	0.9779	0.9607	0.9344	0.8985
9.0	0.9999	0.9997	0.9994	0.9987	0.9973	0.9943	0.9888	0.9786	0.9616	0.9354	0.8995
9.1	0.9999	0.9997	0.9994	0.9988	0.9974	0.9946	0.9892	0.9793	0.9624	0.9364	0.9005
9.2	0.9999	0.9998	0.9995	0.9989	0.9976	0.9949	0.9896	0.9799	0.9633	0.9374	0.9015
9.3	0.9999	0.9998	0.9995	0.9989	0.9977	0.9951	0.9900	0.9805	0.9641	0.9384	0.9024
9.4	0.9999	0.9998	0.9996	0.9990	0.9978	0.9954	0.9904	0.9811	0.9649	0.9393	0.9033
9.5	0.9999	0.9998	0.9996	0.9991	0.9980	0.9956	0.9908	0.9817	0.9657	0.9402	0.9042
9.6	0.9999	0.9998	0.9996	0.9991	0.9981	0.9958	0.9912	0.9822	0.9664	0.9411	0.9051
9.7	0.9999	0.9999	0.9997	0.9992	0.9982	0.9960	0.9915	0.9828	0.9672	0.9420	0.9060
9.8	0.9999	0.9999	0.9997	0.9993	0.9983	0.9962	0.9918	0.9833	0.9679	0.9428	0.9069
9.9	1.0000	0.9999	0.9997	0.9993	0.9984	0.9964	0.9922	0.9838	0.9686	0.9437	0.9077
10.0	1.0000	0.9999	0.9997	0.9994	0.9985	0.9966	0.9925	0.9843	0.9693	0.9445	0.9085

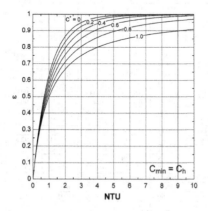

CONFIGURATION 28

1TUBE100ROW100PASS1CIR(CC)

(a) $C_{min} = C_c$

NTU	Heat Exchanger Effectiveness										
	C^* ($C_{min} = C_c$)										
	0	0.1	0.2	0.3	0.4	0.5	0.6	0.7	0.8	0.9	1.0
0	0.0000	0.0000	0.0000	0.0000	0.0000	0.0000	0.0000	0.0000	0.0000	0.0000	0.0000
0.1	0.0952	0.0947	0.0943	0.0939	0.0934	0.0930	0.0926	0.0922	0.0917	0.0913	0.0909
0.2	0.1813	0.1797	0.1782	0.1767	0.1753	0.1738	0.1723	0.1709	0.1695	0.1681	0.1667
0.3	0.2592	0.2562	0.2532	0.2503	0.2474	0.2445	0.2417	0.2389	0.2362	0.2334	0.2308
0.4	0.3297	0.3250	0.3204	0.3158	0.3113	0.3069	0.3025	0.2982	0.2940	0.2898	0.2857
0.5	0.3935	0.3871	0.3807	0.3745	0.3683	0.3623	0.3563	0.3504	0.3446	0.3389	0.3333
0.6	0.4512	0.4431	0.4351	0.4271	0.4194	0.4117	0.4041	0.3966	0.3893	0.3821	0.3750
0.7	0.5034	0.4937	0.4841	0.4746	0.4652	0.4560	0.4468	0.4379	0.4290	0.4203	0.4118
0.8	0.5507	0.5395	0.5284	0.5175	0.5066	0.4959	0.4853	0.4748	0.4645	0.4544	0.4444
0.9	0.5934	0.5810	0.5686	0.5563	0.5441	0.5320	0.5200	0.5082	0.4965	0.4850	0.4737
1.0	0.6321	0.6186	0.6050	0.5915	0.5781	0.5647	0.5515	0.5384	0.5254	0.5126	0.5000
1.1	0.6671	0.6527	0.6382	0.6236	0.6091	0.5946	0.5801	0.5658	0.5516	0.5376	0.5238
1.2	0.6988	0.6836	0.6683	0.6528	0.6373	0.6218	0.6063	0.5909	0.5756	0.5604	0.5454
1.3	0.7275	0.7117	0.6957	0.6795	0.6632	0.6468	0.6303	0.6139	0.5975	0.5813	0.5652
1.4	0.7534	0.7373	0.7208	0.7039	0.6869	0.6697	0.6524	0.6350	0.6177	0.6004	0.5833
1.5	0.7769	0.7605	0.7436	0.7263	0.7087	0.6908	0.6727	0.6545	0.6363	0.6181	0.6000
1.6	0.7981	0.7816	0.7645	0.7468	0.7287	0.7102	0.6915	0.6725	0.6534	0.6344	0.6154
1.7	0.8173	0.8008	0.7836	0.7657	0.7472	0.7282	0.7088	0.6892	0.6694	0.6495	0.6296
1.8	0.8347	0.8183	0.8010	0.7830	0.7642	0.7448	0.7250	0.7047	0.6842	0.6635	0.6428
1.9	0.8504	0.8342	0.8170	0.7989	0.7800	0.7603	0.7400	0.7192	0.6980	0.6766	0.6552
2.0	0.8647	0.8487	0.8317	0.8136	0.7945	0.7746	0.7539	0.7326	0.7109	0.6889	0.6667
2.1	0.8775	0.8619	0.8451	0.8271	0.8080	0.7879	0.7669	0.7452	0.7230	0.7003	0.6774
2.2	0.8892	0.8740	0.8575	0.8396	0.8205	0.8003	0.7791	0.7570	0.7343	0.7110	0.6875
2.3	0.8997	0.8850	0.8688	0.8511	0.8322	0.8119	0.7905	0.7681	0.7449	0.7211	0.6970
2.4	0.9093	0.8950	0.8792	0.8618	0.8430	0.8227	0.8012	0.7785	0.7549	0.7306	0.7059
2.5	0.9179	0.9041	0.8887	0.8717	0.8530	0.8328	0.8112	0.7883	0.7643	0.7396	0.7143
2.6	0.9257	0.9125	0.8975	0.8808	0.8623	0.8422	0.8206	0.7975	0.7732	0.7480	0.7222
2.7	0.9328	0.9201	0.9056	0.8892	0.8710	0.8511	0.8294	0.8062	0.7816	0.7561	0.7297
2.8	0.9392	0.9270	0.9130	0.8970	0.8792	0.8593	0.8377	0.8144	0.7896	0.7636	0.7368
2.9	0.9450	0.9333	0.9198	0.9043	0.8867	0.8671	0.8455	0.8221	0.7971	0.7708	0.7436
3.0	0.9502	0.9391	0.9261	0.9110	0.8938	0.8744	0.8529	0.8295	0.8043	0.7777	0.7500
3.1	0.9550	0.9444	0.9319	0.9172	0.9004	0.8813	0.8599	0.8364	0.8111	0.7842	0.7561
3.2	0.9592	0.9492	0.9372	0.9230	0.9065	0.8877	0.8665	0.8430	0.8176	0.7904	0.7619
3.3	0.9631	0.9536	0.9421	0.9284	0.9123	0.8938	0.8727	0.8493	0.8237	0.7963	0.7674
3.4	0.9666	0.9576	0.9466	0.9334	0.9177	0.8995	0.8786	0.8553	0.8296	0.8019	0.7727
3.5	0.9698	0.9613	0.9508	0.9380	0.9227	0.9048	0.8842	0.8609	0.8352	0.8073	0.7777
3.6	0.9727	0.9646	0.9546	0.9423	0.9274	0.9099	0.8895	0.8663	0.8405	0.8125	0.7826
3.7	0.9753	0.9677	0.9581	0.9463	0.9319	0.9147	0.8945	0.8715	0.8456	0.8174	0.7872
3.8	0.9776	0.9705	0.9614	0.9500	0.9360	0.9192	0.8993	0.8764	0.8505	0.8221	0.7916
3.9	0.9798	0.9730	0.9644	0.9534	0.9399	0.9234	0.9038	0.8810	0.8552	0.8266	0.7959
4.0	0.9817	0.9753	0.9671	0.9566	0.9435	0.9274	0.9081	0.8855	0.8597	0.8310	0.8000
4.1	0.9834	0.9775	0.9697	0.9596	0.9469	0.9312	0.9122	0.8897	0.8640	0.8352	0.8039
4.2	0.9850	0.9794	0.9720	0.9624	0.9501	0.9348	0.9160	0.8938	0.8681	0.8392	0.8076
4.3	0.9864	0.9812	0.9742	0.9650	0.9531	0.9381	0.9197	0.8977	0.8720	0.8430	0.8113
4.4	0.9877	0.9828	0.9762	0.9674	0.9559	0.9413	0.9232	0.9014	0.8758	0.8467	0.8148
4.5	0.9889	0.9843	0.9780	0.9696	0.9586	0.9444	0.9266	0.9050	0.8794	0.8503	0.8181
4.6	0.9899	0.9856	0.9797	0.9717	0.9610	0.9472	0.9298	0.9084	0.8830	0.8538	0.8214
4.7	0.9909	0.9869	0.9813	0.9736	0.9634	0.9499	0.9328	0.9116	0.8863	0.8571	0.8245
4.8	0.9918	0.9880	0.9827	0.9754	0.9655	0.9525	0.9357	0.9148	0.8896	0.8603	0.8275
4.9	0.9926	0.9890	0.9841	0.9771	0.9676	0.9549	0.9384	0.9178	0.8927	0.8634	0.8305
5.0	0.9933	0.9900	0.9853	0.9787	0.9695	0.9572	0.9411	0.9206	0.8957	0.8664	0.8333
5.1	0.9939	0.9909	0.9864	0.9801	0.9713	0.9594	0.9436	0.9234	0.8986	0.8693	0.8360
5.2	0.9945	0.9916	0.9875	0.9815	0.9730	0.9614	0.9460	0.9261	0.9014	0.8721	0.8387
5.3	0.9950	0.9924	0.9884	0.9827	0.9746	0.9634	0.9482	0.9286	0.9041	0.8748	0.8412
5.4	0.9955	0.9930	0.9893	0.9839	0.9761	0.9652	0.9504	0.9310	0.9067	0.8774	0.8437
5.5	0.9959	0.9936	0.9902	0.9850	0.9775	0.9670	0.9525	0.9334	0.9092	0.8799	0.8461
5.6	0.9963	0.9942	0.9909	0.9860	0.9789	0.9686	0.9545	0.9357	0.9116	0.8824	0.8484
5.7	0.9967	0.9947	0.9916	0.9870	0.9801	0.9702	0.9564	0.9378	0.9140	0.8848	0.8507
5.8	0.9970	0.9951	0.9923	0.9879	0.9813	0.9717	0.9582	0.9399	0.9163	0.8871	0.8529
5.9	0.9973	0.9956	0.9929	0.9887	0.9824	0.9731	0.9599	0.9419	0.9185	0.8893	0.8550
6.0	0.9975	0.9959	0.9934	0.9895	0.9834	0.9745	0.9616	0.9439	0.9206	0.8915	0.8571

NTU	Heat Exchanger Effectiveness										
	$C^* (C_{min} = C_c)$										
	0	0.1	0.2	0.3	0.4	0.5	0.6	0.7	0.8	0.9	1.0
6.1	0.9978	0.9963	0.9939	0.9902	0.9844	0.9757	0.9632	0.9457	0.9226	0.8936	0.8591
6.2	0.9980	0.9966	0.9944	0.9908	0.9853	0.9769	0.9647	0.9475	0.9246	0.8957	0.8610
6.3	0.9982	0.9969	0.9948	0.9915	0.9862	0.9781	0.9662	0.9493	0.9266	0.8976	0.8629
6.4	0.9983	0.9972	0.9952	0.9920	0.9870	0.9792	0.9675	0.9509	0.9284	0.8996	0.8648
6.5	0.9985	0.9974	0.9956	0.9926	0.9878	0.9802	0.9689	0.9526	0.9302	0.9015	0.8666
6.6	0.9986	0.9976	0.9959	0.9931	0.9885	0.9812	0.9702	0.9541	0.9320	0.9033	0.8683
6.7	0.9988	0.9978	0.9962	0.9936	0.9892	0.9821	0.9714	0.9556	0.9337	0.9051	0.8701
6.8	0.9989	0.9980	0.9965	0.9940	0.9898	0.9830	0.9725	0.9570	0.9353	0.9068	0.8717
6.9	0.9990	0.9982	0.9968	0.9944	0.9904	0.9839	0.9737	0.9584	0.9369	0.9085	0.8733
7.0	0.9991	0.9984	0.9970	0.9948	0.9910	0.9847	0.9747	0.9598	0.9385	0.9101	0.8749
7.1	0.9992	0.9985	0.9973	0.9951	0.9915	0.9854	0.9758	0.9611	0.9400	0.9117	0.8765
7.2	0.9993	0.9986	0.9975	0.9955	0.9920	0.9861	0.9767	0.9623	0.9415	0.9133	0.8780
7.3	0.9993	0.9987	0.9977	0.9958	0.9925	0.9868	0.9777	0.9635	0.9429	0.9148	0.8794
7.4	0.9994	0.9989	0.9979	0.9961	0.9929	0.9875	0.9786	0.9647	0.9443	0.9163	0.8809
7.5	0.9994	0.9990	0.9980	0.9963	0.9933	0.9881	0.9795	0.9658	0.9456	0.9177	0.8823
7.6	0.9995	0.9990	0.9982	0.9966	0.9937	0.9887	0.9803	0.9669	0.9469	0.9192	0.8836
7.7	0.9995	0.9991	0.9983	0.9968	0.9941	0.9892	0.9811	0.9680	0.9482	0.9205	0.8850
7.8	0.9996	0.9992	0.9985	0.9970	0.9944	0.9898	0.9818	0.9690	0.9494	0.9219	0.8863
7.9	0.9996	0.9993	0.9986	0.9972	0.9947	0.9903	0.9826	0.9700	0.9506	0.9232	0.8875
8.0	0.9997	0.9993	0.9987	0.9974	0.9951	0.9908	0.9833	0.9709	0.9518	0.9245	0.8888
8.1	0.9997	0.9994	0.9988	0.9976	0.9953	0.9912	0.9839	0.9718	0.9529	0.9257	0.8900
8.2	0.9997	0.9994	0.9989	0.9978	0.9956	0.9916	0.9846	0.9727	0.9540	0.9269	0.8912
8.3	0.9998	0.9995	0.9990	0.9979	0.9959	0.9921	0.9852	0.9736	0.9551	0.9281	0.8924
8.4	0.9998	0.9995	0.9990	0.9981	0.9961	0.9924	0.9858	0.9744	0.9561	0.9293	0.8935
8.5	0.9998	0.9996	0.9991	0.9982	0.9963	0.9928	0.9864	0.9752	0.9571	0.9304	0.8946
8.6	0.9998	0.9996	0.9992	0.9983	0.9966	0.9932	0.9869	0.9760	0.9581	0.9316	0.8957
8.7	0.9998	0.9997	0.9993	0.9984	0.9968	0.9935	0.9874	0.9767	0.9591	0.9326	0.8968
8.8	0.9998	0.9997	0.9993	0.9985	0.9970	0.9938	0.9879	0.9774	0.9600	0.9337	0.8978
8.9	0.9999	0.9997	0.9994	0.9986	0.9971	0.9941	0.9884	0.9781	0.9609	0.9348	0.8989
9.0	0.9999	0.9997	0.9994	0.9987	0.9973	0.9944	0.9889	0.9788	0.9618	0.9358	0.8999
9.1	0.9999	0.9998	0.9995	0.9988	0.9975	0.9947	0.9893	0.9795	0.9627	0.9368	0.9009
9.2	0.9999	0.9998	0.9995	0.9989	0.9976	0.9950	0.9897	0.9801	0.9635	0.9378	0.9018
9.3	0.9999	0.9998	0.9996	0.9990	0.9978	0.9952	0.9902	0.9807	0.9644	0.9387	0.9028
9.4	0.9999	0.9998	0.9996	0.9991	0.9979	0.9954	0.9905	0.9813	0.9652	0.9397	0.9037
9.5	0.9999	0.9998	0.9996	0.9991	0.9980	0.9957	0.9909	0.9819	0.9659	0.9406	0.9046
9.6	0.9999	0.9999	0.9997	0.9992	0.9981	0.9959	0.9913	0.9824	0.9667	0.9415	0.9055
9.7	0.9999	0.9999	0.9997	0.9992	0.9982	0.9961	0.9916	0.9830	0.9675	0.9424	0.9064
9.8	0.9999	0.9999	0.9997	0.9993	0.9983	0.9963	0.9920	0.9835	0.9682	0.9432	0.9073
9.9	1.0000	0.9999	0.9997	0.9993	0.9984	0.9965	0.9923	0.9840	0.9689	0.9441	0.9081
10.0	1.0000	0.9999	0.9998	0.9994	0.9985	0.9966	0.9926	0.9845	0.9696	0.9449	0.9090

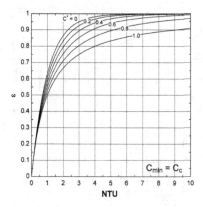

CONFIGURATION 28

1TUBE100ROW100PASS1CIR(CC)

(b) $C_{min} = C_h$

NTU	Heat Exchanger Effectiveness										
	$C^* (C_{min} = C_h)$										
	0	0.1	0.2	0.3	0.4	0.5	0.6	0.7	0.8	0.9	1.0
0	0.0000	0.0000	0.0000	0.0000	0.0000	0.0000	0.0000	0.0000	0.0000	0.0000	0.0000
0.1	0.0952	0.0947	0.0943	0.0939	0.0934	0.0930	0.0926	0.0922	0.0917	0.0913	0.0909
0.2	0.1813	0.1797	0.1782	0.1767	0.1753	0.1738	0.1723	0.1709	0.1695	0.1681	0.1667
0.3	0.2592	0.2562	0.2532	0.2503	0.2474	0.2445	0.2417	0.2389	0.2362	0.2334	0.2308
0.4	0.3297	0.3250	0.3204	0.3158	0.3113	0.3069	0.3025	0.2982	0.2940	0.2898	0.2857
0.5	0.3935	0.3871	0.3807	0.3745	0.3683	0.3623	0.3563	0.3504	0.3446	0.3389	0.3333
0.6	0.4512	0.4431	0.4351	0.4271	0.4194	0.4117	0.4041	0.3966	0.3893	0.3821	0.3750
0.7	0.5034	0.4937	0.4841	0.4746	0.4652	0.4560	0.4468	0.4379	0.4290	0.4203	0.4118
0.8	0.5507	0.5395	0.5284	0.5175	0.5066	0.4959	0.4853	0.4748	0.4645	0.4544	0.4444
0.9	0.5934	0.5810	0.5686	0.5563	0.5441	0.5320	0.5200	0.5082	0.4965	0.4850	0.4737
1.0	0.6321	0.6186	0.6050	0.5915	0.5781	0.5647	0.5515	0.5384	0.5254	0.5126	0.5000
1.1	0.6671	0.6527	0.6382	0.6236	0.6091	0.5946	0.5801	0.5658	0.5516	0.5376	0.5238
1.2	0.6988	0.6836	0.6683	0.6528	0.6373	0.6218	0.6063	0.5909	0.5756	0.5604	0.5454
1.3	0.7275	0.7117	0.6957	0.6795	0.6632	0.6468	0.6303	0.6139	0.5975	0.5813	0.5652
1.4	0.7534	0.7373	0.7208	0.7039	0.6869	0.6697	0.6524	0.6350	0.6177	0.6004	0.5833
1.5	0.7769	0.7605	0.7436	0.7263	0.7087	0.6908	0.6727	0.6545	0.6363	0.6181	0.6000
1.6	0.7981	0.7816	0.7645	0.7468	0.7287	0.7102	0.6915	0.6725	0.6534	0.6344	0.6154
1.7	0.8173	0.8008	0.7836	0.7657	0.7472	0.7282	0.7088	0.6892	0.6694	0.6495	0.6296
1.8	0.8347	0.8183	0.8010	0.7830	0.7642	0.7448	0.7250	0.7047	0.6842	0.6635	0.6428
1.9	0.8504	0.8342	0.8170	0.7989	0.7800	0.7603	0.7400	0.7192	0.6980	0.6766	0.6552
2.0	0.8647	0.8487	0.8317	0.8136	0.7945	0.7746	0.7539	0.7326	0.7109	0.6889	0.6667
2.1	0.8775	0.8619	0.8451	0.8271	0.8080	0.7879	0.7669	0.7452	0.7230	0.7003	0.6774
2.2	0.8892	0.8740	0.8575	0.8396	0.8205	0.8003	0.7791	0.7570	0.7343	0.7110	0.6875
2.3	0.8997	0.8850	0.8688	0.8511	0.8322	0.8119	0.7905	0.7681	0.7449	0.7211	0.6970
2.4	0.9093	0.8950	0.8792	0.8618	0.8430	0.8227	0.8011	0.7785	0.7549	0.7306	0.7059
2.5	0.9179	0.9041	0.8887	0.8717	0.8530	0.8328	0.8112	0.7883	0.7643	0.7396	0.7143
2.6	0.9257	0.9125	0.8975	0.8808	0.8623	0.8422	0.8205	0.7975	0.7732	0.7480	0.7222
2.7	0.9328	0.9201	0.9056	0.8892	0.8710	0.8511	0.8294	0.8062	0.7816	0.7561	0.7297
2.8	0.9392	0.9270	0.9130	0.8970	0.8792	0.8593	0.8377	0.8144	0.7896	0.7636	0.7368
2.9	0.9450	0.9333	0.9198	0.9043	0.8867	0.8671	0.8455	0.8221	0.7971	0.7708	0.7436
3.0	0.9502	0.9391	0.9261	0.9110	0.8938	0.8744	0.8529	0.8295	0.8043	0.7777	0.7500
3.1	0.9550	0.9444	0.9319	0.9172	0.9004	0.8813	0.8599	0.8364	0.8111	0.7842	0.7561
3.2	0.9592	0.9492	0.9372	0.9230	0.9065	0.8877	0.8665	0.8430	0.8176	0.7904	0.7619
3.3	0.9631	0.9536	0.9421	0.9284	0.9123	0.8938	0.8727	0.8493	0.8237	0.7963	0.7674
3.4	0.9666	0.9576	0.9466	0.9334	0.9177	0.8995	0.8786	0.8553	0.8296	0.8019	0.7727
3.5	0.9698	0.9613	0.9507	0.9380	0.9227	0.9048	0.8842	0.8609	0.8352	0.8073	0.7777
3.6	0.9727	0.9646	0.9546	0.9423	0.9274	0.9099	0.8895	0.8663	0.8405	0.8125	0.7826
3.7	0.9753	0.9677	0.9581	0.9463	0.9319	0.9147	0.8945	0.8715	0.8456	0.8174	0.7872
3.8	0.9776	0.9705	0.9614	0.9500	0.9360	0.9192	0.8993	0.8763	0.8505	0.8221	0.7916
3.9	0.9798	0.9730	0.9644	0.9534	0.9399	0.9234	0.9038	0.8810	0.8552	0.8266	0.7959
4.0	0.9817	0.9753	0.9671	0.9566	0.9435	0.9274	0.9081	0.8855	0.8597	0.8310	0.8000
4.1	0.9834	0.9775	0.9697	0.9596	0.9469	0.9312	0.9122	0.8897	0.8640	0.8352	0.8039
4.2	0.9850	0.9794	0.9720	0.9624	0.9501	0.9348	0.9160	0.8938	0.8681	0.8392	0.8076
4.3	0.9864	0.9812	0.9742	0.9650	0.9531	0.9381	0.9197	0.8977	0.8720	0.8430	0.8113
4.4	0.9877	0.9828	0.9762	0.9674	0.9559	0.9413	0.9232	0.9014	0.8758	0.8467	0.8148
4.5	0.9889	0.9843	0.9780	0.9696	0.9585	0.9443	0.9266	0.9049	0.8794	0.8503	0.8181
4.6	0.9899	0.9856	0.9797	0.9717	0.9610	0.9472	0.9298	0.9084	0.8829	0.8538	0.8214
4.7	0.9909	0.9869	0.9813	0.9736	0.9633	0.9499	0.9328	0.9116	0.8863	0.8571	0.8245
4.8	0.9918	0.9880	0.9827	0.9754	0.9655	0.9525	0.9357	0.9148	0.8896	0.8603	0.8275
4.9	0.9926	0.9890	0.9841	0.9771	0.9676	0.9549	0.9384	0.9178	0.8927	0.8634	0.8305
5.0	0.9933	0.9900	0.9853	0.9787	0.9695	0.9572	0.9411	0.9206	0.8957	0.8664	0.8333
5.1	0.9939	0.9909	0.9864	0.9801	0.9713	0.9594	0.9436	0.9234	0.8986	0.8693	0.8360
5.2	0.9945	0.9916	0.9875	0.9815	0.9730	0.9614	0.9459	0.9260	0.9014	0.8721	0.8386
5.3	0.9950	0.9924	0.9884	0.9827	0.9746	0.9634	0.9482	0.9286	0.9041	0.8748	0.8412
5.4	0.9955	0.9930	0.9893	0.9839	0.9761	0.9652	0.9504	0.9310	0.9067	0.8774	0.8437
5.5	0.9959	0.9936	0.9902	0.9850	0.9775	0.9670	0.9525	0.9334	0.9092	0.8799	0.8461
5.6	0.9963	0.9942	0.9909	0.9860	0.9788	0.9686	0.9545	0.9357	0.9116	0.8824	0.8484
5.7	0.9967	0.9947	0.9916	0.9870	0.9801	0.9702	0.9564	0.9378	0.9140	0.8848	0.8507
5.8	0.9970	0.9951	0.9923	0.9879	0.9813	0.9717	0.9582	0.9399	0.9163	0.8871	0.8529
5.9	0.9973	0.9955	0.9929	0.9887	0.9824	0.9731	0.9599	0.9419	0.9185	0.8893	0.8550
6.0	0.9975	0.9959	0.9934	0.9894	0.9834	0.9744	0.9616	0.9439	0.9206	0.8915	0.8571

NTU	Heat Exchanger Effectiveness										
	C^* ($C_{min} = C_h$)										
	0	0.1	0.2	0.3	0.4	0.5	0.6	0.7	0.8	0.9	1.0
6.1	0.9978	0.9963	0.9939	0.9902	0.9844	0.9757	0.9632	0.9457	0.9226	0.8936	0.8591
6.2	0.9980	0.9966	0.9944	0.9908	0.9853	0.9769	0.9647	0.9475	0.9246	0.8956	0.8610
6.3	0.9982	0.9969	0.9948	0.9915	0.9862	0.9781	0.9661	0.9493	0.9266	0.8976	0.8629
6.4	0.9983	0.9972	0.9952	0.9920	0.9870	0.9792	0.9675	0.9509	0.9284	0.8996	0.8648
6.5	0.9985	0.9974	0.9956	0.9926	0.9877	0.9802	0.9689	0.9525	0.9302	0.9015	0.8666
6.6	0.9986	0.9976	0.9959	0.9931	0.9885	0.9812	0.9701	0.9541	0.9320	0.9033	0.8683
6.7	0.9988	0.9978	0.9962	0.9935	0.9891	0.9821	0.9714	0.9556	0.9337	0.9051	0.8700
6.8	0.9989	0.9980	0.9965	0.9940	0.9898	0.9830	0.9725	0.9570	0.9353	0.9068	0.8717
6.9	0.9990	0.9982	0.9968	0.9944	0.9904	0.9838	0.9736	0.9584	0.9369	0.9085	0.8733
7.0	0.9991	0.9983	0.9970	0.9948	0.9909	0.9846	0.9747	0.9598	0.9385	0.9101	0.8749
7.1	0.9992	0.9985	0.9973	0.9951	0.9915	0.9854	0.9757	0.9611	0.9400	0.9117	0.8765
7.2	0.9993	0.9986	0.9975	0.9955	0.9920	0.9861	0.9767	0.9623	0.9415	0.9133	0.8780
7.3	0.9993	0.9987	0.9977	0.9958	0.9924	0.9868	0.9777	0.9635	0.9429	0.9148	0.8794
7.4	0.9994	0.9988	0.9978	0.9960	0.9929	0.9875	0.9786	0.9647	0.9443	0.9163	0.8809
7.5	0.9994	0.9989	0.9980	0.9963	0.9933	0.9881	0.9794	0.9658	0.9456	0.9177	0.8823
7.6	0.9995	0.9990	0.9982	0.9966	0.9937	0.9887	0.9803	0.9669	0.9469	0.9192	0.8836
7.7	0.9995	0.9991	0.9983	0.9968	0.9941	0.9892	0.9811	0.9679	0.9482	0.9205	0.8850
7.8	0.9996	0.9992	0.9984	0.9970	0.9944	0.9898	0.9818	0.9690	0.9494	0.9219	0.8863
7.9	0.9996	0.9993	0.9986	0.9972	0.9947	0.9903	0.9826	0.9699	0.9506	0.9232	0.8875
8.0	0.9997	0.9993	0.9987	0.9974	0.9950	0.9907	0.9833	0.9709	0.9518	0.9245	0.8888
8.1	0.9997	0.9994	0.9988	0.9976	0.9953	0.9912	0.9839	0.9718	0.9529	0.9257	0.8900
8.2	0.9997	0.9994	0.9989	0.9977	0.9956	0.9916	0.9846	0.9727	0.9540	0.9269	0.8912
8.3	0.9998	0.9995	0.9990	0.9979	0.9959	0.9920	0.9852	0.9735	0.9551	0.9281	0.8924
8.4	0.9998	0.9995	0.9990	0.9980	0.9961	0.9924	0.9858	0.9744	0.9561	0.9293	0.8935
8.5	0.9998	0.9996	0.9991	0.9982	0.9963	0.9928	0.9863	0.9752	0.9571	0.9304	0.8946
8.6	0.9998	0.9996	0.9992	0.9983	0.9965	0.9932	0.9869	0.9759	0.9581	0.9316	0.8957
8.7	0.9998	0.9996	0.9992	0.9984	0.9967	0.9935	0.9874	0.9767	0.9591	0.9326	0.8968
8.8	0.9998	0.9997	0.9993	0.9985	0.9969	0.9938	0.9879	0.9774	0.9600	0.9337	0.8978
8.9	0.9999	0.9997	0.9994	0.9986	0.9971	0.9941	0.9884	0.9781	0.9609	0.9348	0.8989
9.0	0.9999	0.9997	0.9994	0.9987	0.9973	0.9944	0.9889	0.9788	0.9618	0.9358	0.8999
9.1	0.9999	0.9998	0.9994	0.9988	0.9974	0.9947	0.9893	0.9794	0.9627	0.9368	0.9009
9.2	0.9999	0.9998	0.9995	0.9989	0.9976	0.9949	0.9897	0.9801	0.9635	0.9377	0.9018
9.3	0.9999	0.9998	0.9995	0.9990	0.9977	0.9952	0.9901	0.9807	0.9644	0.9387	0.9028
9.4	0.9999	0.9998	0.9996	0.9990	0.9979	0.9954	0.9905	0.9813	0.9652	0.9396	0.9037
9.5	0.9999	0.9998	0.9996	0.9991	0.9980	0.9956	0.9909	0.9819	0.9659	0.9406	0.9046
9.6	0.9999	0.9998	0.9996	0.9992	0.9981	0.9959	0.9913	0.9824	0.9667	0.9415	0.9055
9.7	0.9999	0.9999	0.9997	0.9992	0.9982	0.9961	0.9916	0.9830	0.9674	0.9423	0.9064
9.8	0.9999	0.9999	0.9997	0.9993	0.9983	0.9962	0.9919	0.9835	0.9682	0.9432	0.9073
9.9	1.0000	0.9999	0.9997	0.9993	0.9984	0.9964	0.9923	0.9840	0.9689	0.9441	0.9081
10.0	1.0000	0.9999	0.9997	0.9994	0.9985	0.9966	0.9926	0.9845	0.9696	0.9449	0.9090

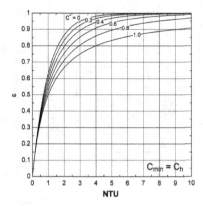

CONFIGURATION 29

2TUBE2ROW2PASS2CIR(IDO)

(a) $C_{min} = C_c$

NTU	Heat Exchanger Effectiveness										
	C^* ($C_{min} = C_c$)										
	0	0.1	0.2	0.3	0.4	0.5	0.6	0.7	0.8	0.9	1.0
0	0.0000	0.0000	0.0000	0.0000	0.0000	0.0000	0.0000	0.0000	0.0000	0.0000	0.0000
0.1	0.0952	0.0947	0.0943	0.0939	0.0934	0.0930	0.0926	0.0921	0.0917	0.0913	0.0909
0.2	0.1813	0.1797	0.1782	0.1767	0.1752	0.1737	0.1722	0.1707	0.1693	0.1679	0.1664
0.3	0.2592	0.2561	0.2531	0.2500	0.2471	0.2442	0.2413	0.2384	0.2356	0.2329	0.2302
0.4	0.3297	0.3248	0.3200	0.3153	0.3107	0.3062	0.3017	0.2973	0.2929	0.2887	0.2845
0.5	0.3935	0.3867	0.3801	0.3736	0.3673	0.3610	0.3548	0.3488	0.3428	0.3370	0.3313
0.6	0.4512	0.4426	0.4342	0.4259	0.4177	0.4097	0.4019	0.3942	0.3866	0.3792	0.3719
0.7	0.5034	0.4930	0.4828	0.4728	0.4629	0.4532	0.4437	0.4344	0.4253	0.4163	0.4076
0.8	0.5507	0.5386	0.5267	0.5150	0.5035	0.4922	0.4811	0.4703	0.4596	0.4492	0.4390
0.9	0.5934	0.5798	0.5664	0.5531	0.5401	0.5273	0.5147	0.5024	0.4903	0.4784	0.4668
1.0	0.6321	0.6171	0.6023	0.5876	0.5732	0.5589	0.5450	0.5312	0.5178	0.5046	0.4917
1.1	0.6671	0.6509	0.6348	0.6189	0.6031	0.5876	0.5723	0.5573	0.5425	0.5281	0.5139
1.2	0.6988	0.6815	0.6643	0.6473	0.6304	0.6136	0.5972	0.5809	0.5650	0.5493	0.5340
1.3	0.7275	0.7093	0.6912	0.6731	0.6551	0.6373	0.6197	0.6024	0.5853	0.5685	0.5521
1.4	0.7534	0.7345	0.7156	0.6966	0.6777	0.6590	0.6403	0.6220	0.6038	0.5860	0.5685
1.5	0.7769	0.7574	0.7378	0.7181	0.6984	0.6787	0.6592	0.6398	0.6207	0.6019	0.5835
1.6	0.7981	0.7782	0.7580	0.7377	0.7173	0.6968	0.6765	0.6562	0.6362	0.6165	0.5971
1.7	0.8173	0.7971	0.7765	0.7557	0.7346	0.7135	0.6923	0.6713	0.6505	0.6299	0.6096
1.8	0.8347	0.8143	0.7934	0.7721	0.7505	0.7288	0.7070	0.6852	0.6636	0.6422	0.6211
1.9	0.8504	0.8299	0.8088	0.7872	0.7652	0.7429	0.7204	0.6980	0.6756	0.6535	0.6317
2.0	0.8647	0.8442	0.8229	0.8010	0.7786	0.7559	0.7329	0.7098	0.6868	0.6639	0.6414
2.1	0.8775	0.8571	0.8358	0.8137	0.7910	0.7679	0.7444	0.7208	0.6971	0.6736	0.6504
2.2	0.8892	0.8689	0.8476	0.8254	0.8025	0.7790	0.7551	0.7309	0.7067	0.6826	0.6587
2.3	0.8997	0.8797	0.8585	0.8362	0.8131	0.7893	0.7650	0.7403	0.7156	0.6909	0.6664
2.4	0.9093	0.8895	0.8684	0.8461	0.8229	0.7988	0.7742	0.7491	0.7238	0.6986	0.6736
2.5	0.9179	0.8984	0.8775	0.8553	0.8320	0.8077	0.7827	0.7573	0.7316	0.7058	0.6802
2.6	0.9257	0.9066	0.8859	0.8638	0.8404	0.8160	0.7907	0.7649	0.7387	0.7125	0.6864
2.7	0.9328	0.9140	0.8936	0.8716	0.8482	0.8236	0.7982	0.7720	0.7454	0.7188	0.6922
2.8	0.9392	0.9208	0.9007	0.8788	0.8555	0.8308	0.8051	0.7786	0.7517	0.7246	0.6976
2.9	0.9450	0.9270	0.9072	0.8855	0.8622	0.8375	0.8116	0.7848	0.7576	0.7301	0.7026
3.0	0.9502	0.9327	0.9132	0.8917	0.8685	0.8437	0.8177	0.7907	0.7631	0.7352	0.7073
3.1	0.9550	0.9379	0.9187	0.8975	0.8743	0.8495	0.8233	0.7961	0.7682	0.7400	0.7117
3.2	0.9592	0.9427	0.9238	0.9028	0.8798	0.8550	0.8287	0.8012	0.7730	0.7445	0.7159
3.3	0.9631	0.9470	0.9285	0.9078	0.8849	0.8600	0.8336	0.8060	0.7776	0.7487	0.7197
3.4	0.9666	0.9510	0.9329	0.9124	0.8896	0.8648	0.8383	0.8105	0.7818	0.7527	0.7234
3.5	0.9698	0.9546	0.9369	0.9167	0.8940	0.8693	0.8427	0.8148	0.7859	0.7564	0.7268
3.6	0.9727	0.9580	0.9406	0.9206	0.8982	0.8735	0.8468	0.8188	0.7896	0.7599	0.7300
3.7	0.9753	0.9610	0.9441	0.9243	0.9020	0.8774	0.8507	0.8225	0.7932	0.7632	0.7330
3.8	0.9776	0.9638	0.9472	0.9278	0.9057	0.8811	0.8544	0.8260	0.7965	0.7663	0.7359
3.9	0.9798	0.9664	0.9502	0.9310	0.9091	0.8845	0.8578	0.8294	0.7997	0.7692	0.7385
4.0	0.9817	0.9688	0.9529	0.9340	0.9122	0.8878	0.8610	0.8325	0.8026	0.7720	0.7411
4.1	0.9834	0.9709	0.9554	0.9368	0.9152	0.8908	0.8641	0.8355	0.8054	0.7746	0.7434
4.2	0.9850	0.9729	0.9578	0.9395	0.9180	0.8937	0.8670	0.8382	0.8081	0.7771	0.7457
4.3	0.9864	0.9748	0.9600	0.9419	0.9206	0.8964	0.8697	0.8409	0.8106	0.7794	0.7478
4.4	0.9877	0.9765	0.9620	0.9442	0.9231	0.8990	0.8722	0.8434	0.8129	0.7816	0.7498
4.5	0.9889	0.9780	0.9639	0.9463	0.9254	0.9014	0.8746	0.8457	0.8152	0.7836	0.7517
4.6	0.9899	0.9795	0.9656	0.9483	0.9276	0.9037	0.8769	0.8479	0.8173	0.7856	0.7534
4.7	0.9909	0.9808	0.9673	0.9502	0.9296	0.9058	0.8791	0.8500	0.8193	0.7874	0.7551
4.8	0.9918	0.9820	0.9688	0.9520	0.9316	0.9078	0.8811	0.8520	0.8212	0.7892	0.7567
4.9	0.9926	0.9831	0.9702	0.9536	0.9334	0.9097	0.8830	0.8539	0.8229	0.7908	0.7582
5.0	0.9933	0.9842	0.9715	0.9552	0.9351	0.9115	0.8848	0.8557	0.8246	0.7924	0.7596
5.1	0.9939	0.9851	0.9728	0.9566	0.9367	0.9132	0.8866	0.8574	0.8262	0.7939	0.7610
5.2	0.9945	0.9860	0.9739	0.9580	0.9382	0.9148	0.8882	0.8589	0.8277	0.7953	0.7623
5.3	0.9950	0.9868	0.9750	0.9593	0.9397	0.9163	0.8897	0.8605	0.8292	0.7966	0.7635
5.4	0.9955	0.9876	0.9760	0.9605	0.9410	0.9178	0.8912	0.8619	0.8305	0.7979	0.7646
5.5	0.9959	0.9883	0.9769	0.9616	0.9423	0.9191	0.8926	0.8632	0.8318	0.7991	0.7657
5.6	0.9963	0.9889	0.9778	0.9627	0.9435	0.9204	0.8939	0.8645	0.8330	0.8002	0.7667
5.7	0.9967	0.9895	0.9786	0.9637	0.9446	0.9216	0.8951	0.8657	0.8342	0.8013	0.7677
5.8	0.9970	0.9901	0.9794	0.9646	0.9457	0.9228	0.8963	0.8669	0.8353	0.8023	0.7686
5.9	0.9973	0.9906	0.9801	0.9655	0.9467	0.9238	0.8974	0.8680	0.8363	0.8033	0.7695
6.0	0.9975	0.9911	0.9808	0.9664	0.9477	0.9249	0.8984	0.8690	0.8373	0.8042	0.7703

NTU	Heat Exchanger Effectiveness										
	C^* ($C_{min} = C_c$)										
	0	0.1	0.2	0.3	0.4	0.5	0.6	0.7	0.8	0.9	1.0
6.1	0.9978	0.9915	0.9814	0.9672	0.9486	0.9258	0.8994	0.8700	0.8383	0.8050	0.7711
6.2	0.9980	0.9919	0.9820	0.9679	0.9494	0.9268	0.9004	0.8709	0.8391	0.8059	0.7718
6.3	0.9982	0.9923	0.9826	0.9686	0.9502	0.9276	0.9013	0.8718	0.8400	0.8066	0.7725
6.4	0.9983	0.9927	0.9831	0.9693	0.9510	0.9285	0.9021	0.8726	0.8408	0.8074	0.7732
6.5	0.9985	0.9930	0.9836	0.9699	0.9517	0.9293	0.9029	0.8734	0.8415	0.8081	0.7738
6.6	0.9986	0.9933	0.9841	0.9705	0.9524	0.9300	0.9037	0.8742	0.8423	0.8087	0.7744
6.7	0.9988	0.9936	0.9845	0.9711	0.9531	0.9307	0.9044	0.8749	0.8429	0.8094	0.7750
6.8	0.9989	0.9939	0.9849	0.9716	0.9537	0.9314	0.9051	0.8756	0.8436	0.8100	0.7756
6.9	0.9990	0.9941	0.9853	0.9721	0.9543	0.9320	0.9058	0.8762	0.8442	0.8105	0.7761
7.0	0.9991	0.9944	0.9857	0.9726	0.9548	0.9326	0.9064	0.8768	0.8448	0.8111	0.7766
7.1	0.9992	0.9946	0.9860	0.9730	0.9553	0.9332	0.9070	0.8774	0.8454	0.8116	0.7770
7.2	0.9993	0.9948	0.9864	0.9734	0.9558	0.9337	0.9075	0.8780	0.8459	0.8121	0.7775
7.3	0.9993	0.9950	0.9867	0.9738	0.9563	0.9342	0.9081	0.8785	0.8464	0.8126	0.7779
7.4	0.9994	0.9952	0.9870	0.9742	0.9567	0.9347	0.9086	0.8790	0.8469	0.8130	0.7783
7.5	0.9994	0.9954	0.9872	0.9746	0.9572	0.9352	0.9091	0.8795	0.8473	0.8134	0.7787
7.6	0.9995	0.9955	0.9875	0.9749	0.9576	0.9356	0.9095	0.8799	0.8477	0.8138	0.7790
7.7	0.9995	0.9957	0.9877	0.9752	0.9579	0.9360	0.9099	0.8804	0.8481	0.8142	0.7794
7.8	0.9996	0.9958	0.9880	0.9755	0.9583	0.9364	0.9103	0.8808	0.8485	0.8145	0.7797
7.9	0.9996	0.9959	0.9882	0.9758	0.9586	0.9368	0.9107	0.8811	0.8489	0.8149	0.7800
8.0	0.9997	0.9961	0.9884	0.9761	0.9590	0.9372	0.9111	0.8815	0.8492	0.8152	0.7803
8.1	0.9997	0.9962	0.9886	0.9764	0.9593	0.9375	0.9115	0.8819	0.8496	0.8155	0.7806
8.2	0.9997	0.9963	0.9888	0.9766	0.9596	0.9378	0.9118	0.8822	0.8499	0.8158	0.7808
8.3	0.9998	0.9964	0.9889	0.9768	0.9599	0.9381	0.9121	0.8825	0.8502	0.8161	0.7811
8.4	0.9998	0.9965	0.9891	0.9771	0.9601	0.9384	0.9124	0.8828	0.8505	0.8164	0.7813
8.5	0.9998	0.9966	0.9893	0.9773	0.9604	0.9387	0.9127	0.8831	0.8508	0.8166	0.7815
8.6	0.9998	0.9967	0.9894	0.9775	0.9606	0.9390	0.9130	0.8834	0.8510	0.8168	0.7817
8.7	0.9998	0.9968	0.9895	0.9776	0.9608	0.9392	0.9132	0.8836	0.8513	0.8171	0.7820
8.8	0.9998	0.9968	0.9897	0.9778	0.9610	0.9394	0.9135	0.8839	0.8515	0.8173	0.7821
8.9	0.9999	0.9969	0.9898	0.9780	0.9612	0.9397	0.9137	0.8841	0.8517	0.8175	0.7823
9.0	0.9999	0.9970	0.9899	0.9782	0.9614	0.9399	0.9139	0.8843	0.8519	0.8177	0.7825
9.1	0.9999	0.9970	0.9900	0.9783	0.9616	0.9401	0.9141	0.8845	0.8521	0.8179	0.7827
9.2	0.9999	0.9971	0.9901	0.9785	0.9618	0.9403	0.9143	0.8847	0.8523	0.8180	0.7828
9.3	0.9999	0.9971	0.9902	0.9786	0.9620	0.9405	0.9145	0.8849	0.8525	0.8182	0.7830
9.4	0.9999	0.9972	0.9903	0.9787	0.9621	0.9406	0.9147	0.8851	0.8527	0.8184	0.7831
9.5	0.9999	0.9973	0.9904	0.9788	0.9623	0.9408	0.9149	0.8853	0.8528	0.8185	0.7833
9.6	0.9999	0.9973	0.9905	0.9790	0.9624	0.9410	0.9150	0.8854	0.8530	0.8187	0.7834
9.7	0.9999	0.9973	0.9906	0.9791	0.9625	0.9411	0.9152	0.8856	0.8531	0.8188	0.7835
9.8	0.9999	0.9974	0.9907	0.9792	0.9627	0.9412	0.9154	0.8857	0.8533	0.8189	0.7836
9.9	1.0000	0.9974	0.9907	0.9793	0.9628	0.9414	0.9155	0.8859	0.8534	0.8191	0.7837
10.0	1.0000	0.9975	0.9908	0.9794	0.9629	0.9415	0.9156	0.8860	0.8535	0.8192	0.7838

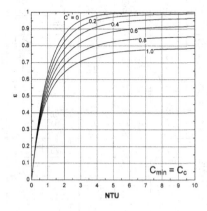

CONFIGURATION 29

2TUBE2ROW2PASS2CIR(IDO)

(b) $C_{min} = C_h$

NTU	Heat Exchanger Effectiveness										
	$C^* (C_{min} = C_h)$										
	0	0.1	0.2	0.3	0.4	0.5	0.6	0.7	0.8	0.9	1.0
0	0.0000	0.0000	0.0000	0.0000	0.0000	0.0000	0.0000	0.0000	0.0000	0.0000	0.0000
0.1	0.0952	0.0947	0.0943	0.0939	0.0934	0.0930	0.0926	0.0921	0.0917	0.0913	0.0909
0.2	0.1813	0.1797	0.1782	0.1767	0.1752	0.1737	0.1722	0.1707	0.1693	0.1679	0.1664
0.3	0.2592	0.2561	0.2531	0.2501	0.2471	0.2442	0.2413	0.2385	0.2356	0.2329	0.2302
0.4	0.3297	0.3248	0.3201	0.3154	0.3107	0.3062	0.3017	0.2973	0.2930	0.2887	0.2845
0.5	0.3935	0.3868	0.3802	0.3737	0.3673	0.3611	0.3549	0.3488	0.3429	0.3370	0.3313
0.6	0.4512	0.4427	0.4342	0.4260	0.4178	0.4098	0.4020	0.3943	0.3867	0.3792	0.3719
0.7	0.5034	0.4931	0.4830	0.4730	0.4631	0.4534	0.4439	0.4345	0.4254	0.4164	0.4076
0.8	0.5507	0.5387	0.5269	0.5153	0.5038	0.4925	0.4814	0.4705	0.4598	0.4493	0.4390
0.9	0.5934	0.5800	0.5667	0.5535	0.5405	0.5277	0.5151	0.5027	0.4905	0.4785	0.4668
1.0	0.6321	0.6174	0.6027	0.5882	0.5737	0.5595	0.5455	0.5316	0.5181	0.5047	0.4917
1.1	0.6671	0.6512	0.6354	0.6196	0.6039	0.5883	0.5730	0.5578	0.5429	0.5283	0.5139
1.2	0.6988	0.6820	0.6651	0.6482	0.6313	0.6145	0.5980	0.5816	0.5654	0.5495	0.5340
1.3	0.7275	0.7098	0.6920	0.6742	0.6563	0.6384	0.6207	0.6032	0.5859	0.5688	0.5521
1.4	0.7534	0.7352	0.7166	0.6979	0.6791	0.6603	0.6415	0.6229	0.6045	0.5863	0.5685
1.5	0.7769	0.7582	0.7390	0.7196	0.7000	0.6803	0.6606	0.6410	0.6215	0.6023	0.5835
1.6	0.7981	0.7791	0.7595	0.7395	0.7192	0.6986	0.6781	0.6575	0.6371	0.6170	0.5971
1.7	0.8173	0.7981	0.7782	0.7577	0.7367	0.7155	0.6942	0.6728	0.6515	0.6304	0.6096
1.8	0.8347	0.8154	0.7952	0.7744	0.7529	0.7311	0.7090	0.6868	0.6647	0.6428	0.6211
1.9	0.8504	0.8312	0.8109	0.7897	0.7679	0.7455	0.7227	0.6998	0.6769	0.6541	0.6317
2.0	0.8647	0.8455	0.8252	0.8038	0.7816	0.7588	0.7354	0.7119	0.6882	0.6647	0.6414
2.1	0.8775	0.8586	0.8383	0.8168	0.7943	0.7711	0.7472	0.7230	0.6987	0.6744	0.6504
2.2	0.8892	0.8705	0.8503	0.8288	0.8061	0.7825	0.7582	0.7334	0.7084	0.6834	0.6587
2.3	0.8997	0.8814	0.8614	0.8399	0.8170	0.7931	0.7683	0.7430	0.7174	0.6918	0.6664
2.4	0.9093	0.8913	0.8715	0.8501	0.8271	0.8029	0.7778	0.7520	0.7259	0.6996	0.6736
2.5	0.9179	0.9004	0.8809	0.8595	0.8365	0.8121	0.7867	0.7604	0.7337	0.7069	0.6802
2.6	0.9257	0.9087	0.8895	0.8683	0.8452	0.8207	0.7949	0.7683	0.7411	0.7137	0.6864
2.7	0.9328	0.9162	0.8974	0.8763	0.8534	0.8287	0.8027	0.7756	0.7480	0.7200	0.6922
2.8	0.9392	0.9231	0.9046	0.8838	0.8609	0.8362	0.8099	0.7825	0.7544	0.7260	0.6976
2.9	0.9450	0.9294	0.9114	0.8908	0.8680	0.8431	0.8167	0.7890	0.7604	0.7315	0.7026
3.0	0.9502	0.9352	0.9175	0.8973	0.8746	0.8497	0.8230	0.7950	0.7661	0.7367	0.7073
3.1	0.9550	0.9405	0.9232	0.9033	0.8807	0.8558	0.8290	0.8007	0.7714	0.7416	0.7117
3.2	0.9592	0.9453	0.9285	0.9089	0.8865	0.8616	0.8346	0.8061	0.7764	0.7462	0.7159
3.3	0.9631	0.9497	0.9334	0.9141	0.8918	0.8670	0.8399	0.8111	0.7811	0.7505	0.7197
3.4	0.9666	0.9538	0.9379	0.9189	0.8969	0.8721	0.8449	0.8159	0.7856	0.7546	0.7234
3.5	0.9698	0.9575	0.9421	0.9234	0.9016	0.8769	0.8496	0.8204	0.7898	0.7584	0.7268
3.6	0.9727	0.9609	0.9459	0.9276	0.9060	0.8814	0.8541	0.8246	0.7937	0.7620	0.7300
3.7	0.9753	0.9640	0.9495	0.9315	0.9102	0.8856	0.8582	0.8287	0.7975	0.7654	0.7330
3.8	0.9776	0.9668	0.9528	0.9352	0.9141	0.8896	0.8622	0.8324	0.8010	0.7686	0.7359
3.9	0.9798	0.9695	0.9558	0.9386	0.9178	0.8934	0.8660	0.8360	0.8043	0.7716	0.7385
4.0	0.9817	0.9719	0.9587	0.9418	0.9212	0.8970	0.8695	0.8394	0.8075	0.7745	0.7411
4.1	0.9834	0.9741	0.9613	0.9448	0.9245	0.9003	0.8729	0.8427	0.8105	0.7772	0.7434
4.2	0.9850	0.9761	0.9638	0.9476	0.9275	0.9035	0.8760	0.8457	0.8133	0.7797	0.7457
4.3	0.9864	0.9779	0.9660	0.9503	0.9304	0.9065	0.8791	0.8486	0.8160	0.7821	0.7478
4.4	0.9877	0.9796	0.9681	0.9527	0.9331	0.9094	0.8819	0.8514	0.8186	0.7844	0.7498
4.5	0.9889	0.9812	0.9701	0.9551	0.9357	0.9121	0.8847	0.8540	0.8210	0.7866	0.7517
4.6	0.9899	0.9826	0.9719	0.9572	0.9381	0.9147	0.8872	0.8565	0.8233	0.7886	0.7534
4.7	0.9909	0.9840	0.9736	0.9593	0.9404	0.9171	0.8897	0.8588	0.8255	0.7906	0.7551
4.8	0.9918	0.9852	0.9752	0.9612	0.9426	0.9195	0.8920	0.8611	0.8276	0.7924	0.7567
4.9	0.9926	0.9863	0.9767	0.9630	0.9447	0.9217	0.8943	0.8632	0.8295	0.7942	0.7582
5.0	0.9933	0.9873	0.9781	0.9647	0.9466	0.9238	0.8964	0.8653	0.8314	0.7958	0.7596
5.1	0.9939	0.9883	0.9794	0.9663	0.9485	0.9257	0.8984	0.8672	0.8332	0.7974	0.7610
5.2	0.9945	0.9892	0.9806	0.9678	0.9502	0.9276	0.9004	0.8691	0.8349	0.7989	0.7623
5.3	0.9950	0.9900	0.9817	0.9692	0.9519	0.9294	0.9022	0.8709	0.8365	0.8003	0.7635
5.4	0.9955	0.9907	0.9827	0.9705	0.9535	0.9312	0.9039	0.8725	0.8381	0.8017	0.7646
5.5	0.9959	0.9914	0.9837	0.9718	0.9550	0.9328	0.9056	0.8742	0.8395	0.8030	0.7657
5.6	0.9963	0.9920	0.9846	0.9730	0.9564	0.9344	0.9072	0.8757	0.8409	0.8042	0.7667
5.7	0.9967	0.9926	0.9855	0.9741	0.9577	0.9359	0.9088	0.8772	0.8423	0.8054	0.7677
5.8	0.9970	0.9931	0.9863	0.9752	0.9590	0.9373	0.9102	0.8786	0.8435	0.8065	0.7686
5.9	0.9973	0.9936	0.9870	0.9762	0.9602	0.9387	0.9116	0.8799	0.8448	0.8075	0.7695
6.0	0.9975	0.9941	0.9877	0.9772	0.9614	0.9400	0.9130	0.8812	0.8459	0.8085	0.7703

NTU	Heat Exchanger Effectiveness										
	C^* ($C_{min} = C_h$)										
	0	0.1	0.2	0.3	0.4	0.5	0.6	0.7	0.8	0.9	1.0
6.1	0.9978	0.9945	0.9884	0.9781	0.9625	0.9412	0.9142	0.8824	0.8470	0.8095	0.7711
6.2	0.9980	0.9949	0.9890	0.9789	0.9636	0.9424	0.9155	0.8836	0.8481	0.8104	0.7718
6.3	0.9982	0.9953	0.9896	0.9797	0.9646	0.9435	0.9166	0.8847	0.8491	0.8112	0.7725
6.4	0.9983	0.9956	0.9901	0.9805	0.9656	0.9446	0.9178	0.8858	0.8501	0.8120	0.7732
6.5	0.9985	0.9959	0.9906	0.9812	0.9665	0.9457	0.9189	0.8868	0.8510	0.8128	0.7738
6.6	0.9986	0.9962	0.9911	0.9819	0.9674	0.9467	0.9199	0.8878	0.8519	0.8136	0.7744
6.7	0.9988	0.9965	0.9915	0.9826	0.9682	0.9476	0.9209	0.8888	0.8527	0.8143	0.7750
6.8	0.9989	0.9967	0.9920	0.9832	0.9690	0.9486	0.9218	0.8897	0.8535	0.8150	0.7756
6.9	0.9990	0.9969	0.9924	0.9838	0.9698	0.9494	0.9228	0.8906	0.8543	0.8156	0.7761
7.0	0.9991	0.9971	0.9927	0.9844	0.9705	0.9503	0.9236	0.8914	0.8550	0.8162	0.7766
7.1	0.9992	0.9973	0.9931	0.9849	0.9712	0.9511	0.9245	0.8922	0.8557	0.8168	0.7770
7.2	0.9993	0.9975	0.9934	0.9854	0.9719	0.9519	0.9253	0.8930	0.8564	0.8174	0.7775
7.3	0.9993	0.9977	0.9937	0.9859	0.9725	0.9526	0.9261	0.8937	0.8570	0.8179	0.7779
7.4	0.9994	0.9978	0.9940	0.9864	0.9732	0.9534	0.9268	0.8944	0.8577	0.8184	0.7783
7.5	0.9994	0.9980	0.9943	0.9868	0.9738	0.9541	0.9276	0.8951	0.8583	0.8189	0.7787
7.6	0.9995	0.9981	0.9946	0.9872	0.9743	0.9547	0.9282	0.8957	0.8588	0.8193	0.7790
7.7	0.9995	0.9983	0.9948	0.9876	0.9749	0.9554	0.9289	0.8964	0.8594	0.8198	0.7794
7.8	0.9996	0.9984	0.9951	0.9880	0.9754	0.9560	0.9296	0.8970	0.8599	0.8202	0.7797
7.9	0.9996	0.9985	0.9953	0.9884	0.9759	0.9566	0.9302	0.8976	0.8604	0.8206	0.7800
8.0	0.9997	0.9986	0.9955	0.9887	0.9764	0.9572	0.9308	0.8981	0.8609	0.8210	0.7803
8.1	0.9997	0.9987	0.9957	0.9891	0.9769	0.9577	0.9313	0.8986	0.8613	0.8214	0.7806
8.2	0.9997	0.9988	0.9959	0.9894	0.9773	0.9583	0.9319	0.8992	0.8618	0.8217	0.7808
8.3	0.9998	0.9988	0.9961	0.9897	0.9777	0.9588	0.9324	0.8996	0.8622	0.8220	0.7811
8.4	0.9998	0.9989	0.9962	0.9900	0.9782	0.9593	0.9329	0.9001	0.8626	0.8224	0.7813
8.5	0.9998	0.9990	0.9964	0.9903	0.9786	0.9597	0.9334	0.9006	0.8630	0.8227	0.7815
8.6	0.9998	0.9990	0.9966	0.9905	0.9789	0.9602	0.9339	0.9010	0.8633	0.8229	0.7817
8.7	0.9998	0.9991	0.9967	0.9908	0.9793	0.9606	0.9344	0.9014	0.8637	0.8232	0.7820
8.8	0.9998	0.9992	0.9968	0.9911	0.9797	0.9611	0.9348	0.9018	0.8640	0.8235	0.7821
8.9	0.9999	0.9992	0.9970	0.9913	0.9800	0.9615	0.9352	0.9022	0.8644	0.8237	0.7823
9.0	0.9999	0.9993	0.9971	0.9915	0.9804	0.9619	0.9357	0.9026	0.8647	0.8240	0.7825
9.1	0.9999	0.9993	0.9972	0.9917	0.9807	0.9623	0.9361	0.9030	0.8650	0.8242	0.7827
9.2	0.9999	0.9994	0.9973	0.9920	0.9810	0.9626	0.9364	0.9033	0.8652	0.8244	0.7828
9.3	0.9999	0.9994	0.9974	0.9922	0.9813	0.9630	0.9368	0.9037	0.8655	0.8246	0.7830
9.4	0.9999	0.9994	0.9975	0.9923	0.9816	0.9634	0.9372	0.9040	0.8658	0.8248	0.7831
9.5	0.9999	0.9995	0.9976	0.9925	0.9819	0.9637	0.9375	0.9043	0.8660	0.8250	0.7833
9.6	0.9999	0.9995	0.9977	0.9927	0.9821	0.9640	0.9378	0.9046	0.8663	0.8252	0.7834
9.7	0.9999	0.9995	0.9978	0.9929	0.9824	0.9643	0.9382	0.9049	0.8665	0.8254	0.7835
9.8	0.9999	0.9996	0.9979	0.9931	0.9826	0.9646	0.9385	0.9052	0.8667	0.8256	0.7836
9.9	1.0000	0.9996	0.9980	0.9932	0.9829	0.9649	0.9388	0.9054	0.8670	0.8257	0.7837
10.0	1.0000	0.9996	0.9980	0.9934	0.9831	0.9652	0.9391	0.9057	0.8672	0.8259	0.7838

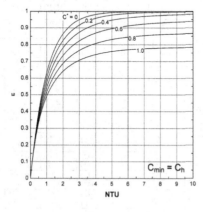

CONFIGURATION 30

2TUBE3ROW3PASS2CIR(IDO)

(a) $C_{min} = C_c$

NTU	Heat Exchanger Effectiveness $C^* (C_{min} = C_c)$										
	0	0.1	0.2	0.3	0.4	0.5	0.6	0.7	0.8	0.9	1.0
0	0.0000	0.0000	0.0000	0.0000	0.0000	0.0000	0.0000	0.0000	0.0000	0.0000	0.0000
0.1	0.0952	0.0947	0.0943	0.0939	0.0934	0.0930	0.0926	0.0921	0.0917	0.0913	0.0909
0.2	0.1813	0.1797	0.1782	0.1767	0.1752	0.1737	0.1723	0.1708	0.1694	0.1680	0.1666
0.3	0.2592	0.2561	0.2531	0.2502	0.2473	0.2444	0.2415	0.2387	0.2359	0.2332	0.2305
0.4	0.3297	0.3249	0.3202	0.3156	0.3111	0.3066	0.3022	0.2978	0.2935	0.2893	0.2852
0.5	0.3935	0.3869	0.3805	0.3741	0.3679	0.3617	0.3556	0.3497	0.3438	0.3381	0.3324
0.6	0.4512	0.4429	0.4347	0.4266	0.4186	0.4108	0.4031	0.3955	0.3881	0.3808	0.3736
0.7	0.5034	0.4934	0.4835	0.4738	0.4642	0.4547	0.4455	0.4363	0.4273	0.4185	0.4099
0.8	0.5507	0.5391	0.5277	0.5164	0.5052	0.4942	0.4834	0.4728	0.4623	0.4521	0.4420
0.9	0.5934	0.5805	0.5676	0.5549	0.5423	0.5299	0.5176	0.5056	0.4937	0.4820	0.4706
1.0	0.6321	0.6179	0.6038	0.5898	0.5759	0.5622	0.5486	0.5352	0.5220	0.5090	0.4963
1.1	0.6671	0.6519	0.6367	0.6215	0.6064	0.5915	0.5767	0.5620	0.5476	0.5333	0.5194
1.2	0.6988	0.6827	0.6665	0.6504	0.6342	0.6182	0.6022	0.5864	0.5708	0.5554	0.5403
1.3	0.7275	0.7107	0.6937	0.6767	0.6596	0.6426	0.6256	0.6087	0.5920	0.5755	0.5593
1.4	0.7534	0.7361	0.7185	0.7007	0.6828	0.6649	0.6470	0.6292	0.6115	0.5939	0.5766
1.5	0.7769	0.7591	0.7410	0.7227	0.7041	0.6854	0.6667	0.6479	0.6293	0.6108	0.5925
1.6	0.7981	0.7801	0.7616	0.7428	0.7236	0.7043	0.6848	0.6652	0.6457	0.6263	0.6071
1.7	0.8173	0.7992	0.7805	0.7612	0.7416	0.7216	0.7015	0.6812	0.6609	0.6406	0.6206
1.8	0.8347	0.8165	0.7977	0.7782	0.7581	0.7377	0.7169	0.6959	0.6749	0.6539	0.6330
1.9	0.8504	0.8323	0.8134	0.7937	0.7734	0.7525	0.7312	0.7096	0.6879	0.6661	0.6445
2.0	0.8647	0.8467	0.8278	0.8081	0.7875	0.7663	0.7445	0.7224	0.7000	0.6775	0.6551
2.1	0.8775	0.8599	0.8411	0.8212	0.8005	0.7790	0.7568	0.7342	0.7113	0.6882	0.6651
2.2	0.8892	0.8718	0.8532	0.8334	0.8126	0.7908	0.7683	0.7453	0.7218	0.6981	0.6743
2.3	0.8997	0.8827	0.8643	0.8446	0.8238	0.8019	0.7791	0.7556	0.7316	0.7073	0.6829
2.4	0.9093	0.8926	0.8745	0.8550	0.8341	0.8121	0.7891	0.7652	0.7408	0.7160	0.6910
2.5	0.9179	0.9017	0.8839	0.8645	0.8438	0.8217	0.7984	0.7743	0.7494	0.7241	0.6986
2.6	0.9257	0.9099	0.8925	0.8734	0.8527	0.8306	0.8072	0.7828	0.7575	0.7318	0.7057
2.7	0.9328	0.9175	0.9004	0.8816	0.8610	0.8389	0.8154	0.7907	0.7652	0.7389	0.7124
2.8	0.9392	0.9244	0.9077	0.8892	0.8688	0.8467	0.8231	0.7982	0.7723	0.7457	0.7187
2.9	0.9450	0.9307	0.9144	0.8962	0.8760	0.8540	0.8304	0.8053	0.7791	0.7521	0.7246
3.0	0.9502	0.9364	0.9206	0.9027	0.8828	0.8609	0.8372	0.8119	0.7855	0.7581	0.7302
3.1	0.9550	0.9417	0.9263	0.9088	0.8891	0.8673	0.8436	0.8182	0.7915	0.7638	0.7354
3.2	0.9592	0.9465	0.9316	0.9144	0.8950	0.8733	0.8496	0.8241	0.7972	0.7692	0.7404
3.3	0.9631	0.9509	0.9364	0.9196	0.9004	0.8790	0.8553	0.8297	0.8026	0.7742	0.7451
3.4	0.9666	0.9549	0.9409	0.9245	0.9056	0.8843	0.8607	0.8350	0.8077	0.7791	0.7496
3.5	0.9698	0.9585	0.9450	0.9290	0.9104	0.8893	0.8658	0.8401	0.8125	0.7837	0.7538
3.6	0.9727	0.9619	0.9488	0.9332	0.9149	0.8940	0.8705	0.8448	0.8171	0.7880	0.7579
3.7	0.9753	0.9650	0.9523	0.9371	0.9191	0.8984	0.8751	0.8493	0.8215	0.7921	0.7617
3.8	0.9776	0.9678	0.9556	0.9407	0.9231	0.9026	0.8794	0.8536	0.8257	0.7961	0.7653
3.9	0.9798	0.9704	0.9586	0.9441	0.9268	0.9066	0.8834	0.8576	0.8296	0.7998	0.7688
4.0	0.9817	0.9727	0.9614	0.9473	0.9303	0.9103	0.8873	0.8615	0.8334	0.8033	0.7720
4.1	0.9834	0.9749	0.9639	0.9503	0.9336	0.9138	0.8909	0.8652	0.8369	0.8067	0.7752
4.2	0.9850	0.9769	0.9663	0.9530	0.9367	0.9171	0.8944	0.8687	0.8403	0.8100	0.7781
4.3	0.9864	0.9787	0.9685	0.9556	0.9396	0.9203	0.8977	0.8720	0.8436	0.8130	0.7810
4.4	0.9877	0.9803	0.9706	0.9580	0.9423	0.9232	0.9008	0.8751	0.8467	0.8160	0.7837
4.5	0.9889	0.9819	0.9725	0.9603	0.9449	0.9261	0.9038	0.8781	0.8496	0.8188	0.7863
4.6	0.9899	0.9833	0.9742	0.9624	0.9473	0.9287	0.9066	0.8810	0.8525	0.8215	0.7888
4.7	0.9909	0.9846	0.9759	0.9644	0.9496	0.9312	0.9093	0.8838	0.8552	0.8240	0.7911
4.8	0.9918	0.9857	0.9774	0.9662	0.9517	0.9336	0.9118	0.8864	0.8577	0.8265	0.7934
4.9	0.9926	0.9868	0.9788	0.9679	0.9538	0.9359	0.9142	0.8889	0.8602	0.8288	0.7955
5.0	0.9933	0.9878	0.9801	0.9696	0.9557	0.9381	0.9165	0.8913	0.8626	0.8311	0.7976
5.1	0.9939	0.9887	0.9813	0.9711	0.9575	0.9401	0.9187	0.8935	0.8648	0.8332	0.7996
5.2	0.9945	0.9896	0.9825	0.9725	0.9592	0.9420	0.9209	0.8957	0.8670	0.8353	0.8015
5.3	0.9950	0.9904	0.9835	0.9739	0.9608	0.9439	0.9229	0.8978	0.8690	0.8373	0.8033
5.4	0.9955	0.9911	0.9845	0.9751	0.9623	0.9456	0.9248	0.8998	0.8710	0.8391	0.8050
5.5	0.9959	0.9917	0.9854	0.9763	0.9638	0.9473	0.9266	0.9017	0.8729	0.8410	0.8067
5.6	0.9963	0.9923	0.9863	0.9774	0.9651	0.9489	0.9283	0.9035	0.8747	0.8427	0.8083
5.7	0.9967	0.9929	0.9870	0.9785	0.9664	0.9504	0.9300	0.9052	0.8765	0.8444	0.8098
5.8	0.9970	0.9934	0.9878	0.9794	0.9676	0.9518	0.9316	0.9069	0.8782	0.8460	0.8113
5.9	0.9973	0.9939	0.9885	0.9804	0.9688	0.9532	0.9331	0.9085	0.8798	0.8475	0.8127
6.0	0.9975	0.9943	0.9891	0.9812	0.9699	0.9545	0.9346	0.9101	0.8813	0.8490	0.8140

NTU	Heat Exchanger Effectiveness										
	C^* ($C_{min} = C_c$)										
	0	0.1	0.2	0.3	0.4	0.5	0.6	0.7	0.8	0.9	1.0
6.1	0.9978	0.9947	0.9897	0.9821	0.9709	0.9557	0.9360	0.9115	0.8828	0.8504	0.8153
6.2	0.9980	0.9951	0.9903	0.9828	0.9719	0.9569	0.9373	0.9129	0.8842	0.8518	0.8166
6.3	0.9982	0.9954	0.9908	0.9836	0.9729	0.9580	0.9386	0.9143	0.8856	0.8531	0.8178
6.4	0.9983	0.9957	0.9913	0.9842	0.9738	0.9591	0.9398	0.9156	0.8869	0.8543	0.8190
6.5	0.9985	0.9960	0.9918	0.9849	0.9746	0.9601	0.9409	0.9168	0.8881	0.8555	0.8201
6.6	0.9986	0.9963	0.9922	0.9855	0.9754	0.9611	0.9421	0.9180	0.8893	0.8567	0.8211
6.7	0.9988	0.9965	0.9926	0.9861	0.9762	0.9621	0.9431	0.9192	0.8905	0.8578	0.8222
6.8	0.9989	0.9968	0.9930	0.9866	0.9769	0.9629	0.9441	0.9203	0.8916	0.8589	0.8231
6.9	0.9990	0.9970	0.9933	0.9872	0.9776	0.9638	0.9451	0.9213	0.8927	0.8599	0.8241
7.0	0.9991	0.9972	0.9936	0.9876	0.9783	0.9646	0.9461	0.9224	0.8937	0.8609	0.8250
7.1	0.9992	0.9974	0.9940	0.9881	0.9789	0.9654	0.9470	0.9233	0.8947	0.8619	0.8259
7.2	0.9993	0.9975	0.9943	0.9885	0.9795	0.9661	0.9478	0.9243	0.8957	0.8628	0.8267
7.3	0.9993	0.9977	0.9945	0.9890	0.9800	0.9669	0.9487	0.9252	0.8966	0.8637	0.8275
7.4	0.9994	0.9978	0.9948	0.9894	0.9806	0.9675	0.9495	0.9261	0.8975	0.8646	0.8283
7.5	0.9994	0.9980	0.9950	0.9897	0.9811	0.9682	0.9502	0.9269	0.8984	0.8654	0.8291
7.6	0.9995	0.9981	0.9952	0.9901	0.9816	0.9688	0.9510	0.9277	0.8992	0.8662	0.8298
7.7	0.9995	0.9982	0.9955	0.9904	0.9821	0.9694	0.9517	0.9285	0.9000	0.8669	0.8305
7.8	0.9996	0.9983	0.9957	0.9907	0.9825	0.9700	0.9524	0.9292	0.9007	0.8677	0.8312
7.9	0.9996	0.9984	0.9959	0.9910	0.9829	0.9705	0.9530	0.9299	0.9015	0.8684	0.8318
8.0	0.9997	0.9985	0.9960	0.9913	0.9833	0.9711	0.9536	0.9306	0.9022	0.8691	0.8325
8.1	0.9997	0.9987	0.9964	0.9918	0.9841	0.9721	0.9548	0.9319	0.9035	0.8704	0.8336
8.2	0.9998	0.9988	0.9965	0.9921	0.9845	0.9725	0.9554	0.9325	0.9042	0.8710	0.8342
8.3	0.9998	0.9988	0.9967	0.9923	0.9848	0.9730	0.9559	0.9331	0.9048	0.8716	0.8347
8.4	0.9998	0.9989	0.9968	0.9925	0.9851	0.9734	0.9564	0.9337	0.9054	0.8722	0.8353
8.5	0.9998	0.9990	0.9969	0.9928	0.9854	0.9738	0.9569	0.9343	0.9059	0.8727	0.8358
8.6	0.9998	0.9990	0.9970	0.9930	0.9857	0.9742	0.9574	0.9348	0.9065	0.8732	0.8363
8.7	0.9998	0.9991	0.9971	0.9932	0.9860	0.9746	0.9579	0.9353	0.9070	0.8737	0.8367
8.8	0.9999	0.9991	0.9972	0.9933	0.9863	0.9749	0.9583	0.9358	0.9075	0.8742	0.8372
8.9	0.9999	0.9992	0.9973	0.9935	0.9866	0.9753	0.9587	0.9363	0.9080	0.8747	0.8376
9.0	0.9999	0.9992	0.9974	0.9937	0.9868	0.9756	0.9591	0.9367	0.9085	0.8752	0.8380
9.1	0.9999	0.9993	0.9975	0.9938	0.9870	0.9759	0.9595	0.9372	0.9090	0.8756	0.8384
9.2	0.9999	0.9993	0.9976	0.9940	0.9873	0.9763	0.9599	0.9376	0.9094	0.8760	0.8388
9.3	0.9999	0.9993	0.9977	0.9941	0.9875	0.9766	0.9603	0.9380	0.9098	0.8765	0.8392
9.4	0.9999	0.9994	0.9978	0.9943	0.9877	0.9768	0.9606	0.9384	0.9102	0.8769	0.8396
9.5	0.9999	0.9994	0.9978	0.9944	0.9879	0.9771	0.9610	0.9388	0.9106	0.8772	0.8399
9.6	0.9999	0.9994	0.9979	0.9945	0.9881	0.9774	0.9613	0.9391	0.9110	0.8776	0.8403
9.7	0.9999	0.9995	0.9980	0.9946	0.9883	0.9776	0.9616	0.9395	0.9114	0.8780	0.8406
9.8	1.0000	0.9995	0.9980	0.9948	0.9885	0.9779	0.9619	0.9398	0.9117	0.8783	0.8409
9.9	1.0000	0.9995	0.9981	0.9949	0.9886	0.9781	0.9622	0.9402	0.9121	0.8787	0.8412
10.0	0.9997	0.9987	0.9964	0.9918	0.9841	0.9721	0.9548	0.9319	0.9035	0.8704	0.8336

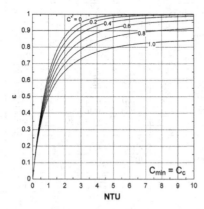

CONFIGURATION 30

2TUBE3ROW3PASS2CIR(IDO)

(b) $C_{min} = C_h$

NTU	Heat Exchanger Effectiveness										
	$C^* (C_{min} = C_h)$										
	0	0.1	0.2	0.3	0.4	0.5	0.6	0.7	0.8	0.9	1.0
0	0.0000	0.0000	0.0000	0.0000	0.0000	0.0000	0.0000	0.0000	0.0000	0.0000	0.0000
0.1	0.0952	0.0947	0.0943	0.0939	0.0934	0.0930	0.0926	0.0921	0.0917	0.0913	0.0909
0.2	0.1813	0.1797	0.1782	0.1767	0.1752	0.1737	0.1723	0.1708	0.1694	0.1680	0.1666
0.3	0.2592	0.2561	0.2531	0.2502	0.2473	0.2444	0.2415	0.2387	0.2359	0.2332	0.2305
0.4	0.3297	0.3249	0.3202	0.3156	0.3111	0.3066	0.3022	0.2978	0.2935	0.2893	0.2852
0.5	0.3935	0.3869	0.3805	0.3741	0.3679	0.3617	0.3557	0.3497	0.3438	0.3381	0.3324
0.6	0.4512	0.4429	0.4347	0.4266	0.4187	0.4109	0.4032	0.3956	0.3881	0.3808	0.3736
0.7	0.5034	0.4934	0.4836	0.4739	0.4643	0.4548	0.4455	0.4364	0.4274	0.4186	0.4099
0.8	0.5507	0.5392	0.5278	0.5165	0.5054	0.4944	0.4835	0.4729	0.4624	0.4521	0.4420
0.9	0.5934	0.5805	0.5677	0.5551	0.5425	0.5301	0.5178	0.5057	0.4938	0.4821	0.4706
1.0	0.6321	0.6180	0.6040	0.5900	0.5761	0.5624	0.5488	0.5353	0.5221	0.5091	0.4963
1.1	0.6671	0.6520	0.6369	0.6218	0.6067	0.5918	0.5769	0.5622	0.5477	0.5334	0.5194
1.2	0.6988	0.6829	0.6668	0.6507	0.6346	0.6186	0.6026	0.5867	0.5710	0.5555	0.5403
1.3	0.7275	0.7109	0.6941	0.6771	0.6601	0.6430	0.6260	0.6091	0.5923	0.5757	0.5593
1.4	0.7534	0.7363	0.7189	0.7012	0.6834	0.6655	0.6475	0.6295	0.6117	0.5941	0.5766
1.5	0.7769	0.7594	0.7415	0.7233	0.7048	0.6860	0.6672	0.6484	0.6296	0.6110	0.5925
1.6	0.7981	0.7805	0.7622	0.7435	0.7244	0.7050	0.6854	0.6657	0.6461	0.6265	0.6071
1.7	0.8173	0.7996	0.7811	0.7621	0.7425	0.7225	0.7022	0.6818	0.6613	0.6408	0.6206
1.8	0.8347	0.8170	0.7984	0.7791	0.7591	0.7386	0.7177	0.6966	0.6753	0.6541	0.6330
1.9	0.8504	0.8328	0.8142	0.7948	0.7745	0.7536	0.7321	0.7104	0.6884	0.6664	0.6445
2.0	0.8647	0.8473	0.8287	0.8092	0.7887	0.7674	0.7455	0.7232	0.7006	0.6778	0.6551
2.1	0.8775	0.8604	0.8420	0.8225	0.8018	0.7803	0.7580	0.7351	0.7119	0.6885	0.6651
2.2	0.8892	0.8724	0.8542	0.8347	0.8140	0.7922	0.7696	0.7462	0.7224	0.6984	0.6743
2.3	0.8997	0.8834	0.8654	0.8460	0.8253	0.8033	0.7804	0.7566	0.7323	0.7077	0.6829
2.4	0.9093	0.8933	0.8757	0.8565	0.8358	0.8137	0.7905	0.7664	0.7416	0.7164	0.6910
2.5	0.9179	0.9024	0.8852	0.8661	0.8455	0.8234	0.7999	0.7755	0.7503	0.7245	0.6986
2.6	0.9257	0.9107	0.8938	0.8751	0.8546	0.8324	0.8088	0.7841	0.7584	0.7322	0.7057
2.7	0.9328	0.9183	0.9018	0.8834	0.8630	0.8408	0.8171	0.7921	0.7661	0.7394	0.7124
2.8	0.9392	0.9252	0.9092	0.8910	0.8708	0.8487	0.8249	0.7997	0.7733	0.7462	0.7187
2.9	0.9450	0.9316	0.9160	0.8982	0.8782	0.8561	0.8323	0.8068	0.7802	0.7526	0.7246
3.0	0.9502	0.9373	0.9222	0.9048	0.8850	0.8631	0.8392	0.8136	0.7866	0.7587	0.7302
3.1	0.9550	0.9426	0.9279	0.9109	0.8914	0.8696	0.8457	0.8199	0.7927	0.7644	0.7354
3.2	0.9592	0.9474	0.9332	0.9166	0.8974	0.8757	0.8518	0.8259	0.7984	0.7698	0.7404
3.3	0.9631	0.9518	0.9381	0.9219	0.9030	0.8815	0.8576	0.8316	0.8039	0.7749	0.7451
3.4	0.9666	0.9559	0.9426	0.9268	0.9082	0.8869	0.8631	0.8370	0.8091	0.7798	0.7496
3.5	0.9698	0.9595	0.9468	0.9314	0.9131	0.8920	0.8682	0.8421	0.8140	0.7844	0.7538
3.6	0.9727	0.9629	0.9506	0.9356	0.9177	0.8968	0.8731	0.8469	0.8186	0.7888	0.7579
3.7	0.9753	0.9660	0.9542	0.9396	0.9220	0.9013	0.8778	0.8515	0.8231	0.7929	0.7617
3.8	0.9776	0.9688	0.9575	0.9433	0.9260	0.9056	0.8821	0.8559	0.8273	0.7969	0.7653
3.9	0.9798	0.9714	0.9605	0.9467	0.9298	0.9096	0.8863	0.8600	0.8313	0.8007	0.7688
4.0	0.9817	0.9737	0.9633	0.9499	0.9334	0.9134	0.8902	0.8640	0.8351	0.8042	0.7720
4.1	0.9834	0.9759	0.9659	0.9529	0.9367	0.9170	0.8940	0.8677	0.8388	0.8077	0.7752
4.2	0.9850	0.9779	0.9683	0.9557	0.9398	0.9205	0.8975	0.8713	0.8422	0.8109	0.7781
4.3	0.9864	0.9797	0.9705	0.9583	0.9428	0.9237	0.9009	0.8747	0.8455	0.8140	0.7810
4.4	0.9877	0.9814	0.9725	0.9608	0.9456	0.9267	0.9041	0.8780	0.8487	0.8170	0.7837
4.5	0.9889	0.9829	0.9744	0.9630	0.9482	0.9296	0.9072	0.8811	0.8517	0.8199	0.7863
4.6	0.9899	0.9843	0.9762	0.9652	0.9507	0.9324	0.9101	0.8840	0.8546	0.8226	0.7888
4.7	0.9909	0.9855	0.9778	0.9672	0.9530	0.9350	0.9128	0.8868	0.8574	0.8252	0.7911
4.8	0.9918	0.9867	0.9793	0.9690	0.9552	0.9374	0.9155	0.8895	0.8600	0.8277	0.7934
4.9	0.9926	0.9878	0.9807	0.9708	0.9573	0.9398	0.9180	0.8921	0.8626	0.8301	0.7955
5.0	0.9933	0.9888	0.9821	0.9724	0.9592	0.9420	0.9204	0.8946	0.8650	0.8323	0.7976
5.1	0.9939	0.9897	0.9833	0.9740	0.9611	0.9441	0.9227	0.8970	0.8673	0.8345	0.7996
5.2	0.9945	0.9905	0.9844	0.9754	0.9628	0.9461	0.9249	0.8992	0.8695	0.8366	0.8015
5.3	0.9950	0.9913	0.9854	0.9767	0.9645	0.9480	0.9270	0.9014	0.8717	0.8386	0.8033
5.4	0.9955	0.9920	0.9864	0.9780	0.9660	0.9498	0.9290	0.9035	0.8737	0.8406	0.8050
5.5	0.9959	0.9926	0.9873	0.9792	0.9675	0.9516	0.9309	0.9055	0.8757	0.8424	0.8067
5.6	0.9963	0.9932	0.9881	0.9803	0.9689	0.9532	0.9327	0.9074	0.8776	0.8442	0.8083
5.7	0.9967	0.9937	0.9889	0.9813	0.9702	0.9548	0.9345	0.9092	0.8794	0.8459	0.8098
5.8	0.9970	0.9942	0.9896	0.9823	0.9715	0.9563	0.9362	0.9110	0.8812	0.8476	0.8113
5.9	0.9973	0.9947	0.9903	0.9833	0.9727	0.9577	0.9378	0.9127	0.8829	0.8491	0.8127
6.0	0.9975	0.9951	0.9909	0.9841	0.9738	0.9591	0.9393	0.9143	0.8845	0.8507	0.8140

NTU	Heat Exchanger Effectiveness										
	C^* ($C_{min} = C_h$)										
	0	0.1	0.2	0.3	0.4	0.5	0.6	0.7	0.8	0.9	1.0
6.1	0.9978	0.9955	0.9915	0.9849	0.9749	0.9604	0.9408	0.9159	0.8860	0.8521	0.8153
6.2	0.9980	0.9958	0.9920	0.9857	0.9759	0.9616	0.9422	0.9174	0.8875	0.8535	0.8166
6.3	0.9982	0.9962	0.9925	0.9864	0.9768	0.9628	0.9436	0.9188	0.8890	0.8549	0.8178
6.4	0.9983	0.9965	0.9930	0.9871	0.9778	0.9639	0.9449	0.9202	0.8904	0.8562	0.8190
6.5	0.9985	0.9967	0.9934	0.9878	0.9786	0.9650	0.9461	0.9216	0.8917	0.8574	0.8201
6.6	0.9986	0.9970	0.9939	0.9884	0.9795	0.9661	0.9473	0.9229	0.8930	0.8586	0.8211
6.7	0.9988	0.9972	0.9942	0.9889	0.9802	0.9670	0.9485	0.9241	0.8942	0.8598	0.8222
6.8	0.9989	0.9974	0.9946	0.9895	0.9810	0.9680	0.9496	0.9253	0.8954	0.8609	0.8231
6.9	0.9990	0.9976	0.9949	0.9900	0.9817	0.9689	0.9506	0.9264	0.8966	0.8620	0.8241
7.0	0.9991	0.9978	0.9952	0.9905	0.9824	0.9698	0.9517	0.9276	0.8977	0.8630	0.8250
7.1	0.9992	0.9980	0.9955	0.9909	0.9830	0.9706	0.9527	0.9286	0.8988	0.8640	0.8259
7.2	0.9993	0.9981	0.9958	0.9913	0.9836	0.9714	0.9536	0.9297	0.8998	0.8650	0.8267
7.3	0.9993	0.9983	0.9960	0.9917	0.9842	0.9722	0.9545	0.9307	0.9008	0.8659	0.8275
7.4	0.9994	0.9984	0.9963	0.9921	0.9848	0.9729	0.9554	0.9316	0.9018	0.8668	0.8283
7.5	0.9994	0.9985	0.9965	0.9925	0.9853	0.9736	0.9563	0.9326	0.9027	0.8677	0.8291
7.6	0.9995	0.9986	0.9967	0.9928	0.9858	0.9743	0.9571	0.9335	0.9036	0.8685	0.8298
7.7	0.9995	0.9987	0.9969	0.9932	0.9863	0.9750	0.9579	0.9343	0.9045	0.8693	0.8305
7.8	0.9996	0.9988	0.9971	0.9935	0.9868	0.9756	0.9586	0.9352	0.9053	0.8701	0.8312
7.9	0.9996	0.9989	0.9972	0.9937	0.9872	0.9762	0.9594	0.9360	0.9061	0.8709	0.8318
8.0	0.9997	0.9990	0.9974	0.9940	0.9876	0.9768	0.9601	0.9367	0.9069	0.8716	0.8325
8.1	0.9997	0.9991	0.9976	0.9943	0.9880	0.9773	0.9607	0.9375	0.9077	0.8723	0.8331
8.2	0.9997	0.9991	0.9977	0.9945	0.9884	0.9778	0.9614	0.9382	0.9084	0.8730	0.8336
8.3	0.9998	0.9992	0.9978	0.9948	0.9888	0.9784	0.9620	0.9389	0.9091	0.8736	0.8342
8.4	0.9998	0.9993	0.9979	0.9950	0.9891	0.9789	0.9627	0.9396	0.9098	0.8743	0.8347
8.5	0.9998	0.9993	0.9981	0.9952	0.9895	0.9793	0.9633	0.9403	0.9105	0.8749	0.8353
8.6	0.9998	0.9994	0.9982	0.9954	0.9898	0.9798	0.9638	0.9409	0.9111	0.8755	0.8358
8.7	0.9998	0.9994	0.9983	0.9956	0.9901	0.9802	0.9644	0.9415	0.9117	0.8760	0.8363
8.8	0.9998	0.9995	0.9984	0.9958	0.9904	0.9807	0.9649	0.9422	0.9123	0.8766	0.8367
8.9	0.9999	0.9995	0.9985	0.9959	0.9907	0.9811	0.9654	0.9427	0.9129	0.8771	0.8372
9.0	0.9999	0.9995	0.9985	0.9961	0.9910	0.9815	0.9659	0.9433	0.9135	0.8776	0.8376
9.1	0.9999	0.9996	0.9986	0.9963	0.9912	0.9818	0.9664	0.9438	0.9140	0.8781	0.8380
9.2	0.9999	0.9996	0.9987	0.9964	0.9915	0.9822	0.9669	0.9444	0.9146	0.8786	0.8384
9.3	0.9999	0.9996	0.9988	0.9965	0.9917	0.9826	0.9674	0.9449	0.9151	0.8791	0.8388
9.4	0.9999	0.9997	0.9988	0.9967	0.9919	0.9829	0.9678	0.9454	0.9156	0.8795	0.8392
9.5	0.9999	0.9997	0.9989	0.9968	0.9922	0.9832	0.9682	0.9459	0.9161	0.8800	0.8396
9.6	0.9999	0.9997	0.9989	0.9969	0.9924	0.9836	0.9686	0.9463	0.9165	0.8804	0.8399
9.7	0.9999	0.9997	0.9990	0.9970	0.9926	0.9839	0.9690	0.9468	0.9170	0.8808	0.8403
9.8	0.9999	0.9997	0.9991	0.9972	0.9928	0.9842	0.9694	0.9472	0.9174	0.8812	0.8406
9.9	1.0000	0.9998	0.9991	0.9973	0.9930	0.9845	0.9698	0.9476	0.9179	0.8816	0.8409
10.0	1.0000	0.9998	0.9991	0.9974	0.9932	0.9847	0.9702	0.9481	0.9183	0.8820	0.8412

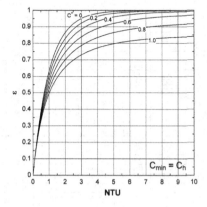

CONFIGURATION 31

2TUBE4ROW4PASS2CIR(IDO)

(a) $C_{min} = C_c$

NTU	Heat Exchanger Effectiveness										
	C^* ($C_{min} = C_c$)										
	0	0.1	0.2	0.3	0.4	0.5	0.6	0.7	0.8	0.9	1.0
0	0.0000	0.0000	0.0000	0.0000	0.0000	0.0000	0.0000	0.0000	0.0000	0.0000	0.0000
0.1	0.0952	0.0947	0.0943	0.0939	0.0934	0.0930	0.0926	0.0922	0.0917	0.0913	0.0909
0.2	0.1813	0.1797	0.1782	0.1767	0.1752	0.1738	0.1723	0.1709	0.1694	0.1680	0.1666
0.3	0.2592	0.2562	0.2532	0.2502	0.2473	0.2444	0.2416	0.2388	0.2360	0.2333	0.2306
0.4	0.3297	0.3250	0.3203	0.3157	0.3112	0.3067	0.3023	0.2980	0.2937	0.2895	0.2854
0.5	0.3935	0.3870	0.3806	0.3743	0.3681	0.3619	0.3559	0.3500	0.3442	0.3385	0.3328
0.6	0.4512	0.4430	0.4348	0.4268	0.4189	0.4112	0.4035	0.3960	0.3886	0.3814	0.3742
0.7	0.5034	0.4935	0.4838	0.4741	0.4646	0.4553	0.4461	0.4370	0.4281	0.4193	0.4107
0.8	0.5507	0.5393	0.5280	0.5168	0.5058	0.4950	0.4842	0.4737	0.4633	0.4531	0.4431
0.9	0.5934	0.5807	0.5680	0.5555	0.5431	0.5308	0.5187	0.5067	0.4949	0.4833	0.4719
1.0	0.6321	0.6182	0.6044	0.5906	0.5769	0.5633	0.5498	0.5366	0.5235	0.5106	0.4979
1.1	0.6671	0.6522	0.6373	0.6224	0.6076	0.5928	0.5782	0.5637	0.5493	0.5352	0.5213
1.2	0.6988	0.6831	0.6673	0.6515	0.6356	0.6198	0.6040	0.5884	0.5729	0.5576	0.5425
1.3	0.7275	0.7111	0.6946	0.6779	0.6612	0.6444	0.6277	0.6110	0.5944	0.5780	0.5619
1.4	0.7534	0.7366	0.7195	0.7021	0.6846	0.6670	0.6493	0.6317	0.6142	0.5968	0.5795
1.5	0.7769	0.7597	0.7422	0.7243	0.7061	0.6878	0.6693	0.6508	0.6323	0.6140	0.5958
1.6	0.7981	0.7808	0.7629	0.7446	0.7259	0.7069	0.6877	0.6684	0.6491	0.6298	0.6107
1.7	0.8173	0.7999	0.7818	0.7632	0.7440	0.7245	0.7047	0.6847	0.6646	0.6445	0.6245
1.8	0.8347	0.8173	0.7992	0.7803	0.7608	0.7408	0.7204	0.6998	0.6789	0.6581	0.6372
1.9	0.8504	0.8332	0.8150	0.7960	0.7763	0.7559	0.7350	0.7138	0.6923	0.6707	0.6491
2.0	0.8647	0.8476	0.8295	0.8105	0.7906	0.7699	0.7486	0.7268	0.7047	0.6824	0.6601
2.1	0.8775	0.8608	0.8429	0.8238	0.8038	0.7829	0.7613	0.7390	0.7163	0.6934	0.6704
2.2	0.8892	0.8728	0.8551	0.8361	0.8161	0.7950	0.7730	0.7504	0.7272	0.7037	0.6800
2.3	0.8997	0.8837	0.8663	0.8475	0.8274	0.8063	0.7841	0.7610	0.7374	0.7133	0.6890
2.4	0.9093	0.8937	0.8766	0.8580	0.8380	0.8167	0.7944	0.7710	0.7469	0.7223	0.6974
2.5	0.9179	0.9028	0.8860	0.8677	0.8478	0.8265	0.8040	0.7804	0.7559	0.7308	0.7053
2.6	0.9257	0.9111	0.8947	0.8767	0.8570	0.8357	0.8130	0.7892	0.7643	0.7388	0.7128
2.7	0.9328	0.9186	0.9027	0.8850	0.8655	0.8443	0.8215	0.7975	0.7723	0.7463	0.7198
2.8	0.9392	0.9256	0.9101	0.8927	0.8734	0.8523	0.8295	0.8053	0.7798	0.7534	0.7265
2.9	0.9450	0.9319	0.9168	0.8998	0.8808	0.8598	0.8370	0.8126	0.7869	0.7602	0.7327
3.0	0.9502	0.9376	0.9231	0.9064	0.8877	0.8668	0.8441	0.8196	0.7936	0.7665	0.7387
3.1	0.9550	0.9429	0.9288	0.9125	0.8941	0.8734	0.8507	0.8262	0.8000	0.7726	0.7443
3.2	0.9592	0.9477	0.9341	0.9182	0.9001	0.8797	0.8570	0.8324	0.8060	0.7783	0.7496
3.3	0.9631	0.9521	0.9390	0.9235	0.9057	0.8855	0.8630	0.8383	0.8117	0.7837	0.7547
3.4	0.9666	0.9561	0.9434	0.9284	0.9110	0.8910	0.8686	0.8439	0.8172	0.7889	0.7595
3.5	0.9698	0.9598	0.9476	0.9330	0.9159	0.8962	0.8739	0.8491	0.8223	0.7938	0.7640
3.6	0.9727	0.9631	0.9514	0.9373	0.9205	0.9010	0.8789	0.8542	0.8273	0.7985	0.7684
3.7	0.9753	0.9662	0.9549	0.9412	0.9248	0.9056	0.8836	0.8590	0.8319	0.8030	0.7726
3.8	0.9776	0.9690	0.9582	0.9449	0.9289	0.9100	0.8881	0.8635	0.8364	0.8072	0.7765
3.9	0.9798	0.9716	0.9612	0.9483	0.9327	0.9141	0.8924	0.8678	0.8407	0.8113	0.7803
4.0	0.9817	0.9739	0.9640	0.9515	0.9363	0.9179	0.8965	0.8720	0.8447	0.8152	0.7839
4.1	0.9834	0.9761	0.9666	0.9545	0.9396	0.9216	0.9003	0.8759	0.8486	0.8189	0.7873
4.2	0.9850	0.9780	0.9689	0.9573	0.9428	0.9250	0.9040	0.8796	0.8523	0.8224	0.7906
4.3	0.9864	0.9798	0.9711	0.9599	0.9457	0.9283	0.9074	0.8832	0.8558	0.8258	0.7938
4.4	0.9877	0.9815	0.9732	0.9623	0.9485	0.9314	0.9107	0.8866	0.8592	0.8291	0.7968
4.5	0.9889	0.9830	0.9751	0.9646	0.9511	0.9343	0.9139	0.8899	0.8625	0.8322	0.7997
4.6	0.9899	0.9844	0.9768	0.9667	0.9536	0.9371	0.9169	0.8930	0.8656	0.8352	0.8025
4.7	0.9909	0.9856	0.9784	0.9687	0.9559	0.9397	0.9197	0.8960	0.8686	0.8381	0.8052
4.8	0.9918	0.9868	0.9799	0.9705	0.9581	0.9422	0.9224	0.8988	0.8714	0.8409	0.8077
4.9	0.9926	0.9879	0.9813	0.9722	0.9601	0.9445	0.9250	0.9015	0.8742	0.8435	0.8102
5.0	0.9933	0.9888	0.9826	0.9738	0.9621	0.9468	0.9275	0.9041	0.8768	0.8460	0.8126
5.1	0.9939	0.9897	0.9837	0.9753	0.9639	0.9489	0.9299	0.9066	0.8793	0.8485	0.8148
5.2	0.9945	0.9906	0.9848	0.9767	0.9656	0.9509	0.9321	0.9090	0.8818	0.8509	0.8170
5.3	0.9950	0.9913	0.9859	0.9781	0.9673	0.9528	0.9342	0.9113	0.8841	0.8531	0.8191
5.4	0.9955	0.9920	0.9868	0.9793	0.9688	0.9546	0.9363	0.9135	0.8864	0.8553	0.8212
5.5	0.9959	0.9926	0.9877	0.9805	0.9702	0.9564	0.9383	0.9156	0.8885	0.8574	0.8231
5.6	0.9963	0.9932	0.9885	0.9815	0.9716	0.9580	0.9401	0.9177	0.8906	0.8594	0.8250
5.7	0.9967	0.9938	0.9893	0.9826	0.9729	0.9596	0.9419	0.9196	0.8926	0.8614	0.8268
5.8	0.9970	0.9942	0.9900	0.9835	0.9741	0.9611	0.9436	0.9215	0.8945	0.8633	0.8285
5.9	0.9973	0.9947	0.9906	0.9844	0.9753	0.9625	0.9453	0.9233	0.8964	0.8651	0.8302
6.0	0.9975	0.9951	0.9912	0.9853	0.9764	0.9638	0.9469	0.9250	0.8982	0.8668	0.8319

NTU	Heat Exchanger Effectiveness										
	C^* ($C_{min} = C_c$)										
	0	0.1	0.2	0.3	0.4	0.5	0.6	0.7	0.8	0.9	1.0
6.1	0.9978	0.9955	0.9918	0.9860	0.9774	0.9651	0.9484	0.9267	0.8999	0.8685	0.8334
6.2	0.9980	0.9958	0.9923	0.9868	0.9784	0.9664	0.9498	0.9282	0.9016	0.8702	0.8349
6.3	0.9982	0.9962	0.9928	0.9875	0.9793	0.9675	0.9512	0.9298	0.9032	0.8717	0.8364
6.4	0.9983	0.9965	0.9933	0.9881	0.9802	0.9686	0.9525	0.9313	0.9047	0.8733	0.8378
6.5	0.9985	0.9967	0.9937	0.9888	0.9811	0.9697	0.9538	0.9327	0.9062	0.8747	0.8392
6.6	0.9986	0.9970	0.9941	0.9893	0.9819	0.9707	0.9550	0.9341	0.9076	0.8761	0.8405
6.7	0.9988	0.9972	0.9945	0.9899	0.9826	0.9717	0.9562	0.9354	0.9090	0.8775	0.8418
6.8	0.9989	0.9974	0.9948	0.9904	0.9833	0.9726	0.9573	0.9366	0.9104	0.8788	0.8430
6.9	0.9990	0.9976	0.9951	0.9909	0.9840	0.9735	0.9584	0.9379	0.9117	0.8801	0.8442
7.0	0.9991	0.9978	0.9954	0.9913	0.9846	0.9743	0.9594	0.9390	0.9129	0.8814	0.8454
7.1	0.9992	0.9980	0.9957	0.9917	0.9852	0.9751	0.9604	0.9402	0.9141	0.8826	0.8465
7.2	0.9993	0.9981	0.9960	0.9922	0.9858	0.9759	0.9614	0.9413	0.9153	0.8837	0.8476
7.3	0.9993	0.9982	0.9962	0.9925	0.9864	0.9767	0.9623	0.9423	0.9164	0.8849	0.8486
7.4	0.9994	0.9984	0.9964	0.9929	0.9869	0.9774	0.9632	0.9434	0.9175	0.8860	0.8496
7.5	0.9994	0.9985	0.9966	0.9932	0.9874	0.9780	0.9640	0.9444	0.9186	0.8870	0.8506
7.6	0.9995	0.9986	0.9968	0.9935	0.9879	0.9787	0.9648	0.9453	0.9196	0.8880	0.8515
7.7	0.9995	0.9987	0.9970	0.9938	0.9883	0.9793	0.9656	0.9462	0.9206	0.8890	0.8525
7.8	0.9996	0.9988	0.9972	0.9941	0.9887	0.9799	0.9664	0.9471	0.9216	0.8900	0.8534
7.9	0.9996	0.9989	0.9973	0.9944	0.9891	0.9805	0.9671	0.9480	0.9225	0.8909	0.8542
8.0	0.9997	0.9990	0.9975	0.9947	0.9895	0.9810	0.9678	0.9488	0.9234	0.8918	0.8551
8.1	0.9997	0.9990	0.9976	0.9949	0.9899	0.9815	0.9685	0.9496	0.9243	0.8927	0.8559
8.2	0.9997	0.9991	0.9978	0.9951	0.9903	0.9820	0.9691	0.9504	0.9251	0.8935	0.8567
8.3	0.9998	0.9992	0.9979	0.9953	0.9906	0.9825	0.9697	0.9511	0.9259	0.8944	0.8574
8.4	0.9998	0.9992	0.9980	0.9955	0.9909	0.9830	0.9703	0.9518	0.9267	0.8952	0.8582
8.5	0.9998	0.9993	0.9981	0.9957	0.9912	0.9834	0.9709	0.9525	0.9275	0.8959	0.8589
8.6	0.9998	0.9993	0.9982	0.9959	0.9915	0.9838	0.9715	0.9532	0.9283	0.8967	0.8596
8.7	0.9998	0.9994	0.9983	0.9961	0.9918	0.9842	0.9720	0.9539	0.9290	0.8974	0.8603
8.8	0.9998	0.9994	0.9984	0.9962	0.9921	0.9846	0.9726	0.9545	0.9297	0.8981	0.8609
8.9	0.9999	0.9995	0.9985	0.9964	0.9923	0.9850	0.9731	0.9551	0.9304	0.8988	0.8616
9.0	0.9999	0.9995	0.9986	0.9965	0.9925	0.9854	0.9735	0.9557	0.9310	0.8995	0.8622
9.1	0.9999	0.9995	0.9986	0.9967	0.9928	0.9857	0.9740	0.9563	0.9317	0.9001	0.8628
9.2	0.9999	0.9996	0.9987	0.9968	0.9930	0.9860	0.9745	0.9568	0.9323	0.9008	0.8634
9.3	0.9999	0.9996	0.9988	0.9969	0.9932	0.9864	0.9749	0.9574	0.9329	0.9014	0.8639
9.4	0.9999	0.9996	0.9988	0.9971	0.9934	0.9867	0.9753	0.9579	0.9335	0.9020	0.8645
9.5	0.9999	0.9996	0.9989	0.9972	0.9936	0.9870	0.9757	0.9584	0.9340	0.9025	0.8650
9.6	0.9999	0.9997	0.9990	0.9973	0.9938	0.9872	0.9761	0.9589	0.9346	0.9031	0.8655
9.7	0.9999	0.9997	0.9990	0.9974	0.9940	0.9875	0.9765	0.9594	0.9351	0.9036	0.8661
9.8	0.9999	0.9997	0.9991	0.9975	0.9941	0.9878	0.9769	0.9598	0.9356	0.9042	0.8665
9.9	1.0000	0.9997	0.9991	0.9976	0.9943	0.9880	0.9772	0.9603	0.9361	0.9047	0.8670
10.0	1.0000	0.9997	0.9991	0.9977	0.9945	0.9883	0.9776	0.9607	0.9366	0.9052	0.8675

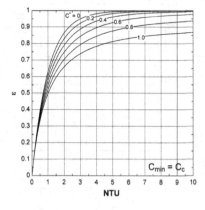

CONFIGURATION 31

2TUBE4ROW4PASS2CIR(IDO)

(b) $C_{min} = C_h$

NTU	Heat Exchanger Effectiveness C^* ($C_{min} = C_h$)										
	0	0.1	0.2	0.3	0.4	0.5	0.6	0.7	0.8	0.9	1.0
0	0.0000	0.0000	0.0000	0.0000	0.0000	0.0000	0.0000	0.0000	0.0000	0.0000	0.0000
0.1	0.0952	0.0947	0.0943	0.0939	0.0934	0.0930	0.0926	0.0922	0.0917	0.0913	0.0909
0.2	0.1813	0.1797	0.1782	0.1767	0.1752	0.1738	0.1723	0.1709	0.1694	0.1680	0.1666
0.3	0.2592	0.2562	0.2532	0.2502	0.2473	0.2444	0.2416	0.2388	0.2360	0.2333	0.2306
0.4	0.3297	0.3250	0.3203	0.3157	0.3112	0.3067	0.3023	0.2980	0.2937	0.2895	0.2854
0.5	0.3935	0.3870	0.3806	0.3743	0.3681	0.3620	0.3559	0.3500	0.3442	0.3385	0.3328
0.6	0.4512	0.4430	0.4349	0.4269	0.4190	0.4112	0.4036	0.3960	0.3886	0.3814	0.3742
0.7	0.5034	0.4936	0.4838	0.4742	0.4647	0.4553	0.4461	0.4370	0.4281	0.4193	0.4107
0.8	0.5507	0.5393	0.5281	0.5169	0.5059	0.4950	0.4843	0.4737	0.4633	0.4531	0.4431
0.9	0.5934	0.5807	0.5681	0.5556	0.5432	0.5309	0.5188	0.5068	0.4950	0.4834	0.4719
1.0	0.6321	0.6183	0.6045	0.5907	0.5770	0.5634	0.5500	0.5367	0.5235	0.5106	0.4979
1.1	0.6671	0.6523	0.6375	0.6226	0.6078	0.5930	0.5783	0.5638	0.5494	0.5353	0.5213
1.2	0.6988	0.6832	0.6675	0.6517	0.6358	0.6200	0.6042	0.5885	0.5730	0.5577	0.5425
1.3	0.7275	0.7112	0.6948	0.6782	0.6614	0.6447	0.6279	0.6112	0.5946	0.5781	0.5619
1.4	0.7534	0.7367	0.7197	0.7024	0.6849	0.6673	0.6496	0.6319	0.6143	0.5968	0.5795
1.5	0.7769	0.7599	0.7424	0.7246	0.7065	0.6881	0.6696	0.6510	0.6325	0.6140	0.5958
1.6	0.7981	0.7810	0.7632	0.7450	0.7263	0.7073	0.6880	0.6687	0.6493	0.6299	0.6107
1.7	0.8173	0.8001	0.7822	0.7636	0.7445	0.7250	0.7051	0.6850	0.6648	0.6446	0.6245
1.8	0.8347	0.8176	0.7996	0.7808	0.7613	0.7413	0.7209	0.7001	0.6792	0.6582	0.6372
1.9	0.8504	0.8334	0.8155	0.7966	0.7769	0.7565	0.7355	0.7142	0.6926	0.6708	0.6491
2.0	0.8647	0.8479	0.8300	0.8111	0.7912	0.7705	0.7492	0.7273	0.7050	0.6826	0.6601
2.1	0.8775	0.8611	0.8434	0.8245	0.8045	0.7836	0.7618	0.7395	0.7167	0.6936	0.6704
2.2	0.8892	0.8731	0.8556	0.8368	0.8168	0.7957	0.7737	0.7509	0.7276	0.7039	0.6800
2.3	0.8997	0.8841	0.8669	0.8483	0.8283	0.8070	0.7847	0.7616	0.7378	0.7135	0.6890
2.4	0.9093	0.8941	0.8772	0.8588	0.8389	0.8176	0.7951	0.7716	0.7473	0.7225	0.6974
2.5	0.9179	0.9032	0.8867	0.8685	0.8487	0.8274	0.8048	0.7810	0.7563	0.7310	0.7053
2.6	0.9257	0.9115	0.8954	0.8776	0.8579	0.8366	0.8139	0.7898	0.7648	0.7390	0.7128
2.7	0.9328	0.9191	0.9035	0.8859	0.8665	0.8453	0.8224	0.7982	0.7728	0.7466	0.7198
2.8	0.9392	0.9260	0.9108	0.8936	0.8744	0.8533	0.8304	0.8060	0.7803	0.7537	0.7265
2.9	0.9450	0.9323	0.9176	0.9008	0.8819	0.8609	0.8380	0.8134	0.7875	0.7604	0.7327
3.0	0.9502	0.9381	0.9239	0.9075	0.8888	0.8680	0.8451	0.8204	0.7942	0.7668	0.7387
3.1	0.9550	0.9434	0.9296	0.9136	0.8953	0.8746	0.8518	0.8270	0.8006	0.7729	0.7443
3.2	0.9592	0.9482	0.9349	0.9194	0.9013	0.8809	0.8581	0.8333	0.8066	0.7786	0.7496
3.3	0.9631	0.9526	0.9398	0.9247	0.9070	0.8868	0.8641	0.8392	0.8124	0.7841	0.7547
3.4	0.9666	0.9566	0.9443	0.9296	0.9123	0.8923	0.8698	0.8448	0.8179	0.7892	0.7595
3.5	0.9698	0.9603	0.9485	0.9342	0.9173	0.8975	0.8751	0.8502	0.8231	0.7942	0.7640
3.6	0.9727	0.9636	0.9523	0.9385	0.9219	0.9024	0.8802	0.8552	0.8280	0.7989	0.7684
3.7	0.9753	0.9667	0.9559	0.9425	0.9263	0.9071	0.8850	0.8601	0.8327	0.8034	0.7726
3.8	0.9776	0.9695	0.9591	0.9462	0.9303	0.9114	0.8895	0.8646	0.8372	0.8076	0.7765
3.9	0.9798	0.9721	0.9622	0.9496	0.9342	0.9156	0.8938	0.8690	0.8415	0.8117	0.7803
4.0	0.9817	0.9744	0.9649	0.9528	0.9378	0.9195	0.8979	0.8732	0.8456	0.8156	0.7839
4.1	0.9834	0.9766	0.9675	0.9558	0.9411	0.9231	0.9018	0.8771	0.8495	0.8193	0.7873
4.2	0.9850	0.9785	0.9699	0.9586	0.9443	0.9266	0.9055	0.8809	0.8532	0.8229	0.7906
4.3	0.9864	0.9803	0.9721	0.9612	0.9473	0.9299	0.9090	0.8845	0.8568	0.8263	0.7938
4.4	0.9877	0.9820	0.9741	0.9636	0.9501	0.9330	0.9123	0.8880	0.8602	0.8296	0.7968
4.5	0.9889	0.9835	0.9760	0.9659	0.9527	0.9360	0.9155	0.8913	0.8635	0.8327	0.7997
4.6	0.9899	0.9849	0.9777	0.9680	0.9552	0.9388	0.9185	0.8944	0.8666	0.8357	0.8025
4.7	0.9909	0.9861	0.9793	0.9700	0.9575	0.9414	0.9214	0.8974	0.8696	0.8386	0.8052
4.8	0.9918	0.9873	0.9808	0.9718	0.9597	0.9439	0.9242	0.9003	0.8725	0.8414	0.8077
4.9	0.9926	0.9883	0.9822	0.9735	0.9618	0.9463	0.9268	0.9030	0.8753	0.8441	0.8102
5.0	0.9933	0.9893	0.9835	0.9751	0.9637	0.9486	0.9293	0.9057	0.8780	0.8466	0.8126
5.1	0.9939	0.9902	0.9846	0.9767	0.9656	0.9507	0.9317	0.9082	0.8805	0.8491	0.8148
5.2	0.9945	0.9910	0.9857	0.9781	0.9673	0.9528	0.9340	0.9107	0.8830	0.8515	0.8170
5.3	0.9950	0.9917	0.9867	0.9794	0.9689	0.9547	0.9361	0.9130	0.8854	0.8538	0.8191
5.4	0.9955	0.9924	0.9877	0.9806	0.9705	0.9565	0.9382	0.9152	0.8876	0.8560	0.8212
5.5	0.9959	0.9930	0.9885	0.9817	0.9719	0.9583	0.9402	0.9174	0.8898	0.8581	0.8231
5.6	0.9963	0.9936	0.9893	0.9828	0.9733	0.9599	0.9421	0.9194	0.8919	0.8601	0.8250
5.7	0.9967	0.9941	0.9901	0.9838	0.9746	0.9615	0.9439	0.9214	0.8940	0.8621	0.8268
5.8	0.9970	0.9946	0.9908	0.9848	0.9758	0.9630	0.9457	0.9233	0.8959	0.8640	0.8285
5.9	0.9973	0.9951	0.9914	0.9857	0.9770	0.9645	0.9474	0.9251	0.8978	0.8658	0.8302
6.0	0.9975	0.9955	0.9920	0.9865	0.9781	0.9658	0.9490	0.9269	0.8996	0.8676	0.8319

NTU	Heat Exchanger Effectiveness										
	C^* $(C_{min} = C_h)$										
	0	0.1	0.2	0.3	0.4	0.5	0.6	0.7	0.8	0.9	1.0
6.1	0.9978	0.9958	0.9926	0.9873	0.9791	0.9671	0.9505	0.9286	0.9014	0.8693	0.8334
6.2	0.9980	0.9962	0.9931	0.9880	0.9801	0.9684	0.9520	0.9302	0.9031	0.8710	0.8349
6.3	0.9982	0.9965	0.9935	0.9887	0.9810	0.9695	0.9534	0.9318	0.9047	0.8726	0.8364
6.4	0.9983	0.9968	0.9940	0.9893	0.9819	0.9707	0.9547	0.9333	0.9063	0.8741	0.8378
6.5	0.9985	0.9970	0.9944	0.9899	0.9827	0.9718	0.9560	0.9348	0.9078	0.8756	0.8392
6.6	0.9986	0.9973	0.9948	0.9905	0.9835	0.9728	0.9573	0.9362	0.9093	0.8770	0.8405
6.7	0.9988	0.9975	0.9951	0.9910	0.9843	0.9738	0.9585	0.9375	0.9107	0.8784	0.8418
6.8	0.9989	0.9977	0.9955	0.9915	0.9850	0.9747	0.9596	0.9389	0.9121	0.8798	0.8430
6.9	0.9990	0.9979	0.9958	0.9920	0.9856	0.9756	0.9607	0.9401	0.9134	0.8811	0.8442
7.0	0.9991	0.9980	0.9960	0.9924	0.9863	0.9764	0.9618	0.9413	0.9147	0.8824	0.8454
7.1	0.9992	0.9982	0.9963	0.9928	0.9869	0.9773	0.9628	0.9425	0.9160	0.8836	0.8465
7.2	0.9993	0.9983	0.9966	0.9932	0.9874	0.9780	0.9638	0.9436	0.9172	0.8848	0.8476
7.3	0.9993	0.9985	0.9968	0.9936	0.9880	0.9788	0.9647	0.9447	0.9184	0.8859	0.8486
7.4	0.9994	0.9986	0.9970	0.9939	0.9885	0.9795	0.9656	0.9458	0.9195	0.8870	0.8496
7.5	0.9994	0.9987	0.9972	0.9943	0.9890	0.9802	0.9665	0.9468	0.9206	0.8881	0.8506
7.6	0.9995	0.9988	0.9974	0.9946	0.9895	0.9808	0.9674	0.9478	0.9216	0.8892	0.8515
7.7	0.9995	0.9989	0.9975	0.9948	0.9899	0.9815	0.9682	0.9488	0.9227	0.8902	0.8525
7.8	0.9996	0.9990	0.9977	0.9951	0.9903	0.9821	0.9689	0.9497	0.9237	0.8911	0.8534
7.9	0.9996	0.9991	0.9979	0.9954	0.9907	0.9826	0.9697	0.9506	0.9246	0.8921	0.8542
8.0	0.9997	0.9991	0.9980	0.9956	0.9911	0.9832	0.9704	0.9515	0.9256	0.8930	0.8551
8.1	0.9997	0.9992	0.9981	0.9958	0.9915	0.9837	0.9711	0.9523	0.9265	0.8939	0.8559
8.2	0.9997	0.9993	0.9982	0.9961	0.9918	0.9842	0.9718	0.9531	0.9274	0.8948	0.8567
8.3	0.9998	0.9993	0.9984	0.9963	0.9921	0.9847	0.9725	0.9539	0.9282	0.8956	0.8574
8.4	0.9998	0.9994	0.9985	0.9964	0.9925	0.9852	0.9731	0.9547	0.9291	0.8965	0.8582
8.5	0.9998	0.9994	0.9986	0.9966	0.9928	0.9856	0.9737	0.9554	0.9299	0.8973	0.8589
8.6	0.9998	0.9995	0.9986	0.9968	0.9930	0.9861	0.9743	0.9561	0.9306	0.8980	0.8596
8.7	0.9998	0.9995	0.9987	0.9970	0.9933	0.9865	0.9748	0.9568	0.9314	0.8988	0.8603
8.8	0.9998	0.9996	0.9988	0.9971	0.9936	0.9869	0.9754	0.9575	0.9321	0.8995	0.8609
8.9	0.9999	0.9996	0.9989	0.9972	0.9938	0.9873	0.9759	0.9581	0.9329	0.9002	0.8616
9.0	0.9999	0.9996	0.9990	0.9974	0.9940	0.9876	0.9764	0.9588	0.9336	0.9009	0.8622
9.1	0.9999	0.9996	0.9990	0.9975	0.9943	0.9880	0.9769	0.9594	0.9342	0.9016	0.8628
9.2	0.9999	0.9997	0.9991	0.9976	0.9945	0.9883	0.9774	0.9600	0.9349	0.9022	0.8634
9.3	0.9999	0.9997	0.9991	0.9977	0.9947	0.9886	0.9778	0.9605	0.9355	0.9029	0.8639
9.4	0.9999	0.9997	0.9992	0.9979	0.9949	0.9890	0.9783	0.9611	0.9362	0.9035	0.8645
9.5	0.9999	0.9997	0.9992	0.9980	0.9951	0.9893	0.9787	0.9616	0.9368	0.9041	0.8650
9.6	0.9999	0.9998	0.9993	0.9980	0.9952	0.9895	0.9791	0.9622	0.9374	0.9047	0.8655
9.7	0.9999	0.9998	0.9993	0.9981	0.9954	0.9898	0.9795	0.9627	0.9379	0.9052	0.8661
9.8	0.9999	0.9998	0.9994	0.9982	0.9956	0.9901	0.9799	0.9632	0.9385	0.9058	0.8665
9.9	1.0000	0.9998	0.9994	0.9983	0.9957	0.9904	0.9803	0.9637	0.9390	0.9063	0.8670
10.0	1.0000	0.9998	0.9994	0.9984	0.9959	0.9906	0.9807	0.9641	0.9396	0.9069	0.8675

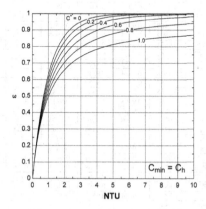

CONFIGURATION 32

2TUBE5ROW5PASS2CIR(IDO)

(a) $C_{min} = C_c$

NTU	Heat Exchanger Effectiveness										
	C^* ($C_{min} = C_c$)										
	0	0.1	0.2	0.3	0.4	0.5	0.6	0.7	0.8	0.9	1.0
0	0.0000	0.0000	0.0000	0.0000	0.0000	0.0000	0.0000	0.0000	0.0000	0.0000	0.0000
0.1	0.0952	0.0947	0.0943	0.0939	0.0934	0.0930	0.0926	0.0922	0.0917	0.0913	0.0909
0.2	0.1813	0.1797	0.1782	0.1767	0.1752	0.1738	0.1723	0.1709	0.1694	0.1680	0.1666
0.3	0.2592	0.2562	0.2532	0.2502	0.2473	0.2445	0.2416	0.2388	0.2361	0.2334	0.2307
0.4	0.3297	0.3250	0.3203	0.3157	0.3112	0.3068	0.3024	0.2981	0.2938	0.2896	0.2855
0.5	0.3935	0.3870	0.3806	0.3743	0.3682	0.3621	0.3561	0.3502	0.3443	0.3386	0.3330
0.6	0.4512	0.4430	0.4349	0.4269	0.4191	0.4114	0.4037	0.3962	0.3889	0.3816	0.3745
0.7	0.5034	0.4936	0.4839	0.4743	0.4649	0.4555	0.4463	0.4373	0.4284	0.4197	0.4111
0.8	0.5507	0.5394	0.5282	0.5171	0.5061	0.4953	0.4846	0.4741	0.4637	0.4536	0.4436
0.9	0.5934	0.5808	0.5682	0.5558	0.5434	0.5312	0.5191	0.5072	0.4955	0.4839	0.4726
1.0	0.6321	0.6183	0.6046	0.5909	0.5773	0.5638	0.5504	0.5372	0.5242	0.5113	0.4986
1.1	0.6671	0.6524	0.6376	0.6229	0.6081	0.5935	0.5789	0.5644	0.5502	0.5361	0.5222
1.2	0.6988	0.6833	0.6677	0.6520	0.6362	0.6205	0.6048	0.5893	0.5739	0.5586	0.5436
1.3	0.7275	0.7113	0.6950	0.6785	0.6619	0.6453	0.6286	0.6120	0.5955	0.5792	0.5631
1.4	0.7534	0.7368	0.7199	0.7028	0.6854	0.6680	0.6504	0.6329	0.6154	0.5981	0.5809
1.5	0.7769	0.7600	0.7427	0.7250	0.7070	0.6888	0.6705	0.6521	0.6337	0.6154	0.5973
1.6	0.7981	0.7811	0.7635	0.7454	0.7269	0.7081	0.6890	0.6699	0.6506	0.6314	0.6124
1.7	0.8173	0.8002	0.7825	0.7641	0.7452	0.7258	0.7062	0.6863	0.6663	0.6463	0.6263
1.8	0.8347	0.8177	0.7998	0.7813	0.7620	0.7423	0.7221	0.7015	0.6808	0.6600	0.6393
1.9	0.8504	0.8336	0.8157	0.7971	0.7776	0.7575	0.7368	0.7157	0.6943	0.6728	0.6513
2.0	0.8647	0.8480	0.8303	0.8116	0.7920	0.7716	0.7505	0.7289	0.7069	0.6847	0.6625
2.1	0.8775	0.8612	0.8437	0.8250	0.8053	0.7847	0.7633	0.7412	0.7187	0.6959	0.6729
2.2	0.8892	0.8732	0.8559	0.8374	0.8177	0.7969	0.7752	0.7528	0.7297	0.7063	0.6827
2.3	0.8997	0.8842	0.8672	0.8488	0.8291	0.8083	0.7864	0.7636	0.7401	0.7161	0.6918
2.4	0.9093	0.8942	0.8775	0.8594	0.8398	0.8189	0.7968	0.7737	0.7498	0.7253	0.7004
2.5	0.9179	0.9033	0.8870	0.8691	0.8497	0.8288	0.8066	0.7832	0.7589	0.7339	0.7085
2.6	0.9257	0.9116	0.8957	0.8782	0.8589	0.8381	0.8157	0.7922	0.7675	0.7421	0.7162
2.7	0.9328	0.9192	0.9037	0.8865	0.8675	0.8467	0.8244	0.8006	0.7756	0.7498	0.7234
2.8	0.9392	0.9261	0.9111	0.8943	0.8755	0.8548	0.8325	0.8085	0.7833	0.7571	0.7302
2.9	0.9450	0.9324	0.9179	0.9014	0.8829	0.8624	0.8401	0.8160	0.7906	0.7640	0.7366
3.0	0.9502	0.9382	0.9242	0.9081	0.8899	0.8696	0.8473	0.8231	0.7975	0.7705	0.7427
3.1	0.9550	0.9434	0.9299	0.9143	0.8964	0.8763	0.8540	0.8299	0.8040	0.7767	0.7485
3.2	0.9592	0.9482	0.9352	0.9200	0.9024	0.8826	0.8604	0.8362	0.8102	0.7826	0.7540
3.3	0.9631	0.9526	0.9401	0.9253	0.9081	0.8885	0.8665	0.8422	0.8160	0.7882	0.7592
3.4	0.9666	0.9567	0.9446	0.9302	0.9134	0.8941	0.8722	0.8480	0.8216	0.7935	0.7642
3.5	0.9698	0.9603	0.9488	0.9348	0.9184	0.8993	0.8776	0.8534	0.8269	0.7986	0.7689
3.6	0.9727	0.9637	0.9526	0.9391	0.9230	0.9043	0.8827	0.8586	0.8320	0.8035	0.7734
3.7	0.9753	0.9667	0.9561	0.9431	0.9274	0.9089	0.8876	0.8635	0.8368	0.8081	0.7777
3.8	0.9776	0.9695	0.9594	0.9468	0.9315	0.9133	0.8922	0.8681	0.8414	0.8125	0.7819
3.9	0.9798	0.9721	0.9624	0.9502	0.9353	0.9175	0.8965	0.8726	0.8459	0.8167	0.7858
4.0	0.9817	0.9744	0.9652	0.9534	0.9389	0.9214	0.9007	0.8768	0.8501	0.8208	0.7896
4.1	0.9834	0.9766	0.9677	0.9564	0.9423	0.9251	0.9046	0.8809	0.8541	0.8247	0.7932
4.2	0.9850	0.9785	0.9701	0.9592	0.9455	0.9286	0.9083	0.8847	0.8579	0.8284	0.7967
4.3	0.9864	0.9803	0.9723	0.9618	0.9484	0.9319	0.9119	0.8884	0.8616	0.8319	0.8000
4.4	0.9877	0.9820	0.9743	0.9642	0.9512	0.9350	0.9153	0.8919	0.8652	0.8354	0.8032
4.5	0.9889	0.9835	0.9762	0.9665	0.9539	0.9380	0.9185	0.8953	0.8686	0.8386	0.8062
4.6	0.9899	0.9849	0.9779	0.9686	0.9563	0.9408	0.9216	0.8985	0.8718	0.8418	0.8092
4.7	0.9909	0.9861	0.9795	0.9705	0.9586	0.9435	0.9245	0.9016	0.8749	0.8448	0.8120
4.8	0.9918	0.9873	0.9810	0.9724	0.9609	0.9460	0.9273	0.9046	0.8779	0.8477	0.8147
4.9	0.9926	0.9883	0.9823	0.9741	0.9629	0.9484	0.9299	0.9074	0.8808	0.8505	0.8173
5.0	0.9933	0.9893	0.9836	0.9757	0.9649	0.9506	0.9325	0.9101	0.8836	0.8532	0.8198
5.1	0.9939	0.9902	0.9848	0.9772	0.9667	0.9528	0.9349	0.9127	0.8862	0.8558	0.8223
5.2	0.9945	0.9910	0.9858	0.9785	0.9684	0.9548	0.9372	0.9152	0.8888	0.8584	0.8246
5.3	0.9950	0.9917	0.9869	0.9798	0.9700	0.9568	0.9394	0.9176	0.8913	0.8608	0.8269
5.4	0.9955	0.9924	0.9878	0.9811	0.9716	0.9586	0.9415	0.9199	0.8936	0.8631	0.8290
5.5	0.9959	0.9930	0.9886	0.9822	0.9730	0.9603	0.9435	0.9221	0.8959	0.8654	0.8311
5.6	0.9963	0.9936	0.9894	0.9833	0.9744	0.9620	0.9454	0.9242	0.8981	0.8675	0.8332
5.7	0.9967	0.9941	0.9902	0.9843	0.9757	0.9636	0.9473	0.9262	0.9002	0.8696	0.8351
5.8	0.9970	0.9946	0.9909	0.9852	0.9769	0.9651	0.9490	0.9282	0.9023	0.8716	0.8370
5.9	0.9973	0.9950	0.9915	0.9861	0.9780	0.9665	0.9507	0.9301	0.9043	0.8736	0.8389
6.0	0.9975	0.9954	0.9921	0.9869	0.9791	0.9679	0.9523	0.9319	0.9062	0.8755	0.8406

NTU	Heat Exchanger Effectiveness										
	C^* ($C_{min} = C_c$)										
	0	0.1	0.2	0.3	0.4	0.5	0.6	0.7	0.8	0.9	1.0
6.1	0.9978	0.9958	0.9926	0.9877	0.9801	0.9692	0.9539	0.9336	0.9080	0.8773	0.8423
6.2	0.9980	0.9961	0.9931	0.9884	0.9811	0.9704	0.9554	0.9353	0.9098	0.8791	0.8440
6.3	0.9982	0.9964	0.9936	0.9891	0.9820	0.9716	0.9568	0.9369	0.9115	0.8808	0.8456
6.4	0.9983	0.9967	0.9940	0.9897	0.9829	0.9727	0.9582	0.9385	0.9132	0.8825	0.8472
6.5	0.9985	0.9970	0.9944	0.9903	0.9837	0.9738	0.9595	0.9400	0.9148	0.8841	0.8487
6.6	0.9986	0.9972	0.9948	0.9908	0.9845	0.9748	0.9607	0.9414	0.9164	0.8857	0.8501
6.7	0.9988	0.9975	0.9952	0.9913	0.9852	0.9757	0.9619	0.9428	0.9179	0.8872	0.8515
6.8	0.9989	0.9977	0.9955	0.9918	0.9859	0.9767	0.9631	0.9441	0.9193	0.8886	0.8529
6.9	0.9990	0.9978	0.9958	0.9923	0.9865	0.9775	0.9642	0.9454	0.9207	0.8900	0.8542
7.0	0.9991	0.9980	0.9961	0.9927	0.9872	0.9784	0.9652	0.9467	0.9221	0.8914	0.8555
7.1	0.9992	0.9982	0.9963	0.9931	0.9877	0.9792	0.9663	0.9479	0.9234	0.8928	0.8568
7.2	0.9993	0.9983	0.9966	0.9935	0.9883	0.9799	0.9672	0.9491	0.9247	0.8941	0.8580
7.3	0.9993	0.9984	0.9968	0.9939	0.9888	0.9807	0.9682	0.9502	0.9259	0.8953	0.8592
7.4	0.9994	0.9986	0.9970	0.9942	0.9893	0.9814	0.9691	0.9513	0.9271	0.8965	0.8603
7.5	0.9994	0.9987	0.9972	0.9945	0.9898	0.9820	0.9700	0.9523	0.9283	0.8977	0.8614
7.6	0.9995	0.9988	0.9974	0.9948	0.9903	0.9827	0.9708	0.9533	0.9294	0.8989	0.8625
7.7	0.9995	0.9989	0.9976	0.9951	0.9907	0.9833	0.9716	0.9543	0.9305	0.9000	0.8636
7.8	0.9996	0.9990	0.9977	0.9954	0.9911	0.9839	0.9724	0.9553	0.9316	0.9011	0.8646
7.9	0.9996	0.9990	0.9979	0.9956	0.9915	0.9844	0.9731	0.9562	0.9326	0.9021	0.8656
8.0	0.9997	0.9991	0.9980	0.9958	0.9918	0.9850	0.9738	0.9571	0.9336	0.9032	0.8666
8.1	0.9997	0.9992	0.9981	0.9960	0.9922	0.9855	0.9745	0.9579	0.9346	0.9042	0.8675
8.2	0.9997	0.9993	0.9982	0.9963	0.9925	0.9860	0.9752	0.9587	0.9355	0.9051	0.8684
8.3	0.9998	0.9993	0.9984	0.9964	0.9928	0.9864	0.9758	0.9595	0.9364	0.9061	0.8693
8.4	0.9998	0.9994	0.9985	0.9966	0.9931	0.9869	0.9764	0.9603	0.9373	0.9070	0.8702
8.5	0.9998	0.9994	0.9986	0.9968	0.9934	0.9873	0.9770	0.9611	0.9382	0.9079	0.8710
8.6	0.9998	0.9995	0.9986	0.9970	0.9937	0.9877	0.9776	0.9618	0.9390	0.9087	0.8718
8.7	0.9998	0.9995	0.9987	0.9971	0.9939	0.9881	0.9782	0.9625	0.9398	0.9096	0.8726
8.8	0.9998	0.9995	0.9988	0.9973	0.9942	0.9885	0.9787	0.9632	0.9406	0.9104	0.8734
8.9	0.9999	0.9996	0.9989	0.9974	0.9944	0.9888	0.9792	0.9638	0.9414	0.9112	0.8741
9.0	0.9999	0.9996	0.9989	0.9975	0.9946	0.9892	0.9797	0.9645	0.9421	0.9120	0.8749
9.1	0.9999	0.9996	0.9990	0.9976	0.9948	0.9895	0.9802	0.9651	0.9428	0.9128	0.8756
9.2	0.9999	0.9997	0.9991	0.9978	0.9950	0.9898	0.9806	0.9657	0.9435	0.9135	0.8763
9.3	0.9999	0.9997	0.9991	0.9979	0.9952	0.9902	0.9811	0.9663	0.9442	0.9142	0.8770
9.4	0.9999	0.9997	0.9992	0.9980	0.9954	0.9904	0.9815	0.9669	0.9449	0.9149	0.8776
9.5	0.9999	0.9997	0.9992	0.9981	0.9956	0.9907	0.9819	0.9674	0.9455	0.9156	0.8783
9.6	0.9999	0.9997	0.9993	0.9982	0.9958	0.9910	0.9823	0.9679	0.9462	0.9163	0.8789
9.7	0.9999	0.9998	0.9993	0.9982	0.9959	0.9913	0.9827	0.9685	0.9468	0.9169	0.8795
9.8	0.9999	0.9998	0.9994	0.9983	0.9961	0.9915	0.9831	0.9690	0.9474	0.9176	0.8801
9.9	1.0000	0.9998	0.9994	0.9984	0.9962	0.9917	0.9835	0.9694	0.9480	0.9182	0.8807
10.0	1.0000	0.9998	0.9994	0.9985	0.9963	0.9920	0.9838	0.9699	0.9485	0.9188	0.8813

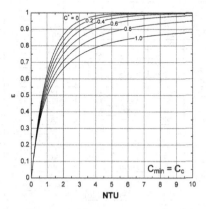

CONFIGURATION 32

2TUBE5ROW5PASS2CIR(IDO)

(b) $C_{min} = C_h$

NTU	Heat Exchanger Effectiveness										
	$C^* (C_{min} = C_h)$										
	0	0.1	0.2	0.3	0.4	0.5	0.6	0.7	0.8	0.9	1.0
0	0.0000	0.0000	0.0000	0.0000	0.0000	0.0000	0.0000	0.0000	0.0000	0.0000	0.0000
0.1	0.0952	0.0947	0.0943	0.0939	0.0934	0.0930	0.0926	0.0922	0.0917	0.0913	0.0909
0.2	0.1813	0.1797	0.1782	0.1767	0.1752	0.1738	0.1723	0.1709	0.1694	0.1680	0.1666
0.3	0.2592	0.2562	0.2532	0.2502	0.2473	0.2445	0.2416	0.2388	0.2361	0.2334	0.2307
0.4	0.3297	0.3250	0.3203	0.3158	0.3112	0.3068	0.3024	0.2981	0.2938	0.2896	0.2855
0.5	0.3935	0.3870	0.3806	0.3744	0.3682	0.3621	0.3561	0.3502	0.3443	0.3386	0.3330
0.6	0.4512	0.4430	0.4349	0.4270	0.4191	0.4114	0.4038	0.3963	0.3889	0.3816	0.3745
0.7	0.5034	0.4936	0.4839	0.4743	0.4649	0.4556	0.4464	0.4373	0.4284	0.4197	0.4111
0.8	0.5507	0.5394	0.5282	0.5171	0.5062	0.4953	0.4847	0.4741	0.4638	0.4536	0.4436
0.9	0.5934	0.5808	0.5683	0.5558	0.5435	0.5313	0.5192	0.5073	0.4955	0.4839	0.4726
1.0	0.6321	0.6184	0.6047	0.5910	0.5774	0.5639	0.5505	0.5373	0.5242	0.5113	0.4986
1.1	0.6671	0.6524	0.6377	0.6230	0.6082	0.5936	0.5790	0.5645	0.5502	0.5361	0.5222
1.2	0.6988	0.6834	0.6678	0.6521	0.6364	0.6206	0.6050	0.5894	0.5739	0.5587	0.5436
1.3	0.7275	0.7114	0.6951	0.6787	0.6621	0.6454	0.6288	0.6121	0.5956	0.5792	0.5631
1.4	0.7534	0.7369	0.7201	0.7030	0.6856	0.6682	0.6506	0.6330	0.6155	0.5981	0.5809
1.5	0.7769	0.7601	0.7429	0.7252	0.7073	0.6891	0.6707	0.6523	0.6338	0.6155	0.5973
1.6	0.7981	0.7812	0.7637	0.7456	0.7271	0.7083	0.6893	0.6700	0.6508	0.6315	0.6124
1.7	0.8173	0.8004	0.7827	0.7644	0.7455	0.7261	0.7064	0.6865	0.6664	0.6463	0.6263
1.8	0.8347	0.8178	0.8001	0.7816	0.7624	0.7426	0.7223	0.7018	0.6810	0.6601	0.6393
1.9	0.8504	0.8337	0.8160	0.7974	0.7780	0.7578	0.7371	0.7160	0.6945	0.6729	0.6513
2.0	0.8647	0.8482	0.8306	0.8120	0.7924	0.7720	0.7509	0.7292	0.7071	0.6848	0.6625
2.1	0.8775	0.8614	0.8440	0.8254	0.8058	0.7851	0.7637	0.7415	0.7189	0.6960	0.6729
2.2	0.8892	0.8734	0.8563	0.8378	0.8182	0.7974	0.7756	0.7531	0.7300	0.7064	0.6827
2.3	0.8997	0.8844	0.8676	0.8493	0.8297	0.8088	0.7868	0.7639	0.7403	0.7162	0.6918
2.4	0.9093	0.8944	0.8779	0.8599	0.8403	0.8194	0.7973	0.7741	0.7500	0.7254	0.7004
2.5	0.9179	0.9035	0.8874	0.8697	0.8503	0.8293	0.8071	0.7836	0.7592	0.7341	0.7085
2.6	0.9257	0.9118	0.8962	0.8787	0.8595	0.8386	0.8163	0.7926	0.7678	0.7422	0.7162
2.7	0.9328	0.9194	0.9042	0.8871	0.8681	0.8473	0.8249	0.8010	0.7759	0.7500	0.7234
2.8	0.9392	0.9264	0.9116	0.8949	0.8761	0.8555	0.8330	0.8090	0.7836	0.7572	0.7302
2.9	0.9450	0.9327	0.9184	0.9021	0.8836	0.8631	0.8407	0.8165	0.7909	0.7641	0.7366
3.0	0.9502	0.9385	0.9247	0.9087	0.8906	0.8703	0.8479	0.8236	0.7978	0.7707	0.7427
3.1	0.9550	0.9437	0.9304	0.9149	0.8971	0.8770	0.8547	0.8304	0.8043	0.7769	0.7485
3.2	0.9592	0.9485	0.9357	0.9207	0.9032	0.8833	0.8611	0.8368	0.8105	0.7828	0.7540
3.3	0.9631	0.9529	0.9406	0.9260	0.9089	0.8893	0.8672	0.8428	0.8164	0.7884	0.7592
3.4	0.9666	0.9570	0.9452	0.9310	0.9142	0.8949	0.8729	0.8486	0.8220	0.7937	0.7642
3.5	0.9698	0.9606	0.9493	0.9356	0.9192	0.9001	0.8784	0.8540	0.8274	0.7988	0.7689
3.6	0.9727	0.9640	0.9531	0.9399	0.9239	0.9051	0.8835	0.8592	0.8324	0.8037	0.7734
3.7	0.9753	0.9671	0.9567	0.9438	0.9283	0.9098	0.8884	0.8641	0.8373	0.8083	0.7777
3.8	0.9776	0.9699	0.9599	0.9475	0.9324	0.9142	0.8930	0.8688	0.8419	0.8128	0.7819
3.9	0.9798	0.9724	0.9629	0.9510	0.9362	0.9184	0.8974	0.8733	0.8463	0.8170	0.7858
4.0	0.9817	0.9748	0.9657	0.9542	0.9398	0.9223	0.9015	0.8775	0.8506	0.8210	0.7896
4.1	0.9834	0.9769	0.9683	0.9572	0.9432	0.9260	0.9055	0.8816	0.8546	0.8249	0.7932
4.2	0.9850	0.9789	0.9707	0.9600	0.9464	0.9295	0.9092	0.8855	0.8585	0.8287	0.7967
4.3	0.9864	0.9806	0.9728	0.9626	0.9494	0.9329	0.9128	0.8892	0.8622	0.8322	0.8000
4.4	0.9877	0.9823	0.9749	0.9650	0.9522	0.9360	0.9162	0.8927	0.8657	0.8356	0.8032
4.5	0.9889	0.9838	0.9767	0.9672	0.9548	0.9390	0.9194	0.8961	0.8691	0.8389	0.8062
4.6	0.9899	0.9851	0.9784	0.9693	0.9573	0.9418	0.9225	0.8993	0.8724	0.8421	0.8092
4.7	0.9909	0.9864	0.9800	0.9713	0.9596	0.9445	0.9255	0.9024	0.8755	0.8451	0.8120
4.8	0.9918	0.9875	0.9815	0.9731	0.9618	0.9470	0.9283	0.9054	0.8785	0.8481	0.8147
4.9	0.9926	0.9886	0.9829	0.9748	0.9639	0.9494	0.9309	0.9083	0.8814	0.8509	0.8173
5.0	0.9933	0.9895	0.9841	0.9764	0.9658	0.9517	0.9335	0.9110	0.8842	0.8536	0.8198
5.1	0.9939	0.9904	0.9853	0.9779	0.9676	0.9538	0.9359	0.9136	0.8869	0.8562	0.8223
5.2	0.9945	0.9912	0.9864	0.9793	0.9694	0.9559	0.9382	0.9161	0.8895	0.8587	0.8246
5.3	0.9950	0.9920	0.9873	0.9806	0.9710	0.9578	0.9405	0.9185	0.8919	0.8611	0.8269
5.4	0.9955	0.9926	0.9883	0.9818	0.9725	0.9597	0.9426	0.9208	0.8943	0.8635	0.8290
5.5	0.9959	0.9932	0.9891	0.9829	0.9739	0.9614	0.9446	0.9230	0.8966	0.8657	0.8311
5.6	0.9963	0.9938	0.9899	0.9840	0.9753	0.9631	0.9465	0.9252	0.8989	0.8679	0.8332
5.7	0.9967	0.9943	0.9906	0.9850	0.9766	0.9647	0.9484	0.9272	0.9010	0.8700	0.8351
5.8	0.9970	0.9948	0.9913	0.9859	0.9778	0.9662	0.9502	0.9292	0.9031	0.8721	0.8370
5.9	0.9973	0.9952	0.9919	0.9868	0.9789	0.9676	0.9519	0.9311	0.9051	0.8740	0.8389
6.0	0.9975	0.9956	0.9925	0.9876	0.9800	0.9689	0.9535	0.9329	0.9070	0.8759	0.8406

NTU	Heat Exchanger Effectiveness										
	$C^* (C_{min} = C_h)$										
	0	0.1	0.2	0.3	0.4	0.5	0.6	0.7	0.8	0.9	1.0
6.1	0.9978	0.9960	0.9930	0.9883	0.9810	0.9702	0.9550	0.9347	0.9088	0.8778	0.8423
6.2	0.9980	0.9963	0.9935	0.9890	0.9820	0.9715	0.9565	0.9364	0.9106	0.8796	0.8440
6.3	0.9982	0.9966	0.9940	0.9897	0.9829	0.9727	0.9580	0.9380	0.9124	0.8813	0.8456
6.4	0.9983	0.9969	0.9944	0.9903	0.9838	0.9738	0.9593	0.9396	0.9141	0.8830	0.8472
6.5	0.9985	0.9972	0.9948	0.9909	0.9846	0.9748	0.9607	0.9411	0.9157	0.8846	0.8487
6.6	0.9986	0.9974	0.9952	0.9914	0.9853	0.9759	0.9619	0.9426	0.9172	0.8861	0.8501
6.7	0.9988	0.9976	0.9955	0.9919	0.9861	0.9768	0.9631	0.9440	0.9188	0.8877	0.8515
6.8	0.9989	0.9978	0.9958	0.9924	0.9867	0.9777	0.9643	0.9453	0.9202	0.8891	0.8529
6.9	0.9990	0.9980	0.9961	0.9929	0.9874	0.9786	0.9654	0.9466	0.9217	0.8906	0.8542
7.0	0.9991	0.9982	0.9964	0.9933	0.9880	0.9795	0.9665	0.9479	0.9230	0.8920	0.8555
7.1	0.9992	0.9983	0.9967	0.9937	0.9886	0.9803	0.9675	0.9491	0.9244	0.8933	0.8568
7.2	0.9993	0.9984	0.9969	0.9941	0.9891	0.9810	0.9685	0.9503	0.9257	0.8946	0.8580
7.3	0.9993	0.9986	0.9971	0.9944	0.9896	0.9818	0.9694	0.9514	0.9269	0.8959	0.8592
7.4	0.9994	0.9987	0.9973	0.9947	0.9901	0.9825	0.9704	0.9525	0.9282	0.8971	0.8603
7.5	0.9994	0.9988	0.9975	0.9950	0.9906	0.9831	0.9712	0.9536	0.9293	0.8983	0.8614
7.6	0.9995	0.9989	0.9977	0.9953	0.9910	0.9838	0.9721	0.9546	0.9305	0.8995	0.8625
7.7	0.9995	0.9990	0.9978	0.9956	0.9915	0.9844	0.9729	0.9556	0.9316	0.9006	0.8636
7.8	0.9996	0.9991	0.9980	0.9958	0.9919	0.9849	0.9737	0.9566	0.9327	0.9017	0.8646
7.9	0.9996	0.9991	0.9981	0.9961	0.9922	0.9855	0.9744	0.9575	0.9337	0.9028	0.8656
8.0	0.9997	0.9992	0.9982	0.9963	0.9926	0.9860	0.9751	0.9584	0.9347	0.9038	0.8666
8.1	0.9997	0.9993	0.9984	0.9965	0.9929	0.9865	0.9758	0.9593	0.9357	0.9048	0.8675
8.2	0.9997	0.9993	0.9985	0.9967	0.9933	0.9870	0.9765	0.9601	0.9367	0.9058	0.8684
8.3	0.9998	0.9994	0.9986	0.9969	0.9936	0.9875	0.9772	0.9610	0.9376	0.9068	0.8693
8.4	0.9998	0.9994	0.9987	0.9971	0.9939	0.9879	0.9778	0.9618	0.9385	0.9077	0.8702
8.5	0.9998	0.9995	0.9988	0.9972	0.9941	0.9884	0.9784	0.9625	0.9394	0.9086	0.8710
8.6	0.9998	0.9995	0.9988	0.9974	0.9944	0.9888	0.9790	0.9633	0.9403	0.9095	0.8718
8.7	0.9998	0.9996	0.9989	0.9975	0.9946	0.9892	0.9795	0.9640	0.9411	0.9103	0.8726
8.8	0.9998	0.9996	0.9990	0.9977	0.9949	0.9895	0.9801	0.9647	0.9419	0.9112	0.8734
8.9	0.9999	0.9996	0.9991	0.9978	0.9951	0.9899	0.9806	0.9654	0.9427	0.9120	0.8741
9.0	0.9999	0.9997	0.9991	0.9979	0.9953	0.9902	0.9811	0.9660	0.9434	0.9128	0.8749
9.1	0.9999	0.9997	0.9992	0.9980	0.9955	0.9906	0.9816	0.9667	0.9442	0.9135	0.8756
9.2	0.9999	0.9997	0.9992	0.9981	0.9957	0.9909	0.9821	0.9673	0.9449	0.9143	0.8763
9.3	0.9999	0.9997	0.9993	0.9982	0.9959	0.9912	0.9825	0.9679	0.9456	0.9150	0.8770
9.4	0.9999	0.9998	0.9993	0.9983	0.9961	0.9915	0.9829	0.9685	0.9463	0.9158	0.8776
9.5	0.9999	0.9998	0.9994	0.9984	0.9962	0.9918	0.9834	0.9690	0.9470	0.9164	0.8783
9.6	0.9999	0.9998	0.9994	0.9985	0.9964	0.9920	0.9838	0.9696	0.9476	0.9171	0.8789
9.7	0.9999	0.9998	0.9995	0.9986	0.9965	0.9923	0.9842	0.9701	0.9483	0.9178	0.8795
9.8	0.9999	0.9998	0.9995	0.9986	0.9967	0.9925	0.9845	0.9706	0.9489	0.9184	0.8801
9.9	1.0000	0.9998	0.9995	0.9987	0.9968	0.9928	0.9849	0.9711	0.9495	0.9191	0.8807
10.0	1.0000	0.9999	0.9996	0.9988	0.9970	0.9930	0.9853	0.9716	0.9501	0.9197	0.8813

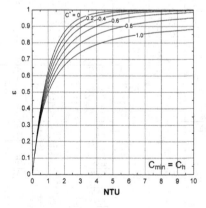

CONFIGURATION 33

2TUBE6ROW6PASS2CIR(IDO)

(a) $C_{min} = C_c$

NTU	Heat Exchanger Effectiveness										
	C^* ($C_{min} = C_c$)										
	0	0.1	0.2	0.3	0.4	0.5	0.6	0.7	0.8	0.9	1.0
0	0.0000	0.0000	0.0000	0.0000	0.0000	0.0000	0.0000	0.0000	0.0000	0.0000	0.0000
0.1	0.0952	0.0947	0.0943	0.0939	0.0934	0.0930	0.0926	0.0922	0.0917	0.0913	0.0909
0.2	0.1813	0.1797	0.1782	0.1767	0.1752	0.1738	0.1723	0.1709	0.1695	0.1680	0.1666
0.3	0.2592	0.2562	0.2532	0.2503	0.2473	0.2445	0.2417	0.2389	0.2361	0.2334	0.2307
0.4	0.3297	0.3250	0.3203	0.3158	0.3113	0.3068	0.3024	0.2981	0.2939	0.2897	0.2856
0.5	0.3935	0.3870	0.3807	0.3744	0.3682	0.3621	0.3561	0.3502	0.3444	0.3387	0.3331
0.6	0.4512	0.4430	0.4350	0.4270	0.4192	0.4114	0.4038	0.3964	0.3890	0.3818	0.3747
0.7	0.5034	0.4936	0.4840	0.4744	0.4650	0.4557	0.4465	0.4375	0.4286	0.4199	0.4113
0.8	0.5507	0.5394	0.5282	0.5172	0.5063	0.4955	0.4848	0.4743	0.4640	0.4538	0.4438
0.9	0.5934	0.5809	0.5684	0.5559	0.5436	0.5314	0.5194	0.5075	0.4958	0.4843	0.4729
1.0	0.6321	0.6184	0.6047	0.5911	0.5775	0.5641	0.5508	0.5376	0.5245	0.5117	0.4991
1.1	0.6671	0.6525	0.6378	0.6231	0.6084	0.5938	0.5793	0.5649	0.5506	0.5366	0.5227
1.2	0.6988	0.6834	0.6678	0.6522	0.6366	0.6209	0.6053	0.5898	0.5744	0.5592	0.5441
1.3	0.7275	0.7115	0.6952	0.6788	0.6623	0.6457	0.6291	0.6126	0.5961	0.5798	0.5637
1.4	0.7534	0.7370	0.7202	0.7031	0.6859	0.6685	0.6510	0.6335	0.6161	0.5988	0.5816
1.5	0.7769	0.7601	0.7430	0.7254	0.7075	0.6894	0.6712	0.6528	0.6345	0.6162	0.5981
1.6	0.7981	0.7812	0.7638	0.7458	0.7274	0.7087	0.6898	0.6707	0.6515	0.6323	0.6133
1.7	0.8173	0.8004	0.7828	0.7646	0.7458	0.7266	0.7070	0.6872	0.6672	0.6473	0.6273
1.8	0.8347	0.8179	0.8002	0.7818	0.7627	0.7431	0.7230	0.7025	0.6819	0.6611	0.6403
1.9	0.8504	0.8338	0.8161	0.7976	0.7783	0.7583	0.7378	0.7168	0.6955	0.6740	0.6525
2.0	0.8647	0.8482	0.8307	0.8122	0.7928	0.7725	0.7516	0.7301	0.7082	0.6860	0.6637
2.1	0.8775	0.8614	0.8441	0.8257	0.8062	0.7857	0.7644	0.7425	0.7200	0.6972	0.6743
2.2	0.8892	0.8735	0.8564	0.8381	0.8186	0.7980	0.7764	0.7541	0.7311	0.7078	0.6841
2.3	0.8997	0.8844	0.8677	0.8495	0.8301	0.8094	0.7876	0.7650	0.7416	0.7176	0.6934
2.4	0.9093	0.8944	0.8780	0.8601	0.8408	0.8201	0.7981	0.7752	0.7513	0.7269	0.7021
2.5	0.9179	0.9035	0.8875	0.8699	0.8507	0.8300	0.8080	0.7848	0.7606	0.7357	0.7103
2.6	0.9257	0.9118	0.8963	0.8790	0.8600	0.8393	0.8172	0.7938	0.7693	0.7439	0.7180
2.7	0.9328	0.9194	0.9043	0.8873	0.8686	0.8481	0.8259	0.8023	0.7775	0.7517	0.7253
2.8	0.9392	0.9264	0.9117	0.8951	0.8766	0.8562	0.8341	0.8103	0.7852	0.7591	0.7322
2.9	0.9450	0.9327	0.9185	0.9023	0.8841	0.8639	0.8418	0.8179	0.7926	0.7661	0.7387
3.0	0.9502	0.9385	0.9248	0.9090	0.8911	0.8711	0.8490	0.8251	0.7995	0.7727	0.7449
3.1	0.9550	0.9437	0.9305	0.9152	0.8976	0.8778	0.8558	0.8319	0.8061	0.7790	0.7508
3.2	0.9592	0.9485	0.9358	0.9209	0.9037	0.8842	0.8623	0.8383	0.8124	0.7850	0.7564
3.3	0.9631	0.9529	0.9407	0.9263	0.9094	0.8901	0.8684	0.8444	0.8184	0.7907	0.7617
3.4	0.9666	0.9569	0.9452	0.9312	0.9147	0.8957	0.8742	0.8502	0.8241	0.7961	0.7668
3.5	0.9698	0.9606	0.9494	0.9358	0.9197	0.9010	0.8796	0.8557	0.8295	0.8013	0.7716
3.6	0.9727	0.9640	0.9532	0.9401	0.9244	0.9060	0.8848	0.8609	0.8346	0.8062	0.7762
3.7	0.9753	0.9670	0.9567	0.9441	0.9288	0.9107	0.8897	0.8659	0.8395	0.8109	0.7806
3.8	0.9776	0.9698	0.9600	0.9478	0.9329	0.9151	0.8944	0.8706	0.8442	0.8154	0.7848
3.9	0.9798	0.9724	0.9630	0.9512	0.9367	0.9193	0.8988	0.8752	0.8487	0.8198	0.7889
4.0	0.9817	0.9747	0.9658	0.9544	0.9403	0.9232	0.9030	0.8795	0.8530	0.8239	0.7927
4.1	0.9834	0.9769	0.9683	0.9574	0.9437	0.9270	0.9069	0.8836	0.8571	0.8279	0.7964
4.2	0.9850	0.9788	0.9707	0.9602	0.9469	0.9305	0.9107	0.8875	0.8610	0.8317	0.8000
4.3	0.9864	0.9806	0.9729	0.9628	0.9499	0.9338	0.9143	0.8913	0.8648	0.8353	0.8034
4.4	0.9877	0.9822	0.9749	0.9652	0.9527	0.9370	0.9177	0.8948	0.8684	0.8388	0.8067
4.5	0.9889	0.9837	0.9767	0.9674	0.9553	0.9400	0.9210	0.8983	0.8719	0.8422	0.8098
4.6	0.9899	0.9851	0.9785	0.9695	0.9578	0.9428	0.9241	0.9015	0.8752	0.8454	0.8128
4.7	0.9909	0.9864	0.9801	0.9715	0.9601	0.9455	0.9271	0.9047	0.8784	0.8485	0.8158
4.8	0.9918	0.9875	0.9815	0.9733	0.9623	0.9480	0.9299	0.9077	0.8815	0.8515	0.8186
4.9	0.9926	0.9885	0.9829	0.9750	0.9644	0.9504	0.9326	0.9106	0.8844	0.8544	0.8213
5.0	0.9933	0.9895	0.9841	0.9766	0.9663	0.9527	0.9351	0.9133	0.8873	0.8572	0.8239
5.1	0.9939	0.9904	0.9853	0.9781	0.9682	0.9548	0.9376	0.9160	0.8900	0.8599	0.8264
5.2	0.9945	0.9912	0.9864	0.9795	0.9699	0.9569	0.9399	0.9185	0.8926	0.8625	0.8288
5.3	0.9950	0.9919	0.9874	0.9808	0.9715	0.9588	0.9422	0.9210	0.8952	0.8650	0.8312
5.4	0.9955	0.9926	0.9883	0.9820	0.9730	0.9607	0.9443	0.9233	0.8976	0.8674	0.8334
5.5	0.9959	0.9932	0.9891	0.9831	0.9745	0.9624	0.9463	0.9256	0.9000	0.8698	0.8356
5.6	0.9963	0.9938	0.9899	0.9842	0.9758	0.9641	0.9483	0.9277	0.9022	0.8720	0.8377
5.7	0.9967	0.9943	0.9906	0.9851	0.9771	0.9657	0.9501	0.9298	0.9044	0.8742	0.8398
5.8	0.9970	0.9948	0.9913	0.9861	0.9783	0.9672	0.9519	0.9318	0.9066	0.8763	0.8418
5.9	0.9973	0.9952	0.9919	0.9869	0.9794	0.9686	0.9536	0.9337	0.9086	0.8783	0.8437
6.0	0.9975	0.9956	0.9925	0.9877	0.9805	0.9700	0.9553	0.9356	0.9106	0.8803	0.8455

NTU	Heat Exchanger Effectiveness										
	C^* ($C_{min} = C_c$)										
	0	0.1	0.2	0.3	0.4	0.5	0.6	0.7	0.8	0.9	1.0
6.1	0.9978	0.9960	0.9930	0.9885	0.9815	0.9713	0.9568	0.9374	0.9125	0.8822	0.8473
6.2	0.9980	0.9963	0.9935	0.9892	0.9825	0.9725	0.9583	0.9391	0.9143	0.8841	0.8491
6.3	0.9982	0.9966	0.9940	0.9898	0.9834	0.9737	0.9598	0.9408	0.9161	0.8859	0.8508
6.4	0.9983	0.9969	0.9944	0.9905	0.9842	0.9748	0.9611	0.9423	0.9178	0.8876	0.8524
6.5	0.9985	0.9971	0.9948	0.9910	0.9850	0.9758	0.9624	0.9439	0.9195	0.8893	0.8540
6.6	0.9986	0.9974	0.9952	0.9916	0.9858	0.9768	0.9637	0.9454	0.9211	0.8910	0.8556
6.7	0.9988	0.9976	0.9955	0.9921	0.9865	0.9778	0.9649	0.9468	0.9227	0.8925	0.8570
6.8	0.9989	0.9978	0.9958	0.9926	0.9872	0.9787	0.9661	0.9482	0.9242	0.8941	0.8585
6.9	0.9990	0.9980	0.9961	0.9930	0.9878	0.9796	0.9672	0.9495	0.9257	0.8956	0.8599
7.0	0.9991	0.9981	0.9964	0.9934	0.9884	0.9804	0.9683	0.9508	0.9271	0.8970	0.8613
7.1	0.9992	0.9983	0.9967	0.9938	0.9890	0.9812	0.9693	0.9520	0.9285	0.8985	0.8626
7.2	0.9993	0.9984	0.9969	0.9942	0.9895	0.9820	0.9703	0.9532	0.9298	0.8998	0.8639
7.3	0.9993	0.9985	0.9971	0.9945	0.9900	0.9827	0.9712	0.9544	0.9311	0.9012	0.8652
7.4	0.9994	0.9987	0.9973	0.9948	0.9905	0.9834	0.9721	0.9555	0.9324	0.9025	0.8664
7.5	0.9994	0.9988	0.9975	0.9951	0.9910	0.9840	0.9730	0.9566	0.9336	0.9037	0.8676
7.6	0.9995	0.9989	0.9977	0.9954	0.9914	0.9847	0.9739	0.9576	0.9348	0.9049	0.8688
7.7	0.9995	0.9990	0.9978	0.9957	0.9918	0.9853	0.9747	0.9586	0.9359	0.9061	0.8699
7.8	0.9996	0.9990	0.9980	0.9959	0.9922	0.9858	0.9754	0.9596	0.9370	0.9073	0.8710
7.9	0.9996	0.9991	0.9981	0.9962	0.9926	0.9864	0.9762	0.9605	0.9381	0.9084	0.8721
8.0	0.9997	0.9992	0.9982	0.9964	0.9929	0.9869	0.9769	0.9614	0.9392	0.9095	0.8731
8.1	0.9997	0.9993	0.9984	0.9966	0.9933	0.9874	0.9776	0.9623	0.9402	0.9106	0.8741
8.2	0.9997	0.9993	0.9985	0.9968	0.9936	0.9879	0.9783	0.9632	0.9412	0.9116	0.8751
8.3	0.9998	0.9994	0.9986	0.9970	0.9939	0.9883	0.9789	0.9640	0.9421	0.9127	0.8761
8.4	0.9998	0.9994	0.9987	0.9971	0.9942	0.9888	0.9795	0.9648	0.9431	0.9136	0.8770
8.5	0.9998	0.9995	0.9988	0.9973	0.9944	0.9892	0.9801	0.9656	0.9440	0.9146	0.8779
8.6	0.9998	0.9995	0.9988	0.9974	0.9947	0.9896	0.9807	0.9663	0.9449	0.9156	0.8788
8.7	0.9998	0.9995	0.9989	0.9976	0.9949	0.9900	0.9812	0.9670	0.9457	0.9165	0.8797
8.8	0.9998	0.9996	0.9990	0.9977	0.9952	0.9903	0.9818	0.9677	0.9466	0.9174	0.8805
8.9	0.9999	0.9996	0.9991	0.9978	0.9954	0.9907	0.9823	0.9684	0.9474	0.9182	0.8814
9.0	0.9999	0.9996	0.9991	0.9980	0.9956	0.9910	0.9828	0.9691	0.9482	0.9191	0.8822
9.1	0.9999	0.9997	0.9992	0.9981	0.9958	0.9913	0.9833	0.9697	0.9490	0.9199	0.8830
9.2	0.9999	0.9997	0.9992	0.9982	0.9960	0.9916	0.9837	0.9703	0.9497	0.9207	0.8837
9.3	0.9999	0.9997	0.9993	0.9983	0.9961	0.9919	0.9842	0.9710	0.9504	0.9215	0.8845
9.4	0.9999	0.9997	0.9993	0.9984	0.9963	0.9922	0.9846	0.9715	0.9511	0.9223	0.8852
9.5	0.9999	0.9998	0.9994	0.9985	0.9965	0.9925	0.9850	0.9721	0.9518	0.9230	0.8859
9.6	0.9999	0.9998	0.9994	0.9985	0.9966	0.9927	0.9854	0.9726	0.9525	0.9238	0.8866
9.7	0.9999	0.9998	0.9994	0.9986	0.9968	0.9930	0.9858	0.9732	0.9532	0.9245	0.8873
9.8	0.9999	0.9998	0.9995	0.9987	0.9969	0.9932	0.9861	0.9737	0.9538	0.9252	0.8880
9.9	1.0000	0.9998	0.9995	0.9988	0.9970	0.9934	0.9865	0.9742	0.9544	0.9258	0.8886
10.0	1.0000	0.9998	0.9995	0.9988	0.9972	0.9937	0.9868	0.9747	0.9550	0.9265	0.8893

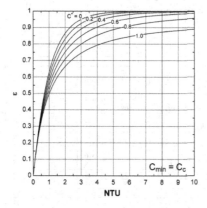

CONFIGURATION 33

2TUBE6ROW6PASS2CIR(IDO)

(b) $C_{min} = C_h$

NTU	Heat Exchanger Effectiveness										
	C^* ($C_{min} = C_h$)										
	0	0.1	0.2	0.3	0.4	0.5	0.6	0.7	0.8	0.9	1.0
0	0.0000	0.0000	0.0000	0.0000	0.0000	0.0000	0.0000	0.0000	0.0000	0.0000	0.0000
0.1	0.0952	0.0947	0.0943	0.0939	0.0934	0.0930	0.0926	0.0922	0.0917	0.0913	0.0909
0.2	0.1813	0.1797	0.1782	0.1767	0.1752	0.1738	0.1723	0.1709	0.1695	0.1680	0.1666
0.3	0.2592	0.2562	0.2532	0.2503	0.2474	0.2445	0.2417	0.2389	0.2361	0.2334	0.2307
0.4	0.3297	0.3250	0.3203	0.3158	0.3113	0.3068	0.3024	0.2981	0.2939	0.2897	0.2856
0.5	0.3935	0.3870	0.3807	0.3744	0.3682	0.3621	0.3561	0.3502	0.3444	0.3387	0.3331
0.6	0.4512	0.4430	0.4350	0.4270	0.4192	0.4115	0.4039	0.3964	0.3890	0.3818	0.3747
0.7	0.5034	0.4936	0.4840	0.4744	0.4650	0.4557	0.4465	0.4375	0.4286	0.4199	0.4113
0.8	0.5507	0.5394	0.5283	0.5172	0.5063	0.4955	0.4848	0.4743	0.4640	0.4538	0.4438
0.9	0.5934	0.5809	0.5684	0.5560	0.5437	0.5315	0.5194	0.5075	0.4958	0.4843	0.4729
1.0	0.6321	0.6184	0.6048	0.5912	0.5776	0.5641	0.5508	0.5376	0.5246	0.5117	0.4991
1.1	0.6671	0.6525	0.6378	0.6232	0.6085	0.5939	0.5793	0.5649	0.5507	0.5366	0.5227
1.2	0.6988	0.6834	0.6679	0.6523	0.6367	0.6210	0.6054	0.5898	0.5744	0.5592	0.5441
1.3	0.7275	0.7115	0.6953	0.6789	0.6624	0.6458	0.6292	0.6127	0.5962	0.5799	0.5637
1.4	0.7534	0.7370	0.7203	0.7033	0.6860	0.6686	0.6511	0.6336	0.6162	0.5988	0.5816
1.5	0.7769	0.7602	0.7431	0.7256	0.7077	0.6896	0.6713	0.6530	0.6346	0.6163	0.5981
1.6	0.7981	0.7813	0.7639	0.7460	0.7276	0.7089	0.6899	0.6708	0.6516	0.6324	0.6133
1.7	0.8173	0.8005	0.7829	0.7647	0.7460	0.7268	0.7072	0.6873	0.6673	0.6473	0.6273
1.8	0.8347	0.8180	0.8004	0.7820	0.7629	0.7433	0.7231	0.7027	0.6820	0.6611	0.6403
1.9	0.8504	0.8339	0.8163	0.7979	0.7786	0.7586	0.7380	0.7169	0.6956	0.6740	0.6525
2.0	0.8647	0.8484	0.8309	0.8125	0.7930	0.7728	0.7518	0.7302	0.7083	0.6861	0.6637
2.1	0.8775	0.8616	0.8443	0.8259	0.8065	0.7860	0.7647	0.7427	0.7202	0.6973	0.6743
2.2	0.8892	0.8736	0.8566	0.8384	0.8189	0.7983	0.7767	0.7543	0.7313	0.7078	0.6841
2.3	0.8997	0.8846	0.8679	0.8499	0.8304	0.8097	0.7879	0.7652	0.7417	0.7177	0.6934
2.4	0.9093	0.8946	0.8783	0.8605	0.8411	0.8204	0.7984	0.7754	0.7515	0.7270	0.7021
2.5	0.9179	0.9037	0.8878	0.8703	0.8511	0.8304	0.8083	0.7850	0.7608	0.7358	0.7103
2.6	0.9257	0.9120	0.8966	0.8793	0.8604	0.8397	0.8176	0.7941	0.7695	0.7440	0.7180
2.7	0.9328	0.9196	0.9046	0.8877	0.8690	0.8485	0.8263	0.8026	0.7777	0.7518	0.7253
2.8	0.9392	0.9266	0.9120	0.8955	0.8771	0.8567	0.8345	0.8106	0.7855	0.7592	0.7322
2.9	0.9450	0.9329	0.9188	0.9027	0.8846	0.8643	0.8422	0.8182	0.7928	0.7662	0.7387
3.0	0.9502	0.9387	0.9251	0.9094	0.8916	0.8715	0.8494	0.8254	0.7998	0.7728	0.7449
3.1	0.9550	0.9439	0.9309	0.9156	0.8981	0.8783	0.8563	0.8322	0.8064	0.7791	0.7508
3.2	0.9592	0.9487	0.9362	0.9214	0.9042	0.8847	0.8628	0.8387	0.8127	0.7851	0.7564
3.3	0.9631	0.9531	0.9411	0.9267	0.9099	0.8906	0.8689	0.8448	0.8186	0.7908	0.7617
3.4	0.9666	0.9572	0.9456	0.9317	0.9153	0.8963	0.8747	0.8506	0.8243	0.7962	0.7668
3.5	0.9698	0.9608	0.9497	0.9363	0.9203	0.9016	0.8801	0.8561	0.8297	0.8014	0.7716
3.6	0.9727	0.9642	0.9536	0.9406	0.9250	0.9066	0.8853	0.8614	0.8349	0.8064	0.7762
3.7	0.9753	0.9672	0.9571	0.9446	0.9294	0.9113	0.8902	0.8663	0.8398	0.8111	0.7806
3.8	0.9776	0.9700	0.9604	0.9483	0.9335	0.9157	0.8949	0.8711	0.8445	0.8156	0.7848
3.9	0.9798	0.9726	0.9634	0.9517	0.9373	0.9199	0.8993	0.8756	0.8490	0.8199	0.7889
4.0	0.9817	0.9749	0.9661	0.9549	0.9409	0.9239	0.9035	0.8799	0.8533	0.8241	0.7927
4.1	0.9834	0.9771	0.9687	0.9579	0.9443	0.9276	0.9075	0.8841	0.8574	0.8280	0.7964
4.2	0.9850	0.9790	0.9711	0.9607	0.9475	0.9311	0.9113	0.8880	0.8614	0.8318	0.8000
4.3	0.9864	0.9808	0.9732	0.9633	0.9505	0.9345	0.9149	0.8918	0.8652	0.8355	0.8034
4.4	0.9877	0.9824	0.9753	0.9657	0.9533	0.9376	0.9183	0.8954	0.8688	0.8390	0.8067
4.5	0.9889	0.9839	0.9771	0.9680	0.9559	0.9406	0.9216	0.8988	0.8722	0.8424	0.8098
4.6	0.9899	0.9853	0.9788	0.9701	0.9584	0.9434	0.9247	0.9021	0.8756	0.8456	0.8128
4.7	0.9909	0.9865	0.9804	0.9720	0.9608	0.9461	0.9277	0.9052	0.8788	0.8487	0.8158
4.8	0.9918	0.9877	0.9819	0.9738	0.9629	0.9487	0.9305	0.9082	0.8819	0.8517	0.8186
4.9	0.9926	0.9887	0.9832	0.9755	0.9650	0.9511	0.9332	0.9111	0.8848	0.8546	0.8213
5.0	0.9933	0.9897	0.9845	0.9771	0.9669	0.9533	0.9358	0.9139	0.8877	0.8574	0.8239
5.1	0.9939	0.9906	0.9856	0.9786	0.9688	0.9555	0.9382	0.9166	0.8904	0.8601	0.8264
5.2	0.9945	0.9914	0.9867	0.9800	0.9705	0.9576	0.9406	0.9191	0.8931	0.8627	0.8288
5.3	0.9950	0.9921	0.9877	0.9812	0.9721	0.9595	0.9428	0.9216	0.8956	0.8652	0.8312
5.4	0.9955	0.9928	0.9886	0.9824	0.9736	0.9614	0.9450	0.9239	0.8981	0.8677	0.8334
5.5	0.9959	0.9934	0.9894	0.9836	0.9750	0.9631	0.9470	0.9262	0.9004	0.8700	0.8356
5.6	0.9963	0.9939	0.9902	0.9846	0.9764	0.9648	0.9490	0.9284	0.9027	0.8723	0.8377
5.7	0.9967	0.9944	0.9909	0.9856	0.9777	0.9664	0.9508	0.9304	0.9049	0.8744	0.8398
5.8	0.9970	0.9949	0.9916	0.9865	0.9789	0.9679	0.9526	0.9325	0.9070	0.8766	0.8418
5.9	0.9973	0.9953	0.9922	0.9874	0.9800	0.9693	0.9543	0.9344	0.9091	0.8786	0.8437
6.0	0.9975	0.9957	0.9928	0.9881	0.9811	0.9706	0.9560	0.9362	0.9111	0.8806	0.8455

NTU	Heat Exchanger Effectiveness										
	C^* ($C_{min} = C_h$)										
	0	0.1	0.2	0.3	0.4	0.5	0.6	0.7	0.8	0.9	1.0
6.1	0.9978	0.9961	0.9933	0.9889	0.9821	0.9719	0.9575	0.9380	0.9130	0.8825	0.8473
6.2	0.9980	0.9964	0.9938	0.9896	0.9830	0.9732	0.9590	0.9398	0.9148	0.8844	0.8491
6.3	0.9982	0.9967	0.9943	0.9902	0.9839	0.9743	0.9605	0.9414	0.9166	0.8862	0.8508
6.4	0.9983	0.9970	0.9947	0.9908	0.9848	0.9754	0.9618	0.9430	0.9184	0.8879	0.8524
6.5	0.9985	0.9972	0.9951	0.9914	0.9856	0.9765	0.9632	0.9446	0.9201	0.8896	0.8540
6.6	0.9986	0.9975	0.9954	0.9919	0.9863	0.9775	0.9644	0.9461	0.9217	0.8913	0.8556
6.7	0.9988	0.9977	0.9957	0.9924	0.9870	0.9785	0.9657	0.9475	0.9233	0.8929	0.8570
6.8	0.9989	0.9979	0.9961	0.9929	0.9877	0.9794	0.9668	0.9489	0.9248	0.8944	0.8585
6.9	0.9990	0.9980	0.9963	0.9933	0.9883	0.9802	0.9679	0.9502	0.9263	0.8959	0.8599
7.0	0.9991	0.9982	0.9966	0.9938	0.9889	0.9811	0.9690	0.9515	0.9277	0.8974	0.8613
7.1	0.9992	0.9984	0.9968	0.9941	0.9895	0.9819	0.9700	0.9528	0.9291	0.8988	0.8626
7.2	0.9993	0.9985	0.9971	0.9945	0.9900	0.9826	0.9710	0.9540	0.9304	0.9002	0.8639
7.3	0.9993	0.9986	0.9973	0.9948	0.9905	0.9833	0.9720	0.9551	0.9317	0.9015	0.8652
7.4	0.9994	0.9987	0.9975	0.9951	0.9910	0.9840	0.9729	0.9562	0.9330	0.9028	0.8664
7.5	0.9994	0.9988	0.9977	0.9954	0.9915	0.9847	0.9738	0.9573	0.9342	0.9041	0.8676
7.6	0.9995	0.9989	0.9978	0.9957	0.9919	0.9853	0.9746	0.9584	0.9354	0.9053	0.8688
7.7	0.9995	0.9990	0.9980	0.9960	0.9923	0.9859	0.9754	0.9594	0.9366	0.9065	0.8699
7.8	0.9996	0.9991	0.9981	0.9962	0.9927	0.9865	0.9762	0.9604	0.9377	0.9077	0.8710
7.9	0.9996	0.9992	0.9983	0.9964	0.9930	0.9870	0.9770	0.9613	0.9388	0.9088	0.8721
8.0	0.9997	0.9992	0.9984	0.9966	0.9934	0.9875	0.9777	0.9622	0.9398	0.9099	0.8731
8.1	0.9997	0.9993	0.9985	0.9968	0.9937	0.9880	0.9784	0.9631	0.9409	0.9110	0.8741
8.2	0.9997	0.9994	0.9986	0.9970	0.9940	0.9885	0.9790	0.9640	0.9419	0.9120	0.8751
8.3	0.9998	0.9994	0.9987	0.9972	0.9943	0.9889	0.9797	0.9648	0.9429	0.9131	0.8761
8.4	0.9998	0.9995	0.9988	0.9974	0.9946	0.9894	0.9803	0.9656	0.9438	0.9141	0.8770
8.5	0.9998	0.9995	0.9989	0.9975	0.9948	0.9898	0.9809	0.9664	0.9447	0.9150	0.8779
8.6	0.9998	0.9996	0.9990	0.9977	0.9951	0.9902	0.9815	0.9672	0.9456	0.9160	0.8788
8.7	0.9998	0.9996	0.9990	0.9978	0.9953	0.9906	0.9820	0.9679	0.9465	0.9169	0.8797
8.8	0.9998	0.9996	0.9991	0.9979	0.9955	0.9909	0.9826	0.9686	0.9473	0.9178	0.8805
8.9	0.9999	0.9997	0.9992	0.9981	0.9958	0.9913	0.9831	0.9693	0.9482	0.9187	0.8814
9.0	0.9999	0.9997	0.9992	0.9982	0.9960	0.9916	0.9836	0.9700	0.9490	0.9195	0.8822
9.1	0.9999	0.9997	0.9993	0.9983	0.9961	0.9919	0.9840	0.9706	0.9498	0.9204	0.8830
9.2	0.9999	0.9997	0.9993	0.9984	0.9963	0.9922	0.9845	0.9712	0.9505	0.9212	0.8837
9.3	0.9999	0.9998	0.9994	0.9985	0.9965	0.9925	0.9849	0.9719	0.9513	0.9220	0.8845
9.4	0.9999	0.9998	0.9994	0.9986	0.9967	0.9928	0.9854	0.9724	0.9520	0.9228	0.8852
9.5	0.9999	0.9998	0.9994	0.9986	0.9968	0.9930	0.9858	0.9730	0.9527	0.9235	0.8859
9.6	0.9999	0.9998	0.9995	0.9987	0.9970	0.9933	0.9862	0.9736	0.9534	0.9243	0.8866
9.7	0.9999	0.9998	0.9995	0.9988	0.9971	0.9935	0.9866	0.9741	0.9540	0.9250	0.8873
9.8	0.9999	0.9998	0.9996	0.9989	0.9972	0.9938	0.9869	0.9746	0.9547	0.9257	0.8880
9.9	1.0000	0.9999	0.9996	0.9989	0.9974	0.9940	0.9873	0.9752	0.9553	0.9264	0.8886
10.0	1.0000	0.9999	0.9996	0.9990	0.9975	0.9942	0.9876	0.9756	0.9559	0.9271	0.8893

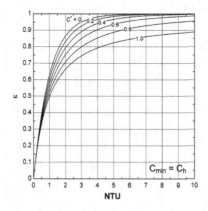

CONFIGURATION 34

2TUBE4ROW2PASS4CIR (Guo et al., 2002)

(a) $C_{min} = C_c$

NTU	Heat Exchanger Effectiveness										
	C^* ($C_{min} = C_c$)										
	0	0.1	0.2	0.3	0.4	0.5	0.6	0.7	0.8	0.9	1.0
0	0.0000	0.0000	0.0000	0.0000	0.0000	0.0000	0.0000	0.0000	0.0000	0.0000	0.0000
0.1	0.0952	0.0947	0.0943	0.0939	0.0934	0.0930	0.0926	0.0921	0.0917	0.0913	0.0909
0.2	0.1813	0.1797	0.1782	0.1767	0.1752	0.1737	0.1722	0.1707	0.1693	0.1679	0.1665
0.3	0.2592	0.2561	0.2531	0.2501	0.2471	0.2442	0.2413	0.2385	0.2357	0.2329	0.2302
0.4	0.3297	0.3248	0.3201	0.3154	0.3108	0.3062	0.3017	0.2973	0.2930	0.2888	0.2846
0.5	0.3935	0.3868	0.3802	0.3737	0.3674	0.3611	0.3550	0.3489	0.3430	0.3372	0.3315
0.6	0.4512	0.4426	0.4342	0.4260	0.4179	0.4099	0.4021	0.3944	0.3869	0.3795	0.3723
0.7	0.5034	0.4931	0.4830	0.4730	0.4632	0.4535	0.4441	0.4348	0.4257	0.4168	0.4081
0.8	0.5507	0.5387	0.5269	0.5153	0.5039	0.4927	0.4817	0.4709	0.4603	0.4499	0.4397
0.9	0.5934	0.5800	0.5667	0.5536	0.5407	0.5280	0.5155	0.5032	0.4912	0.4794	0.4679
1.0	0.6321	0.6173	0.6027	0.5882	0.5740	0.5599	0.5460	0.5324	0.5191	0.5059	0.4931
1.1	0.6671	0.6512	0.6354	0.6197	0.6042	0.5888	0.5737	0.5588	0.5442	0.5299	0.5158
1.2	0.6988	0.6819	0.6651	0.6483	0.6317	0.6152	0.5989	0.5829	0.5671	0.5515	0.5363
1.3	0.7275	0.7098	0.6920	0.6744	0.6567	0.6392	0.6219	0.6048	0.5879	0.5713	0.5549
1.4	0.7534	0.7351	0.7166	0.6981	0.6797	0.6613	0.6430	0.6248	0.6069	0.5893	0.5719
1.5	0.7769	0.7581	0.7390	0.7199	0.7007	0.6815	0.6623	0.6433	0.6244	0.6058	0.5875
1.6	0.7981	0.7790	0.7595	0.7398	0.7200	0.7000	0.6801	0.6602	0.6405	0.6210	0.6018
1.7	0.8173	0.7980	0.7782	0.7581	0.7377	0.7171	0.6965	0.6759	0.6554	0.6351	0.6150
1.8	0.8347	0.8153	0.7953	0.7748	0.7540	0.7330	0.7117	0.6904	0.6692	0.6481	0.6272
1.9	0.8504	0.8310	0.8109	0.7902	0.7691	0.7476	0.7258	0.7039	0.6820	0.6602	0.6385
2.0	0.8647	0.8453	0.8252	0.8044	0.7830	0.7611	0.7389	0.7164	0.6939	0.6714	0.6490
2.1	0.8775	0.8584	0.8384	0.8175	0.7959	0.7737	0.7511	0.7281	0.7050	0.6819	0.6588
2.2	0.8892	0.8703	0.8504	0.8296	0.8078	0.7854	0.7624	0.7390	0.7154	0.6916	0.6680
2.3	0.8997	0.8812	0.8615	0.8407	0.8189	0.7963	0.7730	0.7492	0.7251	0.7008	0.6766
2.4	0.9093	0.8911	0.8717	0.8510	0.8292	0.8065	0.7829	0.7588	0.7342	0.7094	0.6846
2.5	0.9179	0.9002	0.8810	0.8605	0.8388	0.8160	0.7922	0.7678	0.7428	0.7175	0.6922
2.6	0.9257	0.9084	0.8896	0.8693	0.8477	0.8248	0.8009	0.7762	0.7509	0.7251	0.6993
2.7	0.9328	0.9160	0.8975	0.8775	0.8560	0.8332	0.8091	0.7842	0.7585	0.7323	0.7060
2.8	0.9392	0.9229	0.9048	0.8851	0.8638	0.8409	0.8168	0.7916	0.7656	0.7391	0.7123
2.9	0.9450	0.9292	0.9116	0.8921	0.8710	0.8482	0.8241	0.7987	0.7724	0.7455	0.7183
3.0	0.9502	0.9349	0.9178	0.8987	0.8778	0.8551	0.8309	0.8054	0.7789	0.7516	0.7240
3.1	0.9550	0.9402	0.9235	0.9048	0.8841	0.8616	0.8374	0.8117	0.7850	0.7574	0.7293
3.2	0.9592	0.9450	0.9288	0.9104	0.8900	0.8676	0.8435	0.8177	0.7907	0.7629	0.7344
3.3	0.9631	0.9494	0.9337	0.9157	0.8956	0.8734	0.8492	0.8234	0.7962	0.7681	0.7393
3.4	0.9666	0.9535	0.9382	0.9206	0.9008	0.8788	0.8547	0.8288	0.8014	0.7730	0.7439
3.5	0.9698	0.9572	0.9424	0.9252	0.9057	0.8838	0.8598	0.8339	0.8064	0.7777	0.7483
3.6	0.9727	0.9606	0.9462	0.9295	0.9103	0.8887	0.8647	0.8388	0.8112	0.7822	0.7525
3.7	0.9753	0.9637	0.9498	0.9335	0.9146	0.8932	0.8694	0.8434	0.8157	0.7865	0.7565
3.8	0.9776	0.9665	0.9531	0.9372	0.9187	0.8975	0.8738	0.8478	0.8200	0.7906	0.7603
3.9	0.9798	0.9692	0.9562	0.9407	0.9225	0.9016	0.8780	0.8521	0.8241	0.7945	0.7639
4.0	0.9817	0.9716	0.9591	0.9440	0.9261	0.9054	0.8820	0.8561	0.8280	0.7983	0.7674
4.1	0.9834	0.9738	0.9617	0.9470	0.9295	0.9091	0.8858	0.8599	0.8318	0.8019	0.7707
4.2	0.9850	0.9758	0.9642	0.9499	0.9327	0.9125	0.8894	0.8636	0.8354	0.8053	0.7739
4.3	0.9864	0.9776	0.9665	0.9526	0.9357	0.9158	0.8929	0.8671	0.8388	0.8086	0.7770
4.4	0.9877	0.9794	0.9686	0.9551	0.9386	0.9189	0.8961	0.8704	0.8421	0.8118	0.7799
4.5	0.9889	0.9809	0.9706	0.9574	0.9413	0.9219	0.8993	0.8736	0.8453	0.8148	0.7827
4.6	0.9899	0.9824	0.9724	0.9596	0.9438	0.9247	0.9023	0.8767	0.8483	0.8177	0.7854
4.7	0.9909	0.9837	0.9741	0.9617	0.9462	0.9274	0.9051	0.8796	0.8512	0.8205	0.7880
4.8	0.9918	0.9849	0.9757	0.9637	0.9485	0.9299	0.9079	0.8824	0.8540	0.8232	0.7905
4.9	0.9926	0.9860	0.9772	0.9655	0.9507	0.9323	0.9105	0.8851	0.8567	0.8258	0.7929
5.0	0.9933	0.9871	0.9785	0.9672	0.9527	0.9347	0.9130	0.8877	0.8593	0.8282	0.7952
5.1	0.9939	0.9880	0.9798	0.9689	0.9546	0.9369	0.9154	0.8902	0.8618	0.8306	0.7975
5.2	0.9945	0.9889	0.9810	0.9704	0.9565	0.9390	0.9177	0.8926	0.8642	0.8329	0.7996
5.3	0.9950	0.9897	0.9821	0.9718	0.9582	0.9410	0.9198	0.8949	0.8665	0.8352	0.8017
5.4	0.9955	0.9905	0.9832	0.9732	0.9599	0.9429	0.9220	0.8971	0.8687	0.8373	0.8037
5.5	0.9959	0.9911	0.9842	0.9744	0.9614	0.9447	0.9240	0.8993	0.8709	0.8394	0.8056
5.6	0.9963	0.9918	0.9851	0.9757	0.9629	0.9464	0.9259	0.9013	0.8729	0.8414	0.8074
5.7	0.9967	0.9924	0.9859	0.9768	0.9643	0.9481	0.9278	0.9033	0.8749	0.8433	0.8092
5.8	0.9970	0.9929	0.9867	0.9779	0.9657	0.9497	0.9295	0.9052	0.8768	0.8451	0.8109
5.9	0.9973	0.9934	0.9875	0.9789	0.9670	0.9512	0.9312	0.9070	0.8787	0.8469	0.8126
6.0	0.9975	0.9939	0.9882	0.9798	0.9682	0.9527	0.9329	0.9087	0.8805	0.8487	0.8142

NTU	Heat Exchanger Effectiveness										
	C^* ($C_{min} = C_c$)										
	0	0.1	0.2	0.3	0.4	0.5	0.6	0.7	0.8	0.9	1.0
6.1	0.9978	0.9943	0.9888	0.9807	0.9693	0.9541	0.9345	0.9104	0.8822	0.8503	0.8157
6.2	0.9980	0.9947	0.9894	0.9816	0.9704	0.9554	0.9360	0.9121	0.8838	0.8519	0.8172
6.3	0.9982	0.9951	0.9900	0.9824	0.9715	0.9567	0.9374	0.9136	0.8854	0.8535	0.8187
6.4	0.9983	0.9954	0.9906	0.9832	0.9725	0.9579	0.9388	0.9151	0.8870	0.8550	0.8201
6.5	0.9985	0.9957	0.9911	0.9839	0.9735	0.9591	0.9402	0.9166	0.8885	0.8565	0.8215
6.6	0.9986	0.9960	0.9915	0.9846	0.9744	0.9602	0.9415	0.9180	0.8900	0.8579	0.8228
6.7	0.9988	0.9963	0.9920	0.9852	0.9752	0.9613	0.9427	0.9194	0.8914	0.8593	0.8240
6.8	0.9989	0.9965	0.9924	0.9858	0.9761	0.9623	0.9439	0.9207	0.8927	0.8606	0.8253
6.9	0.9990	0.9968	0.9928	0.9864	0.9768	0.9633	0.9451	0.9220	0.8940	0.8619	0.8264
7.0	0.9991	0.9970	0.9932	0.9870	0.9776	0.9642	0.9462	0.9232	0.8953	0.8631	0.8276
7.1	0.9992	0.9972	0.9935	0.9875	0.9783	0.9651	0.9473	0.9244	0.8965	0.8643	0.8287
7.2	0.9993	0.9973	0.9938	0.9880	0.9790	0.9660	0.9483	0.9255	0.8977	0.8655	0.8298
7.3	0.9993	0.9975	0.9941	0.9885	0.9797	0.9669	0.9493	0.9266	0.8989	0.8666	0.8308
7.4	0.9994	0.9977	0.9944	0.9889	0.9803	0.9677	0.9503	0.9277	0.9000	0.8677	0.8319
7.5	0.9994	0.9978	0.9947	0.9893	0.9809	0.9684	0.9512	0.9287	0.9011	0.8687	0.8328
7.6	0.9995	0.9980	0.9950	0.9897	0.9815	0.9692	0.9521	0.9297	0.9021	0.8698	0.8338
7.7	0.9995	0.9981	0.9952	0.9901	0.9820	0.9699	0.9530	0.9307	0.9031	0.8708	0.8347
7.8	0.9996	0.9982	0.9954	0.9905	0.9825	0.9706	0.9538	0.9317	0.9041	0.8717	0.8356
7.9	0.9996	0.9983	0.9956	0.9909	0.9831	0.9713	0.9546	0.9326	0.9051	0.8727	0.8365
8.0	0.9997	0.9984	0.9958	0.9912	0.9835	0.9719	0.9554	0.9334	0.9060	0.8736	0.8373
8.1	0.9997	0.9985	0.9960	0.9915	0.9840	0.9725	0.9562	0.9343	0.9069	0.8745	0.8381
8.2	0.9997	0.9986	0.9962	0.9918	0.9844	0.9731	0.9569	0.9351	0.9078	0.8753	0.8389
8.3	0.9998	0.9987	0.9964	0.9921	0.9849	0.9737	0.9576	0.9359	0.9086	0.8762	0.8397
8.4	0.9998	0.9988	0.9965	0.9924	0.9853	0.9742	0.9583	0.9367	0.9094	0.8770	0.8404
8.5	0.9998	0.9988	0.9967	0.9926	0.9857	0.9748	0.9589	0.9374	0.9102	0.8777	0.8412
8.6	0.9998	0.9989	0.9968	0.9929	0.9860	0.9753	0.9596	0.9382	0.9110	0.8785	0.8419
8.7	0.9998	0.9990	0.9970	0.9931	0.9864	0.9758	0.9602	0.9389	0.9117	0.8792	0.8426
8.8	0.9998	0.9990	0.9971	0.9933	0.9867	0.9762	0.9608	0.9396	0.9125	0.8800	0.8432
8.9	0.9999	0.9991	0.9972	0.9936	0.9871	0.9767	0.9613	0.9402	0.9132	0.8807	0.8439
9.0	0.9999	0.9991	0.9974	0.9938	0.9874	0.9771	0.9619	0.9409	0.9138	0.8813	0.8445
9.1	0.9999	0.9992	0.9975	0.9940	0.9877	0.9776	0.9624	0.9415	0.9145	0.8820	0.8451
9.2	0.9999	0.9992	0.9976	0.9941	0.9880	0.9780	0.9630	0.9421	0.9152	0.8826	0.8457
9.3	0.9999	0.9993	0.9977	0.9943	0.9883	0.9784	0.9635	0.9427	0.9158	0.8833	0.8463
9.4	0.9999	0.9993	0.9978	0.9945	0.9886	0.9788	0.9639	0.9432	0.9164	0.8839	0.8468
9.5	0.9999	0.9994	0.9978	0.9947	0.9888	0.9791	0.9644	0.9438	0.9170	0.8844	0.8474
9.6	0.9999	0.9994	0.9979	0.9948	0.9891	0.9795	0.9649	0.9443	0.9176	0.8850	0.8479
9.7	0.9999	0.9994	0.9980	0.9950	0.9893	0.9798	0.9653	0.9449	0.9181	0.8856	0.8484
9.8	0.9999	0.9995	0.9981	0.9951	0.9895	0.9802	0.9657	0.9454	0.9187	0.8861	0.8489
9.9	1.0000	0.9995	0.9982	0.9953	0.9898	0.9805	0.9662	0.9458	0.9192	0.8866	0.8494
10.0	1.0000	0.9995	0.9982	0.9954	0.9900	0.9808	0.9666	0.9463	0.9197	0.8871	0.8499

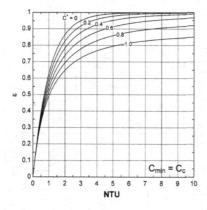

CONFIGURATION 34

2TUBE4ROW2PASS4CIR (Guo et al., 2002)

(b) $C_{min} = C_h$

NTU	Heat Exchanger Effectiveness										
	C^* $(C_{min} = C_h)$										
	0	0.1	0.2	0.3	0.4	0.5	0.6	0.7	0.8	0.9	1.0
0	0.0000	0.0000	0.0000	0.0000	0.0000	0.0000	0.0000	0.0000	0.0000	0.0000	0.0000
0.1	0.0952	0.0947	0.0943	0.0939	0.0934	0.0930	0.0926	0.0921	0.0917	0.0913	0.0909
0.2	0.1813	0.1797	0.1782	0.1767	0.1752	0.1737	0.1722	0.1707	0.1693	0.1679	0.1665
0.3	0.2592	0.2561	0.2531	0.2501	0.2471	0.2442	0.2413	0.2385	0.2357	0.2329	0.2302
0.4	0.3297	0.3248	0.3201	0.3154	0.3108	0.3062	0.3018	0.2973	0.2930	0.2888	0.2846
0.5	0.3935	0.3868	0.3802	0.3737	0.3674	0.3611	0.3550	0.3489	0.3430	0.3372	0.3315
0.6	0.4512	0.4427	0.4343	0.4260	0.4179	0.4099	0.4021	0.3944	0.3869	0.3795	0.3723
0.7	0.5034	0.4931	0.4830	0.4730	0.4632	0.4536	0.4441	0.4349	0.4258	0.4168	0.4081
0.8	0.5507	0.5387	0.5270	0.5154	0.5040	0.4928	0.4818	0.4709	0.4603	0.4499	0.4397
0.9	0.5934	0.5800	0.5668	0.5537	0.5408	0.5281	0.5156	0.5033	0.4913	0.4795	0.4679
1.0	0.6321	0.6174	0.6028	0.5884	0.5741	0.5600	0.5462	0.5325	0.5191	0.5060	0.4931
1.1	0.6671	0.6513	0.6355	0.6199	0.6044	0.5890	0.5739	0.5590	0.5443	0.5299	0.5158
1.2	0.6988	0.6820	0.6652	0.6485	0.6319	0.6154	0.5991	0.5830	0.5672	0.5516	0.5363
1.3	0.7275	0.7099	0.6923	0.6746	0.6570	0.6395	0.6222	0.6050	0.5880	0.5713	0.5549
1.4	0.7534	0.7352	0.7169	0.6985	0.6800	0.6616	0.6433	0.6251	0.6071	0.5894	0.5719
1.5	0.7769	0.7582	0.7393	0.7203	0.7011	0.6818	0.6626	0.6435	0.6246	0.6059	0.5875
1.6	0.7981	0.7792	0.7598	0.7402	0.7204	0.7005	0.6805	0.6606	0.6408	0.6212	0.6018
1.7	0.8173	0.7982	0.7786	0.7586	0.7382	0.7177	0.6970	0.6763	0.6557	0.6352	0.6150
1.8	0.8347	0.8155	0.7957	0.7754	0.7546	0.7335	0.7122	0.6908	0.6695	0.6482	0.6272
1.9	0.8504	0.8313	0.8114	0.7909	0.7698	0.7482	0.7264	0.7044	0.6823	0.6603	0.6385
2.0	0.8647	0.8457	0.8258	0.8051	0.7837	0.7618	0.7395	0.7169	0.6942	0.6716	0.6490
2.1	0.8775	0.8588	0.8390	0.8182	0.7967	0.7745	0.7518	0.7287	0.7054	0.6821	0.6588
2.2	0.8892	0.8707	0.8511	0.8304	0.8087	0.7863	0.7632	0.7396	0.7158	0.6919	0.6680
2.3	0.8997	0.8816	0.8622	0.8416	0.8199	0.7972	0.7738	0.7499	0.7255	0.7011	0.6766
2.4	0.9093	0.8916	0.8724	0.8519	0.8302	0.8075	0.7838	0.7595	0.7347	0.7097	0.6846
2.5	0.9179	0.9006	0.8818	0.8615	0.8399	0.8170	0.7932	0.7685	0.7433	0.7178	0.6922
2.6	0.9257	0.9089	0.8905	0.8704	0.8489	0.8260	0.8020	0.7770	0.7514	0.7254	0.6993
2.7	0.9328	0.9165	0.8984	0.8786	0.8572	0.8344	0.8102	0.7850	0.7591	0.7326	0.7060
2.8	0.9392	0.9234	0.9058	0.8863	0.8651	0.8422	0.8180	0.7926	0.7663	0.7394	0.7123
2.9	0.9450	0.9297	0.9125	0.8934	0.8724	0.8496	0.8253	0.7997	0.7731	0.7459	0.7183
3.0	0.9502	0.9355	0.9188	0.9000	0.8792	0.8565	0.8322	0.8064	0.7796	0.7520	0.7240
3.1	0.9550	0.9408	0.9245	0.9061	0.8856	0.8631	0.8387	0.8128	0.7857	0.7578	0.7293
3.2	0.9592	0.9456	0.9299	0.9118	0.8916	0.8692	0.8449	0.8189	0.7915	0.7633	0.7344
3.3	0.9631	0.9501	0.9348	0.9172	0.8972	0.8750	0.8507	0.8246	0.7971	0.7685	0.7393
3.4	0.9666	0.9541	0.9393	0.9221	0.9025	0.8804	0.8562	0.8301	0.8023	0.7735	0.7439
3.5	0.9698	0.9578	0.9435	0.9268	0.9074	0.8856	0.8614	0.8352	0.8073	0.7782	0.7483
3.6	0.9727	0.9612	0.9474	0.9311	0.9121	0.8905	0.8664	0.8402	0.8121	0.7827	0.7525
3.7	0.9753	0.9644	0.9510	0.9351	0.9165	0.8951	0.8711	0.8449	0.8167	0.7870	0.7565
3.8	0.9776	0.9672	0.9544	0.9389	0.9206	0.8994	0.8756	0.8493	0.8210	0.7912	0.7603
3.9	0.9798	0.9698	0.9575	0.9424	0.9244	0.9036	0.8799	0.8536	0.8252	0.7951	0.7639
4.0	0.9817	0.9722	0.9603	0.9457	0.9281	0.9075	0.8839	0.8577	0.8291	0.7989	0.7674
4.1	0.9834	0.9745	0.9630	0.9488	0.9315	0.9112	0.8878	0.8616	0.8329	0.8025	0.7707
4.2	0.9850	0.9765	0.9655	0.9517	0.9348	0.9147	0.8914	0.8653	0.8366	0.8059	0.7739
4.3	0.9864	0.9783	0.9678	0.9544	0.9378	0.9180	0.8950	0.8688	0.8401	0.8092	0.7770
4.4	0.9877	0.9800	0.9699	0.9569	0.9407	0.9212	0.8983	0.8722	0.8434	0.8124	0.7799
4.5	0.9889	0.9816	0.9719	0.9593	0.9435	0.9242	0.9015	0.8755	0.8466	0.8155	0.7827
4.6	0.9899	0.9830	0.9737	0.9615	0.9461	0.9271	0.9046	0.8786	0.8497	0.8184	0.7854
4.7	0.9909	0.9844	0.9754	0.9636	0.9485	0.9298	0.9075	0.8816	0.8527	0.8212	0.7880
4.8	0.9918	0.9856	0.9770	0.9656	0.9508	0.9324	0.9103	0.8845	0.8555	0.8239	0.7905
4.9	0.9926	0.9867	0.9785	0.9674	0.9530	0.9349	0.9129	0.8872	0.8582	0.8265	0.7929
5.0	0.9933	0.9877	0.9799	0.9692	0.9551	0.9373	0.9155	0.8899	0.8609	0.8290	0.7952
5.1	0.9939	0.9887	0.9812	0.9708	0.9571	0.9395	0.9179	0.8924	0.8634	0.8315	0.7975
5.2	0.9945	0.9895	0.9824	0.9723	0.9589	0.9416	0.9203	0.8949	0.8658	0.8338	0.7996
5.3	0.9950	0.9903	0.9835	0.9738	0.9607	0.9437	0.9225	0.8972	0.8682	0.8360	0.8017
5.4	0.9955	0.9911	0.9845	0.9751	0.9624	0.9456	0.9247	0.8995	0.8705	0.8382	0.8037
5.5	0.9959	0.9918	0.9855	0.9764	0.9640	0.9475	0.9268	0.9017	0.8726	0.8403	0.8056
5.6	0.9963	0.9924	0.9864	0.9776	0.9655	0.9493	0.9287	0.9038	0.8747	0.8423	0.8074
5.7	0.9967	0.9930	0.9872	0.9788	0.9669	0.9510	0.9307	0.9058	0.8768	0.8443	0.8092
5.8	0.9970	0.9935	0.9880	0.9798	0.9683	0.9526	0.9325	0.9077	0.8787	0.8461	0.8109
5.9	0.9973	0.9940	0.9888	0.9809	0.9696	0.9542	0.9342	0.9096	0.8806	0.8480	0.8126
6.0	0.9975	0.9944	0.9894	0.9818	0.9708	0.9557	0.9359	0.9114	0.8825	0.8497	0.8142

NTU	Heat Exchanger Effectiveness										
	C^* ($C_{min} = C_h$)										
	0	0.1	0.2	0.3	0.4	0.5	0.6	0.7	0.8	0.9	1.0
6.1	0.9978	0.9949	0.9901	0.9827	0.9720	0.9571	0.9376	0.9132	0.8842	0.8514	0.8157
6.2	0.9980	0.9952	0.9907	0.9836	0.9731	0.9585	0.9391	0.9149	0.8859	0.8530	0.8172
6.3	0.9982	0.9956	0.9912	0.9844	0.9742	0.9598	0.9406	0.9165	0.8876	0.8546	0.8187
6.4	0.9983	0.9959	0.9918	0.9851	0.9752	0.9610	0.9421	0.9181	0.8892	0.8562	0.8201
6.5	0.9985	0.9962	0.9923	0.9859	0.9761	0.9622	0.9435	0.9196	0.8907	0.8577	0.8215
6.6	0.9986	0.9965	0.9927	0.9865	0.9771	0.9634	0.9448	0.9211	0.8922	0.8591	0.8228
6.7	0.9988	0.9968	0.9931	0.9872	0.9779	0.9645	0.9461	0.9225	0.8937	0.8605	0.8240
6.8	0.9989	0.9970	0.9936	0.9878	0.9788	0.9656	0.9474	0.9238	0.8951	0.8618	0.8253
6.9	0.9990	0.9972	0.9939	0.9884	0.9796	0.9666	0.9486	0.9252	0.8964	0.8631	0.8264
7.0	0.9991	0.9974	0.9943	0.9889	0.9803	0.9676	0.9498	0.9265	0.8978	0.8644	0.8276
7.1	0.9992	0.9976	0.9946	0.9894	0.9811	0.9685	0.9509	0.9277	0.8990	0.8656	0.8287
7.2	0.9993	0.9978	0.9949	0.9899	0.9818	0.9694	0.9520	0.9289	0.9003	0.8668	0.8298
7.3	0.9993	0.9979	0.9952	0.9904	0.9824	0.9703	0.9530	0.9301	0.9015	0.8680	0.8308
7.4	0.9994	0.9981	0.9955	0.9908	0.9831	0.9711	0.9540	0.9312	0.9026	0.8691	0.8319
7.5	0.9994	0.9982	0.9957	0.9912	0.9837	0.9719	0.9550	0.9323	0.9038	0.8702	0.8328
7.6	0.9995	0.9984	0.9960	0.9916	0.9842	0.9727	0.9560	0.9333	0.9049	0.8713	0.8338
7.7	0.9995	0.9985	0.9962	0.9920	0.9848	0.9734	0.9569	0.9344	0.9059	0.8723	0.8347
7.8	0.9996	0.9986	0.9964	0.9924	0.9853	0.9742	0.9577	0.9354	0.9070	0.8733	0.8356
7.9	0.9996	0.9987	0.9966	0.9927	0.9858	0.9749	0.9586	0.9363	0.9080	0.8742	0.8365
8.0	0.9997	0.9988	0.9968	0.9930	0.9863	0.9755	0.9594	0.9373	0.9089	0.8752	0.8373
8.1	0.9997	0.9989	0.9970	0.9933	0.9868	0.9762	0.9602	0.9382	0.9099	0.8761	0.8381
8.2	0.9997	0.9989	0.9972	0.9936	0.9872	0.9768	0.9610	0.9391	0.9108	0.8770	0.8389
8.3	0.9998	0.9990	0.9973	0.9939	0.9877	0.9774	0.9617	0.9399	0.9117	0.8778	0.8397
8.4	0.9998	0.9991	0.9975	0.9941	0.9881	0.9779	0.9625	0.9407	0.9126	0.8787	0.8404
8.5	0.9998	0.9991	0.9976	0.9944	0.9885	0.9785	0.9632	0.9415	0.9134	0.8795	0.8412
8.6	0.9998	0.9992	0.9977	0.9946	0.9888	0.9790	0.9638	0.9423	0.9142	0.8803	0.8419
8.7	0.9998	0.9993	0.9979	0.9949	0.9892	0.9795	0.9645	0.9431	0.9150	0.8810	0.8426
8.8	0.9998	0.9993	0.9980	0.9951	0.9895	0.9800	0.9651	0.9438	0.9158	0.8818	0.8432
8.9	0.9999	0.9994	0.9981	0.9953	0.9899	0.9805	0.9658	0.9445	0.9166	0.8825	0.8439
9.0	0.9999	0.9994	0.9982	0.9955	0.9902	0.9810	0.9664	0.9452	0.9173	0.8832	0.8445
9.1	0.9999	0.9994	0.9983	0.9957	0.9905	0.9814	0.9669	0.9459	0.9180	0.8839	0.8451
9.2	0.9999	0.9995	0.9984	0.9958	0.9908	0.9819	0.9675	0.9466	0.9187	0.8846	0.8457
9.3	0.9999	0.9995	0.9984	0.9960	0.9911	0.9823	0.9680	0.9472	0.9194	0.8852	0.8463
9.4	0.9999	0.9996	0.9985	0.9962	0.9914	0.9827	0.9686	0.9478	0.9200	0.8858	0.8468
9.5	0.9999	0.9996	0.9986	0.9963	0.9916	0.9831	0.9691	0.9484	0.9207	0.8865	0.8474
9.6	0.9999	0.9996	0.9987	0.9965	0.9919	0.9834	0.9696	0.9490	0.9213	0.8871	0.8479
9.7	0.9999	0.9996	0.9987	0.9966	0.9921	0.9838	0.9701	0.9496	0.9219	0.8876	0.8484
9.8	0.9999	0.9997	0.9988	0.9967	0.9923	0.9842	0.9705	0.9502	0.9225	0.8882	0.8489
9.9	1.0000	0.9997	0.9989	0.9969	0.9926	0.9845	0.9710	0.9507	0.9231	0.8888	0.8494
10.0	1.0000	0.9997	0.9989	0.9970	0.9928	0.9848	0.9715	0.9512	0.9236	0.8893	0.8499

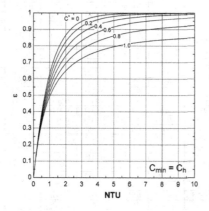

CONFIGURATION 35

1TUBE6ROW3PASS2CIR

(a) $C_{min} = C_c$

NTU	Heat Exchanger Effectiveness										
	C^* $(C_{min} = C_c)$										
	0	0.1	0.2	0.3	0.4	0.5	0.6	0.7	0.8	0.9	1.0
0	0.0000	0.0000	0.0000	0.0000	0.0000	0.0000	0.0000	0.0000	0.0000	0.0000	0.0000
0.1	0.0952	0.0947	0.0943	0.0939	0.0934	0.0930	0.0926	0.0921	0.0917	0.0913	0.0909
0.2	0.1813	0.1797	0.1782	0.1767	0.1752	0.1737	0.1723	0.1708	0.1694	0.1680	0.1666
0.3	0.2592	0.2561	0.2531	0.2502	0.2473	0.2444	0.2415	0.2387	0.2359	0.2332	0.2305
0.4	0.3297	0.3249	0.3202	0.3156	0.3111	0.3066	0.3022	0.2978	0.2935	0.2893	0.2852
0.5	0.3935	0.3869	0.3805	0.3741	0.3679	0.3617	0.3557	0.3497	0.3438	0.3381	0.3324
0.6	0.4512	0.4429	0.4347	0.4266	0.4187	0.4108	0.4031	0.3956	0.3881	0.3808	0.3736
0.7	0.5034	0.4934	0.4836	0.4738	0.4642	0.4548	0.4455	0.4363	0.4274	0.4185	0.4098
0.8	0.5507	0.5391	0.5277	0.5165	0.5053	0.4943	0.4835	0.4728	0.4623	0.4520	0.4419
0.9	0.5934	0.5805	0.5677	0.5550	0.5424	0.5300	0.5177	0.5056	0.4937	0.4820	0.4705
1.0	0.6321	0.6180	0.6039	0.5900	0.5761	0.5623	0.5487	0.5352	0.5220	0.5090	0.4962
1.1	0.6671	0.6520	0.6368	0.6217	0.6066	0.5916	0.5768	0.5621	0.5476	0.5333	0.5193
1.2	0.6988	0.6828	0.6667	0.6506	0.6345	0.6184	0.6024	0.5865	0.5709	0.5554	0.5402
1.3	0.7275	0.7108	0.6940	0.6770	0.6599	0.6428	0.6258	0.6089	0.5921	0.5755	0.5591
1.4	0.7534	0.7362	0.7188	0.7011	0.6832	0.6652	0.6472	0.6293	0.6115	0.5939	0.5765
1.5	0.7769	0.7594	0.7414	0.7231	0.7045	0.6858	0.6669	0.6481	0.6293	0.6107	0.5923
1.6	0.7981	0.7804	0.7621	0.7433	0.7241	0.7047	0.6851	0.6654	0.6458	0.6262	0.6069
1.7	0.8173	0.7995	0.7809	0.7618	0.7421	0.7221	0.7018	0.6814	0.6609	0.6405	0.6203
1.8	0.8347	0.8169	0.7982	0.7788	0.7587	0.7382	0.7173	0.6962	0.6750	0.6538	0.6327
1.9	0.8504	0.8327	0.8140	0.7944	0.7741	0.7531	0.7317	0.7099	0.6880	0.6661	0.6442
2.0	0.8647	0.8471	0.8285	0.8088	0.7882	0.7669	0.7450	0.7227	0.7001	0.6775	0.6549
2.1	0.8775	0.8603	0.8417	0.8220	0.8013	0.7797	0.7574	0.7346	0.7114	0.6881	0.6648
2.2	0.8892	0.8723	0.8539	0.8343	0.8134	0.7916	0.7689	0.7457	0.7219	0.6980	0.6741
2.3	0.8997	0.8832	0.8651	0.8455	0.8247	0.8027	0.7797	0.7560	0.7318	0.7073	0.6827
2.4	0.9093	0.8931	0.8753	0.8560	0.8351	0.8130	0.7898	0.7657	0.7410	0.7160	0.6908
2.5	0.9179	0.9022	0.8848	0.8656	0.8448	0.8226	0.7992	0.7748	0.7497	0.7241	0.6984
2.6	0.9257	0.9105	0.8934	0.8745	0.8538	0.8316	0.8080	0.7833	0.7578	0.7318	0.7055
2.7	0.9328	0.9181	0.9014	0.8827	0.8622	0.8400	0.8163	0.7913	0.7655	0.7390	0.7122
2.8	0.9392	0.9250	0.9087	0.8904	0.8700	0.8478	0.8240	0.7989	0.7727	0.7458	0.7185
2.9	0.9450	0.9313	0.9155	0.8975	0.8773	0.8552	0.8313	0.8060	0.7795	0.7522	0.7245
3.0	0.9502	0.9371	0.9217	0.9040	0.8841	0.8621	0.8382	0.8127	0.7859	0.7583	0.7301
3.1	0.9550	0.9423	0.9274	0.9101	0.8905	0.8686	0.8447	0.8190	0.7920	0.7640	0.7354
3.2	0.9592	0.9472	0.9327	0.9158	0.8964	0.8747	0.8508	0.8250	0.7978	0.7694	0.7405
3.3	0.9631	0.9516	0.9376	0.9210	0.9020	0.8804	0.8565	0.8307	0.8032	0.7746	0.7452
3.4	0.9666	0.9556	0.9421	0.9259	0.9071	0.8857	0.8619	0.8360	0.8084	0.7795	0.7498
3.5	0.9698	0.9593	0.9462	0.9305	0.9120	0.8908	0.8671	0.8411	0.8133	0.7841	0.7541
3.6	0.9727	0.9626	0.9500	0.9347	0.9166	0.8956	0.8719	0.8459	0.8180	0.7885	0.7582
3.7	0.9753	0.9657	0.9536	0.9387	0.9208	0.9001	0.8765	0.8505	0.8224	0.7927	0.7621
3.8	0.9776	0.9685	0.9568	0.9423	0.9248	0.9043	0.8809	0.8548	0.8266	0.7968	0.7658
3.9	0.9798	0.9711	0.9599	0.9458	0.9286	0.9083	0.8850	0.8590	0.8307	0.8006	0.7693
4.0	0.9817	0.9735	0.9627	0.9490	0.9321	0.9121	0.8889	0.8629	0.8345	0.8042	0.7727
4.1	0.9834	0.9756	0.9652	0.9519	0.9354	0.9156	0.8926	0.8667	0.8382	0.8077	0.7759
4.2	0.9850	0.9776	0.9676	0.9547	0.9386	0.9190	0.8962	0.8702	0.8417	0.8111	0.7790
4.3	0.9864	0.9794	0.9698	0.9573	0.9415	0.9222	0.8995	0.8736	0.8450	0.8142	0.7820
4.4	0.9877	0.9811	0.9719	0.9597	0.9443	0.9252	0.9027	0.8769	0.8482	0.8173	0.7848
4.5	0.9889	0.9826	0.9738	0.9620	0.9469	0.9281	0.9057	0.8800	0.8513	0.8202	0.7875
4.6	0.9899	0.9840	0.9756	0.9641	0.9493	0.9308	0.9086	0.8830	0.8542	0.8230	0.7901
4.7	0.9909	0.9853	0.9772	0.9661	0.9516	0.9334	0.9114	0.8858	0.8570	0.8257	0.7926
4.8	0.9918	0.9864	0.9787	0.9680	0.9538	0.9359	0.9140	0.8885	0.8597	0.8283	0.7950
4.9	0.9926	0.9875	0.9801	0.9697	0.9559	0.9382	0.9165	0.8911	0.8623	0.8308	0.7973
5.0	0.9933	0.9885	0.9814	0.9714	0.9578	0.9404	0.9189	0.8936	0.8648	0.8332	0.7995
5.1	0.9939	0.9894	0.9826	0.9729	0.9596	0.9425	0.9212	0.8960	0.8672	0.8355	0.8017
5.2	0.9945	0.9902	0.9837	0.9743	0.9614	0.9445	0.9234	0.8983	0.8695	0.8377	0.8037
5.3	0.9950	0.9910	0.9848	0.9757	0.9630	0.9464	0.9255	0.9005	0.8717	0.8398	0.8057
5.4	0.9955	0.9917	0.9858	0.9769	0.9646	0.9482	0.9275	0.9026	0.8738	0.8419	0.8076
5.5	0.9959	0.9924	0.9867	0.9781	0.9660	0.9499	0.9294	0.9046	0.8759	0.8438	0.8094
5.6	0.9963	0.9929	0.9875	0.9792	0.9674	0.9516	0.9313	0.9066	0.8778	0.8457	0.8112
5.7	0.9967	0.9935	0.9883	0.9803	0.9688	0.9531	0.9330	0.9084	0.8797	0.8476	0.8129
5.8	0.9970	0.9940	0.9890	0.9813	0.9700	0.9546	0.9347	0.9102	0.8816	0.8494	0.8146
5.9	0.9973	0.9945	0.9897	0.9822	0.9712	0.9560	0.9363	0.9120	0.8833	0.8511	0.8162
6.0	0.9975	0.9949	0.9903	0.9831	0.9723	0.9574	0.9379	0.9136	0.8850	0.8527	0.8177

NTU	Heat Exchanger Effectiveness										
	C^* ($C_{min} = C_c$)										
	0	0.1	0.2	0.3	0.4	0.5	0.6	0.7	0.8	0.9	1.0
6.1	0.9978	0.9953	0.9909	0.9839	0.9734	0.9587	0.9394	0.9152	0.8867	0.8543	0.8192
6.2	0.9980	0.9956	0.9914	0.9846	0.9744	0.9599	0.9408	0.9168	0.8883	0.8559	0.8206
6.3	0.9982	0.9960	0.9920	0.9854	0.9754	0.9611	0.9422	0.9183	0.8898	0.8574	0.8220
6.4	0.9983	0.9963	0.9924	0.9861	0.9763	0.9623	0.9435	0.9197	0.8913	0.8588	0.8234
6.5	0.9985	0.9965	0.9929	0.9867	0.9771	0.9633	0.9447	0.9211	0.8927	0.8602	0.8247
6.6	0.9986	0.9968	0.9933	0.9873	0.9780	0.9644	0.9460	0.9225	0.8941	0.8616	0.8259
6.7	0.9988	0.9970	0.9937	0.9879	0.9788	0.9654	0.9471	0.9237	0.8954	0.8629	0.8272
6.8	0.9989	0.9972	0.9940	0.9884	0.9795	0.9663	0.9483	0.9250	0.8967	0.8642	0.8284
6.9	0.9990	0.9975	0.9944	0.9890	0.9802	0.9672	0.9494	0.9262	0.8980	0.8654	0.8295
7.0	0.9991	0.9976	0.9947	0.9894	0.9809	0.9681	0.9504	0.9274	0.8992	0.8666	0.8306
7.1	0.9992	0.9978	0.9950	0.9899	0.9815	0.9690	0.9514	0.9285	0.9004	0.8678	0.8317
7.2	0.9993	0.9980	0.9953	0.9903	0.9822	0.9698	0.9524	0.9296	0.9015	0.8689	0.8328
7.3	0.9993	0.9981	0.9955	0.9908	0.9828	0.9706	0.9533	0.9307	0.9027	0.8700	0.8338
7.4	0.9994	0.9982	0.9958	0.9911	0.9833	0.9713	0.9542	0.9317	0.9037	0.8711	0.8348
7.5	0.9994	0.9984	0.9960	0.9915	0.9839	0.9720	0.9551	0.9327	0.9048	0.8721	0.8358
7.6	0.9995	0.9985	0.9962	0.9919	0.9844	0.9727	0.9560	0.9336	0.9058	0.8731	0.8367
7.7	0.9995	0.9986	0.9964	0.9922	0.9849	0.9734	0.9568	0.9346	0.9068	0.8741	0.8376
7.8	0.9996	0.9987	0.9966	0.9925	0.9853	0.9740	0.9576	0.9355	0.9077	0.8750	0.8385
7.9	0.9996	0.9988	0.9968	0.9928	0.9858	0.9746	0.9583	0.9363	0.9087	0.8760	0.8394
8.0	0.9997	0.9989	0.9970	0.9931	0.9862	0.9752	0.9591	0.9372	0.9096	0.8769	0.8402
8.1	0.9997	0.9989	0.9971	0.9934	0.9866	0.9758	0.9598	0.9380	0.9105	0.8777	0.8410
8.2	0.9997	0.9990	0.9973	0.9936	0.9870	0.9763	0.9605	0.9388	0.9113	0.8786	0.8418
8.3	0.9998	0.9991	0.9974	0.9939	0.9874	0.9768	0.9611	0.9396	0.9121	0.8794	0.8426
8.4	0.9998	0.9992	0.9975	0.9941	0.9878	0.9773	0.9618	0.9403	0.9129	0.8802	0.8434
8.5	0.9998	0.9992	0.9977	0.9943	0.9881	0.9778	0.9624	0.9411	0.9137	0.8810	0.8441
8.6	0.9998	0.9993	0.9978	0.9945	0.9884	0.9783	0.9630	0.9418	0.9145	0.8818	0.8448
8.7	0.9998	0.9993	0.9979	0.9947	0.9888	0.9788	0.9636	0.9425	0.9152	0.8825	0.8455
8.8	0.9998	0.9994	0.9980	0.9949	0.9891	0.9792	0.9642	0.9431	0.9160	0.8833	0.8462
8.9	0.9999	0.9994	0.9981	0.9951	0.9894	0.9796	0.9647	0.9438	0.9167	0.8840	0.8469
9.0	0.9999	0.9994	0.9982	0.9953	0.9896	0.9800	0.9652	0.9444	0.9174	0.8847	0.8475
9.1	0.9999	0.9995	0.9983	0.9954	0.9899	0.9804	0.9658	0.9450	0.9180	0.8853	0.8482
9.2	0.9999	0.9995	0.9983	0.9956	0.9902	0.9808	0.9663	0.9456	0.9187	0.8860	0.8488
9.3	0.9999	0.9995	0.9984	0.9957	0.9904	0.9812	0.9667	0.9462	0.9193	0.8866	0.8494
9.4	0.9999	0.9996	0.9985	0.9959	0.9907	0.9815	0.9672	0.9468	0.9200	0.8873	0.8500
9.5	0.9999	0.9996	0.9986	0.9960	0.9909	0.9819	0.9677	0.9473	0.9206	0.8879	0.8506
9.6	0.9999	0.9996	0.9986	0.9962	0.9911	0.9822	0.9681	0.9479	0.9211	0.8885	0.8511
9.7	0.9999	0.9996	0.9987	0.9963	0.9913	0.9825	0.9686	0.9484	0.9217	0.8891	0.8517
9.8	0.9999	0.9997	0.9987	0.9964	0.9916	0.9829	0.9690	0.9489	0.9223	0.8896	0.8522
9.9	1.0000	0.9997	0.9988	0.9965	0.9918	0.9832	0.9694	0.9494	0.9228	0.8902	0.8527
10.0	1.0000	0.9997	0.9988	0.9966	0.9919	0.9835	0.9698	0.9499	0.9234	0.8907	0.8532

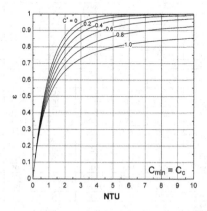

CONFIGURATION 35

1TUBE6ROW3PASS2CIR

(b) $C_{min} = C_h$

NTU	Heat Exchanger Effectiveness										
	C^* ($C_{min} = C_h$)										
	0	0.1	0.2	0.3	0.4	0.5	0.6	0.7	0.8	0.9	1.0
0	0.0000	0.0000	0.0000	0.0000	0.0000	0.0000	0.0000	0.0000	0.0000	0.0000	0.0000
0.1	0.0952	0.0947	0.0943	0.0939	0.0934	0.0930	0.0926	0.0921	0.0917	0.0913	0.0909
0.2	0.1813	0.1797	0.1782	0.1767	0.1752	0.1737	0.1723	0.1708	0.1694	0.1680	0.1666
0.3	0.2592	0.2561	0.2531	0.2502	0.2473	0.2444	0.2415	0.2387	0.2359	0.2332	0.2305
0.4	0.3297	0.3249	0.3202	0.3156	0.3111	0.3066	0.3022	0.2978	0.2935	0.2893	0.2852
0.5	0.3935	0.3869	0.3805	0.3741	0.3679	0.3617	0.3557	0.3497	0.3438	0.3381	0.3324
0.6	0.4512	0.4429	0.4347	0.4266	0.4187	0.4108	0.4031	0.3956	0.3881	0.3808	0.3736
0.7	0.5034	0.4934	0.4836	0.4738	0.4642	0.4548	0.4455	0.4363	0.4274	0.4185	0.4098
0.8	0.5507	0.5391	0.5277	0.5165	0.5053	0.4943	0.4835	0.4728	0.4623	0.4520	0.4419
0.9	0.5934	0.5805	0.5677	0.5550	0.5424	0.5300	0.5177	0.5056	0.4937	0.4820	0.4705
1.0	0.6321	0.6180	0.6039	0.5900	0.5761	0.5623	0.5487	0.5352	0.5220	0.5090	0.4962
1.1	0.6671	0.6520	0.6368	0.6217	0.6066	0.5916	0.5768	0.5621	0.5476	0.5333	0.5193
1.2	0.6988	0.6828	0.6667	0.6506	0.6345	0.6184	0.6024	0.5865	0.5709	0.5554	0.5402
1.3	0.7275	0.7108	0.6940	0.6770	0.6599	0.6428	0.6258	0.6089	0.5921	0.5755	0.5591
1.4	0.7534	0.7362	0.7188	0.7011	0.6832	0.6652	0.6472	0.6293	0.6115	0.5939	0.5765
1.5	0.7769	0.7594	0.7414	0.7231	0.7045	0.6858	0.6669	0.6481	0.6293	0.6107	0.5923
1.6	0.7981	0.7804	0.7621	0.7433	0.7241	0.7047	0.6851	0.6654	0.6458	0.6262	0.6069
1.7	0.8173	0.7995	0.7809	0.7618	0.7421	0.7221	0.7018	0.6814	0.6609	0.6405	0.6203
1.8	0.8347	0.8169	0.7982	0.7788	0.7587	0.7382	0.7173	0.6962	0.6750	0.6538	0.6327
1.9	0.8504	0.8327	0.8140	0.7944	0.7741	0.7531	0.7317	0.7099	0.6880	0.6661	0.6442
2.0	0.8647	0.8471	0.8285	0.8088	0.7882	0.7669	0.7450	0.7227	0.7001	0.6775	0.6549
2.1	0.8775	0.8603	0.8417	0.8220	0.8013	0.7797	0.7574	0.7346	0.7114	0.6881	0.6648
2.2	0.8892	0.8723	0.8539	0.8343	0.8134	0.7916	0.7689	0.7457	0.7219	0.6980	0.6741
2.3	0.8997	0.8832	0.8651	0.8455	0.8247	0.8027	0.7797	0.7560	0.7318	0.7073	0.6827
2.4	0.9093	0.8931	0.8753	0.8560	0.8351	0.8130	0.7898	0.7657	0.7410	0.7160	0.6908
2.5	0.9179	0.9022	0.8848	0.8656	0.8448	0.8226	0.7992	0.7748	0.7497	0.7241	0.6984
2.6	0.9257	0.9105	0.8934	0.8745	0.8538	0.8316	0.8080	0.7833	0.7578	0.7318	0.7055
2.7	0.9328	0.9181	0.9014	0.8827	0.8622	0.8400	0.8163	0.7913	0.7655	0.7390	0.7122
2.8	0.9392	0.9250	0.9087	0.8904	0.8700	0.8478	0.8240	0.7989	0.7727	0.7458	0.7185
2.9	0.9450	0.9313	0.9155	0.8975	0.8773	0.8552	0.8313	0.8060	0.7795	0.7522	0.7245
3.0	0.9502	0.9371	0.9217	0.9040	0.8841	0.8621	0.8382	0.8127	0.7859	0.7583	0.7301
3.1	0.9550	0.9423	0.9274	0.9101	0.8905	0.8686	0.8447	0.8190	0.7920	0.7640	0.7354
3.2	0.9592	0.9472	0.9327	0.9158	0.8964	0.8747	0.8508	0.8250	0.7978	0.7694	0.7405
3.3	0.9631	0.9516	0.9376	0.9210	0.9020	0.8804	0.8565	0.8307	0.8032	0.7746	0.7452
3.4	0.9666	0.9556	0.9421	0.9259	0.9071	0.8857	0.8619	0.8360	0.8084	0.7795	0.7498
3.5	0.9698	0.9593	0.9462	0.9305	0.9120	0.8908	0.8671	0.8411	0.8133	0.7841	0.7541
3.6	0.9727	0.9626	0.9500	0.9347	0.9166	0.8956	0.8719	0.8459	0.8180	0.7885	0.7582
3.7	0.9753	0.9657	0.9536	0.9387	0.9208	0.9001	0.8765	0.8505	0.8224	0.7927	0.7621
3.8	0.9776	0.9685	0.9568	0.9423	0.9248	0.9043	0.8809	0.8548	0.8266	0.7968	0.7658
3.9	0.9798	0.9711	0.9599	0.9458	0.9286	0.9083	0.8850	0.8590	0.8307	0.8006	0.7693
4.0	0.9817	0.9735	0.9627	0.9490	0.9321	0.9121	0.8889	0.8629	0.8345	0.8042	0.7727
4.1	0.9834	0.9756	0.9652	0.9519	0.9354	0.9156	0.8926	0.8667	0.8382	0.8077	0.7759
4.2	0.9850	0.9776	0.9676	0.9547	0.9386	0.9190	0.8962	0.8702	0.8417	0.8111	0.7790
4.3	0.9864	0.9794	0.9698	0.9573	0.9415	0.9222	0.8995	0.8736	0.8450	0.8142	0.7820
4.4	0.9877	0.9811	0.9719	0.9597	0.9443	0.9252	0.9027	0.8769	0.8482	0.8173	0.7848
4.5	0.9889	0.9826	0.9738	0.9620	0.9469	0.9281	0.9057	0.8800	0.8513	0.8202	0.7875
4.6	0.9899	0.9840	0.9756	0.9641	0.9493	0.9308	0.9086	0.8830	0.8542	0.8230	0.7901
4.7	0.9909	0.9853	0.9772	0.9661	0.9516	0.9334	0.9114	0.8858	0.8570	0.8257	0.7926
4.8	0.9918	0.9864	0.9787	0.9680	0.9538	0.9359	0.9140	0.8885	0.8597	0.8283	0.7950
4.9	0.9926	0.9875	0.9801	0.9697	0.9559	0.9382	0.9165	0.8911	0.8623	0.8308	0.7973
5.0	0.9933	0.9885	0.9814	0.9714	0.9578	0.9404	0.9189	0.8936	0.8648	0.8332	0.7995
5.1	0.9939	0.9894	0.9826	0.9729	0.9596	0.9425	0.9212	0.8960	0.8672	0.8355	0.8017
5.2	0.9945	0.9902	0.9837	0.9743	0.9614	0.9445	0.9234	0.8983	0.8695	0.8377	0.8037
5.3	0.9950	0.9910	0.9848	0.9757	0.9630	0.9464	0.9255	0.9005	0.8717	0.8398	0.8057
5.4	0.9955	0.9917	0.9858	0.9769	0.9646	0.9482	0.9275	0.9026	0.8738	0.8419	0.8076
5.5	0.9959	0.9924	0.9867	0.9781	0.9660	0.9499	0.9294	0.9046	0.8759	0.8438	0.8094
5.6	0.9963	0.9929	0.9875	0.9792	0.9674	0.9516	0.9313	0.9066	0.8778	0.8457	0.8112
5.7	0.9967	0.9935	0.9883	0.9803	0.9688	0.9531	0.9330	0.9084	0.8797	0.8476	0.8129
5.8	0.9970	0.9940	0.9890	0.9813	0.9700	0.9546	0.9347	0.9102	0.8816	0.8494	0.8146
5.9	0.9973	0.9945	0.9897	0.9822	0.9712	0.9560	0.9363	0.9120	0.8833	0.8511	0.8162
6.0	0.9975	0.9949	0.9903	0.9831	0.9723	0.9574	0.9379	0.9136	0.8850	0.8527	0.8177

NTU	Heat Exchanger Effectiveness										
	C^* ($C_{min} = C_h$)										
	0	0.1	0.2	0.3	0.4	0.5	0.6	0.7	0.8	0.9	1.0
6.1	0.9978	0.9953	0.9909	0.9839	0.9734	0.9587	0.9394	0.9152	0.8867	0.8543	0.8192
6.2	0.9980	0.9956	0.9914	0.9846	0.9744	0.9599	0.9408	0.9168	0.8883	0.8559	0.8206
6.3	0.9982	0.9960	0.9920	0.9854	0.9754	0.9611	0.9422	0.9183	0.8898	0.8574	0.8220
6.4	0.9983	0.9963	0.9924	0.9861	0.9763	0.9623	0.9435	0.9197	0.8913	0.8588	0.8234
6.5	0.9985	0.9965	0.9929	0.9867	0.9771	0.9633	0.9447	0.9211	0.8927	0.8602	0.8247
6.6	0.9986	0.9968	0.9933	0.9873	0.9780	0.9644	0.9460	0.9225	0.8941	0.8616	0.8259
6.7	0.9988	0.9970	0.9937	0.9879	0.9788	0.9654	0.9471	0.9237	0.8954	0.8629	0.8272
6.8	0.9989	0.9972	0.9940	0.9884	0.9795	0.9663	0.9483	0.9250	0.8967	0.8642	0.8284
6.9	0.9990	0.9975	0.9944	0.9890	0.9802	0.9672	0.9494	0.9262	0.8980	0.8654	0.8295
7.0	0.9991	0.9976	0.9947	0.9894	0.9809	0.9681	0.9504	0.9274	0.8992	0.8666	0.8306
7.1	0.9992	0.9978	0.9950	0.9899	0.9815	0.9690	0.9514	0.9285	0.9004	0.8678	0.8317
7.2	0.9993	0.9980	0.9953	0.9903	0.9822	0.9698	0.9524	0.9296	0.9015	0.8689	0.8328
7.3	0.9993	0.9981	0.9955	0.9908	0.9828	0.9706	0.9533	0.9307	0.9027	0.8700	0.8338
7.4	0.9994	0.9982	0.9958	0.9911	0.9833	0.9713	0.9542	0.9317	0.9037	0.8711	0.8348
7.5	0.9994	0.9984	0.9960	0.9915	0.9839	0.9720	0.9551	0.9327	0.9048	0.8721	0.8358
7.6	0.9995	0.9985	0.9962	0.9919	0.9844	0.9727	0.9560	0.9336	0.9058	0.8731	0.8367
7.7	0.9995	0.9986	0.9964	0.9922	0.9849	0.9734	0.9568	0.9346	0.9068	0.8741	0.8376
7.8	0.9996	0.9987	0.9966	0.9925	0.9853	0.9740	0.9576	0.9355	0.9077	0.8750	0.8385
7.9	0.9996	0.9988	0.9968	0.9928	0.9858	0.9746	0.9583	0.9363	0.9087	0.8760	0.8394
8.0	0.9997	0.9989	0.9970	0.9931	0.9862	0.9752	0.9591	0.9372	0.9096	0.8769	0.8402
8.1	0.9997	0.9989	0.9971	0.9934	0.9866	0.9758	0.9598	0.9380	0.9105	0.8777	0.8410
8.2	0.9997	0.9990	0.9973	0.9936	0.9870	0.9763	0.9605	0.9388	0.9113	0.8786	0.8418
8.3	0.9998	0.9991	0.9974	0.9939	0.9874	0.9768	0.9611	0.9396	0.9121	0.8794	0.8426
8.4	0.9998	0.9992	0.9975	0.9941	0.9878	0.9773	0.9618	0.9403	0.9129	0.8802	0.8434
8.5	0.9998	0.9992	0.9977	0.9943	0.9881	0.9778	0.9624	0.9411	0.9137	0.8810	0.8441
8.6	0.9998	0.9993	0.9978	0.9945	0.9884	0.9783	0.9630	0.9418	0.9145	0.8818	0.8448
8.7	0.9998	0.9993	0.9979	0.9947	0.9888	0.9788	0.9636	0.9425	0.9152	0.8825	0.8455
8.8	0.9998	0.9994	0.9980	0.9949	0.9891	0.9792	0.9642	0.9431	0.9160	0.8833	0.8462
8.9	0.9999	0.9994	0.9981	0.9951	0.9894	0.9796	0.9647	0.9438	0.9167	0.8840	0.8469
9.0	0.9999	0.9994	0.9982	0.9953	0.9896	0.9800	0.9652	0.9444	0.9174	0.8847	0.8475
9.1	0.9999	0.9995	0.9983	0.9954	0.9899	0.9804	0.9658	0.9450	0.9180	0.8853	0.8482
9.2	0.9999	0.9995	0.9983	0.9956	0.9902	0.9808	0.9663	0.9456	0.9187	0.8860	0.8488
9.3	0.9999	0.9995	0.9984	0.9957	0.9904	0.9812	0.9667	0.9462	0.9193	0.8866	0.8494
9.4	0.9999	0.9996	0.9985	0.9959	0.9907	0.9815	0.9672	0.9468	0.9200	0.8873	0.8500
9.5	0.9999	0.9996	0.9986	0.9960	0.9909	0.9819	0.9677	0.9473	0.9206	0.8879	0.8506
9.6	0.9999	0.9996	0.9986	0.9962	0.9911	0.9822	0.9681	0.9479	0.9211	0.8885	0.8511
9.7	0.9999	0.9996	0.9987	0.9963	0.9913	0.9825	0.9686	0.9484	0.9217	0.8891	0.8517
9.8	0.9999	0.9997	0.9987	0.9964	0.9916	0.9829	0.9690	0.9489	0.9223	0.8896	0.8522
9.9	1.0000	0.9997	0.9988	0.9965	0.9918	0.9832	0.9694	0.9494	0.9228	0.8902	0.8527
10.0	1.0000	0.9997	0.9988	0.9966	0.9919	0.9835	0.9698	0.9499	0.9234	0.8907	0.8532

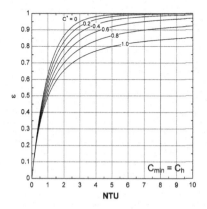

Chapter 8
EES Programs

The EES codes to simulate some simple flow arrangement configurations are shown in this chapter. Details regarding the use of these programs can be found in Sect. 3.2. It must be emphasized that the results from these EES programs coincide with those from the HETE code. They have been introduced to further help the reader in understanding the computational methodology developed with the purpose of evaluating the effectiveness and other heat exchangers related parameters.

8.1 First EES Program: "One_pass_one_row_cross_flow.EES"

```
"!One pass with one row cross-flow heat exchanger"
$UnitSystem SI MASS RAD PA  K J
$TABSTOPS   0.2 0.4 0.6 0.8 3.5 in

"This program allows to compute the HE effectiveness with the EES code.
 The code is programmed following the book material and it is not optimized from the computational point of
view"

"Heat exchanger flow arrangement data"
N_e = 100[-]    "Number of elements in the heat exchanger"
N_t = 1[-]      "Number of tubes per row"
N_r = 1[-]      "Number of rows"
N_c = 1[-]      "Number of in-tube fluid circuits"

"Thermal input data - user"
NTU = 1 [-]"Heat exchanger number of transfer of units"
Cr = 0.5[-]      "Heat capacity rate ratio Cr=C_min/C_max"
K = 1        "Chose of the fluid with the C_min heat capacity rate. K =1 -> C_min = C_c (air), K = 2 -> C_min =
C_h (fluid). To be implemented."

"Thermal input data - code"
UA = 12 [W/K]     "Heat exchanger conductance"
T_h_i = 70 [°C]    "Mean in-tube fluid inlet temperature"
T_c_i = 35 [°C]    "Mean external fluid inlet temperature"
```

© The Author(s) 2015
L. Cabezas-Gómez et al., *Thermal Performance Modeling of Cross-Flow
Heat Exchangers*, SpringerBriefs in Applied Sciences and Technology,
DOI 10.1007/978-3-319-09671-1_8

"Compute the quantities C_c, C_h, UAle, C_cle and C_hle, C_min = C_c"
UAle = UA/(N_e*N_t*N_r)   "Element heat exchanger conductance"
C_c = UA/NTU   "Heat exchanger cold fluid heat capacity rate"
C_h = C_c/Cr   "Heat exchanger hot fluid heat capacity rate"
C_cle = UA/(NTU*N_e*N_t)   "Element cold fluid heat capacity rate"
C_hle = UA/(NTU*Cr*N_c)   "Element hot fluid heat capacity rate"

"Inlet values of temperatures for both fluids"
T_h[0] = T_h_i   "hot fluid"
duplicate j=1,N_e
   T_c[j,0]=T_c_i   "cold fluid"
end

"Element effectiveness Gammale"
GAMMAle = 1-exp(-UAle/C_cle)

{"Heat exchanger temperature field distribution. Use of Eqs. (2.39) and (2.40)"

A = (C_cle*GAMMAle)/C_hle
duplicate j=1, N_e
    T_h[j] = (2-A)/(2+A)*T_h[j-1] + (2*A)/(2+A)*T_c[j,0]   "hot fluid"
    T_c[j,1] = (A+2*(1-GAMMAle))/(2+A)*T_c[j,0] + (2*GAMMAle)/(2+A)*T_h[j-1]   "cold fluid"
end

T_bar_c_o=sum(T_c[k,1],k=1,N_e)/N_e   "mean outlet cold fluid temperature"
epsilon=(T_bar_c_o-T_c[1,0])/(T_h[0]-T_c[1,0])   "element effectiveness"
epsilon_teste = Q_h_total/Q_max   "heat exchanger effectiveness"
epsilon_th = (1/Cr)*(1-exp(-(Cr)*(1-exp(-NTU))))   "theoretical effectiveness"

duplicate j=1, N_e
    Q_dot_hle[j]=-C_hle*(T_h[j]-T_h[j-1])   "hot fluid"
    Q_dot_cle[j]=C_cle*(T_c[j,1]-T_c[j,0])   "cold fluid"
end

Q_h_total=sum(Q_dot_hle[k],k=1,N_e)   "total heat transfer rate, hot fluid"
Q_c_total=sum(Q_dot_cle[k],k=1,N_e)   "total heat transfer rate, cold fluid"
Q_max = min(C_c,C_h)*(T_h_i-T_c_i)   "maximum heat transfer rate"
}

"Heat exchanger temperature distribution. Use of Eqs. (2.35 - 2.38)"
duplicate j=1, N_e
    Q_dotle[j]=-C_hle*(T_h[j]-T_h[j-1])   "hot fluid"
    Q_dotle[j]=C_cle*(T_c[j,1]-T_c[j,0])   "cold fluid"
    (T_c[j,1]-T_c[j,0])/(T_h_m[j]-T_c[j,0]) = GAMMAle   "element effectiveness"
    T_h_m[j] = (T_h[j] + T_h[j-1])/2   "element mean hot fluid temperature"
end

T_bar_c_o=sum(T_c[k,1],k=1,N_e)/N_e   "mean outlet cold fluid temperature"
Q_total = sum(Q_dotle[k],k=1,N_e)   "total heat transfer rate"
Q_max = min(C_c,C_h)*(T_h_i-T_c_i)   "maximum heat transfer rate"

epsilon=(T_bar_c_o-T_c[1,0])/(T_h[0]-T_c[1,0])   "heat exchanger effectiveness"
epsilon_teste = Q_total/Q_max   "heat exchanger effectiveness"
"theoretical effectiveness. One-pass cross-flow with one row"
epsilon_th = (1/Cr)*(1-exp(-(Cr)*(1-exp(-NTU))))

8.2 Second EES Program: "One_pass_two_rows_cross_ flow.EES"

```
"!One pass with two rows cross-flow heat exchanger"
$UnitSystem SI MASS RAD PA  K J
$TABSTOPS   0.2 0.4 0.6 0.8 3.5 in

"This program allows to compute the HE effectiveness with the EES code.
 The code is programmed following the book material and it is not optimized from the computational point of
view "

"Heat exchanger flow arrangement data"
N_e = 100[-]    "Number of elements in the heat exchanger"
N_t = 1[-]      "Number of tubes per row"
N_r = 2[-]      "Number of rows"
N_c = 2[-]      "Number of in-tube fluid circuits"

"Thermal input data - user"
NTU = 1 [-]"Heat exchanger number of transfer of units"
Cr = 0.5[-]      "Heat capacity rate ratio Cr=C_min/C_max"
K = 1            "Chose of the fluid with the C_min heat capacity rate. K =1 -> C_min = C_c (air), K = 2 ->
C_min = C_h (fluid). To be implemented."

"Thermal input data - code"
UA = 12 [W/K]    "Heat exchanger conductance"
T_h_i = 70 [°C]   "Mean in-tube fluid inlet temperature"
T_c_i = 35 [°C]   "Mean external fluid inlet temperature"

"Compute the quantities C_c, C_h, UAle, C_cle and C_hle, C_min = C_c"
UAle = UA/(N_e*N_t*N_r)   "Element heat exchanger conductance"
C_c = UA/NTU              "Heat exchanger cold fluid heat capacity rate"
C_h = C_c/Cr             "Heat exchanger hot fluid heat capacity rate"
C_cle = UA/(NTU*N_e*N_t)  "Element cold fluid heat capacity rate"
C_hle = UA/(NTU*Cr*N_c)   "Element hot fluid heat capacity rate"

"Inlet values of temperatures for both fluids"
duplicate i=1,N_c
    T_h[0,i] = T_h_i        "hot fluid"
end
duplicate j=1,N_e
   T_c[j,0]=T_c_i           "cold fluid"
end

"Element effectiveness Gammale"
GAMMAle = 1-exp(-UAle/C_cle)

"Heat exchanger temperature field distribution. Use of Eqs. (2.39) and (2.40)"
A = (C_cle*GAMMAle)/C_hle
duplicate i=1, N_c
   duplicate j=1, N_e
      T_h[j,i] = (2-A)/(2+A)*T_h[j-1,i] + (2*A)/(2+A)*T_c[j,i-1]                          "hot fluid"
      T_c[j,i] = (A+2*(1-GAMMAle))/(2+A)*T_c[j,i-1] + (2*GAMMAle)/(2+A)*T_h[j-1,i]  "cold fluid"
   end
end

T_bar_c_o=sum(T_c[k,2],k=1,N_e)/N_e                          "mean outlet cold fluid temperature"
epsilon=(T_bar_c_o-T_c_i)/(T_h_i-T_c_i)                      "element effectiveness"
epsilon_teste = Q_h_total/Q_max

"theoretical effectiveness. One pass cross-flow with two rows"
epsilon_th_2row = (1/Cr)*(1-exp(-(2*B2*Cr))*(1+B2^2*Cr))
B2 = 1 - exp(-NTU/2)
```

```
duplicate i=1, N_c                                          "element heat transfer rate"
   duplicate j=1, N_e
      Q_dot_hle[j,i]=-C_hle*(T_h[j,i]-T_h[j-1,i])           "hot fluid"
      Q_dot_cle[j,i]=C_cle*(T_c[j,i]-T_c[j,i-1])            "cold fluid"
   end
end

"total heat transfer rate"
Q_h_total=sum(Q_dot_hle[k,1],k=1,N_e) + sum(Q_dot_hle[k,2],k=1,N_e)   "hot fluid"
Q_c_total=sum(Q_dot_cle[k,1],k=1,N_e) + sum(Q_dot_cle[k,2],k=1,N_e)   "cold fluid"
Q_max = min(C_c,C_h)*(T_h_i-T_c_i)                                    "maximum heat transfer rate"


{"Heat exchanger temperature distribution. Use of Eqs. (2.35 - 2.38)"
duplicate i=1,N_c
   duplicate j=1, N_e
      Q_dotle[j,i]=-C_hle*(T_h[j,i]-T_h[j-1,i])            "hot fluid"
      Q_dotle[j,i]=C_cle*(T_c[j,i]-T_c[j,i-1])             "cold fluid"
      (T_c[j,i]-T_c[j,i-1])/(T_h_m[j,i]-T_c[j,i-1]) = GAMMAle   "element effectiveness"
      T_h_m[j,i] = (T_h[j,i] + T_h[j-1,i])/2               "element mean hot fluid temperature"
   end
end

T_bar_c_o=sum(T_c[k,2],k=1,N_e)/N_e                        "mean outlet cold fluid temperature"
Q_total = sum(Q_dotle[k,1],k=1,N_e)+sum(Q_dotle[k,2],k=1,N_e)   "total heat transfer rate"
Q_max = min(C_c,C_h)*(T_h_i-T_c_i)                        "maximum heat transfer rate"
epsilon=(T_bar_c_o-T_c_i)/(T_h_i-T_c_i)                   "heat exchanger effectiveness"
epsilon_teste = Q_total/Q_max

"theoretical effectiveness. One-pass cross-flow with two rows"
epsilon_th = (1/Cr)*(1-exp(-(2*B*Cr))*(1+B^2*Cr))
B = 1 - exp(-NTU/2)
}
```

8.3 Third EES Program: "One_pass_three_rows_cross_flow.EES"

```
"!One pass with three rows cross-flow heat exchanger"
$UnitSystem SI MASS RAD PA  K J
$TABSTOPS  0.2 0.4 0.6 0.8 3.5 in

"This program allows to compute the HE effectiveness with the EES code.
 The code is programmed following the book material and it is not optimized from the computational point of
view "

"Heat exchanger flow arrangement data"
N_e = 100[-]    "Number of elements in the heat exchanger"
N_t = 1[-]      "Number of tubes per row"
N_r = 3[-]      "Number of rows"
N_c = 3[-]      "Number of in-tube fluid circuits"

"Thermal input data - user"
NTU = 1 [-]"Heat exchanger number of transfer of units"
Cr = 0.5[-]     "Heat capacity rate ratio Cr=C_min/C_max"
K = 1           "Chose of the fluid with the C_min heat capacity rate. K =1 -> C_min = C_c (air), K = 2 ->
C_min = C_h (fluid). To be implemented."
```

```
"Thermal input data - code"
UA = 12 [W/K]     "Heat exchanger conductance"
T_h_i = 70 [°C]   "Mean in-tube fluid inlet temperature"
T_c_i = 35 [°C]   "Mean external fluid inlet temperature"

"Compute the quantities C_c, C_h, UAle, C_cle and C_hle, C_min = C_c"
UAle = UA/(N_e*N_t*N_r)     "Element heat exchanger conductance"
C_c = UA/NTU               "Heat exchanger cold fluid heat capacity rate"
C_h = C_c/Cr               "Heat exchanger hot fluid heat capacity rate"
C_cle = UA/(NTU*N_e*N_t)   "Element cold fluid heat capacity rate"
C_hle = UA/(NTU*Cr*N_c)    "Element hot fluid heat capacity rate"

"Inlet values of temperatures for both fluids"
duplicate i=1,N_c
   T_h[0,i] = T_h_i          "hot fluid"
end
duplicate j=1,N_e
   T_c[j,0]=T_c_i            "cold fluid"
end

"Element effectiveness Gammale"
GAMMAle = 1-exp(-UAle/C_cle)

"Heat exchanger temperature field distribution. Use of Eqs. (2.39) and (2.40)"
A = (C_cle*GAMMAle)/C_hle
duplicate i=1, N_c
   duplicate j=1, N_e
      T_h[j,i] = (2-A)/(2+A)*T_h[j-1,i] + (2*A)/(2+A)*T_c[j,i-1]                        "hot fluid"
      T_c[j,i] = (A+2*(1-GAMMAle))/(2+A)*T_c[j,i-1] + (2*GAMMAle)/(2+A)*T_h[j-1,i] "cold fluid"
   end
end

T_bar_c_o=sum(T_c[k,N_c],k=1,N_e)/N_e                        "mean outlet cold fluid temperature"
epsilon=(T_bar_c_o-T_c_i)/(T_h_i-T_c_i)                      "element effectiveness"
epsilon_teste = Q_h_total/Q_max

"theoretical effectiveness. One pass cross-flow with three rows"
epsilon_th_3row = (1/Cr)*(1-exp(-(3*B3*Cr))*(1+B3^2*(3-B3)*Cr+0.5*3*B3^4*Cr^2))
B3 = 1 - exp(-NTU/3)

duplicate i=1, N_c                                           "element heat transfer rate"
   duplicate j=1, N_e
      Q_dot_hle[j,i]=-C_hle*(T_h[j,i]-T_h[j-1,i])            "hot fluid"
      Q_dot_cle[j,i]=C_cle*(T_c[j,i]-T_c[j,i-1])             "cold fluid"
   end
end

"total heat transfer rate"
Q_h_total=sum(Q_dot_hle[k,1],k=1,N_e) + sum(Q_dot_hle[k,2],k=1,N_e) + sum(Q_dot_hle[k,3],k=1,N_e)
Q_c_total=sum(Q_dot_cle[k,1],k=1,N_e) + sum(Q_dot_cle[k,2],k=1,N_e) + sum(Q_dot_cle[k,3],k=1,N_e)
Q_max = min(C_c,C_h)*(T_h_i-T_c_i)                          "maximum heat transfer rate"

"! This version of the discrete model presents numerical difficulties to converge for more than two rows. Use
update guesses option in the calculate menu"
{"Heat exchanger temperature distribution. Use of Eqs. (2.35 - 2.38)"
duplicate i=1,N_c
   duplicate j=1, N_e
      Q_dotle[j,i]=-C_hle*(T_h[j,i]-T_h[j-1,i])              "hot fluid"
      Q_dotle[j,i]=C_cle*(T_c[j,i]-T_c[j,i-1])              "cold fluid"
      (T_c[j,i]-T_c[j,i-1])/(T_h_m[j,i]-T_c[j,i-1]) = GAMMAle "element effectiveness"
      T_h_m[j,i] = (T_h[j,i] + T_h[j-1,i])/2                "element mean hot fluid temperature"
   end
end
```

```
T_bar_c_o=sum(T_c[k,N_c],k=1,N_e)/N_e                    "mean outlet cold fluid temperature"
epsilon=(T_bar_c_o-T_c_i)/(T_h_i-T_c_i)                  "heat exchanger effectiveness"

"total heat transfer rate"
Q_total = sum(Q_dotle[k,1],k=1,N_e)+sum(Q_dotle[k,2],k=1,N_e)+sum(Q_dotle[k,3],k=1,N_e)
Q_max = min(C_c,C_h)*(T_h_i-T_c_i)                       "maximum heat transfer rate"
epsilon_teste = Q_total/Q_max

"theoretical effectiveness. One-pass cross-flow with three rows"
epsilon_th_3row = (1/Cr)*(1-exp(-(3*B3*Cr))*(1+B3^2*(3-B3)*Cr+0.5*3*B3^4*Cr^2))
B3 = 1 - exp(-NTU/3)
}
```

8.4 Fourth EES Program: "Two_pass_parallel_cross_flow.EES"

```
"!Two-pass parallel cross-flow heat exchanger with one circuit"
$UnitSystem SI MASS RAD PA  K J
$TABSTOPS  0.2 0.4 0.6 0.8 3.5 in

"This program allows to compute the HE effectiveness with the EES code.
 The code is programmed following the book material and it is not optimized from the computational point of
 view "

"Heat exchanger flow arrangement data"
N_e = 10[-] "Number of elements in the heat exchanger"
N_t = 1[-]       "Number of tubes per row"
N_r = 2[-]       "Number of rows"
N_c = 1[-]       "Number of in-tube fluid circuits"

"Thermal input data - user"
NTU = 1 [-]"Heat exchanger number of transfer of units"
Cr = 0.5[-]      "Heat capacity rate ratio Cr=C_min/C_max"
K = 1            "Chose of the fluid with the C_min heat capacity rate. K =1 -> C_min = C_c (air), K = 2 ->
C_min = C_h (fluid). To be implemented."

"Thermal input data - code"
UA = 12 [W/K]    "Heat exchanger conductance"
T_h_i = 70 [°C]  "Mean in-tube fluid inlet temperature"
T_c_i = 35 [°C]  "Mean external fluid inlet temperature"

"Compute the quantities C_c, C_h, UAle, C_cle and C_hle, C_min = C_c"
UAle = UA/((N_e/N_r)*N_t*N_r)      "Element heat exchanger conductance"
C_c = UA/NTU                       "Heat exchanger cold fluid heat capacity rate"
C_h = C_c/Cr                       "Heat exchanger hot fluid heat capacity rate"
C_cle = UA/(NTU*(N_e/N_r)*N_t)  "Element cold fluid heat capacity rate"
C_hle = UA/(NTU*Cr*N_c)           "Element hot fluid heat capacity rate"

"Inlet values of temperatures for both fluids"
T_h[0] = T_h_i         "hot fluid"
duplicate j=1,N_e/N_r
  T_c[j,0]=T_c_i         "cold fluid"
end

"Element effectiveness Gammale"
GAMMAle = 1-exp(-UAle/C_cle)

"Heat exchanger temperature field distribution. Use of Eqs. (2.39) and (2.40)"
A = (C_cle*GAMMAle)/C_hle
duplicate j=1, N_e/N_r
    T_h[j] = (2-A)/(2+A)*T_h[j-1] + (2*A)/(2+A)*T_c[j,0]         "hot fluid"
    T_c[j,1] = (A+2*(1-GAMMAle))/(2+A)*T_c[j,0] + (2*GAMMAle)/(2+A)*T_h[j-1]    "cold fluid"
end
```

```
duplicate j=N_e/N_r+1, N_e
    T_h[j] = (2-A)/(2+A)*T_h[j-1] + (2*A)/(2+A)*T_c[N_e-j+1,1]                    "hot fluid"
    T_c[j,2] = (A+2*(1-GAMMAle))/(2+A)*T_c[N_e-j+1,1] + (2*GAMMAle)/(2+A)*T_h[j-1]   "cold fluid"
end

duplicate j=N_e/N_r+1, N_e
    T_c[j,1] = T_c[N_e-j+1,1]                                                    "cold fluid"
end


T_bar_c_o=sum(T_c[k,2],k=(N_e/N_r+1),N_e)/(N_e/N_r)        "mean outlet cold fluid temperature"


epsilon=(T_bar_c_o-T_c_i)/(T_h_i-T_c_i)                    "element effectiveness"
epsilon_teste = Q_h_total/Q_max                           "heat exchanger effectiveness"

"theoretical effectiveness. Parallel-cross-flow with two passes"
epsilon_th_p = (1/Cr)*(1-B2/2)*(1-exp(-2*B2*Cr))
B2 = 1-exp(-NTU/2)

duplicate j=1, N_e                                        "element heat transfer rate"
    Q_dot_hle[j]=-C_hle*(T_h[j]-T_h[j-1])                 "hot fluid"
end
duplicate j=1,N_e/N_r
    Q_dot_cle[j]=C_cle*(T_c[j,1]-T_c[j,0])                "cold fluid"
end
duplicate j=N_e/N_r+1,N_e
    Q_dot_cle[j]=C_cle*(T_c[j,2]-T_c[j,1])                "cold fluid"
end

Q_h_total=sum(Q_dot_hle[k],k=1,N_e)                       "total heat transfer rate, hot fluid"
Q_c_total=sum(Q_dot_cle[k],k=1,N_e)                       "total heat transfer rate, cold fluid"
Q_max = min(C_c,C_h)*(T_h_i-T_c_i)                        "maximum heat transfer rate"
```

8.5 Fifth EES Program: "Two_pass_counter_cross_flow.EES"

```
"!Two-pass counter cross-flow heat exchanger with one circuit"
$UnitSystem SI MASS RAD PA  K J
$TABSTOPS  0.2 0.4 0.6 0.8 3.5 in

"This program allows to compute the HE effectiveness with the EES code.
 The code is programmed following the book material and it is not optimized from the computational point of
view"

"Heat exchanger flow arrangement data"
N_e = 100[-]    "Number of elements in the heat exchanger"
N_t = 1[-]      "Number of tubes per row"
N_r = 2[-]      "Number of rows"
N_c = 1[-]      "Number of in-tube fluid circuits"

"Thermal input data - user"
NTU = 1 [-]"Heat exchanger number of transfer of units"
Cr = 0.5[-]      "Heat capacity rate ratio Cr=C_min/C_max"
K = 1            "Chose of the fluid with the C_min heat capacity rate. K =1 -> C_min = C_c (air), K = 2 ->
C_min = C_h (fluid). To be implemented."
```

```
"Thermal input data - code"
UA = 12 [W/K]      "Heat exchanger conductance"
T_h_i = 70 [°C]    "Mean in-tube fluid inlet temperature"
T_c_i = 35 [°C]    "Mean external fluid inlet temperature"

"Compute the quantities C_c, C_h, UAle, C_cle and C_hle, C_min = C_c"
UAle = UA/((N_e/N_r)*N_t*N_r)          "Element heat exchanger conductance"
C_c = UA/NTU                           "Heat exchanger cold fluid heat capacity rate"
C_h = C_c/Cr                           "Heat exchanger hot fluid heat capacity rate"
C_cle = UA/(NTU*(N_e/N_r)*N_t)         "Element cold fluid heat capacity rate"
C_hle = UA/(NTU*Cr*N_c)                "Element hot fluid heat capacity rate"

"Inlet values of temperatures for both fluids"
T_h[0] = T_h_i                                                   "hot fluid"
duplicate j=(N_e/N_r+1),N_e
   T_c[j,2]=T_c_i                                                "cold fluid"
end

"Element effectiveness Gammale"
GAMMAle = 1-exp(-UAle/C_cle)

"Heat exchanger temperature field distribution. Use of Eqs. (2.39) and (2.40)"
A = (C_cle*GAMMAle)/C_hle
duplicate j=1, N_e/N_r
   T_h[j] = (2-A)/(2+A)*T_h[j-1] + (2*A)/(2+A)*T_c[N_e-j+1,1]                        "hot fluid"
   T_c[j,0] = (A+2*(1-GAMMAle))/(2+A)*T_c[N_e-j+1,1] + (2*GAMMAle)/(2+A)*T_h[j-1]   "cold fluid"
end

duplicate j=N_e/N_r+1, N_e
   T_h[j] = (2-A)/(2+A)*T_h[j-1] + (2*A)/(2+A)*T_c[j,2]                        "hot fluid"
   T_c[j,1] = (A+2*(1-GAMMAle))/(2+A)*T_c[j,2] + (2*GAMMAle)/(2+A)*T_h[j-1]   "cold fluid"
end

duplicate j=1,N_e/N_r
   T_c[j,1] = T_c[N_e-j+1,1]                                     "cold fluid"
end


T_bar_c_o=sum(T_c[k,0],k=1,(N_e/N_r))/(N_e/N_r)                 "mean outlet cold fluid temperature"


epsilon=(T_bar_c_o-T_c_i)/(T_h_i-T_c_i)                  "element effectiveness"
epsilon_teste = Q_h_total/Q_max                          "heat exchanger effectiveness"

"theoretical effectiveness. Counter-cross-flow with two passes"
epsilon_th_cc = (1/Cr)*(1- ( B2/2+ (1-B2/2)*exp(2*B2*Cr) )^(-1) )
B2 = 1-exp(-NTU/2)

duplicate j=1, N_e                                       "element heat transfer rate"
   Q_dot_hle[j]=-C_hle*(T_h[j]-T_h[j-1])                 "hot fluid"
end
duplicate j=1,N_e/N_r
   Q_dot_cle[j]=C_cle*(T_c[j,0]-T_c[j,1])                "cold fluid"
end
duplicate j=N_e/N_r+1,N_e
   Q_dot_cle[j]=C_cle*(T_c[j,1]-T_c[j,2])                "cold fluid"
end

Q_h_total=sum(Q_dot_hle[k],k=1,N_e)              "total heat transfer rate, hot fluid"
Q_c_total=sum(Q_dot_cle[k],k=1,N_e)             "total heat transfer rate, cold fluid"
Q_max = min(C_c,C_h)*(T_h_i-T_c_i)              "maximum heat transfer rate"
```